This comprehensive book describes the most soundly based methods currently available for evaluating the transport properties, particularly viscosity and thermal conductivity, of pure fluids and fluid mixtures. Particular emphasis is placed on recent theoretical advances in our understanding of fluid transport properties in all the different regions of thermodynamic states.

Following a general introductory section, the important theoretical tools for describing complete transport property surfaces of fluids are presented. Different methods of data representation are then covered, followed by a section which demonstrates the application of selected methods under various specific conditions. Case studies of transport property analysis for real fluids are then given for systems with increasing complexity, and the book concludes with a discussion of various international data banks and prediction packages.

Advanced students of physics, chemistry, and chemical engineering; scientists working in the fields of kinetic theory, experimental determination, and theoretical interpretation of transport properties; as well as engineers involved with the design and optimization of process equipment, will find this book indispensable.

T0236733

Transport Properties of Fluids

Transport Properties of Fluids
Their Correlation, Prediction and Estimation

Edited by

Jürgen Millat
NORDUM Institut für Umwelt und Analytik GmbH, Kessin/Rostock, Germany

J. H. Dymond
The University, Glasgow, UK

C. A. Nieto de Castro
University of Lisbon, Portugal

IUPAC

CAMBRIDGE
UNIVERSITY PRESS

CAMBRIDGE UNIVERSITY PRESS
Cambridge, New York, Melbourne, Madrid, Cape Town, Singapore, São Paulo

Cambridge University Press
The Edinburgh Building, Cambridge CB2 2RU, UK

Published in the United States of America by Cambridge University Press, New York

www.cambridge.org
Information on this title: www.cambridge.org/9780521461788

First published 1996
This digitally printed first paperback version 2005

A catalogue record for this publication is available from the British Library

Library of Congress Cataloguing in Publication data
Transport properties of fluids : their correlation, prediction and estimation / edited by Jürgen Millat,
J.H. Dymond, C.A. Nieto de Castro.
p. cm.
ISBN 0-521-46178-2 (hardcover)
1. Fluid dynamics. 2. Transport theory. 3. Fluids–Thermal properties.
I. Millat, Jürgen. II. Dymond, J. H. (John H.) III. Castro, C.A. Nieto de
QC151.T73 1996
530.4'25–dc20   95-12766

ISBN-13 978-0-521-46178-8 hardback
ISBN-10 0-521-46178-2 hardback

ISBN-13 978-0-521-02290-3 paperback
ISBN-10 0-521-02290-8 paperback

This book is dedicated by its editors and authors to the memory of Professors
Joseph Kestin (1913–1993)
and
Edward A. Mason (1926–1994).
Their outstanding research and inspiration contributed greatly to the concept and
content of this volume.

Contents

Contents

Contributors

M. J. Assael
Faculty of Chemical Engineering, Aristotle University, Univ. Box 453, GR–54006 Thessaloniki, Greece

E. Bich
Universität Rostock, Fachbereich Chemie, D–18055 Rostock, Germany

R. J. B. Craven
IUPAC Thermodynamic Tables Project Centre, Department of Chemical Engineering and Chemical Technology, Imperial College, London SW7 2BY, UK

K. M. de Reuck
IUPAC Thermodynamic Tables Project Centre, Department of Chemical Engineering and Chemical Technology, Imperial College, London SW7 2BY, UK

J. H. Dymond
Chemistry Department, The University, Glasgow G12 8QQ, UK

A. E. Elhassan
IUPAC Thermodynamic Tables Project Centre, Department of Chemical Engineering and Chemical Technology, Imperial College, London SW7 2BY, UK

D. J. Evans
Research School of Chemistry, The Australian National University, Canberra ACT 0200, Australia

J. M. N. A. Fareleira
Technical University of Lisbon, Av. Rovisco Pais, P-1096 Lisbon, Portugal

P. S. Fialho
University of Azores, P-9702 Angra do Heroismo Codex, Portugal

D. G. Friend
National Institute of Standards and Technology, Thermophysics Division, Boulder, CO 80303, USA

H. J. M. Hanley
National Institute of Standards and Technology, Thermophysics Division, Boulder, CO 80303, USA

C. Hoheisel
Ruhr–Universität Bochum, Theoretische Chemie, D–44799 Bochum, Germany

M. L. Huber
National Institute of Standards and Technology, Thermophysics Division, Boulder, CO 80303, USA

A. I. Johns
National Engineering Laboratory, East Kilbride, Glasgow G75 0QU, UK

R. Krauss
Universität Stuttgart, Institut für Technische Thermodynamik und Thermische Verfahrenstechnik, D–70550 Stuttgart, Germany

J. Luettmer–Strathmann
Institute for Physical Science & Technology, University of Maryland, College Park, MD 20742, USA

K. N. Marsh
Thermodynamics Research Center, Texas A&M University System, College Station, TX 77843-3111, USA

E. A. Mason*
Department of Chemistry, Brown University, Providence, R.I., 02912, USA

J. Millat
NORDUM Institut für Umwelt und Analytik GmbH, Gewerbepark Am Weidenbruch, D–18196 Kessin/Rostock, Germany

A. Nagashima
Faculty of Science and Technology, Department of Mechanical Engineering, Keio University, 3–14–1 Hiyoshi, Yokohama 223, Japan

C. A. Nieto de Castro
Chemistry Department, Faculty of Sciences, University of Lisbon, Campo Grande, Ed. C1 – Piso 5, P–1700 Lisbon, Portugal

R. A. Perkins
National Institute of Standards and Technology, Thermophysics Division, Boulder, CO 80303, USA

B. E. Poling
College of Engineering, The University of Toledo, Toledo, Ohio 43606–3390, USA

M. L. V. Ramires
Chemistry Department, Faculty of Sciences, University of Lisbon, Campo Grande, Ed. C1 – Piso 5, P–1700 Lisbon, Portugal

E. P. Sakonidou
Van der Waals–Zeeman Laboratory, University of Amsterdam, Valckenierstraat 65-67, NL–1018 XE Amsterdam, The Netherlands

J. V. Sengers
Institute for Physical Science & Technology, University of Maryland, College Park, MD 20742, USA

* Deceased November 1994.

K. Stephan
Universität Stuttgart, Institut für Technische Thermodynamik und Thermische Verfahrens-technik, D–70550 Stuttgart, Germany

F. J. Uribe
Department of Physics, Universidad Autonoma Metropolitana, Av. Michoacácan y Calz. de la Purisima, 09340 México, D.F., México

H. R. van den Berg
Van der Waals–Zeeman Laboratory, University of Amsterdam, Valckenierstraat 65-67, NL–1018 XE Amsterdam, The Netherlands

A. A. Vasserman
Odessa Institute of Marine Engineers, Mechnikova Street 34, Odessa, Ukraine

V. Vesovic
Department of Mineral Resources Engineering, Imperial College, London SW7 2BP, UK

E. Vogel
Universität Rostock, Fachbereich Chemie, D–18055 Rostock, Germany

W. A. Wakeham
Department of Chemical Engineering and Chemical Technology, Imperial College, London SW7 2BY, UK

J. T. R. Watson
National Engineering Laboratory, East Kilbride, Glasgow G75 OQU, UK

R. C. Wilhoit
Thermodynamics Research Center, Texas A&M University System, College Station, TX 77843-3111, USA

Foreword

The Commission on Thermodynamics of the Physical Chemistry Division of the International Union of Pure and Applied Chemistry is charged by the Union with the duty to define and maintain standards in the general field of thermodynamics. This duty encompasses matters such as the establishment and monitoring of international pressure and temperature scales, recommendations for calorimetric procedures, the selection and evaluation of reference standards for thermodynamic measurements of all types and the standardization of nomenclature and symbols in chemical thermodynamics. One particular aspect of the commission's work from among this set is carried forward by two subcommittees: one on thermodynamic data and the other on transport properties. These two subcommittees are responsible for the critical evaluation of experimental data for the properties of fluids that lie in their respective areas and for the subsequent preparation and dissemination of internationally approved thermodynamic tables of the fluid state and representations of transport properties.

The Subcommittee on Transport Properties has discharged its responsibilities through the work of groups of research workers active in the field drawn from all over the world. These groups have collaborated in the preparation of representations of the viscosity, thermal conductivity and diffusion coefficients of pure fluids and their mixtures over wide ranges of thermodynamic states. The representations have almost always been based upon an extensive body of experimental data for the property in question accumulated over many years by the efforts of laboratories worldwide. The results of this work have been published under the auspices of the subcommittee, with international endorsement, in *Journal of Physical and Chemical Reference Data* and *International Journal of Thermophysics*. The series of papers produced provides equations that describe the properties as a function of temperature and density that can be readily coded to yield transport properties at any prescribed thermodynamic state with a defined uncertainty.

In 1991, in collaboration with the Commission on Thermodynamics, the Subcommittee on Transport Properties brought together a wider group of experts to contribute to Volume III in the series *Experimental Thermodynamics*, which was edited by

W. A. Wakeham, J. V. Sengers and A. Nagashima under the title *Measurement of the Transport Properties of Fluids* (Blackwell Scientific, Oxford). The volume describes the state of the art with respect to the instrumentation for the determination of the transport properties of fluids which has been employed in the acquisition of the data upon which many of the representations of the properties have been based. However, the representations of the transport properties of fluids have also relied heavily on the use of kinetic theory to fill gaps for thermodynamic states where no experimental results exist or for properties where measurements have not been performed. The theory is then employed as a secure means of interpolation or even extrapolation from a smaller set of high–quality information.

The present volume was conceived by the Subcommittee on Transport Properties and the Commission on Thermodynamics to be a complement to the description of experimental techniques. Its purpose is therefore to outline the principles that underlie the statistical mechanical theories of transport processes in fluids and fluid mixtures in a way that leads to results that can be used in practice for their prediction or representation and to give practical examples of how this has been implemented. The brief to the editors of this book from the subcommittee has been admirably fulfilled by the team of authors that they have assembled. The coverage of the theory of transport properties is concise yet comprehensive and is developed in a fashion that leads to useful results. The sections on applications work their way through increasingly complicated archetypal systems from the simplest monatomic species to dense mixtures of polyatomic fluids of industrial significance and always with the emphasis on practical utility. This approach is concluded with examples of practical realizations of the representations of the properties incorporated in computer packages.

The book is intended to be useful for engineers who have to make use of representations of transport properties in order that they should understand the methodology that lies behind published correlations as well as the limitations of the development. It is also intended to be a summary of the status of the field at a particular moment for practitioners of the subject. It is thus a book which is intended to bring the latest state of knowledge to bear on problems of practical importance through international collaboration and thus fulfill one of the main objectives of IUPAC.

W. A. Wakeham
Chairman
Commission on Thermodynamics
International Union of Pure and Applied Chemistry

Part one

GENERAL

1

Introduction

J. MILLAT

NORDUM Institut für Umwelt und Analytik, Kessin/Rostock, Germany

J. H. DYMOND

The University, Glasgow, UK

C. A. NIETO DE CASTRO

University of Lisbon, Portugal

Accurate knowledge of transport properties of pure gases and liquids, and of their mixtures, is essential for the optimum design of the different items of chemical process plants, for determination of intermolecular potential energy functions and for development of accurate theories of transport in dense fluids. A previous IUPAC volume, edited by Wakeham *et al.* (1991), also produced by Commission I.2 through its Subcommittee on Transport Properties, has described experimental methods for the accurate determination of transport properties. However, it is impossible to measure these properties for all industrially important fluids, and their mixtures, at all the thermodynamic states of interest. Measurements therefore need to be supplemented by theoretical calculations.

This present volume, which is complementary to the previous publication, discusses the present state of theory with regard to the dilute–gas state, the initial density dependence, the critical region and the very dense gas and liquid states for pure components and mixtures. In all cases, the intention is to present the theory in usable form and examples are given of its application to nonelectrolyte systems. This will be of particular use to chemical and mechanical engineers. The subtitle of this volume 'Their correlation, prediction and estimation' reflects the preferred order of application to obtain accurate values of transport properties. Careful correlation of accurate experimental data gives reliable values at interpolated temperatures and pressures (densities), and at different compositions when the measurements are for mixtures. Unfortunately, there are only a limited number of systems where data of such accuracy are available. In other cases, sound theoretical methods are necessary to predict the required values. Where information is lacking – for intermolecular forces, for example – estimation methods have to be used. These are of lower accuracy, but usually have more general applicability.

In view of the outstanding need for accurate theoretical prediction, this volume gives a clear presentation of current theory as applicable to fluids and fluid mixtures in different density ranges. As a result of the substantial advances made in recent years it is now

3

possible to describe exactly the low–density transport properties as well as the critical enhancements.

The dilute–gas theory is presented here for the first time in terms of effective collision cross sections in a comprehensive readily usable form which applies to both polyatomic fluids and monatomic fluids. This description should now be used exclusively but, because it is relatively new, expressions are given for the macroscopic quantities in terms of these effective cross sections, and certain simple relationships between these effective cross sections and the previously used collision integrals are also described.

It is possible to account for the initial density dependence (at least for viscosity) and, although the description is not rigorous, it is sufficient for this often relatively small contribution to the transport properties. For higher densities, the modified Enskog theory can be used in a consistent manner, although this does have limitations. This becomes obvious from the fact that different empirical modifications have been proposed and applied to different regions of the transport property surface. Therefore, for certain ranges of thermodynamic states, an empirical estimation scheme based on the density dependence of the excess transport property is frequently to be preferred. For liquids and dense gases under conditions where the critical enhancements are negligible, methods based on hard–sphere theory give the best representation of experimental data.

Although experimental transport properties are measured at different temperatures and pressures, it is the density, or molar volume, which is the theoretically important variable. So, for the prediction of transport properties, it is necessary to convert data at a given temperature and pressure to the corresponding temperature and density, or vice versa, by use of a reliable equation of state. Accordingly, an account is given in this volume of the most useful equations of state to express these relationships for gases and liquids. For dense fluids, it is possible to calculate transport properties directly by molecular simulation techniques under specified conditions when the molecular interactions can be adequately represented. A description is included in this book of these methods, which are significant also for the results which have aided the development of transport theory.

When the above methods fail, estimation methods become important. Schemes based on the Corresponding–States Principle which are particularly important in this respect are described. In order to demonstrate clearly just when the methods of correlation, the theoretical expressions and estimation techniques are applicable, examples are given of transport–property data representation for systems of different complexity: simple monatomic fluids, diatomic fluids, polyatomic fluids (specifically, water and refrigerant R134a), nonreacting mixtures and (dilute) alkali–metal vapors as an example of a reacting mixture.

Rapid access to transport property data is essential for the efficient use of proposed correlation and prediction schemes. As a result, experimental data have been stored in many data banks worldwide and the final section of this volume describes a number of the major data banks, which also incorporate methods for the calculation of transport properties.

The authors are aware of the fact that, although this book demonstrates the significant progress that has been made in this field in the past decade, there is still a need for additional experimental and theoretical work in many parts of the transport–property surface. In the individual chapters, an attempt is made to specify the relevant needs in each density domain. It should be noted that this volume is restricted to a discussion of nonelectrolytes. In spite of their technological importance, ionic systems, including ionized gases and plasmas, molten metals and aqueous electrolyte solutions are not included because of the different nature of the interaction forces. A complete description of the transport properties of these fluids and fluid mixtures would occupy another volume.

The editors acknowledge with thanks the contributions which have been made by all the authors. They have attempted to produce a reasonable uniformity of style and apologize for any gross inconsistencies which remain. It is appreciated that not all theories of transport and estimation methods have been covered. For these omissions, and for all errors, the editors accept full responsibility. Finally, it is with the greatest pleasure that the editors acknowledge the support of members and corresponding members of the IUPAC Subcommittee on Transport Properties of Commission I.2 on Thermodynamics. Particular thanks are extended to its chairman, Professor W.A. Wakeham, for his many constructive comments and his unending enthusiasm and encouragement throughout.

Reference

W.A. Wakeham, A. Nagashima & J.V. Sengers, eds. (1991). *Experimental Thermodynamics, Vol. III: Measurement of the Transport Properties of Fluids.* Oxford: Blackwell Scientific Publications.

2

Technological Importance

W. A. WAKEHAM

Imperial College, London, UK

C. A. NIETO DE CASTRO

University of Lisbon, Portugal

2.1 Introduction

Fluids, that is gases and liquids, are self–evidently prerequisites for normal life. They also play a major role in the production of many artefacts and in the operation of much of the equipment upon which modern life depends. Occasionally, a fluid is the ultimate result of a technological process, such as a liquid or gaseous fuel, so that its existence impinges directly on the public consciousness. More often, fluids are intermediates in processes yielding solid materials or objects, and are then contained within solid objects so that their public image is very much less and their significance not fully appreciated. Nevertheless, every single component of modern life relies upon a fluid at some point and therefore upon our understanding of the fluid state.

The gross behavioral features of a fluid are well understood in the sense that it is easy to grasp that a gas has the property to completely fill any container and that a liquid can be made to flow by the imposition of a very small force. However, beyond these qualitative features lie a wide range of thermophysical and thermochemical properties of fluids that determine their response to external stimuli. This analysis concentrates exclusively on *thermophysical properties* and will not consider any process that involves a change to the molecular entities that comprise the fluid. The most familiar thermophysical properties are those that determine the change in state of a fluid that results from an external stimulus, for example the change of temperature of a mass of fluid that results from the input of a quantity of heat to it. Such properties, which relate to differences between two states of thermodynamic equilibrium, are known as *thermodynamic properties*. On the other hand, those properties which are concerned with the rate of change of the state of a fluid as a result of a change in external conditions, or with the transport of mass, momentum or energy between different parts of a fluid which is not in a uniform state, are known as *transport properties* and form the subject of this volume. The purpose of this chapter is to illustrate the importance of the transport properties of fluids in science and technology.

6

2.2 Areas of technological interest

It will be shown later in the book (see Chapters 4 and 5) that the transport of mass, momentum and energy through a fluid is the consequence of molecular motion and molecular interaction. In the low–density gas phase the mean free path of the molecules is very much greater than a molecular diameter. It is then the free molecular motion that contributes mostly to the transport, and molecular collisions are relatively rare events involving only two molecules at any one time. Such molecular collisions modify the transport process by deflecting molecules from their original course. Thus the nature of the collision, which is determined by the forces exerted between a pair of molecules, necessarily determines the magnitude of the flux, of mass, momentum or energy induced by a gradient of molecular concentration, flow velocity or temperature in the gas. The fluxes, $\mathbf{J}$, of the transported quantities and the imposed gradients, ∇Y, are normally related via simple, phenomenological, linear laws such as those of Fick, Newton and Fourier (Bird *et al.* 1960)

$$\mathbf{J} = -X\nabla Y \tag{2.1}$$

Here, X is the transport property associated with the particular process under consideration. It follows that the transport coefficient, which itself may be a function of the temperature and density of the fluid, will reflect the interactions between the molecules of the dilute gas. For that reason there has been, for approximately 150 years, a purely scientific interest in the transport properties of fluids as a means of probing the forces between pairs of molecules. Within the last twenty years, at least for the interactions of the monatomic, spherically symmetric inert gases, the transport properties have played a significant role in the elucidation of these forces.

As the density of the fluid is increased the free motion of molecules is increasingly dominated in the transport process by the interactions among the molecules and especially groups of them. The mean free path becomes smaller and of the order of several molecular diameters. The details of the interactions between the molecules therefore become less important compared to the fact that so many interactions take place. Thus, when the dense liquid state is attained, it seems that quite simple models of the interaction between molecules are adequate for a description of the behavior of the transport properties (see Chapters 5 and 10). In the extreme case of a fluid near its critical point the specific intermolecular interaction becomes totally irrelevant, since the transport properties of the fluid are determined by the behavior of clusters and their size rather than anything else (see Chapter 6). Thus, the scientific importance of transport properties under these conditions becomes one of seeking to describe the behavior of the property itself through appropriate statistical mechanical theory rather than as a tool to reveal other fundamental information.

The importance of the transport properties of fluids in technology is maintained across the entire spectrum of densities. Almost all chemical–process plants make use of fluids either in process streams or as a means to heat and cool those streams. The process of heat exchange between two fluid streams is conducted in a heat exchanger whose design must be such as to permit the requisite heating or cooling of a process stream to be carried out within prescribed limits of temperature. The rate of heat exchange and, therefore, the design of the heat exchanger is dependent on the physical properties of the fluids involved. A knowledge of these properties is evidently a prerequisite for the design. The design of chemical reactors, particularly those that make use of porous solid catalysts, or of separation equipment, requires a knowledge of the diffusion coefficients for various species in a mixture of fluids in addition to the viscosity and thermal conductivity. Errors in the values of the properties used to design a given item of a chemical plant can produce a significant effect on the capital cost of that item, as well as unexpected increases in the operating costs. Errors of this kind have effects that can propagate throughout the design of the entire plant, sometimes becoming amplified and threatening its operability.

Similarly, the design of refrigeration or air conditioning equipment requires a knowledge of the viscosity and thermal conductivity of the working fluid in the thermodynamic cycle in order to determine the size of the heat–transfer equipment and fluid pumps required to meet a specified duty. Moreover, the viscosity and thermal conductivity of fluid lubricants is of great significance to the process of lubrication. Considerable efforts are expended to select and synthesize fluids with particular characteristics for the viscosity of lubricants as a function of temperature to ensure proper operation of lubricated equipment under a variety of operating conditions. Indeed, it is particularly in this area that the need for some accurate standard reference values for the viscosity of fluids is most acute because of the need to provide meaningful intercomparisons of data obtained by different manufacturers. In addition, most of the equipment used in industry to measure or to control properties of the process streams needs to be calibrated with respect to standard reference data, which, sometimes, require international validation.

Transport coefficients occur in all forms of continuum, hydrodynamic equations concerned with mass, momentum and energy conservation once constitutive equations for the fluids of interest are introduced. Such equations are frequently encountered in trying to model mathematically technological processes with a view to their refinement. Attempts to model such processes mathematically (usually numerically) are frequently limited by a lack of knowledge of the physical properties of the materials involved including the transport coefficients of the fluids.

Increasingly, there is a demand for improved safety of technological processes. The term 'safety' may include environmental damage of various kinds as well as a direct threat to life and property. Here, too, transport properties of fluids have a significant impact. For example, the description of the process of pollutant dispersion contains diffusive and convective components into which the transport coefficients of the gas or liquid medium enter. There is a growing requirement for a demonstration of the pedigree of every number that is employed in a calculation intended to demonstrate a safety case for industrial plant so as to satisfy regulatory or legislative bodies both nationally and internationally. Such requirements dictate that there should be a body of approved, preferably internationally approved, data which can be used as a standard source in all calculations of this kind.

It is clear therefore that the transport properties of fluids are of significance in many areas of technology. At least in some of those areas, the accuracy of the data may be of considerable economic and practical significance. The following sections are intended to provide quantitative examples of this assertion.

2.3 Examples

2.3.1 Intermolecular forces

As was remarked earlier, all of the transport coefficients, as well as other properties, of a dilute gas depend upon the intermolecular forces that exist between the molecules in the gas. Thus, on the one hand a knowledge of the intermolecular forces enables all the dilute–gas properties to be evaluated at an arbitrary temperature even if they have not all been measured. Equally, it might be expected that accurate measurements of the transport properties of the gas might be used to determine the forces between pairs of molecules. It was not until 1970 that such a process was shown to be feasible and then only with the aid of input from other sources of information, but it is interesting to note that one of the factors that contributed to the slow development of this process was the inconsistency of the available data for the viscosity of a gas with independent sources of information. The inconsistency was finally traced to errors in early measurements of the viscosity of gases that were finally eliminated by improved experimental design. The final importance of the transport properties of gases in elucidating the forces between some molecules is best illustrated by means of the example of argon. Argon has always been an archetypal system for this study because of the spherical nature of the molecule and the consequent spherical symmetry of the pair potential and simplicity of the kinetic theory.

The viscosity of a dilute gas composed of spherically symmetric, structureless molecules is related to the pair potential through the equation

$$\eta^{(0)} = \frac{k_B T}{4(k_B T/\pi m)^{1/2} \mathfrak{S}(2000)} f_\eta \tag{2.2}$$

where k_B is the Boltzmann constant, T the absolute temperature and m the molecular mass. In addition, $\mathfrak{S}(2000)$ is an effective cross section that contains all of the dynamical information related to the intermolecular potential that acts between the molecules. It is explicitly related to the pair potential for this interaction in later chapters of this book (see Chapter 4). Finally, f_η is a factor near unity that accounts for kinetic theory approximations beyond the first and is extremely weakly dependent on the nature of the intermolecular interaction.

By 1972 the viscosity of argon had been determined over a range of temperatures from 120 K to 2000 K with an accuracy of better than 2%. At around that time, independent

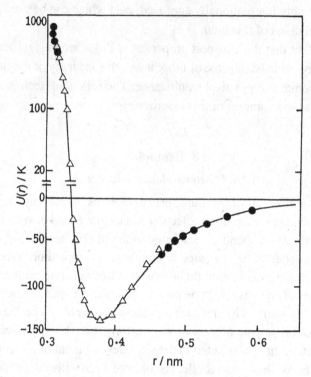

Fig. 2.1. A modern version of the intermolecular pair potential for argon. Solid line: a representation of the full potential; symbols: △ inversion of gas viscosity; ● inversion of second virial coefficient.

measurements of the spectrum of bound argon dimers and molecular–beam–scattering data became available for the first time. When all of the available data for argon were combined it was possible to determine the intermolecular pair potential for argon for the first time (Maitland *et al.* 1987).

Subsequently, and of greater significance in the context of this volume, it was shown that it was possible to determine the pair potential of monatomic species directly from measurements of the viscosity of the dilute gas, by a process of iterative inversion (Maitland *et al.* 1987). As an illustration of the success that can be achieved, Figure 2.1 compares the pair potential that is obtained by application of the inversion process to the viscosity data for argon with that currently thought to be the best available pair potential for argon which is consistent with a wide variety of experimental and theoretical information.

The same techniques have been employed to determine the pair potential for other like and unlike interactions among the monatomic gases (Maitland *et al.* 1987). Attempts have also been made to apply the same sort of techniques to polyatomic gases (Vesovic & Wakeham 1987). However, because of the nonspherical nature of the pair potential and the sheer magnitude of the computational effort required to evaluate the effective cross sections in such a case progress has been limited until recently. The advent of very much faster computers now holds out the hope that such systems may become more amenable to study.

2.3.2 Process–plant design

As an example of the importance of the transport properties of fluids in the design of chemical process plant the catalytic reactor for the synthesis of methanol from hydrogen and carbon monoxide shown in Figure 2.2 is considered. The feed gases consist of a mixture of hydrogen and carbon monoxide which enter the reactor through a gas–gas exchanger, which is an integral part of the pressure vessel for the reactor. In the heat exchanger the incoming gases at 30 MPa are heated from ambient temperature to 610 K by the product gases from the two catalyst beds in the reactor. In the particular design shown, a second heat exchanger is used between the two catalyst beds to provide interstage cooling. Because the two heat exchangers are incorporated into the reactor their design must be specified with only a small safety margin if the size of the entire reactor is to remain within acceptable limits. Furthermore, for the preheater, there is little flexibility in operation by which design deficiencies may be overcome in operation because many of the variables are determined by the requirements of the catalytic reaction zones. In order to study the effect of the transport properties of the gases upon the design of this equipment a standard design methodology has been applied (Armstrong *et al.* 1982) in which a set of realistic, but arbitrary, values for the physical properties of the gas streams has been adopted as a reference case to yield a reference

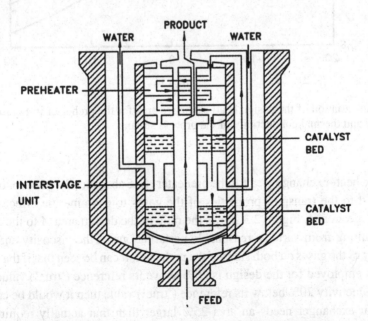

Fig. 2.2. A catalytic reactor for the synthesis of methanol.

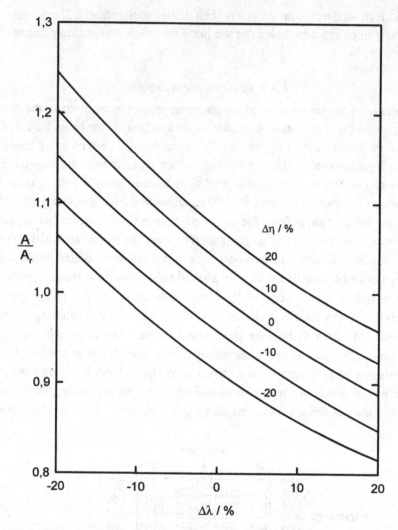

Fig. 2.3. The variation of the design heat–exchange area for the preheater in Figure 2.2 with the viscosity and thermal conductivity of the process streams.

value for the heat–exchange area for the preheater A_r. Subsequently, perturbations have been applied to the transport properties of the gases to examine the effect upon the heat–exchange area A. Figure 2.3 shows the ratio of the design area A to the reference area A_r resulting from various, reasonable perturbations of the viscosity and thermal conductivity of the gases on both sides of the preheater. It can be seen that if the viscosity of the gases employed for the design is 20% above its reference ('true') value and the thermal conductivity 20% below its reference ('true') value then it would be concluded that the heat exchanger needs an area 25% larger than that actually required. As a consequence, if this design were adopted for the preheater, the total reactor system could be constructed 15% larger than necessary with a significant increase in capital costs. On the other hand, if the errors in the two properties were in the opposite sense then the preheater would be underdesigned, leading to a lower feed temperature to the catalytic beds and a subsequent reduction in the efficiency of the overall plant, with an increase in operating costs. For this last reason it is usual to overdesign the heat–transfer equipment and accept a larger capital cost that could be avoided if more accurate values for the transport properties of the fluids were available.

2.3.3 Nuclear reactor safety

Figure 2.4 contains a schematic diagram of a natural uranium graphite–moderated (Magnox) nuclear reactor (Collier & Hewitt 1987). In a nuclear reactor the energy produced by the self–sustaining fission of a material such as ^{235}U is used to generate heat, which is transferred to a coolant circulating through the reactor. In turn, the heat absorbed by the coolant stream is transferred to a steam generator, which is then used to power a turbine for electricity generation. In the Magnox reactor the coolant is carbon dioxide at a pressure of 2 MPa, which leaves the reactor core at a temperature of 670 K. The reactor core itself is typically 14 m in diameter and 8 m high and is contained in a steel or concrete pressure vessel. The safety of such reactors has, quite naturally, caused considerable public concern, and their design, as well as that of other nuclear installations, is the subject of national legislation and international regulation. In particular, the safety audit for such a reactor must consider the circumstances surrounding the failure of one or more components in the entire plant. Thus, in addition to design

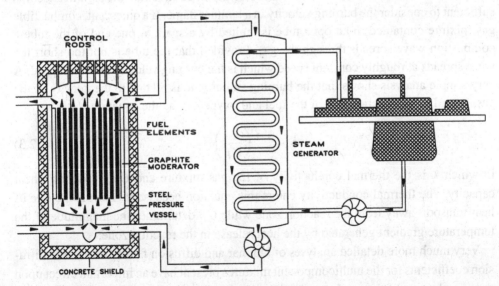

Fig. 2.4. A schematic diagram of a Magnox nuclear power plant.

calculations for the heat–exchange processes, one must consider extreme conditions when, for example, there is a loss of pressure in the coolant cycle or a failure of circulation for some reason. Generally, these latter calculations form a part of the safety case for the installation, and quality assurance then dictates that every single numerical value for a physical property is validated and of an appropriate pedigree. Included in such validated data would be the transport properties of the coolant stream allowing for possible contaminants as a result of a variety of modes of failure. The implication of the need for validated data implies high accuracy and a degree of approval from an appropriate body. In many cases, data produced by a body such as IUPAC (Vesovic *et al.* 1990), which have international approval, would be deemed satisfactory. It seems likely that owing to the internationalization in the trade of plant designs of all kinds, such international approval will become increasingly important in satisfying the demands of national regulatory bodies for quality assurance.

2.3.4 Combustion

Combustion is important in many areas of technology and it remains the single most important process of energy production in the world. While the essential features of combustion processes are well–understood, the details are still the subject of active research. Indeed there is currently great interest in studies to reduce pollutant formation in combustion systems as well as studies of detonations in the area of safety.

The principal feature of any combustion process is the chemical reaction of a fuel with oxygen which releases useful heat. Naturally, therefore, the process is dominated by the reaction kinetics involved, but the transport properties of the fuel and the products of combustion have some significance in its description. To illustrate the significance it is sufficient to consider the burning velocity in a laminar flame. If a quiescent, combustible gas mixture contained in an open tube is ignited by a spark at one end of the tube a combustion wave spreads through the gas. Provided that the tube is not too short the wave spreads at roughly constant speed which is the burning velocity or flame speed. A very simple analysis shows that the burning velocity, v, is related to the density of the gas, ρ, and to the reaction rate of the fuel and oxygen, r, by the equation

$$v = \rho r^{1/2} \left(\frac{\lambda}{C_P} \right)^{1/2} \tag{2.3}$$

in which λ is the thermal conductivity of the gas mixture and C_P its isobaric heat capacity. The thermal conductivity enters this equation because it controls the rate of heat transport away from the reaction zone while C_P determines the magnitude of the temperature gradient generated by the heat release in the reaction zone.

Very much more detailed analyses of laminar and diffusion flames reveal that diffusion coefficients for the multicomponent mixtures present have an important effect upon the complete set of the products of combustion while even the viscosity of the system can be important. Of course, in a particular combustion system it is not possible to alter the transport properties of the fluid so that data on such properties are of significance in the interpretation of combustion experiments and in modeling processes rather than in design.

2.3.5 *Modeling of heat transfer*

There are many industrial processes, frequently operated at high temperature, where the quality of a solid product is crucially dependent on the cooling and solidification from a molten state. The most familiar examples are the continuous casting process for steels and the float–glass process. However, there are more recent developments, such as the production of near–perfect crystals of semiconducting materials, where the degree of perfection of the crystal has a profound effect on the ultimate performance of electronic devices constructed from them.

In the improvement of all of these processes a complete understanding of heat transfer by both conduction and convection is essential. Since the governing hydrodynamic equations are well known, the accuracy of models of such processes depends sensitively on, and is currently limited by, our knowledge of the constitutive equations of the molten materials and, in particular, upon the transport coefficients which enter them. Significant advances in the quality and uniformity of a number of materials might be attainable were accurate data for the thermal conductivity and viscosity of molten materials at high temperature available.

2.4 Standard reference data

For a number of materials it is vital that property values are provided which have a validated accuracy and are identified as standard reference data (SRD), preferably with international approval and recognition. The preparation of such internationally approved data standards is a timeconsuming and delicate activity that requires a critical evaluation of all the available measurements with a detailed assessment of the accuracy of each individual datum reported. For these reasons, only results obtained in instruments characterized by high quality and a complete working equation based upon a sound theory can be employed for the establishment of standard reference data. Whenever possible the results of measurements made with different experimental techniques should be included.

These conditions on the establishment of standard reference data are satisfied by the results for relatively few fluids in the case of transport properties. Examples of cases where the conditions have been satisfied are provided by the standard reference values for the thermal conductivity of three liquids and the transport property correlations for carbon dioxide provided by the International Union of Pure and Applied Chemistry (Nieto de Castro *et al.* 1986; Vesovic *et al.* 1990).

The absence of standard reference data of this kind for calibration or other purposes can cause some industrial activities to be impaired, legal disputes to arise between organizations and limit the assessment of the quality of measurements made in a relative manner.

References

Armstrong, J.B., Li, S.F.Y. & Wakeham, W.A. (1982). The effect of errors in the thermophysical properties of fluids upon plant design. *ASME Winter Annual Meeting, 1982* Paper 82–WAHT–84.

Bird, R.B., Stewart, W.E. & Lightfoot, E.N. (1960). *Transport Phenomena*. New York: Wiley.

Collier, J.G. & Hewitt, G.F. (1987). *Introduction to Nuclear Power*. New York: Hemisphere.

Maitland, G.C., Rigby, M., Smith, E.B. & Wakeham, W.A. (1987). *Intermolecular Forces: Their Origin and Determination*. Oxford: Clarendon Press.

Nieto de Castro, C. A., Li, S. F. Y., Nagashima, A., Trengove R. D. & Wakeham, W. A. (1986). Standard reference data for the thermal conductivity of liquids. *J. Phys. Chem. Ref. Data*, **15**, 1073–1086.

Vesovic, V. & Wakeham, W. A. (1987). An interpretation of intermolecular pair potentials obtained by inversion for non–spherical systems. *Mol. Phys.*, **62**, 1239–1246.

Vesovic, V., Wakeham, W.A., Olchowy, G.A., Sengers, J.V., Watson, J.T.R. & Millat, J. (1990). The transport properties of carbon dioxide. *J. Phys. Chem. Ref. Data*, **19**, 763–808.

3

Methodology

C. A. NIETO DE CASTRO

University of Lisbon, Portugal

W. A. WAKEHAM

Imperial College, London, UK

3.1 Introduction

It is now estimated that there are some 50 million pure chemicals known of which some 20,000 are listed as high–volume, major chemicals by the European Economic Community (Forcheri & de Rijk 1981) some of which may be transported across national borders. For each pure fluid there are approximately 30 properties which are of technological significance of which twelve are functions of temperature and pressure. If just these twelve properties are considered and it is assumed that measurements at only ten pressures and ten temperatures are required then to provide the necessary information for only one pure fluid requires 1200 measurements. If all the pure species and all possible mixtures from among the set of bulk chemicals are included and composition is allowed as a variable then it is rather easy to estimate that, even for a generous estimate of the rate of experimental data acquisition in the world, the total effort required to fulfill the needs identified in Chapter 2 would exceed 100 billion man–years. This figure makes it immediately obvious that industry's needs for physical property data can never be met by measurement alone. It is therefore necessary to replace a complete program of measurements by an alternative strategy designed to meet the same objective. The philosophy and methods for the establishment of such a strategy have been discussed by many authors and have been updated regularly and most recently by Nieto de Castro & Wakeham (1992).

The present chapter is, therefore, devoted to the methodology underlying the alternative strategy which the authors of this book believe to be appropriate. The chapter provides a definition of the levels of a hierarchy of correlation, prediction and estimation procedures that seek to generate the physical properties of fluids and their mixtures by means other than direct measurements. These different methods are then expanded in the subsequent sections of the book and examples of their application given.

3.2 Correlation

Irrespective of any theoretical background it is known from experience that the trans-port properties of a fluid depend upon temperature and density (or pressure). Thus, for a particular fluid, a reasonable means to satisfy the need of industry for the transport properties of a fluid over a wide range of conditions is provided by the empirical cor-relation of available experimental data. In view of the comments made in the previous section this is evidently a route that will only ever be available for a small number of fluids. Moreover, the process of correlating the dependence of the transport property upon the independent state variables is not itself a straightforward one (see Chapter 7).

The dependence of a transport property $X(\rho, T)$ can be always written as the sum of three contributions

$$X(\rho, T) = X^{(0)}(T) + \Delta X(\rho, T) + \Delta X_c(\rho, T) \tag{3.1}$$

where X = viscosity (η), thermal conductivity (λ), the product of molar density and diffusion (ρD) or the product of molar density and thermal diffusivity (ρa). Further-more, $X^{(0)}(T)$ is the dilute–gas value of the property, $\Delta X(\rho, T)$ is the excess property and $\Delta X_c(\rho, T)$ is the property critical enhancement. Equation (3.1) can also be written in the form

$$X(\rho, T) = \bar{X}(\rho, T) + \Delta X_c(\rho, T) \tag{3.2}$$

where $\bar{X}(\rho, T) = X^{(0)}(T) + \Delta X(\rho, T)$ is the background or regular value of the property. Figure 3.1 displays schematically these definitions for an isotherm of the thermal conductivity in the $\lambda(\rho)$ plane.

However, it should be recognized that although the theory of the transport properties of fluids is not completely developed, it can provide some guidance in the process of correlation. For example, all kinetic theories of transport reveal that it is the tempera-ture and density that are the fundamental state variables and that pressure is of no direct significance. Since most measurements are carried out at specified pressures and not specified densities, this automatically means that a single, uniform equation of state must be used to convert any experimental data to (ρ, T) space from the experimental (p, T) space. Furthermore, the dilute–gas kinetic theory reveals a number of relation-ships between different properties of a gas that are exact or nearly exact so that these relationships provide consistency tests for experimental data as well as constraints that must be satisfied by the final correlation of the properties.

One point of significance in the process of correlation is the recognition that not all experimental values are of equal worth. The field of transport properties is littered with examples of quite erroneous measurements made, in good faith, with instruments whose theory was not completely understood. It is therefore always necessary to separate all of the experimental data collected during a literature search into primary and secondary data by means of a thorough study of each paper.

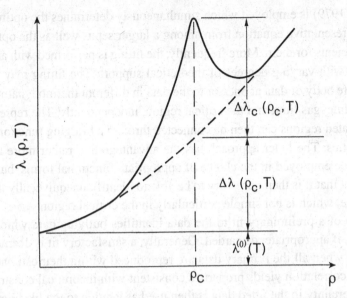

Fig. 3.1. The different contributions to the thermal conductivity of a fluid close to the critical point.

Primary data are those to be used in the development of the correlation and they will satisfy the following conditions:

(i) the measurements will have been carried out in an instrument for which a complete working equation is available together with a complete set of corrections,
(ii) the instrument will have had a high sensitivity to the property to be measured,
(iii) the primary, measured variables will have been determined with a high precision.

Occasionally, experimental data that fail to satisfy these conditions may be included in the primary data set if they are unique in their coverage of a particular region of state and cannot be shown to be inconsistent with theoretical constraints. Their inclusion is encouraged if other measurements made in the same instrument are consistent with independent, nominally more accurate data. Secondary data, excluded by the above conditions, are used for comparison only.

In association with the process of data selection an estimate of the accuracy must be made for each set of results on a basis which is independent of that of the original authors. This estimate of uncertainty usually refers to precision and not accuracy, and a statistical weight is then determined for each datum based upon that uncertainty. This process of weighting cannot be exact, and the relative weight among different data sets is the only relevant parameter; the process is important in cases where a number of data sets are available of different precision.

Finally, the selected data are fitted to an appropriate functional form using least–squares procedures (see Chapter 7). Occasionally a procedure such as SEEQ (de Reuck

& Armstrong 1979) is employed which simultaneously determines the optimum set of terms in a representative equation from among a larger set as well as the optimum numerical coefficients for them. More frequently, the fitting is performed with an equation of fixed form with varying degrees of theoretical support. The fitting may be carried out to the entire body of data at once or to the data in different thermodynamic regions, such as the dilute–gas limit and the critical region, independently. The representations in these separated regions can then be connected through a bridging function for intermediate densities. The latter approach has the advantage that rather more theoretical guidance can be employed in the choice of appropriate functional forms, but it suffers from the defect that it is then necessary to be able to identify unequivocally the various regions of state, which is not simple particularly in the critical region.

Completion of a preliminary fit to the data identifies outlying data which must be examined and, if appropriate, discarded. Generally, a satisfactory fit is deemed to have been achieved when all the primary data are reproduced within their estimated uncertainty and the correlation yields properties consistent with theoretical constraints. The estimated uncertainty in the fitted data is then used as a guide to the likely error in the property arising from the use of the correlation.

Several examples of the use of this procedure for the transport properties of fluids will be found later in this book.

3.3 Prediction

The word 'prediction' is defined to imply the generation of values of a transport property of a fluid or fluid mixture, by means of a method based upon a rigorous theory, in a region of state where direct measurements do not exist.

Undoubtedly the most effective substitute for direct measurement of the transport properties of fluids would be a complete, rigorous statistical–mechanical theory that enabled the calculation of the properties of a macroscopic ensemble of molecules from a knowledge of molecular properties and of the forces between the molecules. Indeed, one might perceive this as the ultimate objective of statistical–mechanical theory. The impossibility of carrying through this program at present rests upon the fact that there exists no rigorous, applicable theory of transport properties except in the dilute–gas state, where only pair interaction potentials matter or, asymptotically close to the critical point, where the details of the intermolecular forces are irrelevant and effects depend upon the behavior of large clusters of molecules. Nevertheless, a procedure of this kind is defined as a *prediction* of the physical properties. It is thus defined because it is possible to evaluate the properties of pure fluids or mixtures without recourse to any measurements of the properties being evaluated. From the comments made above it should be clear that the number of occasions on which this route to properties will be available is exceedingly small.

3.3.1 Semitheoretical prediction

A subset of predictive procedures is formed when a limited set of measurements of some physical property for a particular fluid, either microscopic or macroscopic, has been performed for which a rigorous theory exists, but which is devoid of approximations or contains approximations that are well–characterized. The theory may then be used to allow the available measurements on one property to be used to predict another for which no measurements are available. This may be done either directly through an explicit theoretical relationship between the properties, or through the intermediacy of an intermolecular potential derived from data on one property. In the first case it is unlikely that the temperature range of the prediction can exceed that of the original measurements, whereas in the latter case it is possible that new thermodynamic states may be treated. The scheme is predictive in the sense that no information on the property to be evaluated is required; an example of such a procedure is the evaluation of the thermal conductivity of monatomic gases from the viscosity, which is discussed in Chapters 4 and 11.

A further type of predictive method arises when mixtures are considered. Again, if a complete, rigorous theory for the properties of a fluid containing many components exists it is frequently the case that the mixture properties depend only upon well–defined quantities, characteristic of all of the various binary interactions in the system. In such circumstances, either of the predictive means set out above may be used to evaluate the quantities for each binary interaction, and their combination with theory then leads to values of the property of the multicomponent mixture without the need for data on that property.

3.3.2 Predictive methods and models

While retaining the rigor of theory it is possible to develop less exact predictive schemes that are of considerable utility. Such methods generally invoke a physical model of the interactions between the molecules in the system. Perhaps the best known of these methods, as well as that which departs the least from absolute rigor, is the Extended Law of Corresponding States, discussed in detail in Chapter 11. In this procedure, to the rigorous kinetic theory of dilute monatomic gases and gas mixtures is added the hypothesis that the spherical intermolecular pair potentials for interactions among the species can be rendered conformal by a suitable choice of scaling parameters for energy ϵ and distance σ. It follows from this hypothesis that each functional of the pair potential is a universal function of the reduced temperature, $T^* = k_B T / \epsilon$. Thus, determinations of the functional from measurements of a property for a variety of different species mean that the functional can be determined empirically for the complete set of species over a wider range of reduced temperature than can be investigated for any one species. The hypothesis of universality means that the property of one particular species may then be predicted in a range of conditions outside of those in which it has been measured.

A second, but less satisfactory, version of the same idea is provided by the use of a prescribed intermolecular pair potential such as the Lennard–Jones (12–6) model (Maitland *et al.* 1987). This model, while not an exact representation of any known intermolecular force, has the virtues of mathematical simplicity and the generally correct gross features of real intermolecular potentials. The disadvantages are that when used in the manner outlined above to evaluate the appropriate functionals for a particular property of a fluid, it is unable to represent the experimentally observed behavior correctly even in a limited temperature range. As a consequence, the predictions of properties that follow from its use are intrinsically less accurate than those of the corresponding–states procedure. Indeed, this example serves to illustrate a general point that the more constrained the model attached to a rigorous theory, the less reliable the predictions of the procedure based upon it.

It should be noted that for all of the transport properties of dilute gases an exact theory exists so that for them all, according to the present definition, prediction is possible, given an appropriate model and some experimental information. When no experimental information on a particular substance exists at all and when there is no known intermolecular potential for a particular interaction, it is clear that, to a large extent, the existence of an exact theory is irrelevant and one is forced to use an estimation procedure from among the set discussed below.

3.4 Estimation

Estimation procedures are frequently thought of as those which are simply less accurate than the type of methods discussed above. However, this is a generalization which is neither useful nor, necessarily, correct, especially in the case when prescribed intermolecular pair potentials are used. It is preferable, therefore, to define estimation in terms of the degree of departure from a sound theoretical basis so that it is then possible to distinguish a variety of different levels of procedure.

3.4.1 Semitheoretical estimation

Within the category of estimation procedures those that enjoy the highest level of confidence are those where the exactitude of a rigorous theory is maintained, but the parameters that enter into it are estimated without the benefit of experimental information. The use of an exact theory ensures that the estimated property will possess gross features that are fundamentally correct, although the use of approximate values of the parameters will mean that the values of the property will be less accurate than if experimental data were employed. For example, in the application of the Extended Law of Corresponding States for dilute–gas properties to mixtures it is always a sensible procedure to estimate the scaling parameters of an unlike interaction from those of the two pure component interactions from combination rules if no information on the binary interaction is available from elsewhere. This preserves the exact form of the equation for the transport property and confines the estimation to a quantity that has a relatively minor effect upon the overall property calculation.

3.4.2 Approximate theoretical methods

Another class of estimation procedures may be identified which is founded upon an inexact theory. The theory employed may be inexact because of simplifications made in its development or because it is based upon a particular molecular model. Such procedures may be further subdivided according to the degree of approximation introduced by the model. A familiar example of such a procedure from thermodynamics is the van der Waals equation of state. An extremely simple model of the finite volume of the molecules and of the attractive forces between them leads to an equation of state that possesses many of the gross features of real fluids, although it represents the behavior of none exactly (see Chapter 8). In the field of transport properties a similar type of procedure is the basis of the Enskog theory of transport in dense fluids, discussed in Chapters 5 and 10. In this theory, molecules are modeled as rigid spheres of fixed volume and a simple ansatz used to retain the concept of molecular chaos. Neither of these two elements of the theory is strictly correct, but they render the analytic evaluation of the theory practicable. Once again, the theoretical results in their original form fail to represent the behavior of any real fluid. However, if the parameters and/or functions that arise in the theoretical results are adjusted to fit a limited set of experimental data then a reliable interpolatory and extrapolatory procedure can be developed for a particular fluid or fluid mixture. Naturally, because the theory upon which the procedure is based is approximate, the reliability is reduced when compared with that of the schemes discussed earlier. Furthermore, the empirical determination of parameters in the theory makes it difficult to use procedures of this kind in a predictive manner, that is, without input from experiment unless further, entirely heuristic steps are taken. For example, it may be possible to relate the parameters or functions determined from experiment for some fluids to molecular structure, density or temperature in a manner that might apply to other fluids for which there are no transport property data. In this case the methods may be used to evaluate the property for a given fluid outside the range of conditions for which experimental data were used to derive values for the parameters. In some cases the process may be extended to new fluids for which no measurements exist. In any event, the further down the road of empiricism one proceeds the less reliable the calculations of the properties become.

3.4.3 Semiempirical estimation

When an approximate theory is linked with behavior determined from experiment alone, the further step away from rigor identifies a class of semiempirical estimation schemes.

An example of such a procedure is the generalized corresponding–states method of Hanley *et al.*, described in Chapter 12. Here, a simplified, and therefore approximate, version of the exact theory is combined with dimensional analysis to lead to a corresponding–states procedure in terms of reduced, macroscopic variables. The implementation of this corresponding–states procedure rests upon the availability of the behavior of the transport properties of a reference substance. Since no theoretical means is available for determining the behavior of any reference substance over a wide range of thermodynamic states, the reference behavior is determined from experiment for a particular fluid. This empirical determination of the behavior of a reference substance then provides the means to determine adjustable parameters for a large number of other fluids from a limited set of information.

The advantages of this kind of procedure are their generality and ability to make estimations over a wide range of states and classes of pure substances and mixtures. Indeed, extensions of such procedures by means of further empirical steps relating the requisite parameters of substances to critical point parameters and/or molecular structure then enable estimations of the transport properties of substances to be made for which no direct measurements exist.

3.4.4 Empirical estimation

A final class of estimation properties are those that are entirely empirical. They are not usually based on any theory beyond, possibly, dimensional analysis, although they are sometimes derived from simplistic physical models. Instead these procedures rely upon the observation and representation of the relationship between a property and pressure, density temperature or molecular structure across a wide range of materials or mixtures. Generally, such procedures make use of a large body of experimental data of variable, and frequently unknown, accuracy for a wide range of compounds to establish an empirical, functional relationship between the property and a number of selected variables or combinations of variables. Generally, an attempt is made in the development of such a representation to use as the characteristic parameters for each fluid those that are most frequently known for chemical materials such as the normal boiling point, molecular mass, critical temperature *etc*. In the sense that all of these parameters must, ultimately, be implicitly related to the transport property through the laws governing molecular motion and the interaction between molecules, it is not unreasonable that one should seek simple, explicit relationships of this kind. However, that the same simple relationship should remain exact for a wide range of materials is evidently unlikely. Thus, procedures based on such principles are often not robust in the sense that their extrapolation beyond the range of conditions or type of compound for which they were originally developed can lead to serious errors.

Such procedures will always have a role in the evaluation of the properties of fluids because of their wide range of application in terms of materials. However, they should be viewed as a means of last resort to evaluate a property and they must always be employed with caution and even skepticism. Methods of this kind are dealt with only briefly in this volume (see Chapter 13), since they are exhaustively and admirably reviewed elsewhere (Reid *et al.* 1987).

3.5 Guidelines

It should be clear from the preceding classification of methods of correlation, prediction and estimation of the transport properties of fluids that the list has been presented in the preferred order of application. That is, whenever a correlation of critically evaluated data is available it should be used. Examples of the development of some of these correlations are given in later chapters for different classes of fluids. Wide–ranging correlations of this type are available for only a small subset of the fluids of interest, and the next best means of obtaining the properties is either directly from theory (in rare cases) or from a representation of the results of an exact theory supported by experimental data. This would, in fact, always be the preferred choice of method for the evaluation of the properties of mixtures where wide–ranging correlations in temperature, density and composition are not practicable. This approach is viable at present only for the dilute state of gases and gas mixtures.

If no experimental data are available for the implementation of this approach in its entirety then the use of a model intermolecular pair potential with scaling parameters derived from some other property or by estimation is to be preferred above any other means of evaluation for the dilute–gas state. In the dense fluid state it is at present necessary to make use of a procedure based upon an approximate theory. The Enskog theory in one of its forms is usually the best procedure of this kind. However, its application requires the availability of some experimental data for the property of interest at least for pure components so that again its application is limited.

The generalized corresponding–states procedures based upon an approximate theory can be applied to a much wider range of materials and require much less input information than the methods that place a greater reliance on theory. Since they have a basis in a simplified theory and some experiment, these methods are always to be preferred over entirely empirical schemes.

If all other methods are inapplicable, one must have recourse to entirely empirical methods. Owing to the unreliability of such methods, it is best to employ a number of different methods of this type and to assess their consistency but always to limit their application to fluids structurally similar to those employed in the formulation of the procedure.

The order of preference given here is that of decreasing accuracy of the predicted results and increasing order of generality of application. In computational terms none of the methods for the evaluation of the transport properties of fluids require complicated solution algorithms such as those that arise in the treatment of equilibrium properties.

Thus, there are seldom any reasons to select a particular method on the grounds of computational speed. It is preferable to use the best possible predictive method available in the hierarchy described here for particular applications.

References

de Reuck, K.M. & Armstrong, B.A. (1979). A method of correlation using a search procedure based on a step–wise least–squares technique, and its application to an equation of state for propylene. *Cryogenics*, **19**, 505–512.

Forcheri, S. & de Rijk, J.R., eds. (1981). *Classification of Chemicals in the Customs Tariff of the European Community. Vol. III.* Brussels: Commission of the European Community.

Maitland, G.C., Rigby, M., Smith, E.B. & Wakeham, W.A. (1987). *Intermolecular Forces: Their Origin and Determination.* Oxford: Clarendon Press.

Nieto de Castro, C. A. & Wakeham, W. A. (1992). The prediction of the transport properties of fluids. *Fluid Phase Equil.*, **79**, 265–276.

Reid, R. C., Prausnitz, J. M. & Poling, B. J. (1987). *The Properties of Gases and Liquids.* 4th ed., New York: McGraw-Hill.

Part two
THEORY

4

Transport Properties of Dilute Gases and Gaseous Mixtures

J. MILLAT

NORDUM Institut für Umwelt und Analytik, Kessin/Rostock, Germany

V. VESOVIC and W. A. WAKEHAM

Imperial College, London, UK

4.1 Introduction

Transport coefficients describe the process of relaxation to equilibrium from a state perturbed by application of temperature, pressure, density, velocity or composition gradients. The theoretical description of these phenomena constitutes that part of nonequilibrium statistical mechanics that is known as the *kinetic theory*. The ultimate purpose of this theory is to relate the macroscopic (observable) properties of a system to the microscopic properties of the individual molecules and their interaction potentials.

The kinetic theory of *dilute gases* assumes a macroscopic system at densities low enough so that molecules most of the time move freely and interact through *binary encounters* only. Nevertheless, the densities are high enough to ensure that the effects of molecule–wall collisions can be neglected compared to those from molecule–molecule encounters. The first condition implies that the thermodynamic state of the fluid should be adequately described by a virial expansion up to and including the second virial coefficient. The second condition means that the mean free path of molecules is much smaller than any dimension of the vessel and that Knudsen effects play no significant role.

It is worth noting that in this context the terms 'dilute' or 'low–density gas' represent a real physical situation, whereas the frequently used expression 'zero–density limit' is related to results of a mathematical extrapolation of a density series of a particular transport property at constant temperature to zero density. The derived value is assumed to be identical with the 'true value' for the dilute–gas state, a statement that in most cases turns out to be correct.

This volume is predominantly concerned with the transport properties of fluids and fluid mixtures of practical significance. This means that attention is concentrated upon systems containing polyatomic molecules and upon the 'traditional' transport properties such as the viscosity, thermal conductivity and diffusion coefficients. The development of the kinetic theory of polyatomic gases has led to a unifying methodology for the description of the practical relationships between the microscopic characteristics of

29

molecules and their macroscopic properties (McCourt *et al.* 1990, 1991; Vesovic & Wakeham 1992; McCourt 1992).

The ease of the practical evaluation of the transport properties of a dilute gas by means of these relationships decreases as the complexity of the molecules increases. Thus for a pure monatomic gas, with no internal degrees of freedom, the calculations are now trivial, consuming minutes on a personal computer. For systems involving atoms and rigid rotors the computations are now almost routine and take hours on a work station (see, for instance, Dickinson & Heck 1990). For two rigid rotors the calculations are only just beginning to be carried out and take days on a work station (Heck & Dickinson 1994). For systems that involve molecules other than rigid rotors the theory is still approximate and calculations are heuristic.

The relevant results from these analyses are summarized in this chapter first for pure gases and then for mixtures. The presentation includes exact and approximate relationships derived from the theory between various transport properties which are of considerable value in the evaluation of properties or even for their representation.

Although the contribution is written with an emphasis on polyatomic gases, it also includes results for pure monatomic gases and mixtures containing monatomic species. Chapters 5 and 6 consider fluids at moderate and high densities as well as the critical enhancements to the transport properties and, therefore, complete this brief summary of the statistical theory of fluids.

4.2 Pure gases

4.2.1 Kinetic theory of dilute polyatomic gases

4.2.1.1 General results

There are two essential features that distinguish polyatomic molecules from structureless (monatomic) species. First, they possess internal energy in the form of rotational and vibrational modes of motion. Second, they interact through non–spherically symmetric intermolecular pair potentials. Hence, during molecular encounters, energy may be exchanged between internal states of the two molecules and between internal states and translational motion. In principle, a quantum–mechanical treatment of processes in the gas is necessary to account for the changes of internal state. However, the explicit form of the intermolecular potential does not need to be known in order to carry the theory through.

The fully quantum–mechanical kinetic theory of polyatomic gases is based on the Waldmann–Snider equation (Waldmann 1957; Snider 1960) and was subsequently developed and summarized by McCourt and co–workers (for instance, McCourt *et al.* 1990, 1991). Wang Chang and Uhlenbeck (1951) and independently de Boer (see Wang Chang *et al.* 1964) formulated a semiclassical kinetic theory (Wang Chang–Uhlenbeck– de Boer theory, or WCUB) subsequently developed by Monchick, Mason and their collaborators (Mason & Monchick 1962; Monchick *et al.* 1965). The essential features of the latter theory are to identify each internal quantum state of a particular molecule as a separate chemical species and to treat the translational motion classically.

The quantum–mechanical theory has the advantage that it can treat the degeneracy of rotational energy states and is therefore able to describe the effect of magnetic and electric fields on the transport properties of a gas (McCourt *et al.* 1990, 1991). These effects, known as polarization, arise because a gradient of any state variable, such as temperature or velocity, causes a small, preferential alignment of the molecules through collision and thus a partial polarization of their angular momentum vector. The disadvantage of this theory for practical applications is that it is only formally established for gases with rotational degrees of freedom (rigid rotors). On the other hand, the semiclassical theory has the advantages that it treats all forms of internal energy and is the semiclassical limit of the quantum–mechanical approach, although it has the disadvantage that it describes strictly only the thermophysical properties of a gas in the limit of very high magnetic fields where the rotational states are nondegenerate. For the present purpose of discussing the traditional transport properties of gases, it seems important that it should be possible to treat all forms of internal energy, so that it is preferable to employ the WCUB (semiclassical) results and to correct, where necessary, for the small (order of magnitude of 1%) effects arising from the neglect of degenerate states.

The starting point for the kinetic theory of polyatomic gases is the formulation of the equation of change for the singlet distribution function f_i ($\mathbf{c}, E_i, \mathbf{r}, t$) for each quantum state $\{i\}$, where the single index i refers to the entire set of quantum numbers necessary to characterize the internal state of a molecule. Here $f d\mathbf{c} d\mathbf{r}$ is the number of molecules with internal energy E_i, translational velocity $\mathbf{c}$ and position $\mathbf{r}$ in the range $d\mathbf{c}$ and $d\mathbf{r}$ at time t. For small departures from the equilibrium state the singlet distribution function is represented in terms of the generalized local equilibrium distribution function $f_i^{(0)}$ and a linear perturbation ϕ_i so that (Ferziger and Kaper 1972)

$$f_i = f_i^{(0)}(1 + \phi_i) \tag{4.1}$$

where

$$f_i^{(0)} = \frac{n}{Q}\left(\frac{m}{2\pi k_B T}\right)^{3/2} \exp\left(-C^2 - \epsilon_i\right) \tag{4.2}$$

where m denotes the mass of a molecule, n the local number density, T the temperature, k_B the Boltzmann constant, $Q = \sum_i \exp(-\epsilon_i)$ the internal state partition function, $\epsilon_i = E_i/k_B T$ the reduced energy of the ith internal quantum state and C is the reduced peculiar velocity given by

$$C = (m/2k_B T)^{1/2}(\mathbf{c} - \mathbf{c_0}) \tag{4.3}$$

where $\mathbf{c_0}$ is the average velocity.

In the absence of external fields the evolution of f_i is then governed by the Wang Chang–Uhlenbeck–de Boer equation – a form of generalized Boltzmann equation – that can be written after linearization as

$$\mathcal{D}f_i^{(0)} + f_i^{(0)}\mathcal{D}(\phi_i) = -f_i^{(0)}n\mathcal{R}(\phi_i + \phi_j) \tag{4.4}$$

where $\mathcal{D}$ is the streaming operator and the WCUB collision operator $\mathcal{R}$ is defined by

$$nf_i^{(0)}\mathcal{R}(Y) = \sum_{jkl} \int\int\int f_i^{(0)} f_j^{(0)} [Y - Y'] g\sigma_{ij}^{kl}(g, \chi, \psi)\sin\chi\, d\chi\, d\psi\, d\mathbf{c}_j \tag{4.5}$$

and is related to a binary collision in which the two colliding molecules initially in states i and j approach each other with the relative velocity $\mathbf{g}$. After collision the relative velocity has changed to $\mathbf{g}'$ and is rotated through the polar angle χ and the azimuthal angle ψ while the internal states of the molecules have changed to k and l. Furthermore, $\sigma_{ij}^{kl}(g, \chi, \psi)$ is the *inelastic* differential cross section for the collision process.

A solution to the WCUB equation for $\phi_i(\mathbf{c}, E_i, \mathbf{r}, t)$ can be obtained by an extension of the Chapman–Enskog solution of the classical Boltzmann equation for a monatomic gas (Ferziger & Kaper 1972; Maitland *et al.* 1987). Accordingly, the perturbation ϕ_i has the form

$$\phi_i = -\frac{1}{n}\Gamma\nabla\mathbf{c}_0 - \frac{1}{n}\mathbf{A}\cdot\nabla\ln T - \frac{1}{n}\mathbf{B}:\nabla\mathbf{c}_0 \tag{4.6}$$

with n, $\mathbf{c}_0$ and T being the local values of the number density, mass–average velocity and temperature respectively. The scalar Γ, the vector $\mathbf{A}$ and the second–rank tensor $\mathbf{B}$ are then expanded in series of orthonormal basis functions which are chosen to be scalars, vectors or second–rank tensors respectively. One choice of basis functions, known as the two–flux approach (McCourt *et al.* 1990, 1991; Ross *et al.* 1992), leads to

$$\Gamma = \sum_{s,t} c_{st}\Phi^{00st}; \qquad \mathbf{A} = \sum_{s,t} a_{st}\Phi^{10st}; \qquad \mathbf{B} = \sum_{s,t} b_{st}\Phi^{20st} \tag{4.7}$$

The orthonormal basis functions Φ^{pqst} are tensors of rank p in molecular velocity C and q in molecular angular momentum $\mathbf{j}$. The remaining indices s and t denote the order of polynomials in C^2 (Sonine polynomials) and in $\epsilon_i - \langle\epsilon\rangle_0$ (Wang Chang–Uhlenbeck polynomials) respectively (see Ferziger & Kaper 1972). The symbol $\langle\epsilon\rangle_0$ represents the equilibrium average internal energy, because the operation $\langle X\rangle_0$ is defined by

$$\langle X\rangle_0 \equiv \frac{\sum_i \int f_i^{(0)} X_i\, d\mathbf{c}}{\sum_i \int f_i^{(0)}\, d\mathbf{c}} \tag{4.8}$$

An alternative approach to thermal conductivity is based on a different set of basis functions Φ^{p0X} originally proposed by Thijsse *et al.* (1979). The starting point of this so–called total–energy–flux approach is the natural assumption that the leading

Table 4.1. *Selected basis tensors and their meaning.*

Tensor	Process
Two–flux approach	
$\Phi^{1000} = \sqrt{2}\,C$	particle flux
$\Phi^{1010} = \left(\frac{4}{5}\right)^{1/2}\left(C^2 - \frac{5}{2}\right)C$	translational energy flux
$\Phi^{1001} = \left(\frac{4}{5}\right)^{1/2}\frac{1}{r}\left(\epsilon - \langle\epsilon\rangle_0\right)C$	internal energy flux[a]
$\Phi^{2000} = \sqrt{2}\,\overline{CC}$	momentum flux
$\Phi^{0001} = \left(\frac{2}{5}\right)^{1/2}\frac{1}{r}\left(\epsilon - \langle\epsilon\rangle_0\right)$	internal energy[a]
$\Phi^{0010} = \left(\frac{2}{3}\right)^{1/2}\left(C^2 - \frac{3}{2}\right)$	translational energy
Total–energy flux approach	
$\Phi^{10E} = C\left(\frac{4}{5}\right)^{1/2}\frac{1}{(1+r^2)^{1/2}}\left[C^2 - \frac{5}{2} + \epsilon - \langle\epsilon\rangle_0\right]$	total–energy flux[a]
$\Phi^{10D} = C\left(\frac{4}{5}\right)^{1/2}\frac{1}{(1+r^2)^{1/2}}$ $\left[r\left(C^2 - \frac{5}{2}\right) - \left(\frac{1}{r}\right)\left(\epsilon - \langle\epsilon\rangle_0\right)\right]$	difference energy flux[a]

[a] $r = (2c_{int}/5k_B)^{1/2}$.

expansion term should be proportional to the quantity that is transferred, i.e., to the total energy E. Accordingly, the basis function Φ^{10E} represents the total energy flux, whereas Φ^{10D} is a difference function that is orthogonal to Φ^{10E}. The total–energy basis functions are linear combinations of the basis functions of the two–flux approach introduced before, as can be seen from Table 4.1, where selected basis functions are explicitly defined.

4.2.1.2 Transport coefficients

The explicit evaluation of the transport coefficients of the gas is now performed in a series of steps detailed elsewhere (Hirschfelder *et al.* 1954; Chapman & Cowling 1970; Ferziger & Kaper 1972; Maitland *et al.* 1987; McCourt *et al.* 1990, 1991). First, using the local equilibrium solution $f_i^{(0)}$, Euler's equations of hydrodynamics are constructed and employed to eliminate the time dependence from the left–hand sides of equation (4.4). This then becomes an integral equation for the perturbations ϕ_i. Furthermore, if ϕ_i is assumed to be of the form of equation (4.6) then each of the quantities Γ, $\mathbf{A}$ and $\mathbf{B}$ must independently satisfy similar integral equations. Since the local equilibrium solution $f_i^{(0)}$ was already applied to provide the local values of the molecular number density, mean velocity and temperature, it follows that the quantities Γ, $\mathbf{A}$ and $\mathbf{B}$ must also obey auxiliary conditions, if this assignment is to remain valid. The integral equations, together with the auxiliary conditions, serve to determine Γ, $\mathbf{A}$ and $\mathbf{B}$ uniquely and thereby ϕ_i.

Using the subsequent first–order perturbation solution for f_i, the stress tensor and heat–flux vector in the gas can be evaluated explicitly in terms of Γ, $\mathbf{A}$ and $\mathbf{B}$, and thus the transport coefficients, which are the proportionality factors in the phenomenological constitutive equations for the gas, may be determined. It turns out that Γ is related only to the bulk viscosity of the gas, $\mathbf{A}$ to the thermal conductivity and $\mathbf{B}$ to the viscosity. Specifically the bulk viscosity is given by

$$\kappa = k_B [\Gamma \cdot \mathcal{R}(\Gamma)] \tag{4.9}$$

the thermal conductivity by

$$\lambda = \frac{2k_B^2 T}{3m} [\mathbf{A} \cdot \mathcal{R}(\mathbf{A})] \tag{4.10}$$

and the viscosity by

$$\eta = \frac{k_B T}{10} [\mathbf{B} : \mathcal{R}(\mathbf{B})] \tag{4.11}$$

where the notation $[X \odot \mathcal{R}(X)]$ denotes kinetic theory bracket integrals (Ferziger & Kaper 1972).

If the basis functions Φ^{pqst} of equation (4.7) are employed then these results become remarkably simple, since the orthonormality of the basis functions ensures that the bulk viscosity κ and the dynamic viscosity η depend upon just one of the relevant expansion coefficients while the thermal conductivity depends upon two. Specifically, for bulk viscosity

$$\kappa = -\frac{2(5)^{1/2} k_B T}{3} r c_{01} \tag{4.12}$$

for thermal conductivity

$$\lambda = \frac{2k_B^2 T}{m} \left[-\left(\frac{5}{2}\right)^{1/2} a_{10} + \left(\frac{5}{2}\right)^{1/2} r a_{01} \right] \tag{4.13}$$

and for viscosity

$$\eta = k_B T b_{00} \tag{4.14}$$

where

$$r^2 = \frac{2c_{int}}{5k_B} \tag{4.15}$$

and c_{int} is the ideal–gas isochoric internal heat capacity.

The expansion coefficients c_{st}, a_{st} and b_{st} may themselves be determined from the independent sets of linear algebraic equations that result from substitution of the expressions (4.7) into the auxiliary conditions for Γ, **A** and **B** (Maitland *et al.* 1987; McCourt *et al.* 1991). For example, for the viscosity η

$$\sum_{s'=0}^{\infty} \sum_{t'=0}^{\infty} b_{s't'} \left\langle \Phi^{20st} : \mathcal{R}(\Phi^{20s't'}) \right\rangle_0 = 5\delta_{s0}\delta_{t0} \quad \text{for } s, t = 0 \text{ to } \infty \qquad (4.16)$$

in which the notation $\left\langle \Phi^{20st} : \mathcal{R}(\Phi^{20s't'}) \right\rangle_0$ denotes the equilibrium average defined by equation (4.8).

An approximate solution for b_{00} is obtained by truncating the infinite set of equations in some way. Here, the Chapman–Cowling method (Chapman & Cowling 1970) is employed. A first–order solution is obtained by retaining the smallest number of nonvanishing terms of equivalent order in the expansion of the quantities Φ^{p0st}. This means that for viscosity one retains only $s' = t' = 0$, so that the first–order result for b_{00} is obtained trivially from equation (4.16). For bulk viscosity and thermal conductivity in a first–order approximation the terms $s' = 1$, $t' = 0$ and $s' = 0$, $t' = 1$ are retained because $a_{00} = c_{00} = 0$. In this case one must solve a set of two equations each containing two terms to derive first–order approximations for a_{10}, a_{01} and c_{01} and hence the transport coefficients. In a second–order approximation one must include terms in the expansion of Φ^{p0st} up to $s' = 1$, $t' = 1$ for the viscosity and $s' = 2$, $t' = 0$; $s' = 0$, $t' = 2$ and $s' = t' = 1$ for bulk viscosity and thermal conductivity. It follows that it is possible by means of equations (4.12) to (4.14) to relate the transport coefficients directly to the quantities $\left\langle \Phi^{p0st} \cdot \mathcal{R}(\Phi^{p0s't'}) \right\rangle_0$. It is now, however, usual to employ effective collision cross sections in their place, which are discussed in the next section.

4.2.1.3 Effective collision cross sections

Effective collision cross sections are related to the reduced matrix elements of the linearized collision operator $\mathcal{R}$ and incorporate all of the information about the binary molecular interactions, and therefore, about the intermolecular potential. Effective collision cross sections represent the collisional coupling between microscopic tensor polarizations which depend in general upon the reduced peculiar velocity C and the rotational angular momentum **j**. The meaning of the indices p, p'; q, q'; s, s' and t, t' is the same as already introduced for the basis tensors Φ^{pqst}.

In the two–flux approach only cross sections of equal rank in velocity ($p = p'$) and zero rank in angular momentum ($q = q' = 0$) enter the description of the traditional transport properties. Such cross sections are defined by

$$\mathfrak{S}\begin{pmatrix} p\,0\,s\,t \\ p\,0\,s'\,t' \end{pmatrix} = \frac{\left\langle \Phi^{p0st} \cdot \mathcal{R}\Phi^{p0s't'} \right\rangle_0}{(2p+1)\,\langle c \rangle} \qquad (4.17)$$

where the operation $\langle X \rangle_0$ is defined by equation (4.8) and $\langle c \rangle$ is the mean relative velocity

$$\langle c \rangle = 4 \left(\frac{k_B T}{\pi m} \right)^{1/2} \tag{4.18}$$

The collision cross sections arising from the total–energy flux approach of Thijsse *et al.* (1979) are linearly related to those defined in equation (4.17):

$$\mathfrak{S}(10E) = (1 + r^2)^{-1} \left[\mathfrak{S}(1010) + 2r\mathfrak{S} \begin{pmatrix} 1\,0\,1\,0 \\ 1\,0\,0\,1 \end{pmatrix} + r^2 \mathfrak{S}(1001) \right] \tag{4.19}$$

$$\mathfrak{S}(10D) = (1 + r^2)^{-1} \left[r^2 \mathfrak{S}(1010) - 2r\mathfrak{S} \begin{pmatrix} 1\,0\,1\,0 \\ 1\,0\,0\,1 \end{pmatrix} + \mathfrak{S}(1001) \right] \tag{4.20}$$

$$\mathfrak{S} \begin{pmatrix} 1\,0\,E \\ 1\,0\,D \end{pmatrix} = (1 + r^2)^{-1} \left[r\mathfrak{S}(1010) - (1 - r^2)\mathfrak{S} \begin{pmatrix} 1\,0\,1\,0 \\ 1\,0\,0\,1 \end{pmatrix} - r\mathfrak{S}(1001) \right] \tag{4.21}$$

The effective cross sections, where unprimed and primed indices are different, are generally known as *coupling cross sections*, since they are related to the effects of coupling of different polarizations. The cross sections that are characterized by identical unprimed and primed indices are known as relaxation or transport cross sections and, to simplify the notation, are written as $\mathfrak{S}(pqst)$ or $\mathfrak{S}(pqX)$, respectively.

4.2.2 *Explicit formulas for the transport coefficients of polyatomic gases*

4.2.2.1 *First–order expressions*

As noted earlier, the use of the WCUB semiclassical theory means that the derived equations cannot account for polarization of the angular momentum vector (spin polarization). The theory therefore leads to the so–called isotropic value for the lowest–order approximations to viscosity, bulk viscosity and thermal conductivity.

The single basis tensor Φ^{2000} is sufficient to deduce the formula for the viscosity in the first–order approximation

$$[\eta]_1^{\text{iso}} = \frac{k_B T}{\langle c \rangle \, \mathfrak{S}(2000)} \tag{4.22}$$

As a result of the fact that neither individual energy associated with translational degrees of freedom nor that associated with modes of internal motion is conserved, but only their sum, for polyatomic molecules the coefficient of bulk viscosity (volume viscosity, dilatational viscosity) exists. The first Chapman–Cowling approximation derived from the WCUB theory for this quantity is

$$[\kappa]_1^{\text{iso}} = \frac{k_B c_{\text{int}}}{c_V^2} \frac{k_B T}{\langle c \rangle \, \mathfrak{S}(0001)} = \frac{2}{3} \left(\frac{c_{\text{int}}}{c_V} \right)^2 \frac{k_B T}{\langle c \rangle \, \mathfrak{S}(0010)} \tag{4.23}$$

At present the derivation of this equation from the fully quantum–mechanical theory relies upon a heuristic extension of the semiclassical treatment for rigid rotors in which the rotational heat capacity and rotational collision number are replaced by the total internal heat capacity and the internal collision number respectively. This (approximate) extension seems to be justified by a result of van den Oord & Korving (1988).

In place of the bulk viscosity it is often convenient to introduce collision numbers ζ_{int} for internal energy relaxation. The latter represent approximately the number of collisions necessary to relax an internal energy perturbation to equilibrium. Here, the *definition* that was introduced by Herzfeld and Litovitz (1959)

$$[\zeta_{int}]_1^{iso} = \frac{4k_B T}{\pi \eta \langle c \rangle \, \mathfrak{S}(0001)} \tag{4.24}$$

is used.

The first Chapman–Cowling approximation to the thermal conductivity of a dilute polyatomic gas within the two–flux approach leads to a total value that is the sum of two contributions related to translational and internal degrees of freedom respectively:

$$[\lambda]_1^{iso} = [\lambda_{tr}]_1^{iso} + [\lambda_{int}]_1^{iso} \tag{4.25}$$

This result justifies to some extent the empirical assumption of Eucken (1913) that transport of translational and internal energy occurs independently. Two expansion vectors Φ^{1010} and Φ^{1001} are needed in order to evaluate both contributions, which read

$$[\lambda_{tr}]_1^{iso} = \frac{5k_B^2 T}{2m \langle c \rangle} \left[\frac{\mathfrak{S}(1001) - r\mathfrak{S}\binom{1010}{1001}}{\mathfrak{S}(1010)\,\mathfrak{S}(1001) - \mathfrak{S}^2\binom{1010}{1001}} \right] \tag{4.26}$$

and

$$[\lambda_{int}]_1^{iso} = \frac{5k_B^2 T}{2m \langle c \rangle} \left[\frac{r^2\mathfrak{S}(1010) - r\mathfrak{S}\binom{1010}{1001}}{\mathfrak{S}(1010)\,\mathfrak{S}(1001) - \mathfrak{S}^2\binom{1010}{1001}} \right] \tag{4.27}$$

Accordingly, within this approximation the total thermal conductivity can be written

$$[\lambda]_1^{iso} = \frac{5k_B^2 T}{2m \langle c \rangle} \left[\frac{\mathfrak{S}(1001) - 2r\mathfrak{S}\binom{1010}{1001} + r^2\mathfrak{S}(1010)}{\mathfrak{S}(1010)\,\mathfrak{S}(1001) - \mathfrak{S}^2\binom{1010}{1001}} \right] \tag{4.28}$$

The lowest–order result for the total–energy flux approach is solely based on the use of the expansion vector Φ^{10E} and can be written as

$$[\lambda_T]_1^{iso} = \frac{5k_B^2 T \left(1 + r^2\right)}{2m \langle c \rangle \, \mathfrak{S}(10E)} \tag{4.29}$$

From the discussion in Section 4.2.3.1 and the exact relationships given in equations (4.19)–(4.21) it is clear that (4.29) is not exactly equivalent to (4.28), although it turns out to be a remarkably good approximation, as will be shown in Section 4.2.2.3.

In order to be able to develop practical correlation schemes for the transport properties, as detailed below, it is convenient to introduce the coefficient of self–diffusion that is related to an effective cross section* by

$$[D]_1^{iso} = \frac{k_B T}{mn \langle c \rangle \mathfrak{S}(1000)} \qquad (4.30)$$

A further, sometimes useful, quantity was originally *defined* by Monchick *et al.* (1965) and was defined in terms of effective cross sections by Viehland *et al.* (1978). It represents the transport of energy associated with internal degrees of freedom by a diffusion mechanism. This so–called internal diffusion coefficient is defined by

$$[D_{int}]_1^{iso} = \frac{k_B T}{mn \langle c \rangle} \left[\frac{1}{\mathfrak{S}(1001) - \frac{1}{2}\mathfrak{S}(0001)} \right] \qquad (4.31)$$

Hence, for the isotropic case the frequently used ratio of the two diffusion coefficients is

$$\left[\frac{D_{int}}{D} \right]_1^{iso} = \frac{\mathfrak{S}(1000)}{\mathfrak{S}(1001) - \frac{1}{2}\mathfrak{S}(0001)} \qquad (4.32)$$

4.2.2.2 Second–order kinetic theory corrections

As the development of the kinetic theory outlined above has shown, the transport co-efficients are obtained in different orders of approximations according to the number of terms included in the basis vectors of equation (4.7). Fortunately, the lowest–order approximations, at least for viscosity and thermal conductivity, are remarkably accurate and adequate for many purposes. However, for the most accurate work it is necessary to take account of higher–order kinetic theory corrections. These corrections may be expressed in the form

$$[X]_n^{iso} = [X]_1^{iso} f_X^{(n)} \qquad (X = \kappa, \zeta_{int}, \eta, \lambda_{tr}, \lambda_{int}) \qquad (4.33)$$

where $[X]_n^{iso}$ is the corrected property in the nth order and $f_X^{(n)}$ the relevant correction factor. Formal expressions for higher–order approximations to the transport coefficients for polyatomic gases have been derived by Maitland *et al.* (1983), but their extreme complexity inhibits their inclusion here and has also prevented their evaluation except in particular cases (see Section 14.2).

In the second–order approximation the results for the viscosity and the translational part of the thermal conductivity within the Mason–Monchick approximation are

$$f_\eta^{(2)} = \left[1 - \frac{\mathfrak{S}^2 \begin{pmatrix} 2 0 0 0 \\ 2 0 1 0 \end{pmatrix}}{\mathfrak{S}(2000)\,\mathfrak{S}(2010)} \right]^{-1} \qquad (4.34)$$

* In fact $\mathfrak{S}(1000)$ is zero for identical molecules, so that strictly its so–called self–part denoted by $\mathfrak{S}'(1000)$ should be employed. There is, however, no such difficulty for isotopic species.

$$f_{\lambda_{tr}}^{(2)} = \left[1 - \frac{\mathfrak{S}^2 \begin{pmatrix} 1\,0\,1\,0 \\ 1\,0\,2\,0 \end{pmatrix}}{\mathfrak{S}\,(1010)\,\mathfrak{S}\,(1020)}\right]^{-1} \tag{4.35}$$

These results may be written in terms of the ratios of other effective cross sections as has usually been done for monatomic species (Maitland *et al.* 1987) and has recently been performed for polyatomic molecules (Heck *et al.* 1994). Generally $f_\eta^{(2)}$ and $f_\lambda^{(2)}$ depart from unity by no more than 2%. In the particular case of monatomic gases explicit results for the transport coefficients up to the third–order approximation are available (Maitland *et al.* 1987).

A further case of some interest is the higher–order approximation to the thermal conductivity of a polyatomic gas based on the expansion vectors Φ^{10E} and Φ^{10D}. This leads to

$$[\lambda_T]_2 = [\lambda_T]_1\, f_T = [\lambda_T]_1 \left[1 - \frac{\mathfrak{S}^2 \begin{pmatrix} 1\,0\,E \\ 1\,0\,D \end{pmatrix}}{\mathfrak{S}\,(10E)\,\mathfrak{S}\,(10D)}\right]^{-1} \tag{4.36}$$

The result in the second–order approximation is exactly equivalent to equation (4.28), derived from the two–flux approach, as can be seen by substituting relations (4.19)–(4.21) into equation (4.36). Practical experience (Thijsse *et al.* 1979; Millat *et al.* 1988a; van den Oord & Korving 1988) and model calculations for nitrogen (Heck & Dickinson 1994) demonstrate that

$$\mathfrak{S}^2 \begin{pmatrix} 1\,0\,E \\ 1\,0\,D \end{pmatrix} \ll \mathfrak{S}\,(10E)\,\mathfrak{S}\,(10D) \tag{4.37}$$

In practice, the correction factor f_T usually departs by at most 1% from unity. This means that equation (4.29) provides at the same time a simple and accurate description of the thermal conductivity of a polyatomic gas.

The inequality (4.37) conveys the physical idea that the change in internal energy in an inelastic collision (one in which translational energy is not conserved) is generally much less than the relative kinetic energy and that this small energy change produces negligible distortion of the molecular trajectories. Sometimes this physical notion is termed the Mason–Monchick approximation (MMA) (Mason & Monchick 1962), and it is also consistent with the Infinite Order Sudden Approximation (IOSA) of molecular collision theory (e.g. McLaughlin *et al.* 1986). The same approximation has been employed to make an evaluation of the effect of the higher–order correction factors for all of the transport properties of polyatomic gases (Heck *et al.* 1993, 1994), where it has been shown to be remarkably accurate.

4.2.2.3 Corrections for angular momentum–polarization effects

A further set of corrections to the first–order isotropic transport coefficients arises from the use of the semiclassical WCUB theory. As already mentioned, the WCUB equation fails to account for the presence of degenerate internal states or, in other words, for the

j–dependence of the nonequilibrium distribution function f. This result is characteristic of the isotropic approximation (frequently called the Pidduck approximation) to the field free transport coefficients as given above.

If the equivalent approximation is introduced into the quantum–mechanical Wald-mann–Snider expressions (see Viehland *et al.* 1978; McCourt *et al.* 1990, 1991), the WCUB results are formally recovered. This means that it is possible to derive corrections to the WCUB expressions to account for the neglect of polarization effects. These corrections may be expressed in the form

$$X = [X]_1^{\text{iso}} \cdot S_X \qquad (X = \kappa, \zeta_{int}, \eta, \lambda) \tag{4.38}$$

where X is the corrected property and S_X the approximate correction factor for angular momentum polarization effects. Each of the correction factors may be expressed in terms of effective cross sections that involve tensorial components of different rank. One difficulty in the evaluation of S_X is that slightly different solution methods are needed for various types of polyatomic molecules (McCourt *et al.* 1991). In order to simplify the discussion only the results for diamagnetic rigid rotors are given in the rest of this section. For such molecules,

$$S_\kappa = S_\zeta = 1 \tag{4.39}$$

$$S_\eta \simeq \left[1 - \frac{\mathfrak{S}^2 \left(\begin{smallmatrix} 2\,0\,0\,0 \\ 0\,2\,0\,0 \end{smallmatrix} \right)}{\mathfrak{S}\,(2000)\,\mathfrak{S}\,(0200)} \right]^{-1} \simeq 1 \tag{4.40}$$

and

$$S_\lambda \simeq \left[1 - \frac{5}{3} \frac{\mathfrak{S}^2 \left(\begin{smallmatrix} 1\,2\,E \\ 1\,0\,E \end{smallmatrix} \right)}{\mathfrak{S}\,(12E)\,\mathfrak{S}\,(10E)} \right]^{-1} \tag{4.41}$$

All of these corrections are close to unity, with S_λ being the most significantly different; but even then its effect does not exceed 1.5%. For monatomic species no corrections are of course required, since there is no molecular angular momentum.

4.2.3 Pure monatomic gases

The original Chapman–Enskog theory is based on the Boltzmann equation (Boltzmann 1872) that was developed for structureless particles which exhibit isotropic intermolec-ular potentials. Monatomic species are a subset of polyatomic molecules that do not possess modes of internal motion and interact via elastic collisions only. This simplifies the problem dramatically because the WCUB equation (or Waldmann–Snider equation) can be replaced by the Boltzmann equation that is solved in its linearized form.

The results for the viscosity, $\eta^{\text{iso}} \equiv \eta$, and the self–diffusion coefficient, $D^{\text{iso}} \equiv D$, for monatomic gases are formally identical to those for polyatomic gases (4.22) and

(4.30). However, the effective cross sections for monatomic species are fully elastic and the relevant expressions, therefore, are in fact considerably simpler. The bulk viscosity, κ, does not exist for dilute monatomic gases. The thermal conductivity, on the other hand, is given by a significantly simpler expression compared to (4.28) because of the absence of internal energy and may be written

$$\lambda = \frac{5k_B^2 T}{2m \langle c \rangle \, \mathfrak{S}(1010)} f_\lambda \tag{4.42}$$

Furthermore, the exact relationship

$$\mathfrak{S}(1010) = \frac{2}{3} \mathfrak{S}(2000) \tag{4.43}$$

can be employed to derive the expression

$$\lambda = \frac{15 k_B \eta}{4m} \frac{f_\lambda}{f_\eta} \simeq \frac{15 k_B \eta}{4m} \tag{4.44}$$

This equation is important because it allows the prediction of the thermal conductivity of noble gases from known viscosity data. The ratio $f_\lambda / f_\eta = \xi$ may be evaluated in a number of ways and is largely independent of the manner of the calculation, as will be discussed later. Additionally, it is noted for later use that the self–diffusion coefficient and the viscosity may be related by the equation

$$\frac{nm \, [D]_1}{[\eta]_1} = \frac{6}{5} A^* \tag{4.45}$$

where

$$A^* = \frac{5 \, \mathfrak{S}(2000)}{6 \, \mathfrak{S}(1000)} \tag{4.46}$$

The advantage of this formula is that the values of A^* are insensitive to the pair potential upon which the individual effective cross sections depend.

4.2.4 Applicable formulas

4.2.4.1 Prediction from first principles

All of the effective collision cross sections introduced above can, in principle, be evaluated from a knowledge of the intermolecular pair potential by means of equation (4.17) and the definition of the collision operator (4.5). This process would then make it possible to predict the transport properties of dilute gases from first principles. However, such a procedure would require that it is possible to evaluate the *inelastic* differential scattering cross section σ_{ij}^{kl} or an equivalent to it which enters (4.5). Until very recently this could only be accomplished for dilute monatomic gases. There were two reasons for this: first, only for such systems are accurate intermolecular potentials available (Aziz 1984; Maitland *et al.* 1987; van der Avoird 1992) second, only for such systems was the evaluation of the differential scattering cross section within the capabilities of available computing resources. Only very recently (Heck & Dickinson 1994; Heck & Dickinson, personal communication) has it become possible to evaluate effective cross sections for the collisions of two rigid rotors (nitrogen–nitrogen, carbon monoxide–carbon monoxide) classically. Nevertheless, such calculations are still too time consuming for routine use and the pair potentials are known for very few polyatomic interactions (van der Avoird 1992). Thus, only for monatomic species is the route of complete prediction based on equations (4.22), (4.30) and (4.42) a viable option (Bich *et al.* 1990).

4.2.4.2 *Correlation and prediction using the law of corresponding states*

The law of corresponding states is based upon the hypothesis that, to an acceptable degree of accuracy, the intermolecular pair potentials for a number of fluids can be rendered conformal by the choice of two scaling factors, ϵ_c for energy and σ_c for length, characteristic of each fluid. From this hypothesis it follows that the reduced collision cross sections for this 'universal' potential are 'universal' functions of the reduced temperature $T^* = k_B T / \epsilon_c$

$$\mathfrak{S}^* \begin{pmatrix} p\ 0\ s\ t \\ p'\ 0\ s'\ t' \end{pmatrix} = \frac{\mathfrak{S} \begin{pmatrix} p\ 0\ s\ t \\ p'\ 0\ s'\ t' \end{pmatrix}}{\pi \sigma_c^2} = f(T^*) \tag{4.47}$$

It has proved useful to apply such a scheme for the correlation of viscosity and diffusion coefficients of monatomic as well as polyatomic fluids and fluid mixtures (Maitland *et al.* 1987). In addition, the thermal conductivity has been correlated for monatomic gases and their mixtures and for selected simple polyatomic fluids. Correlations based on the law of corresponding states and its extensions and their predictive power are the subject of Chapter 11 of this volume and are, therefore, not detailed here.

4.2.4.3 *Theoretically based correlation schemes*

Theoretically based correlation schemes must be founded on the kinetic theory outlined above and include as much high–precision experimental information as possible. For this purpose some rearrangement of the exact kinetic theory results is useful.

Viscosity

In order to devise a practical correlation scheme for viscosity any polarization corrections are ignored. Accordingly, the related correction factor is assumed to be $S_\eta = 1$. Furthermore, a reduced functional is introduced by

$$\mathfrak{S}_\eta^* = \frac{\mathfrak{S}\ (2000)}{\pi \sigma^2 f_\eta} = \frac{\mathfrak{S}^*\ (2000)}{f_\eta} \tag{4.48}$$

where f_η is the correction factor for higher–order Chapman–Cowling approximations and σ represents a length scaling parameter that is to be adjusted. This leads to

$$\eta = \frac{k_B T}{\pi \sigma^2 \langle c \rangle \mathfrak{S}_\eta^*} \tag{4.49}$$

Then, $\mathfrak{S}_\eta^*$ is correlated as a function of the reduced temperature $T^* = k_B T / \epsilon$ with the adjustable energy scaling parameter ϵ. Usually, for correlation, one of the following functions is employed

$$\ln \mathfrak{S}_\eta^* = \sum_{i=0}^{n} a_i (\ln T^*)^i \tag{4.50}$$

or alternatively

$$\mathfrak{S}_\eta^* = \sum_{i=0}^{n} b_i \left(\frac{1}{T^*} \right)^i \tag{4.51}$$

As the result of this procedure an 'individual' correlation for a particular fluid is provided. The range of validity and the uncertainty of the correlation strongly depend upon the body of experimental results that was chosen as the primary data set. For examples of the application of this technique see Chapter 14.

Self–diffusion

Because of a lack of experimental data, it will usually not be possible to deduce a correlation for this quantity based on experimental information. However, since viscosity as well as diffusion are almost independent of the inelasticity of molecular encounters, it is assumed that D can be approximated to a good degree of accuracy using equation (4.45) and an estimate of the cross section ratio A^*. This quantity can be taken, for instance, from correlations based on the law of corresponding states (Maitland *et al.* 1987) (see Chapter 11). It is difficult, at present, to estimate the error that may result for polyatomic systems in this way, but preliminary results of Heck *et al.* (1993) and Heck & Dickinson (1994) show that for nitrogen A^* obtained from the corresponding–states correlations may be too low by 2 to 4%. In the worst case, one can apply the rough approximation $A^* = 1.1$ with an error that is only slightly greater.

Thermal conductivity

For monatomic systems, equation (4.44) provides a means of evaluating or correlating the thermal conductivity whether or not there are experimental data available. For this purpose it is simply necessary to have available viscosity data for the substance, or a representation of them, and a suitable approximation for the ratio $\xi(T^*)$. As remarked earlier, $\xi(T^*) = f_\lambda / f_\eta$ is very insensitive to the pair potential used to evaluate it (Maitland *et al.* 1987) so that any suitable pair potential can be used for its calculation.

A useful representation based on corresponding states is

$$\xi = 1 + 0.0042\left[1 - \exp 0.33\left(1 - T^*\right)\right] \tag{4.52}$$

For polyatomic molecules the higher–order correction factors obtained from equations (4.34) and (4.35) are considerably more complicated especially if account is taken of the effects of spin polarization. However, the latter effects in the case of the correction factor S_λ may be related to an experimentally accessible quantity (Viehland *et al.* 1978; McCourt *et al.* 1990) so that

$$S_\lambda \simeq 1 - \frac{5}{3}\left(\frac{\Delta\lambda_\parallel}{\lambda}\right)_{\text{sat}} \tag{4.53}$$

where $\left(\Delta\lambda_\parallel/\lambda\right)_{\text{sat}}$ is the almost temperature–independent saturation value of the fractional change in the thermal conductivity of the gas measured parallel to a magnetic field (McCourt *et al.* 1990; Hermans 1992). Since S_λ leads to systematic corrections of about 1 to 1.5%, it should be applied whenever possible. However, experimental values of $\left(\Delta\lambda_\parallel/\lambda\right)_{\text{sat}}$ are not available for all gases, so that it is then necessary to assign $S_\lambda = 1$.

For the purposes of representation of a body of thermal conductivity data it is convenient to make use of results from the total–energy flux approach of the kinetic theory (Thijsse *et al.* 1979; Millat *et al.* 1988a; van den Oord & Korving 1988). According to this procedure a further reduced cross section is defined by

$$\mathfrak{S}_\lambda^* = \frac{\mathfrak{S}(10E)}{\pi\sigma^2 f_T} \tag{4.54}$$

where f_T again absorbs the effects of higher–order approximations. This leads to

$$\lambda = \frac{5k_B^2 T\left(1 + r^2\right)}{2\pi\sigma^2 m \langle c\rangle \mathfrak{S}_\lambda^*} S_\lambda \tag{4.55}$$

$\mathfrak{S}_\lambda^*$ can now be correlated in a way similar to $\mathfrak{S}_\eta^*$ using the scaling parameters derived from viscosity. Obviously, equation (4.55) has no intrinsic predictive power and requires high–precision experimental information over a large temperature range. In practice, this is seldom available, since high–precision data for λ are restricted to a narrow range and a critical data assessment is much more complicated. However, it is an empirical observation that for a number of gases $\mathfrak{S}_\lambda^*$ is a linear function of $1/T^*$ so that extrapolation over a limited temperature range may be permitted (Millat *et al.* 1989).

On the other hand, the two–flux approach to the thermal conductivity of a polyatomic gas offers a rather greater opportunity for its prediction from viscosity and other experimental data because it is possible to make use of exact and approximate relationships between various effective cross sections and, in some cases, theoretically known behavior in the high temperature limit. The starting point for such an approach is equation (4.28) for the total thermal conductivity, provided that S_λ is approximated by equation (4.53)

$$[\lambda]_1 = \frac{5k_B^2 T}{2m\langle c\rangle}\left[\frac{\mathfrak{S}(1001) - 2r\mathfrak{S}\left(\begin{smallmatrix}1010\\1001\end{smallmatrix}\right) + r^2\mathfrak{S}(1010)}{\mathfrak{S}(1010)\mathfrak{S}(1001) - \mathfrak{S}^2\left(\begin{smallmatrix}1010\\1001\end{smallmatrix}\right)}\right] \cdot S_\lambda \qquad (4.56)$$

In many cases a limited amount of accurate experimental data for λ is available, and one is concerned to extend the temperature range and to check the internal consistency of the calculated results. For this purpose it is useful to eliminate as many cross sections as possible from equation (4.56) in favor of independently measured quantities. Some exact relationships between effective cross sections make this possible in part

$$\mathfrak{S}\left(\begin{matrix}1010\\1001\end{matrix}\right) = -\frac{5r}{6}\mathfrak{S}(0001) \qquad (4.57)$$

and

$$\mathfrak{S}(1010) = \frac{2}{3}\mathfrak{S}(2000) + \frac{25}{18}r^2\mathfrak{S}(0001) \qquad (4.58)$$

The cross section $\mathfrak{S}(2000)$ is obviously available from the viscosity, whereas $\mathfrak{S}(0001)$ is often available from bulk viscosity or direct or indirect collision number measurements (Millat *et al.* 1988b; Millat & Wakeham 1989). This leaves only the cross section $\mathfrak{S}(1001)$ to be determined, which could, in principle, be deduced from measurements of D_{int} according to equation (4.31). Very few measurements of this quantity have yet been performed (see Ferron 1990). It is therefore necessary at present to adopt a different approach, which is to use the limited amount of thermal conductivity data available to determine $\mathfrak{S}(1001)$ and to evaluate D_{int} from it. Such an analysis can be useful because it is then possible to evaluate the 'experimental' value of D_{int}/D as

$$\frac{D_{int}}{D} = \frac{5}{6A^*}\frac{nm D_{int}}{\eta} \qquad (4.59)$$

The virtue of this determination is that there are theoretical analyses (McCourt *et al.* 1990) that indicate that for rigid rotors at high temperatures

$$\mathfrak{S}(1001) \rightarrow \mathfrak{S}(1000) + \frac{35}{12}r^2\mathfrak{S}(0001) \qquad (4.60)$$

Accordingly, D_{int}/D is given by

$$\frac{D_{int}}{D} \rightarrow \frac{\mathfrak{S}(1000)}{\mathfrak{S}(1000) + \left[\frac{35}{12}r^2 - \frac{1}{2}\right]\mathfrak{S}(0001)} \qquad (4.61)$$

so that, in the high temperature limit, this ratio approaches unity from below as $T \rightarrow \infty$.

The ratio D_{int}/D is, in fact, seldom very different from unity. On the one hand, the result of equation (4.61) permits some thermal conductivity data that are inconsistent with this result to be discarded. On the other hand, if the available data are consistent with this limiting behavior, equation (4.61) provides a secure means of extrapolating the thermal conductivity upward in temperature. An example of this approach is given in Section 14.2.

In cases where there are no experimental data for the thermal conductivity, or just a very small number, the assignment

$$\frac{D_{int}}{D} = 1 \tag{4.62}$$

is often adequate for an approximate evaluation of the thermal conductivity and seldom leads to an error of more than 5%.

4.2.4.4 Prediction of thermal conductivity using approximate methods

Van den Oord–Korving scheme

Van den Oord & Korving (1988) argued that to a good approximation Φ^{10E} is nearly an eigenfunction of the collision operator and hence that the cross section $\mathfrak{S}\left(\begin{smallmatrix} 1 & 0 & E \\ 1 & 0 & D \end{smallmatrix}\right)$ is sufficiently small that it can be taken as zero. By means of rearranging (4.19) to (4.21) they derived the formula

$$\mathfrak{S}(10E) \rightarrow \frac{2}{3}\mathfrak{S}(2000) + \frac{5}{9}r^2\mathfrak{S}(0001) \tag{4.63}$$

which relates $\mathfrak{S}(10E)$ to the shear and bulk viscosity cross sections. This result can be used to predict approximately the thermal conductivity if shear and bulk viscosity (or related quantities) are precisely known. Unfortunately, data for the bulk viscosity are often not available. But if the viscosity and thermal conductivity data are available, the equation can be applied to predict $\mathfrak{S}(0001)$ and bulk viscosity (or collision numbers).

Another possible application of equation (4.63) is to evaluate the dimensionless Prandtl number for the dilute gas, which is defined by

$$\mathrm{Pr} = \frac{c_P^0 \eta}{m\lambda} = \frac{\mathfrak{S}(10E)}{\mathfrak{S}(2000)} \approx \frac{2}{3} + \frac{k}{2c_{int}}\frac{\mathfrak{S}(0001)}{\mathfrak{S}(2000)} \tag{4.64}$$

where c_P^0 is the ideal–gas heat capacity at constant pressure. Especially at higher temperatures the Prandtl number was found to be relatively temperature insensitive and, therefore, useful for extrapolation within certain limits (van den Oord & Korving 1988; Hendl et al. 1991).

Eucken formulas

Two frequently used estimation formulas are based on the original paper of Eucken (1913) and can be analyzed here in terms of the full theory. Starting from equations (4.26)–(4.28), that is, from the isotropic value for the thermal conductivity, it is assumed that there is no exchange of energy between translational and internal modes of motion.

Then, one has

$$\mathfrak{S}(0001) \to 0 \tag{4.65}$$

According to equations (4.57) and (4.58), this leads to

$$\mathfrak{S}\begin{pmatrix} 1010 \\ 1001 \end{pmatrix} \to 0 \tag{4.66}$$

and

$$\mathfrak{S}(1010) \to \frac{2}{3}\mathfrak{S}(2000) \tag{4.67}$$

Furthermore, it can be assumed within this context that

$$\mathfrak{S}(1001) \to \mathfrak{S}(1000) \tag{4.68}$$

and, accordingly,

$$\frac{D_{int}}{D} = 1 \tag{4.69}$$

To obtain the original result of Eucken (1913) it is further necessary to assume that

$$\mathfrak{S}(1000) \to \mathfrak{S}(2000) \tag{4.70}$$

which is equivalent to $A^* = 5/6$. It should be noted that this assigns a value to A^* that is about 25% lower than a realistic value. This leads to

$$\lambda = \frac{k_B \eta}{m}\left(\frac{5}{2}\frac{c_{tr}}{k_B} + \frac{c_{int}}{k_B}\right) \tag{4.71}$$

in which the subscripts 'tr' and 'int' refer to contributions to the isochoric heat capacity from the different modes of motion. A less drastic simplification assumes that equations (4.65)–(4.69) are retained, whereas (4.70) is replaced by

$$\frac{\mathfrak{S}(2000)}{\mathfrak{S}(1000)} = \frac{6}{5}A^* = \frac{\rho D}{\eta} \tag{4.72}$$

This leads to the modified Eucken formula

$$\lambda = \frac{k_B \eta}{m}\left(\frac{5}{2}\frac{c_{tr}}{k_B} + \frac{\rho D}{\eta}\frac{c_{int}}{k_B}\right) \tag{4.73}$$

which was derived by Chapman and Cowling (1970). Obviously, both equations are the results of rather crude approximations; the latter is an improvement over the former only at high temperatures, and both should be used with caution.

4.3 Gas mixtures

4.3.1 Kinetic theory of dilute polyatomic gas mixtures

4.3.1.1 General results

Section 4.2 considered the kinetic theory of the transport properties of dilute, pure fluids in some detail; in this section the theory is extended to gas mixtures. Naturally, the theory of mixtures shares many features with that of pure species so that the same pattern will be adopted for the presentation, although duplication will be avoided whenever possible. For these reasons the semiclassical kinetic theory description is immediately adopted here, and similar consequences for the description of some phenomena as they pertain to pure gases are accepted.

In the case of mixtures a singlet distribution function $f_{qi}(\mathbf{c}_{qi}, E_{qi}, \mathbf{r}, t)$ is identified for each component q in an N-component mixture. As a consequence, the theory must seek to solve a set of linearized WCUB equations, one for each species analogous to equation (4.4) so that one has

$$\mathcal{D}_q f_q^{(0)} + f_q^{(0)} \mathcal{D}_q(\phi_{qi}) = -f_q^{(0)} \sum_{q'} n_{q'} \mathcal{R}_{qq'}(\phi_{qi} + \phi_{qj}) \qquad (4.74)$$

The notation remains the same as before except that subscripts q and q' denote independent molecular species and the Wang Chang–Uhlenbeck operator for the mixture is defined by the equation

$$n_{q'} f_{qi}^{(0)} \mathcal{R}_{qq'}(Y) = \sum_{jkl} \int \int \int f_{qi}^{(0)} f_{q'j}^{(0)} [Y - Y'] g\sigma_{ij}^{kl}(g, \chi, \psi) \sin \chi \, d\chi \, d\psi \, d\mathbf{c}_{jq'}$$
$$(4.75)$$

Now, σ_{ij}^{kl} is the *inelastic* differential scattering cross section for a collision in which the internal state of molecule of species q changes from i to k and that of a molecule of species q' changes from j to l.

A solution to the WCUB equation (4.74) for $f_{qi} = f_{qi}^{(0)}(1 + \phi_{qi})$ can be obtained by a generalization of the procedure of Section 4.2.1.1. Thus, the perturbation function ϕ_{qi} has the form

$$\phi_{qi} = -\frac{1}{n}\mathbf{A}_q \cdot \nabla \ln T - \frac{1}{n}\mathbf{B}_q : \nabla \mathbf{c}_0 - \frac{1}{n}\Gamma_q \nabla \cdot \mathbf{c}_0 - \frac{1}{n}\sum_{q'} \mathbf{F}_q^{q'} \cdot \mathbf{d}_{q'} \qquad (4.76)$$

Here it is noted that terms in the perturbation function that allow for departures from the local equilibrium because of temperature gradients and velocity gradients must be taken into account as well as those for the possibility of composition gradients, which is done by including the diffusion driving force $\mathbf{d}_{q'}$. The quantities $\mathbf{A}_q$, $\mathbf{B}_q$, Γ_q and $\mathbf{F}_q^{q'}$ are unknown functions of different tensorial rank that relate the perturbation in the macroscopic variables to the perturbation in the molecular velocity distribution.

By essentially the same techniques as described in Section 4.2 the various transport coefficients of the multicomponent gas mixture can be related to the functions $\mathbf{A}_q$, $\mathbf{B}_q$, Γ_q and $\mathbf{F}_q^{q'}$. In particular, the multicomponent diffusion coefficients are given by

$$D_{qq'} = \frac{1}{3n}[\mathbf{F}^{q'} \cdot \mathcal{R}(\mathbf{F}^q)] \tag{4.77}$$

the thermal diffusion coefficients by

$$D_q^T = \frac{1}{3n}[\mathbf{A} \cdot \mathcal{R}(\mathbf{F}^q)] \tag{4.78}$$

the partial thermal conductivity (in the presence of a net diffusive flux) λ' by

$$\lambda' = \frac{2k_B^2 T}{3m}[\mathbf{A} \cdot \mathcal{R}(\mathbf{A})] \tag{4.79}$$

and the viscosity by

$$\eta = \frac{k_B T}{10}[\mathbf{B} : \mathcal{R}(\mathbf{B})] \tag{4.80}$$

where the notation $[X \odot \mathcal{R}(X)]$ denotes the kinetic theory bracket integrals of Ross *et al.* (1992).

In order to obtain explicit expressions for the various transport coefficients the tensors $\mathbf{A}_q$, $\mathbf{B}_q$ and $\mathbf{F}_q^{q'}$ are expanded in a series of orthogonal basis tensors according to the procedure outlined in Section 4.2.1.1 for pure gases. As noted there, this procedure leads to an infinite set of linear equations for the expansion coefficients, and each of the transport coefficients is itself related to just one of the expansion coefficients. The result is that the transport properties of a multicomponent gas mixture can be expressed formally as the ratio of two infinite determinants. Various orders of approximations to the transport coefficients can then be generated by retaining only a limited number of terms in the polynomial expansion. There are various subtleties associated with the nomenclature of orders of approximation which need to be considered carefully. Here almost exclusively the lowest order of approximation is considered, which is again remarkably accurate. Details of higher–order approximations may be found elsewhere (McCourt *et al.* 1990; Ross *et al.* 1992).

For the expansion tensors of the various functions, Γ_q, $\mathbf{A}_q$, $\mathbf{B}_q$ and $\mathbf{F}_q^{q'}$, the same two choices are available as for the pure fluid. The first is that employed by Wang Chang and Uhlenbeck, so that, for example,

$$\mathbf{A}_q = \sum_s \sum_t a_{qst} \Phi^{10st|q} \tag{4.81}$$

and

$$\mathbf{B}_q = \sum_s \sum_t b_{qst} \Phi^{20sr|q} \tag{4.82}$$

where, apart from the species–distinguishing index q, the polynomials $\Phi^{p0st|q}$ are those listed in Table 4.1.

As for the pure gas case, the transport coefficients are related to individual expansion coefficients of the polynomial series. Thus, for the partial thermal conductivity of a gas mixture, one has

$$\lambda' = \lambda'_{\text{tr}} + \lambda'_{\text{int}} \tag{4.83}$$

where

$$\lambda'_{\text{tr}} = \sum_q x_q \left(\frac{5 k_{\text{B}}^3 T}{2 m_q} \right)^{1/2} a_{q10} \tag{4.84}$$

and

$$\lambda'_{\text{int}} = \sum_q x_q \left(\frac{c_{\text{int},q}}{k_{\text{B}}} \frac{k_{\text{B}}^3 T}{m_q} \right)^{1/2} a_{q01}, \tag{4.85}$$

where x_q is the mole fraction of the qth component and $c_{\text{int},q}$ its internal, isochoric heat capacity.

For systems containing polyatomic components (at least where one component is polyatomic) the alternative set of tensors $\Phi^{p0E|q}$ and $\Phi^{p0D|q}$ proposed by Thijsse *et al.* (1979) and discussed in Section 4.2.1 for pure gases has been applied to the kinetic theory (Ross *et al.* 1992). For the thermal conductivity, these tensors lead to considerably simpler formulas than those of Wang Chang and Uhlenbeck, which are potentially more useful for the purposes that are the subject of this text.

Whichever set of tensors is employed, the methodology for the explicit evaluation of the transport coefficients follows a route analogous to that described in Section 4.2.1.2 for pure gases. Thus, just as for the pure gas, all of the dynamic encounters between molecules in the gas contribute to the transport coefficients through a series of effective cross sections; details of the theoretical development may be found elsewhere (McCourt *et al.* 1990, 1991). Here it suffices to emphasize that theory yields approximations to the transport coefficients of the gas mixture which can be evaluated to an arbitrary order if desired and, just as for pure gases, it is also possible to account for the effects of spin–polarization, as discussed in Section 4.2.2.3. However, in the case of mixtures, for most practical purposes the first nonzero approximation is adequate; accordingly, this is quoted here. The results are given first in a way in which they apply to polyatomic gas mixtures, since the results include monatomic systems as a special case.

4.3.2 Explicit formulas for the transport coefficients of polyatomic gas mixtures

In the limit of zero density the viscosity η_{mix} of an N-component gas mixture is given by the expression

$$\eta_{\text{mix}} = -k_{\text{B}} T \left| \begin{array}{cc} P^{20,20}_{qq'} & x_q \\ x_{q'} & 0 \end{array} \right| \cdot \left| P^{20,20}_{qq'} \right|^{-1} \tag{4.86}$$

where $P_{qq'}^{20,20}$ is itself an $N \times N$ determinant with elements

$$P_{qq'}^{20,20} = \delta_{qq'} \sum_v \frac{n_q n_v}{n^2} < c_{qv} > \mathfrak{S} \begin{pmatrix} 2\,0\,0\,0 & q \\ 2\,0\,0\,0 & q \end{pmatrix}_{qv}$$
$$+ < c_{qq'} > \frac{n_q n_{q'}}{n^2} \mathfrak{S} \begin{pmatrix} 2\,0\,0\,0 & q \\ 2\,0\,0\,0 & q' \end{pmatrix}_{qq'} \tag{4.87}$$

Here, in addition to symbols previously defined,

$$< c_{qq'} > = \left(\frac{8 k_B T}{\pi \mu_{qq'}} \right)^{1/2} \tag{4.88}$$

and

$$\mu_{qq'} = \frac{m_q m_{q'}}{(m_q + m_{q'})} \tag{4.89}$$

is the reduced mass of the molecules q and q'. The symbol $\delta_{qq'}$ represents the Kronecker delta, while $\mathfrak{S} \begin{pmatrix} 2\,0\,0\,0 & q \\ 2\,0\,0\,0 & q' \end{pmatrix}_{qq'}$ represents one of the group of effective cross sections detailed later.

The multicomponent diffusion coefficients, $D_{qq'}$, defined by Waldmann (1957), in an N–component mixture are given to a first approximation by

$$[D_{qq'}]_1 = \frac{m_q m_{q'}}{(\rho/n)^2} [\mathcal{D}_{qq'}]_1 \tag{4.90}$$

where

$$[\mathcal{D}_{qq'}]_1 = \frac{k_B T}{m_q n < c_{qq'} > \mathfrak{S} \begin{pmatrix} 1\,0\,0\,0 & q \\ 1\,0\,0\,0 & q \end{pmatrix}_{qq'}} \tag{4.91}$$

is the first approximation to the binary diffusion coefficient for the gas pair.

The thermal conductivity of a gas mixture which is measured directly is not the quantity λ' introduced in equation (4.79), because measurements are always performed in the absence of a net diffusive flux. In order to evaluate the measured thermal conductivity in the zero–density limit λ, the multicomponent diffusion coefficients are employed (Ross *et al.* 1992) and then one obtains, in a consistent first–order approximation,

$$[\lambda]_1 = [\lambda_{tr}]_1 + [\lambda_{int}]_1 \tag{4.92}$$

where

$$[\lambda_{tr}]_1 = k_B \sum_q \sum_{q'} \left(C_{qq'}^{11} R_q^{10} R_{q'}^{10} + C_{qq'}^{12} R_q^{10} R_{q'}^{01} \right) \tag{4.93}$$

and

$$[\lambda_{int}]_1 = k_B \sum_q \sum_{q'} \left(C_{qq'}^{21} R_q^{01} R_{q'}^{10} + C_{qq'}^{22} R_q^{01} R_{q'}^{01} \right) \tag{4.94}$$

where

$$R_q^{st} = x_q \left(\frac{k_B T}{m_q}\right)^{1/2} \left[\left(\frac{2c_{int,q}}{k_B}\right)^{1/2} \delta_{s0}\delta_{t1} + (5)^{1/2}\delta_{s1}\delta_{t0}\right] \tag{4.95}$$

The elements $C_{qq'}^{ts}$ are related to a further set of effective cross sections by the relationships

$$C_{qq'} = \left[\tilde{S}^{10,10} - \tilde{S}^{10,01}\left[\tilde{S}^{01,01}\right]^{-1}\tilde{S}^{01,10}\right]_{qq'}^{-1} \tag{4.96}$$

$$C_{qq'}^{12} = -[C^{11}\tilde{S}^{10,01}(\tilde{S}^{01,01})^{-1}]_{qq'} \tag{4.97}$$

$$C_{qq'}^{21} = -[(\tilde{S}^{01,01})^{-1}\tilde{S}^{01,10}C^{11}]_{qq'} \tag{4.98}$$

and

$$C_{qq'}^{22} = [(\tilde{S}^{01,01})^{-1} - C^{21}\tilde{S}^{10,01}(\tilde{S}^{01,01})^{-1}]_{qq'} \tag{4.99}$$

Here C^{mn} is the $N \times N$ matrix with elements $C_{qq'}^{mn}$, and the matrices $\tilde{S}_{qq'}^{st,s't'}$ are related to effective cross sections as follows:

$$\tilde{S}_{qq'}^{st,s't'} = S_{qq'}^{st,s't'} - \left(\frac{m_{q'}}{m_q}\right)^{1/2}\frac{x_{q'}}{x_q}S_{qq}^{st,s't'}\delta_{s'0}\delta_{t'0} \tag{4.100}$$

where

$$S_{qq}^{st,s't'} = \sum_{\mu\neq q} x_q x_\mu <c_{q\mu}> \mathfrak{S}\left(\begin{array}{ccc|c} 1 & 0 & s & t \\ 1 & 0 & s' & t' \end{array}\middle|\begin{array}{c} q \\ q \end{array}\right)_{q\mu}$$

$$+ x_q^2 <c_{qq}> \mathfrak{S}\left(\begin{array}{ccc|c} 1 & 0 & s & t \\ 1 & 0 & s' & t' \end{array}\middle|\begin{array}{c} q \\ q \end{array}\right)_{qq} \tag{4.101}$$

$$S_{qq'}^{st,s't'} = x_q x_{q'} <c_{qq'}> \mathfrak{S}\left(\begin{array}{ccc|c} 1 & 0 & s & t \\ 1 & 0 & s' & t' \end{array}\middle|\begin{array}{c} q \\ q' \end{array}\right)_{qq'} \tag{4.102}$$

These particular forms of the kinetic theory results are somewhat cumbersome to write down, although less so than some other formulations (McCourt et al. 1990; Monchick et al. 1965). Their particular virtues are their complete generality, that they are particularly suitable for machine calculations and that they show explicitly that once the effective cross sections are available the transport properties of an arbitrary mixture of components are easily evaluated.

In the case of the thermal conductivity a rather simpler form of result is obtained if the expansion vectors of Thijsse et al. (1979) are employed. The thermal conductivity of the mixture is then given by

$$[\lambda]_T = \frac{5}{2}k_B^2 T \sum_q \sum_{q'} \frac{C_{qq'T}^{11}x_q x_{q'}\left(1+r_q^2\right)\left(1+r_{q'}^2\right)}{(m_q m_{q'})^{1/2}} \tag{4.103}$$

where

$$C_{qq'\,\mathcal{T}}^{11} = (\mathbf{S}^{EE} - \mathbf{S}^{ED}(\mathbf{S}^{DD})^{-1}\mathbf{S}^{DE})_{qq'}^{-1} \tag{4.104}$$

and the subscript $\mathcal{T}$ denotes the use of the Thijsse expansion tensors. Here, $\mathbf{S}^{XY}$ is an $N \times N$ matrix of concentration–dependent effective cross sections in which new cross sections arise that are defined below:

$$
\begin{aligned}
S_{qq'}^{XY} = \left(1 + r_q^2\right)^{1/2} \left(1 + r_{q'}^2\right)^{1/2} \\
\left[\delta_{qq'} \sum_{\mu \neq q} x_\mu x_q < c_{q\mu} > \mathfrak{S} \begin{pmatrix} 1\,0\,X \\ 1\,0\,Y \end{pmatrix} \begin{matrix} q \\ q \end{matrix} \right)_{q\mu} \\
+ x_q x_{q'} < c_{qq'} > \mathfrak{S} \begin{pmatrix} 1\,0\,X \\ 1\,0\,Y \end{pmatrix} \begin{matrix} q \\ q' \end{matrix} \right)_{qq'} \right]
\end{aligned}
\tag{4.105}
$$

4.3.3 Effective cross sections

The theoretical framework for the transport properties of mixtures is completed by the provision of explicit expressions for the effective cross sections introduced above. All of the effective cross sections of interest here are defined by the equations

$$\mathfrak{S}\begin{pmatrix} p\,0\,s\,t \\ p\,0\,s'\,t' \end{pmatrix} \begin{matrix} q \\ q \end{matrix} \Big)_{q\mu} = \frac{<< \Phi^{p0st|q} \cdot \mathcal{R}_{q\mu}(\Phi^{p0s't'|q}) >>}{(2p+1) < c_{q\mu} >} \tag{4.106}$$

$$\mathfrak{S}\begin{pmatrix} p\,0\,s\,t \\ p\,0\,s'\,t' \end{pmatrix} \begin{matrix} q \\ q' \end{matrix} \Big)_{qq'} = \frac{<< \Phi^{p0st|q} \cdot \mathcal{R}_{qq'}(\Phi^{p0s't'|q'}) >>}{(2p+1) < c_{qq'} >} \tag{4.107}$$

where the operation $<< X >>$ is defined for any dynamical variable X as

$$<< X >> = \frac{1}{n_q} \sum_i \int f_{qi}^{(0)} X d c_q. \tag{4.108}$$

The cross sections that arise in the formulation using the expansion vectors of Thijsse *et al.* (1979) are linearly related to those defined above but may be defined independently by

$$\mathfrak{S}\begin{pmatrix} 1\,0\,X \\ 1\,0\,Y \end{pmatrix} \begin{matrix} q \\ q \end{matrix} \Big)_{q\mu} = \frac{<< \Phi^{10X|q} \cdot \mathcal{R}_{q\mu}^{(1)}(\Phi^{10Y|q}) >>}{3 < c_{q\mu} >} \tag{4.109}$$

and

$$\mathfrak{S}\begin{pmatrix} 1\,0\,X \\ 1\,0\,Y \end{pmatrix} \begin{matrix} q \\ q' \end{matrix} \Big)_{qq'} = \frac{<< \Phi^{10X|q} \cdot \mathcal{R}_{qq'}^{(1)}(\Phi^{10Y|q'}) >>}{3 < c_{qq'} >} \tag{4.110}$$

Here, the operator $\mathcal{R}_{qq'}^{(1)}$ is a part of the Wang Chang–Uhlenbeck collision operator, defined by

$$n_{q'} f_{qi}^{(0)} \mathcal{R}_{qq'}^{(1)}(Y) = <Y_{qi} - Y'_{qk}> \tag{4.111}$$

and $R_{qq'}^{(2)}$ is defined in the same way with i and k on the right–hand side replaced by j and l respectively.

The effective cross section $\mathfrak{S} \begin{pmatrix} p\,0\,s\,t & q \\ p\,0\,s\,t & q \end{pmatrix}_{qq}$ is exactly the same as the cross section $\mathfrak{S}(p0st)$ introduced in Section 4.2 for a pure gas. All of the other cross sections introduced depend upon the binary interaction of two molecular species alone, as is made clear by the definition of the Wang Chang–Uhlenbeck or Boltzmann collision operators. From the point of view of the present volume this is a very important result because it means that if the collision cross sections can be determined for each binary interaction, then it becomes possible to evaluate the contribution of that interaction to the transport properties of a multicomponent mixture immediately. If repeated for all possible binary interactions in the mixture then the *prediction* of the transport properties of any multicomponent mixture containing them is possible. Furthermore, if the effective cross sections for the pure components of a binary mixture are known then those characteristic of the single, unlike binary interaction may be deduced from the properties of a mixture. It is worthy of note that this result remains valid even if higher orders of kinetic theory approximation are invoked.

4.3.4 Applicable formulas for mixtures

As for pure gases, the effective cross sections defined above can, in principle, be evaluated from a knowledge of the intermolecular pair potential of the pair of molecules q and q', by integration. This process would then make it possible to predict the transport properties of multicomponent mixtures of dilute gases from first principles. For the same reasons as discussed in Section 4.2.4 this procedure is not yet routinely practicable. It has been possible (Vesovic & Wakeham 1993; Vesovic *et al.* 1995) to evaluate collision cross sections for atom–rigid–rotor collisions both quantum–mechanically and classically, and more recently to treat the collisions of two identical rigid rotors classically (Heck & Dickinson 1994; Heck & Dickinson, personal communication). Nevertheless, such calculations are still very time–consuming, and the pair potentials are known for very few interactions involving polyatomic components. Thus, for most gas mixtures the route of prediction of the transport properties from intermolecular potentials is not yet viable.

The formulation set out above does, however, provide a hierarchy of procedures for the prediction of the transport properties of dilute–gas mixtures in the sense set out in Chapter 3. In the following the formulas given above are detailed in a manner which will enable each level of the hierarchy to be discussed in turn.

4.3.4.1 Viscosity

Equation (4.86) for the viscosity of a dilute–gas mixture may be written in terms of a number of experimentally accessible quantities (Maitland *et al.* 1987; McCourt *et al.* 1990)

$$[\eta_{\mathrm{mix}}]_1 = -\frac{\begin{vmatrix} H_{11} & \cdots & H_{1N} & x_1 \\ \cdot & & \cdot & \cdot \\ \cdot & \cdot & \cdot & \cdot \\ \cdot & & \cdot & \cdot \\ H_{N1} & \cdots & H_{NN} & x_N \\ x_1 & \cdots & x_N & 0 \end{vmatrix}}{\begin{vmatrix} H_{11} & \cdots & H_{1N} \\ \cdot & & \cdot \\ \cdot & \cdot & \cdot \\ \cdot & & \cdot \\ H_{N1} & \cdots & H_{NN} \end{vmatrix}} \tag{4.112}$$

Here,

$$H_{qq} = \frac{x_q^2}{[\eta_q]_1} + \sum_{\mu \neq q} \frac{2 x_q x_\mu}{\eta_{q\mu}} \frac{m_\mu m_q}{(m_q + m_\mu)^2} \left[\frac{5}{3 A_{\mu q}^*} + \frac{m_\mu}{m_q} \right] \tag{4.113}$$

and

$$H_{qq'} = -\frac{2 x_q x_{q'}}{\eta_{qq'}} \frac{m_q m_{q'}}{(m_q + m_{q'})^2} \left[\frac{5}{3 A_{qq'}^*} - 1 \right] \tag{4.114}$$

The symbol $[\eta_q]_1$ represents the first–order approximation to the zero–density viscosity of pure component q, which is defined in terms of an effective cross section for the $q - q$ interaction in equation (4.22). The quantity $\eta_{qq'}$ is the so–called interaction viscosity for the binary pair $q - q'$. It too may be defined in terms of an effective cross section for that interaction as

$$\eta_{qq'} = \frac{k_B T}{< c_{qq'} > \mathfrak{S}\left(\begin{smallmatrix} 2000 \mid q \\ 2000 \mid q' \end{smallmatrix}\right)_{qq'}} \tag{4.115}$$

$$= \frac{5}{6} \rho D_{qq'} / A_{qq'}^* \tag{4.116}$$

in which ρ is the mass density of the gas mixture and $A_{qq'}^*$ is a ratio of two effective cross sections,

$$A_{qq'}^* = \frac{5}{6} \frac{\mathfrak{S}\left(\begin{smallmatrix} 2000 \mid q \\ 2000 \mid q \end{smallmatrix}\right)_{qq'}}{\mathfrak{S}\left(\begin{smallmatrix} 1000 \mid q \\ 1000 \mid q \end{smallmatrix}\right)_{qq'}} \tag{4.117}$$

These equations are exact within the first–order kinetic theory approximation and, indeed, higher–order approximations are available (Maitland *et al.* 1983, 1987) although, for most practical purposes, the first–order formulas, which have an accuracy of better than 2%, are adequate in all cases.

The application of the foregoing fomula to the prediction of the viscosity of a multi-component mixture has two possibilities, depending upon whether experimental measurements of the viscosity of all pure components and of all possible binary mixture viscosities are available or not at the temperature of interest.

Binary data available

First it is assumed that the viscosity data of all binary mixtures from among the components of a multicomponent mixture are available at the temperature of interest. In this case the prediction of the viscosity of a multicomponent mixture proceeds from an analysis of the viscosity data for each binary mixture. The first step is to estimate the ratio of cross sections $A_{qq'}^*$. It turns out that this ratio is remarkably insensitive to temperature, to the intermolecular pair potential chosen for its evaluation or to the occurrence of inelastic collisions (Maitland *et al.* 1987; Vesovic *et al.* 1995). Consequently, it may be estimated from calculations for any reasonable potential model or from the correlations of the extended law of corresponding states discussed elsewhere (see Chapter 11), once a scaling parameter $\epsilon_{qq'}$ for energy is available. If this parameter is not listed for the system of interest (Maitland *et al.* 1987), it may itself be estimated with sufficient accuracy using the combination rule

$$\epsilon_{qq'} = \left(\epsilon_{qq}\epsilon_{q'q'}\right)^{1/2} \tag{4.118}$$

Experimental viscosity data for the binary mixture of species q and q' can then be employed in equation (4.112) for $N = 2$, together with that for the pure gases at the same temperature to yield $\eta_{qq'}$ directly. If this process is repeated for each binary pair in the N–component mixture of interest, then all of the information required to evaluate the viscosity of the mixture through equation (4.112) is available. Practical experience indicates (Kestin *et al.* 1976) that this process generally yields the mixture viscosity with an accuracy of about 1%.

Binary mixture data not available

If there are no data available for the viscosity of some or all of the binary pairs involved in a multicomponent mixture at the temperature of interest, then the interaction viscosity for those missing pairs must be estimated. For those pairs that have been included in the representations of the extended law of corresponding states (Maitland *et al.* 1987; Chapter 11), the interaction viscosity can be estimated from the correlation of the universal, reduced cross section $\mathfrak{S}_\eta^*(T^*)$ defined by the analogue of equation (4.48)

$$\mathfrak{S}_\eta^* = \frac{\mathfrak{S}\left(\begin{smallmatrix}2\,0\,0\,0 & q \\ 2\,0\,0\,0 & q'\end{smallmatrix}\right)_{qq'}}{\pi\sigma_{qq'}^2} \tag{4.119}$$

Here, $\sigma_{qq'}$ is a distance scaling parameter for the specific interaction and the reduced temperature is defined by

$$T^* = k_B T / \epsilon_{qq'} \qquad (4.120)$$

The correlation of the reduced cross section itself was discussed in Section 4.2.4.3 and is contained in Chapter 11. The accuracy of this prediction procedure for the mixture viscosity is, of course, a little worse than that discussed above, but a value of 3% is likely.

For those pair interactions for which scaling parameters are not included in the listing of the extended law of corresponding states (Maitland *et al.* 1987) the parameters themselves must be estimated. If those for the two pure components are known, the combination rule for the energy scaling parameters of equation (4.118) together with the rule

$$\sigma_{qq'} = (\sigma_{qq} + \sigma_{q'q'})/2 \qquad (4.121)$$

is recommended.

If the pure component viscosities are not available, the procedures recommended in Section 4.2 for their evaluation should be followed. This procedure will normally then lead to values of the scaling parameters for the pure gases for use in the procedures outlined above.

It has been argued that in special cases, such as the interactions between strongly polar substances, the representation of the effective cross sections by means of the extended law of corresponding states is inadequate. In such circumstances each interaction will have to be treated individually, in an effort to determine the most appropriate intermolecular pair potential model to represent it, so that a suitable estimate of the cross section can be obtained. Effective cross sections or, more often, collision integrals, for a number of potential models are available.

4.3.4.2 Diffusion coefficients

The binary diffusion coefficient $\mathcal{D}_{qq'}$ depends only on the unlike $q - q'$ interaction and upon one effective cross section, as shown by equation (4.91). If experimental measurements of the diffusion coefficient are unavailable it may be predicted from measurements of the mixture viscosity through equation (4.116). That is, the analysis that led to the interaction viscosity can be applied followed by the application of equation (4.116) to evaluate $\mathcal{D}_{qq'}$ from it. Generally, for viscosity data accurate to within 0.2% this procedure yields diffusion coefficients with an uncertainty of no more than 2%, largely as a result of the error in the evaluation of $A^*_{qq'}$ and $\eta_{qq'}$.

If experimental measurements of neither the diffusion coefficient nor the viscosity exist, then the binary diffusion coefficient can be estimated with the aid of a representation of the reduced effective cross section such as that provided by the correlations of the extended law of corresponding states (Maitland *et al.* 1987; Chapter 11) or the values provided by an appropriate intermolecular potential model.

4.3.4.3 Thermal conductivity

The evaluation of the thermal conductivity of a gas mixture is rather more complicated and difficult than for the other two properties. The difficulty stems from equations (4.92)–(4.102), which make it clear that, in general, many more cross sections are involved for each binary interaction than for the viscosity or diffusion coefficient. This is a result of the presence of internal energy and its relaxation. Of course, if these cross sections could be evaluated from a pair potential, the additional difficulty would be rather minor, since the calculation of extra cross sections is a relatively small additional burden compared with the treatment of the collision dynamics. However, as has been pointed out before, the evaluation of the collision dynamics prevents such calculations from being performed routinely. As a result, the cross sections that enter the expressions for the thermal conductivity must be evaluated by other means; for some of them this is extremely difficult, since there is little guidance from experiment or model calculations.

There is one special case, that of mixtures of exclusively monatomic components, for which rather simple results pertain and where the route of evaluation from a set of pair potentials is practical and straightforward. In other cases it is necessary to resort to the use of simplified and/or approximate methods to make predictions of the thermal conductivity. In what follows some of the hierarchy of procedures available are considered.

Mixtures of monatomic components

In this case there is simply a translational contribution to the thermal conductivity of the mixture, given by equation (4.93), and this can be further simplified owing to the absence of internal energy so that the thermal conductivity of an N–component mixture of monatomic components can be written as (Maitland *et al.* 1987)

$$[\lambda_{\text{mix,tr}}]_1 = [\lambda_{\text{mix}}]_1 = - \frac{\begin{vmatrix} L_{11} & \cdots & L_{1N} & x_1 \\ \cdot & \cdot & \cdot & \cdot \\ \cdot & \cdot & \cdot & \cdot \\ \cdot & \cdot & \cdot & \cdot \\ L_{N1} & \cdots & L_{NN} & x_N \\ x_1 & \cdots & x_N & 0 \end{vmatrix}}{\begin{vmatrix} L_{11} & \cdots & L_{1N} \\ \cdot & \cdot & \cdot \\ \cdot & \cdot & \cdot \\ \cdot & \cdot & \cdot \\ L_{N1} & \cdots & L_{NN} \end{vmatrix}} \qquad (4.122)$$

Here,

$$L_{qq} = \frac{x_q^2}{[\lambda_q]_1} + \sum_{\mu \neq q} \frac{x_q x_\mu}{2\lambda_{q\mu} A_{q\mu}^* (m_q + m_\mu)^2}$$

$$\times \left[\frac{15}{2} m_q^2 + \frac{25}{4} m_\mu^2 - 3m_\mu^2 B_{q\mu}^* + 4m_q m_\mu A_{q\mu}^* \right] \tag{4.123}$$

and

$$L_{qq'} = -\frac{x_q x_{q'}}{2\lambda_{qq'} A_{qq'}^*} \frac{m_q m_{q'}}{(m_q + m_{q'})^2} \left[\frac{55}{4} - 3B_{qq'}^* - 4A_{qq'}^* \right] \tag{4.124}$$

Here, $[\lambda_q]_1$ is the first approximation to the thermal conductivity of pure component q, and $\lambda_{qq'}$ is related to the interaction viscosity and the diffusion coefficient for the $q - q'$ interaction by the equations

$$\lambda_{qq'} = \frac{15}{8} k_B \frac{(m_q + m_{q'})}{m_q m_{q'}} \eta_{qq'} = \frac{25}{8} \frac{k_B}{A_{qq'}^*} n \mathcal{D}_{qq'} \tag{4.125}$$

where n is the molecular number density. The remaining undefined symbol $B_{qq'}^*$ is a further ratio of effective cross sections characteristic of the unlike interaction (Vesovic *et al.* 1995), which is readily evaluated by direct calculation for a spherically symmetric potential by standard methods (Maitland *et al.* 1987).

The thermal conductivity of a multicomponent mixture of monatomic species therefore requires a knowledge of the thermal conductivity of the pure components and of three quantities characteristic of the unlike interaction. The final three quantities may be obtained by direct calculation from intermolecular potentials, whereas the interaction thermal conductivity, $\lambda_{qq'}$, can also be obtained by means of an analysis of viscosity and/or diffusion measurements through equations (4.112) and (4.125) or by the application of equation (4.122) to an analysis of the thermal conductivity data for all possible binary mixtures, or by a combination of both. If experimental data are used in the prediction it may be necessary to estimate both $A_{qq'}^*$ and $B_{qq'}^*$. This is readily done using a realistic model potential or the correlations of the extended law of corresponding states (Maitland *et al.* 1987). Generally, either of these procedures can be expected to yield thermal conductivity predictions with an accuracy of a few percent for monatomic systems. Naturally, all of the methods of evaluating the properties of the pure components and the quantities characteristic of binary interactions that were discussed in the case of viscosity are available for use here too.

Mixtures containing polyatomic components

In order to discuss the evaluation of the thermal conductivity of mixtures containing polyatomic components it is useful to treat the results of the Wang Chang and Uhlenbeck expansion method and that of Thijsse separately. The former is well established but, in its full form, is sufficiently complicated that it has seldom been used for calculations (see, for example, Kestin et al. 1982), although it has been more widely used in a variety of approximate forms. The results of the second expansion method are rather recent (Ross et al. 1992), so that there are few examples of its application. The limited evidence that there is provides encouragement and suggests that this approach may ultimately prove much simpler and more fruitful.

Equations (4.92) to (4.102), which express the thermal conductivity of a gas mixture in terms of the cross sections of the Wang Chang–Uhlenbeck theory have been rewritten by Monchick et al. (1965) in an approximate form, in which so–called complex collisions have been neglected, as

$$[\lambda_{mix}]_1 = \lambda_{HE} + \Delta\lambda \tag{4.126}$$

where

$$\lambda_{HE} = \lambda_{mix,tr} + \sum_q^N [\lambda_q - \lambda_{q,tr}] \left[1 + \sum_{q' \neq q} \frac{x_{q'} D_{q\,int,q}}{x_q D_{q\,int,q'}} \right]^{-1} \tag{4.127}$$

Here, λ_{HE} is the Hirschfelder–Eucken result for the thermal conductivity of the mixture and $\Delta\lambda$ is a relatively small term (of the order of a few percent of the total thermal conductivity) that contains all of the explicit effects of inelastic collisions. The complete expression for $\Delta\lambda$ has been given by Monchick et al. (1965) and Maitland et al. (1987) in a form that contains largely experimentally accessible quantities. However, because there are in fact very few measurements of these quantities it is difficult, if not impossible, to evaluate this term for most systems. Thus, most evaluations of the thermal conductivity of mixtures have been performed with $\Delta\lambda = 0$ and so with the Hirschfelder–Eucken expression (4.127).

Equation (4.127) is written in a form that is suitable for the prediction of the properties of mixtures from independent evaluations, experimental or otherwise, of the thermal conductivity of the pure components, λ_q. It is, in fact, written in such a way that it automatically reproduces the pure component values and acts, therefore, as an interpolatory scheme between them which has its basis in a well–characterized approximation of the rigorous kinetic theory. However, further approximations are almost always required before the Hirschfelder–Eucken equation can be implemented.

The first term of equation (4.127) is an approximation to the translational contribution to the thermal conductivity of the mixture. It is obtained by making use of equations (4.122)–(4.125) for the thermal conductivity of a monatomic gas mixture. For this purpose approximate translational contributions to the thermal conductivity of each pure component $\lambda_{q,tr}$ and an interaction thermal conductivity for each unlike interaction $\lambda_{qq'}$ are evaluated by the heuristic application of equation (4.125) for monatomic species to polyatomic gases. Thus, the technique requires the availability of experimental viscosity data for pure gases and the interaction viscosity for each binary system or estimates of them. As the discussion of Section 4.2 makes clear, the use of this approximation neglects the effects of inelastic collisions on the transport of translational energy and generally overestimates the translational contribution to the thermal conductivity.

The remaining quantities required to evaluate the approximate translational contribution to the thermal conductivity of the mixture are $A^*_{qq'}$ and $B^*_{qq'}$ for each unlike interaction. Almost invariably these quantities must be estimated by means of spherically symmetric potential models or correlations based upon corresponding states (Maitland *et al.* 1987). Recent evidence (Heck & Dickinson 1994a,b) suggests that while such means of evaluating these cross section ratios are not exact, the errors incurred amount to a few percent at most.

The second term of equation (4.127) represents an approximation to the contribution of internal energy transport to the thermal conductivity of the mixture. The numerator of each term in the summation contains the difference between the total thermal conductivity of a pure gas and its translational part, estimated as discussed above. Thus, inasmuch as the latter quantity is approximate, the internal contribution to the thermal conductivity is merely an estimate. However, the combination of the first and second terms of equation (4.127), in the limit of any one mole fraction approaching unity, ensures that the total thermal conductivity of each pure component is reproduced.

Without further approximations, equation (4.127) cannot be used for predictions, since the diffusion coefficients for internal energy of species q within species q', $\mathcal{D}_{q\text{int},q'}$, are not yet accessible to direct measurement and are available from calculation for only one system (Vesovic *et al.* 1995). The most useful approximation is to assume that the diffusion coefficients for internal energy depart very little from those for mass for the same species, so that

$$\mathcal{D}_{q\text{int},q'} = \mathcal{D}_{qq'} \quad \text{and} \quad \mathcal{D}_{q\text{int},q} = \mathcal{D}_{qq} \tag{4.128}$$

This identification means that it is possible to use experimental values of diffusion coefficients or the viscosities of binary mixtures and pure components to estimate the internal energy diffusion coefficients through equation (4.125). What evidence there is for both pure gases (Section 4.2) and gas mixtures (Vesovic *et al.* 1995) suggests that the mass and internal energy diffusion coefficients seldom differ substantially, so that this is a reasonable approximation. In any event, owing to the fact that the approximate theory is used in an interpolatory manner in this formulation, it has usually been possible to predict the thermal conductivity of binary and multicomponent gas mixtures with errors of a few percent.

Recent work (Vesovic & Wakeham 1993) has concentrated on the use of the Thijsse formulation of the thermal conductivity of mixtures because of its relative simplicity and the hope that the accuracy of even simpler approximate forms of it may approach that found for pure gases (Section 4.2.4.3). At present the available results are confined to the special case of a binary mixture of an atomic species, B, and a rigid–rotor species,

A (Vesovic & Wakeham 1993). The result may be written,

$$[\lambda_{\mathrm{mix}}]_T = -\frac{\begin{vmatrix} L_{AA} & L_{AB} & x_A \\ L_{AB} & L_{BB} & x_B \\ x_A & x_B & 0 \end{vmatrix}}{\begin{vmatrix} L_{AA} & L_{AB} \\ L_{AB} & L_{BB} \end{vmatrix}} \tag{4.129}$$

Here,

$$L_{AA} = \frac{x_A^2}{[\lambda_A]_1} + \frac{x_A x_B}{2\lambda_{AB}\left(1 + r_A^2\right)^2 A_{AB}^*(m_A + m_B)^2}$$

$$\times \left[\frac{15}{2}m_A^2 + \frac{25}{4}m_B^2 - 3m_B^2 B_{AB}^* + 4m_A m_B A_{AB}^* \right.$$

$$+ \frac{5}{2}r_A^2 \frac{\mathcal{D}_{AB}}{\mathcal{D}_{Aint,B}}(m_A + m_B)^2 + 10m_B(m_A + m_B)K_{AB}^*$$

$$\left. + \frac{2r_A^2 m_A}{3\pi m_B}(2m_B - 3m_A)^2 \frac{A_{AB}^*}{\zeta_{Aint,B}} \right] \tag{4.130}$$

$$L_{BB} = \frac{x_B^2}{[\lambda_B]_1} + \frac{x_A x_B}{2\lambda_{AB} A_{AB}^*(m_A + m_B)^2}$$

$$\times \left[\frac{15}{2}m_B^2 + \frac{25}{4}m_A^2 - 3m_A^2 B_{AB}^* + 4m_A m_B A_{AB}^* \right.$$

$$\left. + \frac{50}{3\pi}r_A^2 m_A m_B \frac{A_{AB}^*}{\zeta_{Aint,B}} \right] \tag{4.131}$$

$$L_{AB} = -\frac{x_A x_B}{2\lambda_{AB} A_{AB}^*\left(1 + r_A^2\right)} \frac{m_A m_B}{(m_A + m_B)^2} \left[\frac{55}{4} - 3B_{AB}^* - 4A_{AB}^* \right.$$

$$\left. + \frac{5(m_A + m_B)}{m_B}K_{AB}^* - \frac{10r_A^2}{3\pi} \frac{A_{AB}^*}{\zeta_{Aint,B}}\left(\frac{5m_B}{m_A + m_B} - 3\right) \right] \tag{4.132}$$

Equation (4.129) has evident formal similarities to equation (4.122) for the thermal conductivity of a binary mixture of monatomic gases. Indeed, even the expressions for the determinant elements L_{AA}, L_{BB} and L_{AB} have some similarity to those of equations (4.123) and (4.124), which has been emphasized by the way they have been written. The quantity λ_{AB} is, in this formulation, properly *defined* by equation (4.125) and not merely inferred by analogy with the monatomic case. Furthermore, the new quantities which appear in equations (4.130)–(4.132) are all connected with the polyatomic nature of species A , K_{AB}^* being a further ratio of effective cross sections and $\zeta_{Aint,B}$ being the number of collisions with species B necessary to relax the internal energy of species A. For the case when A is also atomic, $r_A = 0$, $K_{AB}^* = 0$ and $\zeta_{Aint,B} = \infty$, so that the result reduces exactly to that for a monatomic gas mixture.

Early calculations with this formulation (Vesovic & Wakeham 1993) suggest that the explicit terms connected with the inelasticity of the pair potential are small, so that it may provide a simpler route to the prediction of the thermal conductivity of mixtures if the result for atom–rigid–rotor systems is more general. The development of this alternative formulation and of methods for the evaluation of effective cross sections for realistic non–spherically symmetric potentials are active lines of research.

Appendix

Relationships between effective collision cross sections and collision integrals

For practical purposes it may be useful to have relationships between effective collision cross section, as introduced in this chapter, and collision integrals, as employed, for instance, in Chapters 11 and 16.

To derive these relationships is not a simple task; therefore, a complete list cannot be found in the literature. But, in order to apply, for instance, the Extended Law of Corresponding States with the coefficients and parameters given by Maitland *et al.* (1987), just three relationships are needed to be able to calculate effective cross sections from expressions for collision integrals. These formulas can be derived from the appropriate equations defining η, D_{11} and A^* and read

$$\frac{\mathfrak{S}(2000)}{\pi\sigma^2} = \mathfrak{S}^*(2000) = \frac{4}{5}\Omega^{(2,2)*} \tag{4.133}$$

$$\frac{\mathfrak{S}(1000)}{\pi\sigma^2} = \mathfrak{S}^*(1000) = \frac{2}{3}\Omega^{(1,1)*} \tag{4.134}$$

$$A^* = \frac{\Omega^{(2,2)*}}{\Omega^{(1,1)*}} = \frac{5}{6}\frac{\mathfrak{S}^*(2000)}{\mathfrak{S}^*(1000)} = \frac{5}{6}\frac{\mathfrak{S}(2000)}{\mathfrak{S}(1000)} \tag{4.135}$$

References

Aziz, R. A. (1984). Interatomic potentials for rare–gases: pure and mixed interactions, in *Inert Gases. Potentials, Dynamics and Energy Transfer in Doped Crystals*, ed. M. Klein, pp. 5–86. Berlin: Springer–Verlag.

Bich, E., Millat, J. & Vogel, E. (1990). The viscosity and thermal conductivity of pure monatomic gases from their normal boiling point up to 5000 K in the limit of zero density and at 0.101325 MPa. *J. Phys. Chem. Ref. Data*, **19**, 1289–1305.

Boltzmann, L. (1872). Weitere Studien über das Wärmegleichgewicht unter Gasmolekülen. *Sitzungsber. Kaiserl. Akad. der Wissensch.*, **66**, 275–370.

Chapman, S. & Cowling, T.G. (1970). *The Mathematical Theory of Non–Uniform Gases*. Cambridge: Cambridge University Press.

Dickinson, A. S. & Heck E. L. (1990). Transport and relaxation cross sections for He–N_2 mixtures. *Mol. Phys.*, **70**, 239–252.

Eucken, A. (1913). Über das Wärmeleitvermögen, die spezifische Wärme und die innere Reibung der Gase. *Phys. Z.*, **14**, 324–332.

Ferron, J. R. (1990). Diffusion coefficients of internal states for the calculation of thermal conductivity. *Physica*, **A 166**, 325–337.

Ferziger, J. H. & Kaper, H. G. (1972). *Mathematical Theory of Transport Processes in Gases*. Amsterdam: North–Holland.

Heck, E. L. & Dickinson A. S. (1994). Transport and relaxation properties of N_2. *Mol. Phys.*, **81**, 1325–1352.

Heck, E.L., Dickinson, A.S. & Vesovic, V. (1993). Testing the Mason–Monchick approximation for transport properties of a pure diatomic gas. *Chem. Phys. Lett.*, **201**, 389–392.

Heck, E.L., Dickinson, A.S. & Vesovic, V. (1994). Second–order corrections for transport properties of pure diatomic gases. *Mol. Phys.*, **83**, 907–932.

Hendl, S., Millat, J., Vesovic, V., Vogel, E. & Wakeham, W. A. (1991). The viscosity and thermal conductivity of ethane in the limit of zero density. *Int. J. Thermophys.*, **12**, 999–1012.

Hermans L. J. F. (1992). Overview on experimental data from Senftleben–Beenakker effects, in *Status and Future Developments in the Study of Transport Properties*, eds. W.A. Wakeham, A.S. Dickinson, F.R.W. McCourt & V. Vesovic, pp. 155–174. Dordrecht: Kluwer.

Herzfeld, K. F. & Litovitz, T. A. (1959). *Absorption and Dispersion of Ultrasonic Waves*. New York: Academic Press.

Hirschfelder, J. O., Curtiss, C. F. & Bird, R. B. (1954). *Molecular Theory of Gases and Liquids*. New York: John Wiley & Sons.

Kestin, J., Khalifa, H.E. & Wakeham, W.A. (1976). Viscosity of multicomponent mixtures of four complex gases. *J. Chem. Phys.*, **65**, 5186–5191.

Kestin, J., Nagasaka, Y. & Wakeham, W.A. (1982). The thermal conductivity of mixtures of carbon dioxide with three noble gases. *Physica*, **113A**, 1–18.

Maitland, G. C., Mustafa, M. & Wakeham, W. A. (1983). Second–order approximations for the transport properties of dilute polyatomic gases. *J. Chem. Soc., Faraday Trans. 2*, **79**, 1425–1441.

Maitland, G. C., Rigby, M., Smith, E. B. & Wakeham, W. A. (1987). *Intermolecular Forces: Their Origin and Determination*. Oxford: Clarendon Press.

Mason E. A. & Monchick, L. (1962). Heat conductivity of polyatomic and polar gases. *J. Chem. Phys.*, **36**, 1622–1639.

McCourt, F. R. W. (1992). Status of kinetic theory, in *Status and Future Developments in the Study of Transport Properties*, eds. W.A. Wakeham, A.S. Dickinson, F.R.W. McCourt & V. Vesovic, pp. 117–153. Dordrecht: Kluwer.

McCourt, F. R. W., Beenakker, J. J. M., Köhler, W. E. & Kuscer, I. (1990). *Nonequilibrium Phenomena in Polyatomic Gases, Volume I*. Oxford: Clarendon Press.

McCourt, F. R. W., Beenakker, J. J. M., Köhler, W. E. & Kuscer, I. (1991). *Nonequilibrium Phenomena in Polyatomic Gases, Volume II*. Oxford: Clarendon Press.

McLaughlin, U. A., Rigby, M., Vesovic, V. & Wakeham, W. A. (1986). Transport properties of diatomic gases – An investigation of the validity of the Mason–Monchick approximation. *Mol. Phys.*, **59**, 579–585.

Millat, J., Plantikow, A., Mathes, A. & Nimz, H. (1988b). Effective collision cross sections for polyatomic gases from transport properties and thermomolecular pressure differences. *Z. Phys. Chem. Leipzig*, **269**, 865–878.

Millat, J., Vesovic, V. & Wakeham, W. A. (1988a). On the validity of the simplified expression for the thermal conductivity of Thijsse *et al*. *Physica*, **148A**, 153–164.

Millat, J., Vesovic, V. & Wakeham, W. A. (1989). Theoretically–based data assessment for the correlation of the thermal conductivity of dilute gases. *Int. J. Thermophys.*, **10**, 805–818.

Millat, J. & Wakeham, W. A. (1989). The correlation and prediction of thermal conductivity and other properties at zero–density. *Int. J. Thermophys.*, **10**, 983–993.

Monchick, L., Pereira, A. N. G. & Mason, E. A. (1965). Heat conductivity of polyatomic and polar gases and gas mixtures. *J. Chem. Phys.*, **42**, 3241–3256.

Ross, M.J., Vesovic, V. & Wakeham, W.A. (1992). Alternative expressions for the thermal conductivity of dilute gas mixtures. *Physica*, **183A**, 519–536.

Snider, R. F. (1960). Quantum–mechanical modified Boltzmann equation for degenerate internal states. *J. Chem. Phys.*, **32**, 1051–1060.

Thijsse, B. J., 't Hooft, G. W., Coombe, D. A., Knaap, H. F. P. & Beenakker, J. J. M. (1979). Some simplified expressions for the thermal conductivity in an external field. *Physica*, **98A**, 307–312.

van den Oord, R. J. & Korving J. (1988). The thermal conductivity of polyatomic molecules. *J. Chem. Phys.*, **89**, 4333–4338.

van der Avoird A. (1992). Overview on intermolecular potentials, in *Status and Future Developments in the Study of Transport Properties*, eds. W.A. Wakeham, A.S. Dickinson, F.R.W. McCourt & V. Vesovic, pp. 1–28. Dordrecht: Kluwer.

Vesovic, V. & Wakeham, W. A. (1992). Traditional transport properties, in *Status and Future Developments in the Study of Transport Properties*, eds. W.A. Wakeham, A.S. Dickinson, F.R.W. McCourt & V. Vesovic, pp. 29–55. Dordrecht: Kluwer.

Vesovic, V. & Wakeham, W.A. (1993). Practical, accurate expressions for the thermal conductivity of atom–diatom gas mixtures. *Physica*, **201A**, 501–514.

Vesovic, V., Wakeham, W.A., Dickinson, A.S., McCourt, F.R.W. & Thachuk, M. (1995). Quantum mechanical calculation of generalised collision cross sections for the He − N_2 interaction. Part II. Thermomagnetic effect. *Mol. Phys.*, submitted.

Viehland, L. A., Mason, E. A. & Sandler, S. I. (1978). Effect of spin polarization on the thermal conductivity of polyatomic gases. *J. Chem. Phys.*, **68**, 5277–5282.

Waldmann, L. (1957). Die Boltzmann–Gleichung für Gase mit rotierenden Molekülen. *Z. Naturforsch.*, **12a**, 660–662.

Wang Chang, C. S. & Uhlenbeck, G. E. (1951). Transport phenomena in polyatomic gases. Univ. of Michigan Eng. Res. Inst. Report CM–681.

Wang Chang C. S., Uhlenbeck G. E. & de Boer, J. (1964). The heat conductivity and viscosity of polyatomic gases, in *Studies in Statistical Mechanics, Vol. II*, eds. J. de Boer & G.E. Uhlenbeck. Amsterdam: North–Holland.

5

Dense Fluids

5.1 Introduction

J. H. DYMOND

The University, Glasgow, UK

One of the triumphs of the simple kinetic theory of gases was the prediction by Maxwell that the viscosity at constant temperature should be independent of density (pressure), for systems of hard–sphere molecules and also molecules which repel with a force proportional to r^{-5} (Maxwell 1860, 1867). This somewhat surprising result was, contrary to expectation, found to be in good agreement with experiment up to moderate pressures. For example, the viscosity of argon at 298.15 K increases just slightly from 22.63 μPa s at 0.1 MPa to 22.81 μPa s at 1 MPa (Kestin *et al.* 1971). At 5 MPa, the viscosity is still only 23.87 μPa s and 25.77 μPa s at 10 MPa. However, as the pressure increases further, the rate of increase in viscosity becomes greater. At the marginally higher temperature of 301.15 K, the viscosity of argon is 104 μPa s at 149.9 MPa and 480 μPa s at 897.1 MPa (Trappeniers *et al.* 1980).

This difference in the pressure dependence of viscosity between a dilute gas and a dense gas arises because in a dilute gas it is the molecules themselves which transport the momentum; in a dense gas, however, transport of momentum occurs over nonzero distances on collision. The same is true for energy transport (thermal conductivity). In addition, the collision rate, which is proportional to the number density of the particles at low densities, increases more rapidly at high densities because the distances traveled between collisions are then significantly reduced by the particle sizes.

In fact it is the density ρ (or molar volume V), rather than pressure, which is the theoretically important independent variable. This is apparent from consideration of Figures 5.1 and 5.2 for the viscosity of argon (Haynes 1973) at four different temperatures, of which one is below the critical temperature. Whereas at low pressure the viscosity increases with increase in temperature as a result of purely kinetic effects, at high pressures the temperature derivative $(\partial \eta / \partial T)_P$ has the opposite sign as collisional

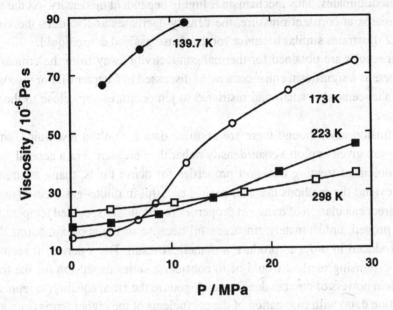

Fig. 5.1. Pressure dependence of the viscosity of argon.

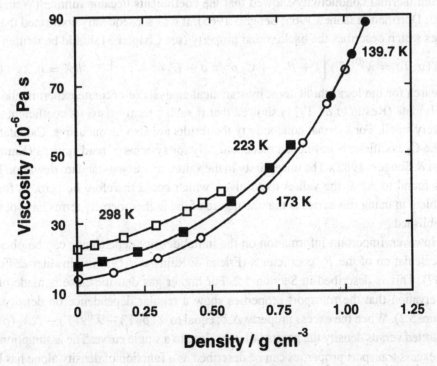

Fig. 5.2. Density dependence of the viscosity of argon.

transfer predominates. This mechanism is highly dependent on density. As the temperature increases at constant pressure, the density decreases and so does the viscosity. Figure 5.2 illustrates similar behavior for the dense gas and dense liquid.

Similar results are obtained for thermal conductivity away from the critical region (where there is a significant enhancement, as discussed in Chapter 6). For viscosity, the critical enhancement is small and restricted to temperatures very close to the critical temperature.

For diffusion also, though there are accurate data for only a few fluids, smoother variations are given in plots versus density rather than pressure. For a general approach to the problem of treating transport properties for dense fluids, many attempts were made to extend the methods that had been successful in dilute–gas theory (see Brush 1972). Direct calculation of transport properties for liquids at specified temperature and pressure proved, unfortunately, unsuccessful because of the approximations that had to be introduced in order to produce a numerical result. For a period, it seemed that the most promising method would be to consider a series expansion for the transport properties in powers of number density (analogous to the virial equation to represent gas imperfection data) with calculation of the coefficients of the higher terms from generalized Boltzmann equations of the type suggested by Bogoliubov (1946), which involved higher–order distribution functions. However, detailed examination of the quadruple collision integrals determining the coefficients of the density–squared term for viscosity and thermal conductivity showed that the coefficients became infinite (Weinstock 1965; Dorfman & Cohen 1965; Sengers 1965). It was subsequently confirmed that the series which describes the background property (see Chapter 3) should be written

$$\bar{X}(\rho, T) = X^{(0)}(T)\left[1 + B_x\rho + C_x\rho^2 \ln \rho + D_x\rho^2 + \dots\right] \qquad (X = \eta, \lambda) \quad (5.1)$$

A search for the logarithmic term by statistical analysis of accurate experimental viscosity data (Kestin *et al.* 1971) showed that if such a term exists, its coefficient must be very small. For thermal conductivity the results are also inconclusive. Calculations of the C_x coefficients have been made, but only for systems of hard spheres (Kamgar–Parsi & Sengers 1982). The uncertainty in the values at the two standard deviation level was found to equal the values themselves, which could therefore be zero. A further problem in using this series is that the nature of the higher–density terms has not been established.

However, important information on the initial density dependence can be obtained by calculation of the B_x coefficients (Friend & Rainwater 1984; Rainwater & Friend 1987). This is described in Section 5.2. For higher gas densities, use is made of the observation that the transport properties show a regular dependence on density (see Figure 5.2). When the excess property ΔX, equal to $X(\rho, T) - X^{(0)}(T) - \Delta X_c(\rho, T)$, is plotted versus density the points fall very close to a single curve. The assumption that the excess transport properties can be described as a function of density alone has been a useful basis for correlating data within stated density and temperature intervals. It should be noted that for liquids the excess properties are not rigorously independent of temperature; where data are available at different temperatures up to elevated pressures, the isotherms do not form a single curve. For a satisfactory representation of transport property data over the complete fluid range, therefore, the excess property must be considered as a function of temperature *and* density (pressure).

Data representation can be considered truly satisfactory only when it has a molecular basis. The first such successful approach was that of Enskog (Enskog 1922) for a system of hard spheres in which he made empirical modifications to the Boltzmann theory to account for the finite size of the molecules. Use of the Boltzmann equation, which considers only binary collisions, is valid for this model, since multiple collisions have a low probability. Enskog obtained expressions relating the diffusion, viscosity and thermal conductivity for the dense system, subscript E, to the dilute–gas values, superscript (0),

$$\frac{n D_E}{[n D]^{(0)}} = \frac{1}{g(\sigma)} \tag{5.2}$$

$$\frac{\eta_E}{\eta^{(0)}} = \left[\frac{1}{g(\sigma)} + \frac{0.8b}{V} + 0.761 g(\sigma) \left(\frac{b}{V} \right)^2 \right] \tag{5.3}$$

$$\frac{\lambda_E}{\lambda^{(0)}} = \left[\frac{1}{g(\sigma)} + \frac{1.2b}{V} + 0.755 g(\sigma) \left(\frac{b}{V} \right)^2 \right] \tag{5.4}$$

where $g(\sigma)$ is the radial distribution function at contact for spheres of diameter σ and $b = 2\pi N_A \sigma^3 / 3$. It is assumed here that a dense hard–sphere system behaves exactly like a low–density system except that the collision rate is higher than expected from consideration of the density change alone. This arises because the distance which two spheres travel between collisions is significantly reduced because of their nonzero diameters. In a dilute system the diameter of the molecule is negligible compared with the interparticle distance. The increase in collision rate can be rigorously calculated from the fact that the pressure, which is proportional to the rate of momentum transfer and hence to the collision rate, is also proportional to the radial distribution function at contact,

$$\frac{PV}{RT} = 1 + \frac{bg(\sigma)}{V} \tag{5.5}$$

where $g(\sigma)$ is unity at low density.

The Enskog theory for diffusion (equation (5.2)) just scales in time the solution of the Boltzmann equation valid at low densities. However, whereas for diffusion the particles themselves must move, in the case of viscosity and thermal conductivity there is the additional mechanism of collisional transfer, which becomes increasingly important as the density increases, whereby momentum and energy can be passed to another molecule upon collision. The Enskog theory for the viscosity η_E and the thermal conductivity

λ_E in terms of the low–density coefficients (equations (5.3) and (5.4)) accordingly contains additional terms. The first, kinetic contribution is the only important term at low densities and scales in time as for diffusion. The final term is the contribution from the potential part alone and the middle term is the cross contribution of the kinetic and potential part. The presence of these terms, and their functional dependence, can be demonstrated simply from a derivation of these expressions by the fluctuation–dissipation theorem, which gives the transport properties in terms of an autocorrelation function of the appropriate flux (see 5.4.1). For thermal conductivity, for example, the flux involves the sum of kinetic and potential energies. The autocorrelation of this flux involves the product of the flux at two different times, producing three different terms which can be shown to have the same dependence on density and $g(\sigma)$ as above.

This theory covers the complete density range from dilute gas to solidification. However, the hard sphere model is inappropriate for a real gas at low and intermediate densities where specific effects of intermolecular forces are significant. It is therefore necessary to modify the theory; this has been done in different ways. In Section 5.2, a method is described for determination of the second viscosity virial coefficient and the translational part of the thermal conductivity coefficient. Although this approach is not rigorous, it does provide a useful estimate for these coefficients – especially for low reduced temperatures.

At higher gas densities, where departures from the dilute–gas values are significantly larger, it is essential to make the radial distribution function consistent with the co–volume b of the molecules. Section 5.3 describes the application and limitations of this approach.

It is for the dense gas, above about the critical density, and the liquid that the hard–sphere model can be considered as a reasonably realistic description of the molecular interactions. For a proper application of the hard–sphere theory, it is necessary to take into account corrections for correlated motions, which are considered in Section 5.4. These were computed using the Einstein expressions for the transport coefficients, which are given in 5.4.1, together with the equivalent autocorrelation function relationships. These are more convenient for direct calculation for dense fluid transport properties using more realistic forms of intermolecular potential energy function, as in recent calculations for molecular fluids (Toxvaerd 1989; Luo & Hoheisel 1992). At the present time, the most successful, general, molecular theory for correlation and prediction of dense fluid transport properties is that based on the hard–sphere model. Application of the smooth hard–sphere theory for atomic fluids is given in 5.4.2, and the rough hard–sphere theory, for molecular fluids, is described in 5.4.3. For dense fluids, other approaches based on corresponding–states theory, absolute reaction–rate theory and free–volume theory have proved useful in the correlation of transport property data. These are presented in Sections 5.4.4–5.4.6. Sections 5.5 and 5.6 consider the application of theory to the calculation of transport properties of dense gas and liquid mixtures.

References

Bogoliubov, N.N. (1946). In *Studies in Statistical Mechanics*, Vol. 1, eds. J. de Boer &
    G. E. Uhlenbeck, (1962), pp. 5–118. Amsterdam: North–Holland.

Brush, S.G. (1972) *Kinetic Theory*, Vol. 3. Oxford: Pergamon Press.

Dorfman, J.R. & Cohen, E.G.D. (1965). On the density expansion of the pair distribution
    function for a dense gas not in equilibrium. *Phys. Lett.*, **16**, 124–125.

Enskog, D. (1922). Kinetische Theorie der Wärmeleitung, Reibung und Selbstdiffusion in
    gewissen verdichteten Gasen und Flüssigkeiten, *Kungl. Sv. Vetenskapsakad. Handl.*, **63**,
    No. 4.

Friend, D.G. & Rainwater, J.C. (1984). Transport properties of a moderately dense gas.
    Chem. Phys. Lett., **107**, 590–594.

Haynes, W.M. (1973). Viscosity of gaseous and liquid argon. *Physica*, **67**, 440–470.

Kamgar–Parsi, B. & Sengers, J.V. (1982). Logarithmic dependence of the transport
    properties of moderately dense gases. *Proc. 8th. Symp. Thermophys. Props.* New York:
    A.S.M.E., 166–171.

Kestin, J., Paykoc, E. & Sengers, J.V. (1971). Density expansion for viscosity in gases.
    Physica, **54**, 1–19.

Luo, H. & Hoheisel, C. (1992). Computation of transport coefficients of liquid benzene and
    cyclohexane using rigid pair interaction models. *J. Chem. Phys.*, **96**, 3173–3176.

Maxwell, J.C. (1860). Illustrations of the dynamical theory of gases. Part 1. On the motions
    and collisions of perfectly elastic spheres. *Phil. Mag.*, **19**, 19–32.

Maxwell, J.C. (1867). On the dynamical theory of gases. *Phil. Trans. Roy. Soc. London*, **157**,
    49–88.

Rainwater, J.C. & Friend, D.G. (1987). Second viscosity and thermal–conductivity virial
    coefficients of gases: Extension to low reduced temperatures. *Phys. Rev.*, **A 36**,
    4062–4066.

Sengers, J.V. (1965). Density expansion of the viscosity of a moderately dense gas. *Phys.
    Rev. Lett.*, **15**, 515–517.

Toxvaerd, S. (1989). Molecular dynamics calculation of the equation of state of liquid
    propane. *J. Chem. Phys.*, **91**, 3716–3720.

Trappeniers, N.J., van der Gulik, P.S. & van den Hooff, H. (1980). The viscosity of argon at
    very high pressure, up to the melting–line. *Chem. Phys. Lett.*, **70**, 438–443.

Weinstock, J. (1965). Nonanalyticity of transport coefficients and the complete density
    expansion of momentum correlation functions. *Phys. Rev.*, **140**, A460–A465.

5.2 Initial density dependence

E. BICH and E. VOGEL

Universität Rostock, Germany

5.2.1 Introduction

Recent advances in the theoretical description of the initial density dependence of the transport properties justify a separate treatment. If moderately dense gases are considered, only the linearized equations (5.1) are needed; that is, the virial form of the density expansion can be truncated after the term linear in density. This means that the deviation from the dilute–gas behavior can be represented by the second transport virial coefficients B_x or alternatively by the initial–density coefficients $(\eta^{(1)}, \lambda^{(1)}, D^{(1)})$, which are related by

$$B_x = \frac{X^{(1)}}{X^{(0)}} \qquad (x = \eta, \lambda, D; X = \eta, \lambda, [\rho D]) \tag{5.6}$$

Since the existence of internal degrees of freedom and their interaction with translational modes of motion has a negligible effect on viscosity, the second viscosity virial coefficient of polyatomic fluids can be formally identified with that of monatomic gases $(B_\eta^{\mathrm{mon}} \equiv B_\eta)$.

On the contrary, the second thermal conductivity virial coefficient originally derived for monatomic gases $(B_\lambda^{\mathrm{mon}})$ represents solely the contribution related to translational degrees of freedom for polyatomic gases $(B_\lambda^{\mathrm{mon}} \equiv B_{\lambda,\mathrm{tr}})$. In order to describe the total initial density contribution to thermal conductivity, therefore, a term related to internal modes of motion has to be included. This can be conveniently done by applying the two–flux approach based on a suggestion of Eucken (Hanley *et al.* 1972; Maitland *et al.* 1987), whereby

$$\lambda(\rho, T) = \lambda_{\mathrm{tr}}(\rho, T) + \lambda_{\mathrm{int}}(\rho, T) \tag{5.7}$$
$$= \lambda_{\mathrm{tr}}^{(0)}(T)\left[1 + B_{\lambda,\mathrm{tr}}(T)\rho\right] + \lambda_{\mathrm{int}}^{(0)}(T)\left[1 + B_{\lambda,\mathrm{int}}(T)\rho\right]$$

In general, there are two methods that can be applied in order to describe $B_x(T)$ $(x = \eta, \lambda)$: The most up–to–date theory, proposed by Friend & Rainwater (1984; Rainwater & Friend 1987), models the moderately dense gas as a mixture of monomers and dimers which interact according to the Lennard–Jones (12–6) potential. Besides the fact that this potential is only a rough approximation of the real physical situation, this model has the disadvantage that it has not yet been extended to describe the internal contribution to the initial density dependence of thermal conductivity.

In this case a less rigorous approach has to be applied which involves a modification of the Enskog hard–sphere theory for real gases (Enskog 1922; Hanley *et al.* 1972; Hanley & Cohen 1976; Ross *et al.* 1986; Nieto de Castro *et al.* 1990). This can be used to find an approximate value for the internal contribution for polyatomic gases, whereas the translational contribution to the second thermal conductivity virial coefficient should be deduced from the Rainwater–Friend theory.

Since this volume concentrates on the application of results derived from kinetic theory, the models behind the working equations are described only briefly in the following subsections in order to enable the reader to understand the advantages and disadvantages of these schemes. For details concerning the theory the reader should refer to the given original papers.

5.2.2 Brief outline of the Rainwater–Friend theory

According to the Rainwater–Friend model, contributions to B_η and $B_{\lambda,\mathrm{tr}}$ are due to interactions between two monomers <2>, three monomers <3>, and a monomer and a dimer <MD>. It is assumed that these contributions can be described independently; that is,

$$B_x = B_x^{<2>} + B_x^{<3>} + B_x^{<MD>} \qquad (x = \eta; (\lambda, \mathrm{tr})) \tag{5.8}$$

Especially for the purpose of comparison with experimental data, it is convenient to introduce reduced quantities, namely,

$$T^* = \frac{k_B T}{\varepsilon} \quad \text{and} \quad B_x^*(T^*) = \frac{B_x}{N_A \sigma^3} \tag{5.9}$$

ε and σ being the Lennard–Jones (12–6) intermolecular potential energy and distance parameters, N_A the Avogadro constant, and k_B the Boltzmann constant.

For the *exact* evaluation of the two–monomer contribution $B_x^{<2>}$ the Rainwater–Friend theory includes results from the kinetic theory of dense gases (see, for example, Cohen 1969), which means that direct kinetic as well as collisional transfer effects are considered. In order to calculate reliable values for $B_x^{<2>}$ Rainwater (1984) extended earlier results of Snider & Curtiss (1958), Hoffman & Curtiss (1965) and Bennett & Curtiss (1969) so that the dynamics of the two particles for a more realistic potential (here the Lennard–Jones (12–6) potential) were included. At the same time, the effects of bound states have been excluded from the evaluation of the complicated set of integrals that defines the two–monomer contribution (see Rainwater 1984).

The analysis of the three–monomer contribution is based on exact results for hard spheres (Sengers *et al.* 1978). Its evaluation to a good *approximation* for realistic potentials was carried out by Hoffman & Curtiss (1965) and by Friend (1983). The effect of a third monomer on the collisional frequency of two monomers – the so–called excluded volume effect – leads to identical expressions for $B_\eta^{<3>}$ and $B_{\lambda,\mathrm{tr}}^{<3>}$, which again depend upon complicated integrals that are not reproduced here.

Whereas the above mentioned contributions can be calculated for the Lennard–Jones (12–6) potential without using any experimental information, the monomer–dimer contribution was derived using selected experimental first density coefficients.

Following a proposal by Stogryn & Hirschfelder (1959) the *approximation* is introduced that the interaction potential between a monomer and a dimer is of the same form as the monomer–monomer potential but characterized by the potential parameter ratios

$$\delta = \frac{\sigma_{MD}}{\sigma_M}, \qquad \theta = \frac{\varepsilon_{MD}}{\varepsilon_M} \qquad\qquad (5.10)$$

In contrast to viscosity, the analysis of thermal conductivity is complicated by the fact that because of the existence of dimers a 'chemical reaction contribution (association correction)' $B_{\lambda,tr,r}^{<MD>*}$ has to be taken into account. This is done using a Eucken–type form that greatly simplifies for monatomic gases.

Because of the complexity of the monomer–dimer collisional process, it seems to be not feasible to derive $B_x^{<MD>}$ solely from theory. However, it is possible to obtain an optimum fit of δ and θ to experimental data of B_η for monatomic as well as polyatomic gases and to obtain results for B_λ for monatomic gases.

The scheme outlined here in a very condensed form leads to tables of total values for B_x^* that refer to a given set of δ and θ. Originally, Friend & Rainwater (1984) proposed

$$\delta = 1.02, \qquad \theta = 1.15 \qquad\qquad (5.11)$$

based on experimental viscosity data for monatomic gases, nitrogen and hydrogen and thermal conductivity data for only the monatomic gases. The experimental data were situated in the reduced temperature range $T^* > 1$, whereas B_x^* is most sensitive to temperature in the range $T^* < 1$. Because $T^* < 1$ is easier to attain for polyatomics, Bich & Vogel (1991) also included B_η data in the range $0.8 < T^* < 3$ for sulfur hexafluoride, neopentane, n–hexane, cyclohexane, and benzene obtained from very accurate viscosity measurements; B_η data for nitrogen, carbon dioxide, ethane, and ethene from literature; and B_λ data for monatomic gases measured with the transient hot–wire technique in order to get improved values of the potential parameter ratios. This led to

$$\delta = 1.04, \qquad \theta = 1.25 \qquad\qquad (5.12)$$

Numerical results for B_x^* are listed in Table 5.1 for the two sets of parameters. This table also includes estimates of the 'chemical reaction contribution' $B_{\lambda,tr,r}^{<MD>*}$ and of the internal contribution $B_{\lambda,int}^*$ (see 5.2.3.2) to the initial density dependence of the thermal conductivity of polyatomic species, which were not included in the original Rainwater–Friend approach. Lennard–Jones (12–6) parameters for these substances, listed in Table 5.2, were determined from the temperature dependence of the (extrapolated) zero–density transport coefficients at low reduced temperatures, where the effect of the initial density dependence is more significant.

Experimental data for the reduced second viscosity virial coefficient are compared in Figure 5.3 with the values calculated by the Rainwater–Friend theory for these two sets of parameters. Polar gases as well as nonpolar gases exhibit much the same temperature dependence (Vogel *et al.* 1986; Hendl & Vogel 1992; Vogel & Hendl 1992). So, to a

Table 5.1. *Reduced second transport virial coefficients and chemical reaction contribution according to the Rainwater–Friend theory for monatomic fluids. Reduced internal contribution for polyatomic gases according to MET–I.*

	Rainwater–Friend			Bich–Vogel		
	$\delta = 1.02$	$\theta = 1.15$		$\delta = 1.04$	$\theta = 1.25$	
T^*	B_η^*	$B_{\lambda,\mathrm{tr}}^*$	B_η^*	$B_{\lambda,\mathrm{tr}}^*$	$B_{\lambda,\mathrm{tr},r}^{<MD>*}$	$B_{\lambda,\mathrm{int}}^*$
0.5	-21.8806	11.3451	-25.2662	5.4659	40.8125	-60.9782
0.6	-9.5862	10.9807	-12.0633	6.9969	25.7191	-28.4576
0.7	-4.5951	9.7170	-6.5113	6.7769	18.0838	-14.8005
0.8	-2.1941	8.5370	-3.7299	6.2558	13.6250	-8.1473
0.9	-.8729	7.5963	-2.1430	5.7581	10.7719	-4.6084
1.0	-.1230	6.8219	-1.1800	5.3102	8.8090	-2.6261
1.1	.3331	6.1885	-.5587	4.9222	7.3885	-1.4793
1.2	.6288	5.6807	-.1438	4.5932	6.3255	-.8087
1.3	.8096	5.2456	.1430	4.3130	5.5048	-.4184
1.4	.9146	4.8699	.3342	4.0574	4.8478	-.1985
1.5	.9736	4.5457	.4620	3.8275	4.3156	-.0788
1.6	1.0075	4.2663	.5507	3.6275	3.8792	-.0226
1.7	1.0233	4.0211	.6143	3.4513	3.5141	-.0078
1.8	1.0348	3.8130	.6678	3.3013	3.2038	-.0189
1.9	1.0265	3.6193	.6955	3.1559	2.9372	-.0442
2.0	1.0108	3.4450	.7103	3.0228	2.7069	-.0760
2.1	.9903	3.2870	.7159	2.9008	2.5061	-.1119
2.2	.9672	3.1433	.7153	2.7889	2.3298	-.1499
2.3	.9427	3.0123	.7105	2.6862	2.1737	-.1885
2.4	.9176	2.8931	.7029	2.5915	2.0346	-.2268
2.5	.8970	2.7891	.6977	2.5089	1.9099	-.2641
2.6	.8715	2.6886	.6861	2.4275	1.7978	-.2995
2.7	.8458	2.5956	.6731	2.3516	1.6964	-.3330
2.8	.8204	2.5091	.6591	2.2806	1.6042	-.3641
2.9	.7955	2.4287	.6443	2.2142	1.5204	-.3928
3.0	.7713	2.3539	.6292	2.1520	1.4439	-.4196
3.5	.6637	2.0466	.5560	1.8935	1.1427	-.5270
4.0	.5721	1.8166	.4871	1.6952	.9341	-.6020
4.5	.4953	1.6387	.4261	1.5394	.7825	-.6531
5.0	.4302	1.4968	.3724	1.4138	.6683	-.6883
6.0	.3267	1.2854	.2841	1.2241	.5089	-.7290
7.0	.2504	1.1382	.2173	1.0906	.4034	-.7463
8.0	.1906	1.0277	.1639	.9892	.3308	-.7526
9.0	.1431	.9416	.1210	.9097	.2774	-.7526
10.0	.1049	.8729	.0862	.8459	.2372	-.7488
15.0	-.0060	.6663	-.0162	.6516	.1323	-.7134
20.0	-.0536	.5605	-.0604	.5507	.0884	-.6754
30.0	-.0870	.4465	-.0905	.4415	.0459	-.6182
40.0	-.0952	.3823	-.0969	.3799	.0221	-.5779
50.0	-.0980	.3401	-.0985	.3394	.0067	-.5476
100.0	-.1359	.2639	-.1365	.2631	.0077	-.4563

Table 5.2. *Lennard–Jones (12–6) parameters.*

Gas or vapor	Collision diameter σ (nm)	Well depth ε/k_B (K)
He	0.263	10.0
Ne	0.272	47.0
Ar	0.341	125.0
Kr	0.362	183.0
Xe	0.396	250.0
H_2	0.287	33.0
N_2	0.368	90.9
CO	0.3698	104.5
CO_2	0.3753	246.1
N_2O	0.3776	248.8
CH_4	0.3706	159.7
CF_4	0.4486	167.3
C_2H_4	0.4155	225.6
C_2H_6	0.4407	227.9
neo $- C_5H_{12}$	0.6160	262.5
C_6H_{14}	0.6136	387.8
C_6H_6	0.5379	411.5

good approximation, $B_\eta(T)$ may be evaluated for other compounds from Table 5.1, given values for the Lennard–Jones (12–6) parameters.

It should be noted that approximations in $B_x^{<3>*}$ and $B_x^{<MD>*}$ preclude the use of more realistic pair potentials at this stage. This approach can be used for helium, hydrogen or neon only with caution because of complications arising from quantum–mechanical effects at low reduced temperatures (Bich & Vogel 1991).

In the case of thermal conductivity, comparison of experimental data with results of this theory is limited to monatomic gases, as shown in Figure 5.4. As for the viscosity (Figure 5.3), so here also it appears that the revised values for δ and θ give a marginally better fit. However, more experimental measurements at $T^* < 1$, especially for the heavier monatomic gases, are required to confirm this.

5.2.3 Modified Enskog theory

5.2.3.1 Viscosity and translational part of thermal conductivity

The expressions for the transport coefficients given by Enskog theory (equations (5.2)–(5.4)) lead to the following results for the second transport virial coefficients

$$B_D = -g^{(1)}, \quad B_\eta = \frac{4b}{5} - g^{(1)}, \quad B_{\lambda,\text{tr}} = \frac{6b}{5} - g^{(1)} \tag{5.13}$$

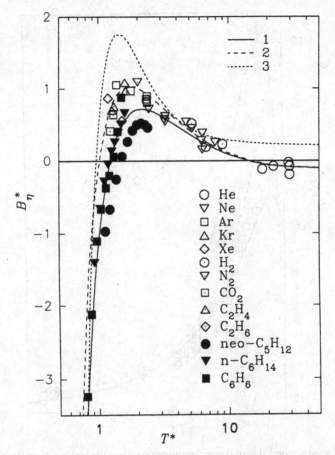

Fig. 5.3. The reduced second viscosity virial coefficient as a function of reduced temperature. Curves: 1 – Rainwater–Friend theory ($\delta = 1.04$ and $\theta = 1.25$); 2 – Rainwater–Friend theory ($\delta = 1.02$ and $\theta = 1.15$); 3 – MET–I.

where $g^{(1)}$, the coefficient of the first density term of the equilibrium radial distribution function at contact for hard spheres, is equal to c/b, the ratio of the third and second pressure virial coefficients.

For application to real gases, this theory has been modified (Enskog 1922; Hanley *et al.* 1972; Hanley & Cohen 1976; Vogel *et al.* 1986; Ross *et al.* 1986). Although this has no rigorous theoretical basis, it does provide an alternative representation of the second viscosity virial coefficient and the translational part of the second thermal conductivity virial coefficient, which is particularly useful at reduced temperatures below $T^* = 0.5$, the lower limit of the coefficients in Table 5.1. On the basis that a real fluid differs from a hard sphere fluid mainly in the temperature dependence of the collision frequency, the pressure P of the hard–sphere fluid is replaced by the thermal pressure $T(\partial P/\partial T)_\rho$ of

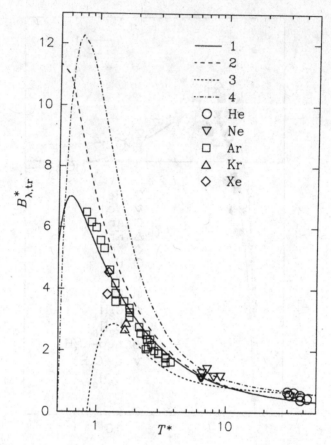

Fig. 5.4. The reduced second thermal conductivity virial coefficient as a function of reduced temperature. Curves: 1 – Rainwater–Friend theory ($\delta = 1.04$ and $\theta = 1.25$); 2 – Rainwater–Friend theory ($\delta = 1.02$ and $\theta = 1.15$); 3 – MET–I (without $B_{\lambda,\mathrm{tr},r}^{<MD>*}$); 4 – MET–I (including $B_{\lambda,\mathrm{tr},r}^{<MD>*}$).

the real fluid. Thus b and c are replaced by

$$b \rightarrow \beta = B + T\frac{dB}{dT}, \qquad c \rightarrow \gamma = C + T\frac{dC}{dT} \qquad (5.14)$$

where B and C are the second and third pressure virial coefficients of the real fluid.

In Figure 5.3, values for B_η^* calculated in this way (MET–I) for the Lennard–Jones (12–6) potential are shown to have the same general temperature dependence as the experimental data (there are other modifications of Enskog's theory which are not considered here.). In Figure 5.4, an analogous comparison is shown for thermal conductivity. The calculated values (MET–I) underestimate the experimental results when the association correction $B_{\lambda,\mathrm{tr},r}^{<MD>*}$ (see Table 5.1) is omitted from the MET approach but are higher than the experiment when this is included.

5.2.3.2 Thermal conductivity of polyatomic fluids

The initial density dependence of the thermal conductivity of a polyatomic gas is given by expression (5.7). Neither the Rainwater–Friend model nor the modified Enskog theory accounts for the contribution of internal degrees of freedom, but it is assumed that this can be modeled as a purely diffusive process following an idea originally introduced by Mason & Monchick (1962) for dilute gases

$$\lambda_{int}(\rho, T) = [\rho D]^{(0)} C^{id}_{V,int} (1 + B_D\rho) \tag{5.15}$$

where $C^{id}_{V,int}$ is the internal contribution to the isochoric molar heat capacity for the dilute gas. Accordingly, the initial density dependence comes entirely from the self–diffusion coefficient; Nieto de Castro *et al.* (1990) proposed that the hard–sphere expression (5.2) should be used for B_D.

According to Stogryn & Hirschfelder (1959) there are two additional terms to be taken into account, one from expansion of the Eucken type correction for polyatomic molecules and a second one due to the actual formation of dimers. Combining all these contributions gives estimates for $B^*_{\lambda,int}$, which are included in Table 5.1.

Calculation of the initial density dependence of the thermal conductivity for polyatomic gases requires knowledge of the translational and internal mode contributions to the dilute–gas thermal conductivity. Thus,

$$B^*_\lambda \lambda^{(0)} = B^*_{\lambda,tr}\lambda^{(0)}_{tr} + B^*_{\lambda,int}\lambda^{(0)}_{int} = B^*_{\lambda,tr}\lambda^{(0)}_{tr} + B^*_{\lambda,int}\left(\lambda^{(0)} - \lambda^{(0)}_{tr}\right) \tag{5.16}$$

The translational part here is obtained from the dilute–gas viscosity according to the approximate Eucken formula

$$\lambda^{(0)}_{tr} = \frac{15R}{4M}\eta^{(0)} \tag{5.17}$$

and $B^*_{\lambda,tr}$ is taken from the results of the Rainwater–Friend theory.

How well this theory accounts for $B^*_{\lambda,tr}$ for polyatomic gases can be judged from Figure 5.5. The $B^*_{\lambda,tr}$ are derived from experiment according to

$$B^*_{\lambda,tr} = B^*_\lambda\frac{\lambda^{(0)}}{\lambda^{(0)}_{tr}} - B^*_{\lambda,int}\left(\frac{\lambda^{(0)}}{\lambda^{(0)}_{tr}} - 1\right) \tag{5.18}$$

with $B^*_{\lambda,int}$ from Table 5.1 and $\lambda^{(0)}_{tr}$ from equation (5.17).

The experimental λ data are mainly those of Nieto de Castro *et al.* (1990) and Ross *et al.* (1986). It is noteworthy that $B^*_{\lambda,tr}$ from methane and ethane measurements agree well at low reduced temperatures, although the temperature dependence is steeper than predicted. There is general agreement between the experimental points for different compounds and theory, but the relatively large scatter reflects mainly the approximate treatment of the effects of internal degrees of freedom.

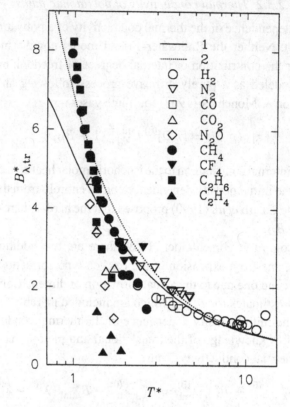

Fig. 5.5. The reduced second thermal conductivity virial coefficient as a function of reduced temperature for polyatomic gases. Curves: 1 – Rainwater–Friend theory ($\delta = 1.04$ and $\theta = 1.25$); 2 – Rainwater–Friend theory ($\delta = 1.02$ and $\theta = 1.15$).

5.2.4 Conclusions

The first density correction for viscosity and for the translational part of the thermal conductivity is best predicted by the Rainwater–Friend model, for which values for the reduced second transport virial coefficients are given in Table 5.1. For computer codes the tabulated values can be approximated using the correlation

$$B_x^* = \sum_{i=0}^{n} a_i \left(\sqrt{T^*} \right)^{-i} , \qquad (x = \eta; \ (\lambda, tr)) \tag{5.19}$$

with the coefficients a_i of Table 5.3 (for the values of Bich & Vogel (1991)) and potential parameters (see Table 5.2). The initial density dependence of thermal conductivity for polyatomic gases can be predicted by means of the Rainwater–Friend theory for the translational part, together with a contribution from the internal modes according to MET–I (including terms due to dimerization of the molecules) or according to the simpler hard–sphere correction.

Table 5.3. *Coefficients for the calculation of the reduced second transport virial coefficients B_η^* and $B_{\lambda,\mathrm{tr}}^*$ according to the Rainwater–Friend theory for monatomic gases ($\delta = 1.04$ and $\theta = 1.25$).*

	Coefficients for B_η^*		Coefficients for $B_{\lambda,\mathrm{tr}}^*$
i	a_i	i	a_i
0	-0.17999496×10^1	0	0.70712663×10^0
1	0.46692621×10^2	1	-0.14929039×10^2
2	-0.53460794×10^3	2	0.17622483×10^3
3	0.33604074×10^4	3	-0.98442060×10^3
4	-0.13019164×10^5	4	0.33417786×10^4
5	0.33414230×10^5	5	-0.71404506×10^4
6	-0.58711743×10^5	6	0.98775397×10^4
7	0.71426686×10^5	7	-0.88362595×10^4
8	-0.59834012×10^5	8	0.49304901×10^4
9	0.33652741×10^5	9	-0.15575002×10^4
10	-0.12027350×10^5	10	0.21212707×10^3
11	0.24348205×10^4		
12	-0.20807957×10^3		

References

Bennett, D. E. & Curtiss, C. F. (1969). Density effects on the transport coefficients of gaseous mixtures. *J. Chem. Phys.*, **51**, 2811–2825.

Bich, E. & Vogel, E. (1991). The initial density dependence of transport properties: Noble gases. *Int. J. Thermophys.*, **12**, 27–42.

Cohen, E. G. D. (1969). The kinetic theory of moderately dense gases, in *Transport Phenomena in Fluids*, ed. H. J. M. Hanley, pp. 157–207. New York: Marcel Dekker.

Enskog, D. (1922). Kinetische Theorie der Wärmeleitung, Reibung und Selbstdiffusion in gewissen verdichteten Gasen und Flüssigkeiten. *Kgl. Svenska Ventensk. Handl.*, **63**, No. 4.

Friend, D. G. (1983). The radial distribution function at low densities: Exact results for small and large separations for smooth potentials. *J. Chem. Phys.*, **79**, 4553–4557.

Friend, D. G. & Rainwater, J. C. (1984). Transport properties of a moderately dense gas. *Chem. Phys. Lett.*, **107**, 590–594.

Hanley, H. J. M. & Cohen, E. G. D. (1976). Analysis of the transport coefficients for simple dense fluids: The diffusion and bulk viscosity coefficients. *Physica*, **A 83**, 215–232.

Hanley, H. J. M., McCarty, R. D. & Cohen, E. G. D. (1972). Analysis of the transport coefficients for simple dense fluids: Application of the modified Enskog theory. *Physica*, **60**, 322–356.

Hendl, S. & Vogel, E. (1992). The viscosity of gaseous ethane and its initial density dependence. *Fluid Phase Equil.*, **76**, 259–272.

Hoffman, D. K. & Curtiss, C. F. (1965). Kinetic theory of dense gases. V. Evaluation of the second transport virial coefficients. *Phys. Fluids*, **8**, 890–895.

Maitland, G. C., Rigby, M., Smith, E. B. & Wakeham, W. A. (1987). *Intermolecular Forces: Their Origin and Determination*, chapter 5. Oxford: Clarendon Press.

Mason, E. A. & Monchick, L. (1962). Heat conductivity of polyatomic and polar gases. *J. Chem. Phys.*, **36**, 1622–1639.

Nieto de Castro, C. A., Friend, D. G., Perkins, R. A. & Rainwater, J. C. (1990). Thermal conductivity of a moderately dense gas. *Chem. Phys.*, **145**, 19–26.

Rainwater, J. C. (1984). On the phase space subdivision of the second virial coefficient and its consequences for kinetic theory. *J. Chem. Phys.*, **81**, 495–510.

Rainwater, J. C. & Friend, D. G. (1987). Second viscosity and thermal–conductivity virial coefficients of gases: Extension to low reduced temperatures. *Phys. Rev.*, **A 36**, 4062–4066.

Ross, M., Szczepanski, R., Trengove, R. D. & Wakeham, W. A. (1986). The initial density dependence of the transport properties of gases. *AIChE Annual Winter Meeting*, paper 86C, held in Miami, Florida, USA.

Sengers, J. V., Gillespie, D. T. & Perez–Esandi, J. J. (1978). Three–particle collision effects in the transport properties of a gas of hard spheres. *Physica*, **A 90**, 365–409.

Snider, R. F. & Curtiss, C. F. (1958). Kinetic theory of moderately dense gases. *Phys. Fluids*, **1**, 122–138.

Stogryn, D. E. & Hirschfelder, J. O. (1959). Initial pressure dependence of thermal conductivity and viscosity. *J. Chem. Phys.*, **31**, 1545–1554.

Vogel, E., Bich, E. & Nimz, R. (1986). The initial density dependence of the viscosity of organic vapours: Benzene and methanol. *Physica*, **A 139**, 188–207.

Vogel, E. & Hendl, S. (1992). Vapor phase viscosity of toluene and *p*–xylene. *Fluid Phase Equil.*, **79**, 313–326.

5.3 Intermediate density range

W. A. WAKEHAM

Imperial College, London, UK

5.3.1 The formal theory

As the density of the fluid is increased above values for which a linear term in the expansion of equation (5.1) is adequate (crudely above values for which a third virial coefficient is adequate to describe the compression factor of a gas), the basis of even the formal kinetic theory is in doubt. In essence, the difficulty arises because it becomes necessary, at higher densities, to consider the distribution function, in configuration and momentum space, of pairs, triplets *etc.* of molecules in order to formulate an equation for the evolution of the single–particle distribution function. Such an equation would be the generalization to higher densities of the Boltzmann equation, discussed in Chapter 4 (Ferziger & Kaper 1972; Dorfman & van Beijeren 1977).

At present this generalization has been accomplished only by means of an unsubstantiated assumption originated by Bogoliubov (1946), who supposed that the higher–order distribution functions could be expressed in terms of time–independent functionals of the single–particle distribution function. As was implied in Section 5.1, this hypothesis leads to a tractable formal theory, but it is one in which the expansion of the multiparticle distribution functions diverge with time in a manner inconsistent with the basis of the theory. The cause of these divergences lies in the fact that, in its original form, the theory treats groups of three or more molecules as if they were interacting among themselves, isolated from all other molecules in the system. The possible encounters among such isolated groups of molecules include sequences of collisions that are not chaotic, in the sense that the initial state before a collision is not independent of the results of previous collisions. Interactions between the groups of molecules and the remainder of the gas can be allowed, by a technique known as resummation, and it is then found that the contributions from four–molecule groups lead to the logarithmic term in the density expansion of the transport properties given in equation (5.1), at least for rigid–sphere molecules (Ferziger & Kaper 1972). The effects of the same considerations for other molecular models or for clusters containing greater numbers of molecules are not known. For densities higher than the critical density, the effects of these correlated collisions may be incorporated into less rigorous theories in a semiempirical manner by making use of the results of computer simulations, as described in the next section. However, in the intermediate range of densities this approach is not available.

The net effect of these considerations is that even the form of the dependence of the transport coefficients on density is not known theoretically. Moreover, of the various coefficients that enter the expansion (5.1), nothing is known about the magnitude of any except the second transport virial coefficients B_x for a Lennard–Jones 12–6 potential (Section 5.2) and the third transport virial coefficients C_x for a rigid–sphere gas (Kamgar–Parsi & Sengers 1982). In the context of the present volume this discussion leads to the conclusion that there is no practical guidance to be derived from rigorous theory for the prediction, estimation or correlation of fluids in the region of states intermediate between the low–density gas and fluids at very high densities.

5.3.2 Modified Enskog Theory

The absence of a rigorous theory for the transport properties of fluids in the intermediate–density range means that it has been necessary to employ methods of evaluation based upon an approximate theory, the principle of corresponding states (Chapter 12) or empiricism (see Section 5.3.3). The only approximate theory to have been used to any extent is the Enskog theory, outlined in Section 5.1 and discussed in a modified form in Section 5.2 in the context of the initial density dependence of the transport properties.

As has been made clear in these earlier sections, the Enskog theory, even for hard–sphere systems, is an *ad hoc* extension of the dilute–gas theory to a dense system. The theory does not, therefore, properly account for the correlations in velocity space discussed above. Furthermore, while a number of methods have been suggested whereby the Enskog theory may be applied to real fluids, all of them require further steps away from a rigorous theory, and, for that reason, none of them is particularly satisfactory. Here just one of the approaches suggested is considered, which has the virtue of self–consistency and is not significantly less accurate than other proposals.

The approach considered is that first proposed by Enskog himself (Chapman & Cowling 1952), who suggested that a pseudo–radial distribution function $\tilde{g}$ for a real gas, to replace the function for hard spheres at contact, and a consistent effective hard–sphere diameter could be obtained from the equation of state for the real gas. Specifically, he proposed that the radial distribution function could be obtained from the equation of state for the real gas by replacing the pressure of the real gas by the thermal pressure P_t through the equation

$$P_t = T \left(\frac{\partial P}{\partial T} \right)_\rho = \frac{RT}{V} (1 + b\tilde{g}\rho) \tag{5.20}$$

which replaces equation (5.5). This substitution is justified by the argument that the real pressure exerted by the molecules of a gas is the sum of the external pressure and an internal pressure arising from the force of cohesion of the molecules. It is therefore supposed to account for a relatively weak, long–range attractive component in the intermolecular forces for real gases.

If a virial expansion is employed for the equation of state and the physical requirement is imposed that $g \to 1$ as $\rho \to 0$, then it is readily shown (Hanley *et al.* 1972) that the co–volume of the molecules b is given by

$$b = B + T \left(\frac{dB}{dT} \right) \tag{5.21}$$

where B is the second virial coefficient of the gas. If the second virial coefficient used in equation (5.21) is that derived from the equation of state for the gas, then the values of b and $\tilde{g}$ are consistent. However, such an approach is not always possible, and the use of virial coefficients from one source and an equation of state from another may have serious consequences.

The viscosity of any gas and the thermal conductivity of a monatomic gas are expressed as

$$\eta(\rho, T) = \eta^{(0)}(T) b\rho \left(\frac{1}{b\rho\tilde{g}(\rho, T)} + 0.8 + 0.761 b\rho\tilde{g}(\rho, T) \right) \tag{5.22}$$

$$\lambda(\rho, T) = \lambda^{(0)}(T) b\rho \left(\frac{1}{b\rho\tilde{g}(\rho, T)} + 1.2 + 0.755 b\rho\tilde{g}(\rho, T) \right) \tag{5.23}$$

Combining these equations with an equation of state yields a predictive procedure for the properties, since equation (5.20) may be used to evaluate $b\rho\tilde{g}$ and b itself is obtained from equation (5.21).

The most extensive calculations and comparisons of this procedure are those reported by Hanley and his collaborators (Hanley *et al.* 1972). Naturally the comparisons have been limited to those substances for which extensive and accurate equations of state and transport property data are available. Figure 5.6 shows the results of the calculation of the viscosity and thermal conductivity of argon as a function of density for several isotherms using this procedure. The results are reported in the form of the excess properties, $\Delta\eta$ and $\Delta\lambda$, defined by equation (3.1).

In the figure, the shadowed region indicates densities above the critical density. Generally, the Enskog theory predicts a more rapid increase in the property than that observed experimentally and a strong dependence of the excess property upon temperature, which is not observed experimentally. Detailed comparisons of this sort show (Hanley *et al.* 1972) that the modified Enskog theory (MET) represents experimental data within $\pm 15\%$ for temperatures below the critical temperature and for densities not exceeding twice the critical density. For temperatures above the critical temperature this degree of agreement can be maintained only up to the critical density itself.

For the thermal conductivity of a polyatomic gas one further extension of the MET is necessary to account for the internal degrees of freedom of the molecule and the transport of internal energy. This is accomplished with the aid of the assumption that the transport of internal energy is diffusive and that there is no relaxation of internal energy. This approximation is similar to that discussed in Chapter 4 for dilute gases and enables one to write

$$\lambda(\rho, T) = \lambda_{tr}(\rho, T) + \lambda_{int}(\rho, T) \tag{5.24}$$

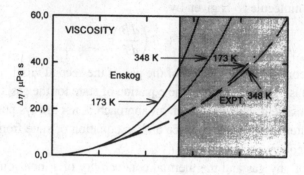

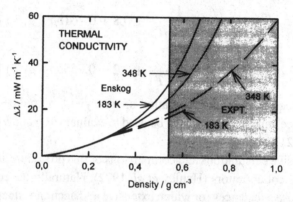

Fig. 5.6. A comparison of the experimental excess viscosity and thermal conductivity of argon as a function of density with the predictions of the modified Enskog theory. The shaded region indicates densities greater than the critical density.

where the translational contribution is given by equation (5.23) for a monatomic gas and the internal contribution by the application of equation (5.2) to the diffusion coefficient of internal energy, so that

$$\lambda_{\text{int}} = \frac{[\rho D]^{(0)} C_{V,\text{int}}}{\tilde{g}} \tag{5.25}$$

where $C_{V,\text{int}}$ is the molar isochoric heat capacity of the gas. Thus, the total thermal conductivity of the gas according to the MET is

$$\lambda(\rho, T) = \lambda_{\text{tr}}^{(0)} b\rho \left(\frac{1}{b\rho\tilde{g}} + 1.2 + 0.755 b\rho\tilde{g} \right) + \frac{[\rho D]^{(0)} C_{V,\text{int}}}{\tilde{g}} \tag{5.26}$$

Notwithstanding the further modification of the Enskog theory required to achieve this result, the procedure yields thermal conductivity predictions for a dense polyatomic gas which are little worse than for monatomic systems.

Despite its considerable limitations, the MET is one of the few means available for the estimation of the transport properties of a gas in the intermediate range of densities when absolutely no experimental data are available, and as such it must occasionally be employed. However, there are demonstrations of its quantitative failure for rather simple molecular models by comparison with computer simulations (Michels & Trappeniers 1980). The procedure should, therefore, always be used with caution, in a consistent manner and with awareness of its limited accuracy.

5.3.3 The excess contribution

The excess transport properties of a pure fluid, that is, the quantities

$$\Delta \eta = \eta(\rho, T) - \eta^{(0)}(T) - \Delta \eta_c(\rho, T) \tag{5.27}$$

$$\Delta \lambda = \lambda(\rho, T) - \lambda^{(0)}(T) - \Delta \lambda_c(\rho, T) \tag{5.28}$$

represent the behavior of the fluid devoid of any explicit effects arising from the critical region itself (Chapter 6). According to the Enskog theory the background and excess properties for monatomic and polyatomic systems are identical, since the theory is unable to account for behavior in the critical region. Thus, equations (5.3) and (5.4) show that for rigid spherical systems the excess transport properties have the relatively strong temperature dependence of the dilute–gas property. As indicated in the previous section, this strong dependence persists in the predictions of the MET. It is in contrast with the experimental behaviour observed for a wide variety of gases as indicated in Figure 5.6, and confirmed in Figure 5.7 where the excess viscosity for methane is plotted as a function of temperature for data in the temperature range 310 K to 475 K. Within the band of uncertainty that characterizes the results of the measurements of the various authors contained in this plot ($\pm 4\%$) it is not possible to discern any significant temperature dependence in the excess property; a similar result pertains for the thermal conductivity.

The contrast between the predictions of the only available theory for the intermediate–density range and experiment casts further doubt on the validity of the MET. At the same time, the empirical observation itself, validated on a large number of pure materials, at least for supercritical temperatures, provides a valuable empirical estimation procedure. This is because, if the density dependence of a transport property of a pure fluid is available along just one isotherm, it is possible to evaluate the property as a function of density along any other isotherm merely assuming that the excess property is temperature independent and by making use of equation (5.27) or (5.28).

Whenever there are any experimental data for the density dependence of a fluid property, this is the preferred estimation technique for the intermediate–density region. Although the temperature dependence of the excess property for a real fluid is weak, it is not entirely negligible, as the discussion of the thermal conductivity of argon in Section 13.1 shows. However, for most practical purposes of estimation the assumption

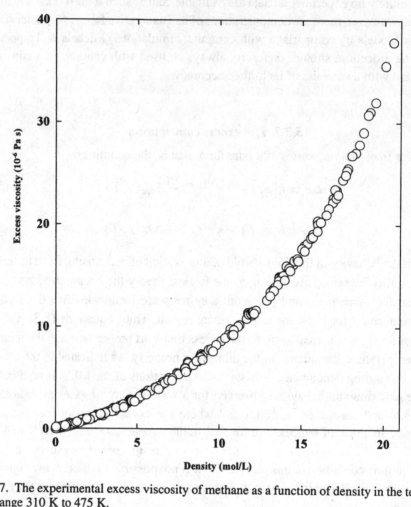

Fig. 5.7. The experimental excess viscosity of methane as a function of density in the temperature range 310 K to 475 K.

of temperature independence in the gas phase is not likely to lead to errors larger than ±5% under any circumstances except possibly near the saturation boundary.

References

Bogoliubov, N. N. (1946). In *Studies in Statistical Mechanics, Vol. I*, eds. J. de Boer & G.E. Uhlenbeck (1962), pp. 5–118. Amsterdam: North–Holland.

Chapman, S. & Cowling, T.G. (1952). *The Mathematical Theory of Non–Uniform Gases. 2nd Edition*. London: Cambridge University Press.

Dorfman, J.R. & van Beijeren, H. (1977). In *Statistical Mechanics, Part B: Time–Dependent Processes*, Chapter 3, ed. B.J. Berne. New York: Plenum.

Ferziger, J. H. & Kaper, H. G. (1972). *Mathematical Theory of Transport Processes in Gases*. Amsterdam: North–Holland.

Hanley, H.J.M., McCarty, R.D. & Cohen, E.G.D. (1972). Analysis of the transport coefficients for simple dense fluids: Application of the modified Enskog theory. *Physica* **60**, 322–356.

Kamgar–Parsi, B. & Sengers, J.V. (1982). Logarithmic dependence of the transport properties of moderately dense gases. *Proc. 8th. Symp. Thermophys. Props.* New York: A.S.M.E., 166–171.

Michels, J.P.J. & Trappeniers, N.J. (1980). Molecular dynamical calculations of the transport properties of a square–well fluid I. The viscosity below critical density. *Physica* **101A**, 156–166.

5.4 Pure fluids at high densities

J. H. DYMOND
The University, Glasgow, UK

5.4.1 Direct computation

There are intractable problems in developing rigorous analytical expressions for the transport properties of dense fluids which can be evaluated for systems of molecules interacting with realistic forms for the intermolecular potential energy function, $U(r)$, without making simplifying assumptions (whose effect on the final result is uncertain). Therefore, it is necessary to consider other approaches. With the advent of fast computers, it became possible to follow each member of a system of particles at a defined temperature and pressure through successive collisions using classical equations of motion. It is necessary just to specify $U(r)$. In this method of Molecular Dynamics (MD; see Chapter 9), knowledge of the positions and velocities of the particles as a function of time allows calculation of transport properties from the Einstein expressions

$$D = \frac{1}{2t} \langle [x_i(t) - x_i(0)] \rangle^2 \tag{5.29}$$

$$\eta = \frac{m^2}{2V k_B T t} \left\langle \left[\sum_{i=1}^{N} \dot{x}_i(t) y_i(t) - \dot{x}_i(0) y_i(0) \right]^2 \right\rangle \tag{5.30}$$

$$\lambda = \frac{1}{2V k_B T^2 t} \left\langle \left[\sum_{i=1}^{N} x_i(t) E_i(t) - x_i(0) E_i(0) \right]^2 \right\rangle \tag{5.31}$$

where x_i is the x–component of the position of particle i and E_i is the sum of its potential and kinetic energies. The angular brackets indicate an average over an equilibrium ensemble or, on the computer, an average starting from several different initial times. For systems of hard spheres, the transport properties were calculated (Alder *et al.* 1970) from $(1/2)(d/dt) < [G(t) - G(0)] >^2$ for a dynamical variable G, which becomes identical to the above expressions in the long–time limit. Equivalent expressions for the transport properties can be written in terms of autocorrelation functions (Helfand 1960), first derived by application of fluctuation–dissipation theory (Green 1954), in which an irreversible process is considered as the return of a perturbed system back to equilibrium. The formulas are

$$D = \int_0^\infty \langle \dot{x}_i(0)\dot{x}_i(t) \rangle \, dt \tag{5.32}$$

$$\eta = \frac{1}{V k_B T} \int_0^\infty \langle J^{xy}(0) J^{xy}(t) \rangle \, dt \tag{5.33}$$

where J^{xy} is the xy component of the microscopic stress tensor.

$$\lambda = \frac{1}{V k_B T^2} \int_0^\infty \langle S(0) S(t) \rangle \, dt \qquad (5.34)$$

where

$$S(t) = \frac{d}{dt} \sum_{i=1}^N x_i E_i \qquad (5.35)$$

The latter formulation of the thermal transport properties is preferable for computer calculations using more realistic forms of intermolecular potential energy function for atomic fluids (Gosling *et al.* 1973), and for molecular fluids (Luo & Hoheisel 1992), as described in Chapter 9. The upper limit for integration is replaced by a time sufficiently long that the autocorrelation function has become zero.

5.4.2 Smooth hard–sphere theory for monatomic fluids

Transport properties for a dense smooth hard–sphere system are given by the approximate Enskog theory (Enskog 1922) in terms of the dilute hard–sphere values by equations (5.2)–(5.4). However, the Enskog theory is not exact, since it is based on the molecular chaos approximation. A sphere is considered as always colliding with other spheres approaching from random directions with random velocities from a Maxwell–Boltzmann distribution for the appropriate temperature. Molecular dynamics calculations (Alder & Wainwright 1963, 1967) have shown that there are correlated molecular motions in hard–sphere systems. The principal correlation effect at high densities is back–scattering, where a sphere closely surrounded by a shell of spheres is most likely to have its velocity reversed on collision with its neighbors. This leads to a decrease in diffusion. At intermediate densities, there is an unexpected persistence of velocities which leads to an enhanced diffusion coefficient. Alder *et al.* (1970) calculated the corrections to the Enskog theory for diffusion, viscosity and thermal conductivity for finite systems of 108 and 500 particles. Their dependence on the ratio (V / V_0) where V_0, given by $N_A \sigma^3 / 2^{1/2}$ or $3b / 2\pi 2^{1/2}$, is the volume of close–packing of spheres is shown in Figure 5.8. In the case of diffusion, values were extrapolated to infinite systems, and the results have been subsequently confirmed (Erpenbeck & Wood 1991). In the case of viscosity and thermal conductivity the correction factors, which are in agreement with more recent values (Michels & Trappeniers 1980, 1981), show larger uncertainties associated with the necessarily larger computing time.

Thus, values for the exact hard–sphere transport properties can be calculated from the very dilute state up to the fluid–solid transition point. Studies have been made on diffusion (Woodcock 1981) that extend the density range to the glass transition.

For calculation of transport properties on the basis of hard–sphere theory, the hard–sphere expressions for the dilute–gas transport properties are required. These are given

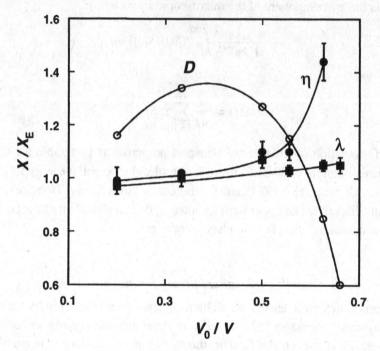

Fig. 5.8. Computed corrections to Enskog theory for transport property X.

to first–order approximation by

$$[nD]^{(0)} = \frac{3}{8\pi\sigma^2}\left(\frac{\pi k_B T}{m}\right)^{1/2} \tag{5.36}$$

$$\eta^{(0)} = \frac{5}{16\pi\sigma^2}(\pi m k_B T)^{1/2} \tag{5.37}$$

$$\lambda^{(0)} = \frac{75}{64\pi\sigma^2}\left(\frac{\pi k_B^3 T}{m}\right)^{1/2} \tag{5.38}$$

for particles of molecular mass m at temperature T and k_B is the Boltzmann constant. Equation (5.38) applies only to monatomics.

It is at the highest densities, above the critical density, that this model can be expected to approximate most closely to real fluids, because there is then a fairly uniform attractive potential energy surface. A comparison of the density dependence of the viscosity and thermal conductivity for argon at 308 K with the exact hard–sphere results (vertical lines) gave close agreement at reduced densities, V_0/V, from 0.2 to 0.5 (Dymond 1987), as shown in Figures 5.9 and 5.10. However, at higher V_0/V, the agreement was outside the estimated uncertainties of the computed corrections to Enskog theory. The optimum V_0 value for argon at 308 K is $13.6 \cdot 10^{-6}$ m^3 mol^{-1}.

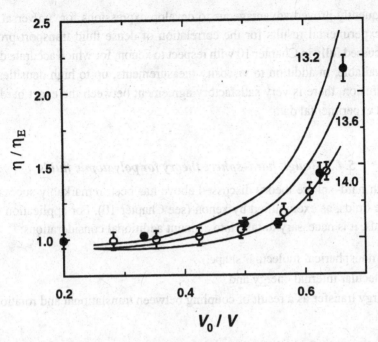

Fig. 5.9. Density dependence of η/η_E derived from argon data at 308 K for different values of $V_0/10^{-6}\,\mathrm{m^3\,mol^{-1}}$.

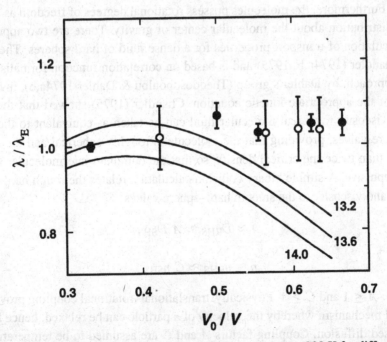

Fig. 5.10. Density dependence of λ/λ_E derived from argon data at 308 K for different values of $V_0/10^{-6}\,\mathrm{m^3\,mol^{-1}}$.

It subsequently proved advantageous to develop expressions for 'universal curves' based on experimental results for the correlation of dense fluid transport properties. This is discussed fully in Chapter 10 with respect to xenon, for which accurate diffusion data are available, in addition to viscosity measurements, up to high densities. In the case of diffusion, there is very satisfactory agreement between the exact hard–sphere results and experimental data.

5.4.3 *Rough hard–sphere theory for polyatomic fluids*

The smooth hard–sphere theory discussed above has been remarkably successful for monatomic fluids, as exemplified by xenon (see Chapter 10). For application to poly-atomic fluids, it is necessary to take into account additional considerations:

(i) the nonspherical molecular shape,
(ii) molecular internal energy and
(iii) energy transfer as a result of coupling between translational and rotational motion.

The simplest model which incorporates some of these features is the rough hard–sphere model. Although it does not account for real molecular shape, there is the possibility of transfer of angular momentum as well as translational momentum upon collision. Furthermore, the molecules possess rotational degrees of freedom as a result of mass distribution about the molecular center of gravity. There are two approaches to the calculation of transport properties for a dense fluid of hard spheres. The first is due to Chandler (1974a,b, 1975) and is based on correlation function formalism. The second approach, by Dahler's group (Theodosopoulou & Dahler 1974a,b), involves a solution of the appropriate kinetic equation. Chandler (1975) showed that the diffusion and viscosity for a real molecular fluid can be taken as equivalent to the rough hard–sphere values, providing that the molecule is roughly spherical and the density is greater than twice the critical density, so that the dominant intermolecular force is harshly repulsive. A simple binary collision calculation relates the rough hard–sphere diffusion and viscosity to the smooth hard–sphere values

$$D \approx D_{\mathrm{RHS}} \approx A \, D_{\mathrm{SHS}} \tag{5.39}$$

$$\eta \approx \eta_{\mathrm{RHS}} \approx C \, \eta_{\mathrm{SHS}} \tag{5.40}$$

where $0 < A \leq 1$ and $C \geq 1$. Physically, translational–rotational coupling provides an additional mechanism whereby the velocity of a particle can be relaxed, hence leading to decreased diffusion. Coupling factors A and C are assumed to be temperature and density independent. Values can be derived only from data fitting – they cannot be determined theoretically.

In Dahler's rough hard–sphere theory (Theodosopoulou & Dahler 1974a,b), the molecules have an internal mass distribution moment characterized by the moment of inertia. Expressions for viscosity and thermal conductivity are given, which are functions only of the reduced volume V/V_0, where V_0 is the volume of close–packing of spheres, and the reduced moment of inertia (Li *et al.* 1985). The Enskog smooth hard–sphere equation is obtained for viscosity in the limit of zero moment of inertia, but, for thermal conductivity the rough hard–sphere theory has an internal energy contribution even when there is no moment of inertia. The model fails to provide an entirely realistic description of inelastic collisions. However, it is interesting to note that calculations of the ratio of the rough hard–sphere viscosity to the corresponding smooth hard–sphere value at different reduced moments of inertia give values which each show a variation of less than 2% for reduced volumes from 1.5 to 2.5 (Li *et al.* 1985). No analytical relation has been obtained for the proportionality, but numerically these results confirm Chandler's expression for viscosity.

In spite of the simplicity of Dahler's model, it is instructive to compare the calculated thermal conductivity for hard spheres having finite moments of inertia with the Enskog smooth hard–sphere values. It is found (Li *et al.* 1985) that the variation is less than 10% over the whole reduced volume range, although the individual contributions from rotation and translation vary to a larger extent, in opposite directions.

This direct proportionality between the rough hard–sphere transport properties and the Enskog coefficients has formed the basis for many correlations of liquid transport properties (Easteal & Woolf 1984; Li *et al.* 1986; Walker *et al.* 1988; Greiner–Schmid *et al.* 1991; Harris *et al.* 1993). For a successful data fit, with unique values for V_0 and the proportionality factors, it is necessary to fit a minimum of two properties simultaneously, with the same V_0 values. This is exemplified in the case of methane in Chapter 10. It is further shown in Chapter 10 that successful correlation of transport property data for nonspherical molecular liquids can be made, based on the assumption that transport properties for these fluids can also be directly related to the smooth hard–sphere values.

5.4.4 Corresponding–states theory

The basis of the simple principle of corresponding states is that

(i) the molecules should interact with a spherically symmetric pairwise additive two–parameter potential energy function $U(r) = \epsilon\varphi(r/\sigma)$, where ϵ is the maximum attractive energy and σ is the separation at zero energy,

(ii) the translational motion can be treated by classical mechanics and

(iii) the internal degrees of freedom of the molecules do not contribute to energy or momentum transport.

Under these conditions, the transport properties, in reduced form, are then universal functions only of reduced state variables. On this basis it is possible to make a reasonable estimate of transport properties. For the simplest application, critical parameters are used as reduction parameters, with values for the (estimated) critical transport properties obtained by extrapolation from outside the critical region, and the ratios of reduced transport property then considered as a function of the reduced pressure, P/P_c, and reduced temperature T/T_c. Transport properties for rare gas fluids and certain polyatomic fluids have been correlated in this way but, in view of the unsuitability of the exact critical transport property values as reduction parameters, molecular parameters are preferred such that the reduced viscosity, thermal conductivity and diffusion become

$$\eta^* = \frac{\eta\sigma^2}{(m\epsilon)^{1/2}}, \qquad \lambda^* = \frac{\lambda m\sigma^2}{k_B(m\epsilon)^{1/2}}, \qquad D^* = \frac{Dm^{1/2}}{\sigma\epsilon^{1/2}} \qquad (5.41)$$

for molecules of mass m. These are considered as functions of reduced temperature $T^* = k_B T/\epsilon$ and reduced density $\rho^* = N_A\sigma^3/V$. In order to include substances that depart from the condition of exact conformality, state–dependent shape factors are introduced into a generalised corresponding–states procedure (Haile, Mo & Gubbins 1976; Hanley 1976; Murad & Gubbins 1977), which leads to the calculation of viscosity and thermal conductivity for a wide range of molecular fluids, and their mixtures, from the dilute gas to the dense liquid using values for a reference fluid plus equation of state data. This approach is described more fully in Chapter 12. An improved fit for viscosity of specific mixtures can be obtained by using a generalized corresponding–states principle based on the properties of two reference fluids (Teja & Turner 1986), especially when a binary interaction coefficient is introduced. For hydrogen–bonded and polar fluids, it is shown (Hwang & Whiting 1986) that introduction of an association parameter and a viscosity acentric factor, respectively, leads to an improved fit for these substances.

5.4.5 Absolute reaction–rate theory

A description of transport processes in dense fluids, based on the absolute reaction–rate theory (Glasstone *et al.* 1941), assumes that molecular motion is confined largely to vibrations of each molecule within a 'cage' formed by nearest neighbors as a result of the close packing. The 'cage' is represented by an activation barrier of height Δ^*G^0, where Δ^*G^0 is the molar energy of activation. In a liquid at rest, molecular rearrangement occurs by one molecule at a time escaping from its 'cage' into a neighboring hole. In a flowing liquid, the frequency of molecular rearrangements is increased, and for Newtonian flow the viscosity is given by the expression

$$\eta = \left(\frac{\delta}{a}\right)^{1/2} \frac{N_A h}{V} \exp\left(\frac{\Delta^*G^0}{RT}\right) \qquad (5.42)$$

where a is the distance traveled per jump, δ is the distance between separate layers in the liquid, h is the Planck constant, V is the molar volume and T is the temperature. In this theory, the ratio δ/a is usually taken as unity, and Δ^*G^0 is then determined from experimental viscosity data. It has been found that Δ^*G^0 is approximately 40% of the enthalpy of vaporization. The equation predicts a linear dependence for the logarithm of the viscosity on reciprocal temperature, strictly at constant molar volume. It had been shown to be obeyed by liquids along the saturation line, providing the temperature range is not too extensive; this is useful as a test of such experimental measurements. However, it is not necessary to assume an activation model to derive a linear relationship between logarithm of the viscosity and reciprocal temperature (Hildebrand & Alder 1973). Indeed, computer simulation studies on simple molecular models have shown (Alder & Einwohner 1965) that the underlying assumption of molecular motion as occurring by a small number of relatively large jumps is in fact incorrect. There is cooperative motion with a large number of small molecular displacements. Nevertheless, this approach is of great interest in that it can readily be extended to mixtures. If the ideal viscosity of a mixture is represented by

$$\Delta^*G^{id} = \sum x_i \Delta^*G_i \qquad (5.43)$$

then a molar excess Gibbs energy of activation for flow Δ^*G^E, given by

$$\Delta^*G^E = \Delta^*G - \Delta^*G^{id} \qquad (5.44)$$

can be derived from experimental viscosity data according to

$$\Delta^*G^E = RT \left[\ln{(\eta V)} - \sum x_i \ln{(\eta_i V_i)} \right] \qquad (5.45)$$

This expression is still frequently used for the description of mixture viscosities (Oswal & Patel 1992).

5.4.6 Free–volume theory

The validity of the simple relationship between viscosity for a liquid at atmospheric pressure and volume, $\eta = c/(v - w)$, where v is specific volume, w is a constant similar to the van der Waals b, and c is another constant, was demonstrated (Batschinski 1913) by linear plots of $1/\eta$ versus v obtained for nonassociated liquids. This expression was later modified (Hildebrand 1971) to emphasize the dependence of fluidity on relative expansion from the intrinsic molar volume, V_∞, where $1/\eta = 0$

$$\frac{1}{\eta} = \frac{E(V - V_\infty)}{V_\infty} \qquad (5.46)$$

Values for V_∞ were found to be approximately equal to 0.3 times the critical volume (Hildebrand 1972), and E was found in the case of n–alkanes to be nearly linearly related to chain length. However, the equation has limited applicability. It does not apply for the saturated liquid at temperatures below 0.46 times the critical temperature, and it does not take into account effects of pressure, since it requires the viscosity to remain constant for constant volume, as temperature and pressure are both varied, contrary to experiment. Furthermore, the corresponding equation for diffusion is not obeyed, as plots of diffusivity versus molar volume show definite curvature (Ertl & Dullien 1973). Nevertheless, by allowing E and V_∞ to vary with temperature, fluidity data for saturated and compressed liquids can be fitted for pressures up to about 30 MPa within about 3%, as for example, for butane and isobutane (Diller & Van Poolen 1985), but this is a purely empirical approach.

An alternative dependence of fluidity on free volume (Doolittle 1951)

$$\frac{1}{\eta} = A \exp \left(\frac{-dV_\infty}{V - V_\infty} \right) \tag{5.47}$$

where d is a constant of order unity, has been given a molecular basis through consideration of the distribution of near–neighbor holes large enough for diffusive displacement, which leads to an equation of the same form for diffusion (Cohen & Turnbull 1959). The importance of the fluidity equation is that it gives an extraordinarily good fit of isothermal data for glycerol, propane-1,2-diol, methanol and dibutylphthalate (Cook et al. 1993) up to high compressions where the viscosity is as much as 108 times the atmospheric pressure viscosity. Values of d for a variety of chemically dissimilar liquids were found to vary between 0.7 and 10.

References

Alder, B.J. & Einwohner, T. (1965). Free–path distribution for hard spheres. *J. Chem. Phys.*, **43**, 3399–3400.

Alder, B.J. & Wainwright, T.E. (1963). *The Many–Body Problem*, ed. J. K. Percus, New York: Interscience Publications Inc.

Alder, B.J. & Wainwright, T.E. (1967). Velocity autocorrelations for hard spheres. *Phys. Rev. Lett.*, **18**, 988–990.

Alder, B.J., Gass, D.M. & Wainwright, T.E. (1970). Studies in molecular dynamics. VIII. The transport coefficients for a hard–sphere fluid. *J. Chem. Phys.*, **53**, 3813–3826.

Batschinski, A.J. (1913). Untersuchungen über die innere Reibung der Flüssigkeiten. *Z. Phys. Chem.*, **84**, 643–706.

Chandler, D. (1974a). Translational and rotational diffusion in liquids I. Translational single–particle correlation functions. *J. Chem. Phys.*, **60**, 3500–3507.

Chandler, D. (1974b). Translational and rotational diffusion in liquids II. Orientational single–particle correlation functions. *J. Chem. Phys.*, **60**, 3508–3512.

Chandler, D. (1975). Rough hard sphere theory of the self–diffusion coefficient for molecular liquids. *J. Chem. Phys.*, **62**, 1358–1363.

Cohen, M.H. & Turnbull, D. (1959). Molecular transport in liquids and glasses. *J. Chem. Phys.*, **31**, 1164–1169.

Cook, R.L., Herbst, C.A. & King, Jr., H.E. (1993). High–pressure viscosity of glass–forming liquids measured by the centrifugal force diamond anvil viscometer. *J. Phys. Chem.*, **97**, 2355–2361.

Diller, D.E. & Van Poolen, L.J. (1985). Measurements of the viscosity of saturated and compressed normal butane and isobutane. *Int. J. Thermophys.*, **6**, 43–62.

Doolittle, A.K. (1951). Studies in Newtonian flow II. The dependence of the viscosity of liquids on free space. *J. Appl. Phys.*, **22**, 1471–1475.

Dymond, J.H. (1987). Corrections to the Enskog theory for viscosity and thermal conductivity. *Physica*, **144B**, 267–276.

Easteal, A.J. & Woolf, L.A. (1984). Developments in the hard–sphere model for self–diffusion and shear viscosity II. Applications based on methane as a model hard sphere fluid. *Physica*, **124B**, 182–192.

Enskog, D. (1922). Kinetische Theorie der Wärmeleitung, Reibung und Selbstdiffusion in gewissen verdichteten Gasen und Flüssigkeiten, *Kungl. Sv. Vetenskapsakad. Handl.*, 63, No. 4.

Erpenbeck, J.J. & Wood, W.W. (1991). Self–diffusion coefficient for the hard–sphere fluid. *Phys. Rev. A*, **43**, 4254–4261.

Ertl, H. & Dullien, F.A.L. (1973). Hildebrand's equations for viscosity and diffusivity. *J. Phys. Chem.*, **77**, 3007–3011.

Glasstone, S., Laidler, K.J. and Eyring, H. (1941). *Theory of Rate Processes*, New York: McGraw–Hill, Chap. 9.

Gosling, E.M., McDonald, I.R. & Singer, K. (1973). On the calculation by molecular dynamics of the shear viscosity of a simple fluid. *Mol. Phys.*, **26**, 1475–1484.

Green, M.S. (1954). Markoff random processes and the statistical mechanics of time–dependent phenomena II. Irreversible processes in liquids. *J. Chem. Phys.*, **22**, 398–413.

Greiner–Schmid, A., Wappmann, S., Has., M. & Lüdemann, H.–D. (1991). Self–diffusion in the compressed fluid lower alkanes: methane, ethane and propane. *J. Chem. Phys.*, **94**, 5643–5649.

Haile, J.M., Mo, K.C. & Gubbins, K.E. (1976). Viscosity of cryogenic liquid mixtures (including LNG) from corresponding states methods. *Adv. Cryogen. Eng.*, **21**, 501–508.

Hanley, H.J.M. (1976). Prediction of the viscosity and thermal conductivity coefficients of mixtures. *Cryogenics*, 643–651.

Harris, K.R., Alexander, J.J., Goscinska, T., Malhotra, R, Woolf, L.A. & Dymond, J.H. (1993). Temperature and density dependence of the self–diffusion coefficients of liquid *n*–octane and toluene. *Mol. Phys.*, **78**, 235–248.

Helfand, E. (1960). Transport coefficients from dissipation in a canonical ensemble. *Phys. Rev.*, **119**, 1–9.

Hildebrand, J.H. (1971). Motions of molecules in liquids: Viscosity and diffusivity. *Science*, **174**, 490–493.

Hildebrand, J.H. (1972). Fluidity: A general theory. *Proc. Nat. Acad. Sci. USA*, **69**, 3428–3431.

Hildebrand, J.H. & Alder, B.J. (1973). Activation energy: Not involved in transport processes in liquids. *I&EC Fundament.*, **12**, 387.

Hwang, M.–J. & Whiting, W.B. (1986). A corresponding–states treatment for the viscosity of polar fluids. *Ind. Eng. Chem. Res.*, **26**, 1758–1766.

Li, S.F.Y., Maitland, G.C. & Wakeham, W.A. (1985). The thermal conductivity of liquid hydrocarbons. *High Temp.–High Press.*, **17**, 241–251.

Li, S.F.Y., Trengove, R.D., Wakeham, W.A. & Zalaf, M. (1986). The transport coefficients of polyatomic liquids. *Int. J. Thermophys.*, **7**, 273–284.

Luo, H. & Hoheisel, C. (1992). Computation of transport coefficients of liquid benzene and cyclohexane using rigid pair interaction models. *J. Chem. Phys.*, **96**, 3173–3176.

Michels, J.P.J. & Trappeniers, N.J. (1980). Molecular dynamical calculations on the transport properties of a square–well fluid II. The viscosity above the critical density. *Physica*, **104A**, 243–254.

Michels, J.P.J. & Trappeniers, N.J. (1981). Molecular dynamical calculations on the transport properties of a square–well fluid III. The thermal conductivity. *Physica*, **107A**, 158–165.

Murad, S. & Gubbins, K.E. (1977). Corresponding states correlation for thermal conductivity of dense gases. *Chem. Eng. Sci.*, **32**, 499–505.

Oswal, S.L. & Patel, S.G. (1992). Viscosity of binary mixtures IV. Triethylamine with alkanes and monoalkylamines at 303.15 K and 313.15 K. *Int. J. Thermophys.*, **13**, 817–825.

Teja, A.S. & Turner, P.A. (1986). The correlation and prediction of viscosities of mixtures over a wide range of temperature and pressure. *Chem. Eng. Commun.*, **49**, 69–79.

Theodosopoulou, M. & Dahler, J.S. (1974a). The kinetic theory of polyatomic liquids I. The generalized moment method. *J. Chem. Phys.*, **60**, 3567–3582.

Theodosopoulou, M. & Dahler, J.S. (1974b). The kinetic theory of polyatomic liquids II. The rough sphere, rigid ellipsoid and square–well ellipsoid models. *J. Chem. Phys.*, **60**, 4048–4057.

Walker, N.A., Lamb, D.M., Adamy, S.T., Jonas, J. & Dare–Edwards, M.P. (1988). Self–diffusion in the compressed highly viscous liquid 2–ethylhexyl benzoate. *J. Phys. Chem.*, **92**, 3675–3679.

Woodcock, L.V. (1981). Glass transition in the hard–sphere model and Kauzman's paradox. *Ann. N.Y. Acad. Sci.*, **371**, 274–298.

5.5 Fluid mixtures at high density – supercritical phase

V. VESOVIC and W. A. WAKEHAM

Imperial College, London, UK

The difficulties that surround the formulation of a rigorous theory of the transport properties of dense pure fluids set out in earlier sections pertain equally strongly to mixtures of fluids with an arbitrary number of components. For that reason the formal theory is not pursued in this section. Instead, attention is concentrated upon an approximate theory that has led to a valuable means of predicting the properties of multicomponent mixtures from a limited amount of experimental information on the pure components.

The most successful approximate theory is founded upon the generalization of the theory of Enskog to multicomponent mixtures carried out by Thorne and reported originally in the monograph by Chapman & Cowling (1952). This generalization is founded upon the Enskog hypothesis that the dominant effect of increasing density is to increase the collision frequency for encounters between every species of rigid–sphere molecule while account is taken of the finite volume of the molecules and of collisional transfer of momentum and energy. For these reasons the generalization of the Enskog theory suffers from all of the defects of the Enskog theory for pure fluids, discussed in Section 5.3. However, in its application to the prediction of the properties of dense fluid mixtures, it is possible to use the Thorne–Enskog equations merely as the framework of a scheme which is interpolatory between the known properties of the pure fluids. In this way it is possible both to reduce the reliance of the procedure on the exactitude of the theory and to devise another means of selecting an appropriate and internally consistent radial distribution function and molecular volume in place of that which characterizes the modified Enskog theory (MET; Section 5.3). Owing to the reduction of the dependence upon the correctness of theory and the ability to use some experimental information, the procedure to be described has proved quite reliable. Attention is confined here to a treatment of viscosity and thermal conductivity where the experimental information on pure components is of direct value. For the diffusion coefficients of mixtures, pure component information is of little value and so the procedure is burdened immediately with the full limitations inherent in the Enskog theory.

The procedure set out below is the culmination of contributions from a number of complementary studies (DiPippo *et al.* 1977; Sandler & Fiszdon 1979; Vesovic & Wakeham 1989a,b, 1991) and they are combined here in the most recent version of the method. It should be emphasized that only the background behavior of the transport properties can be treated by this procedure. The subscript bckg is, however, omitted in the interest of clarity.

5.5.1 Viscosity

The viscosity of a dense gas mixture of rigid, spherical molecules containing N components according to the Thorne–Enskog equation can be written in the form (DiPippo et al. 1977; Vesovic & Wakeham, 1989a,b)

$$\eta(\rho, T) = -\frac{\begin{vmatrix} H_{11} & \cdots & H_{1N} & Y_1 \\ \vdots & \cdots & \vdots & \vdots \\ H_{N1} & \cdots & H_{NN} & Y_N \\ Y_1 & \cdots & Y_N & 0 \end{vmatrix}}{\begin{vmatrix} H_{11} & \cdots & H_{1N} \\ \vdots & \cdots & \vdots \\ H_{N1} & \cdots & H_{NN} \end{vmatrix}} + \kappa_{\text{mix}} \tag{5.48}$$

where

$$Y_i = x_i \left[1 + \sum_{j=i}^{N} \frac{m_j}{(m_i + m_j)} x_j \alpha_{ij} g_{ij} \rho \right] \tag{5.49}$$

$$H_{ii} = \frac{x_i^2 g_{ii}}{\eta_i^{(0)}} + \sum_{j \neq i}^{N} \frac{x_i x_j g_{ij}}{2\eta_{ij}^{(0)} A_{ij}^*} \frac{m_i m_j}{(m_i + m_j)^2} \left(\frac{20}{3} + \frac{4m_j}{m_i} A_{ij}^* \right) \tag{5.50}$$

$$H_{ij}(i \neq j) = -\frac{x_i x_j g_{ij}}{2\eta_{ij}^{(0)} A_{ij}^*} \frac{m_i m_j}{(m_i + m_j)^2} \left(\frac{20}{3} - 4A_{ij}^* \right) \tag{5.51}$$

and

$$\kappa_{\text{mix}} = \left(\frac{16}{5\pi} \right) \frac{15}{16} \rho^2 \sum_{j=1}^{N} \sum_{i=1}^{N} x_i x_j g_{ij} \alpha_{ij}^2 \eta_{ij}^{(0)} \tag{5.52}$$

where ρ is the molar density of the mixture, g_{ij} is the radial distribution function at contact for the hard spherical molecules i and j in the presence of all other species and α_{ij} is a volume that accounts for the mean free–path shortening in the dense gas that is related to the co–volume of the pair of molecules i and j. The superscript zero on $\eta_i^{(0)}$ and $\eta_{ij}^{(0)}$ indicates that these quantities are the viscosity of pure component i and the interaction viscosity for the $i - j$ encounter in the dilute–gas limit; all the other symbols have been defined in Chapter 4. In the dilute–gas limit the radial distribution function, g_{ij}, takes its limiting value of unity for all i, j; equations (5.48)–(5.52) become equivalent to their dilute–gas counterparts in the first–order kinetic theory (Chapter 4).

5.5.1.1 Estimation of the viscosity

In order to apply equations (5.48)–(5.52) to the evaluation of the viscosity of a real–gas mixture it is essential, as it was for pure fluids, to replace the pair distribution function at contact g_{ij} and the mean free–path parameter (or co–volume) with quantities chosen in a consistent fashion to be characteristic of the real gas. This is achieved in a number of steps.

In the first a pseudo–radial distribution function, $\tilde{g}_i$, is identified for each real, pure component at a particular temperature and the density of interest from measurements of the pure component viscosity for each fluid. This can be accomplished by writing equation (5.48) for a pure fluid and solving for the pseudo–radial distribution function at each density (DiPippo *et al.* 1977; Vesovic & Wakeham 1989a,b). Hence,

$$\tilde{g}_i(\rho, T) = \frac{\beta_\eta \left[\eta_i - \rho \alpha_{ii} \eta_i^{(0)} \right]}{2 \rho^2 \alpha_{ii}^2 \eta_i^{(0)}} \pm \beta_\eta \left[\left(\frac{\eta_i - \rho \alpha_{ii} \eta_i^{(0)}}{2 \rho^2 \alpha_{ii}^2 \eta_i^{(0)}} \right)^2 - \frac{1}{\rho^2 \alpha_{ii}^2} \right]^{1/2} \tag{5.53}$$

where

$$\frac{1}{\beta_\eta} = \frac{1}{4} + \left(\frac{16}{5\pi} \right) \frac{15}{16} \tag{5.54}$$

and η_i is the pure–component viscosity at the molar density ρ and temperature T.

There are two possible solutions of equation (5.53), $\tilde{g}_i^+$ and $\tilde{g}_i^-$, corresponding to the positive and negative roots of equation (5.53). Although both solutions for $\tilde{g}_i$ will reproduce the pure component viscosity, only one of the solutions is physically plausible at a given density. At low densities the pseudo–radial distribution function should tend monotonically to unity as the density tends to zero, while at high densities the pseudo–radial distribution function should monotonically increase. The first requirement is met by the $\tilde{g}_i^-$ branch of the solution, while the second one is met only by the $\tilde{g}_i^+$ branch. Thus, in order to construct the total pseudo–radial distribution function, $\tilde{g}_i^-$ is used at low densities up to some switchover density ρ^*, where the switching to the $\tilde{g}_i^+$ branch is performed. In order to ensure a continuous and smooth transition between the two branches, the switchover density must be chosen for each isotherm at the point where the roots $\tilde{g}_i^+$ and $\tilde{g}_i^-$ are equal, so that

$$\tilde{g}_i^+(\rho^*, T) = \tilde{g}_i^-(\rho^*, T) \tag{5.55}$$

which leads to the result that

$$\left(\frac{\eta_i}{\eta_i^{(0)} \alpha_{ii} \rho} \right)_{min} = \frac{\eta_i}{\eta_i^{(0)} \alpha_{ii} \rho^*} = \left(\frac{2}{\sqrt{\beta_\eta}} + 1 \right) = 3.1954 \tag{5.56}$$

This is because the density ρ^* is also the density at which the group $(\eta_i / \eta_i^{(0)} \rho)$ displays its minimum value.

It follows that from measurements of the viscosity of each pure fluid as a function of density at a particular temperature, it is possible to determine both α_{ii} and the pseudo–radial distribution function $\tilde{g}_i$. First, the minimum value of the group $(\eta_i / \eta_i^{(0)} \rho)$ is determined from experimental data, and the switchover density ρ^* is determined from

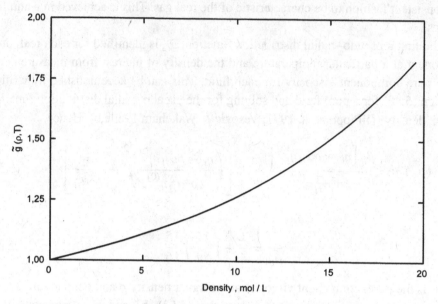

Fig. 5.11. The pseudo–radial distribution function deduced from the viscosity of nitrogen at $T = 200$ K; switching density $\rho^* = 13{,}300$ mol m^{-3}.

the first of equations (5.56). The value of α_{ii} is then also uniquely determined at each temperature from the experimental magnitude of the group $(\eta_i/\eta_i^{(0)}\rho)$ at its minimum and the second of equations (5.56). Finally, the pseudo–radial distribution function can be determined by application of equation (5.53) to the experimental data with the consistent value of α_{ii}.

Figure 5.11 shows the pseudo–radial distribution function for nitrogen at a temperature of 200 K as a function of density determined in this fashion. The resulting distribution function is smooth notwithstanding the fact that the transition between the two roots of equation (5.53) takes place in the density region encompassed by the figure. It is worthwhile emphasizing that this method of determining a pseudo–radial distribution function has the advantages of internal consistency claimed for the MET in Section 5.3. However, in this method, the derived quantities are guaranteed to reproduce the pure component transport property, which was not the case for the MET.

The next step in the procedure of evaluation of the mixture properties is the evaluation of the pseudo–radial distribution functions for all $i - j$ interactions in the mixture as well as the mean free–path parameter α_{ij} for the unlike interaction. It is consistent with the remainder of the procedure to estimate them from mixing rules based upon a rigid–sphere model. Among the many possible mixing rules for the radial distribution function one that has proved successful is based upon the Percus–Yevick equation for the radial distribution function of hard–sphere mixtures (Kestin & Wakeham 1980; Vesovic &

Wakeham 1989a,b). Thus the functions $\tilde{g}_{ij}$ for all i, j are obtained from the equation

$$\tilde{g}_{ij}(\rho, T) = 1 + \frac{2}{5} \sum_{k=1}^{N} x_k (\tilde{g}_k - 1)$$
$$+ \left[\frac{\frac{6}{5} (\tilde{g}_i - 1)^{1/3} (\tilde{g}_j - 1)^{1/3} \sum_{k=1}^{N} x_k (\tilde{g}_k - 1)^{2/3}}{(\tilde{g}_i - 1)^{1/3} + (\tilde{g}_j - 1)^{1/3}} \right]$$

(5.57)

The parameters α_{ij} are also obtained from a rigid-sphere mixing rule,

$$\alpha_{ij} = \frac{1}{8} \left(\alpha_{ii}^{1/3} + \alpha_{jj}^{1/3} \right)^3$$ (5.58)

If, now, the derived values of all of the pseudo–radial distribution functions $\tilde{g}_{ij}$ are used to replace the corresponding rigid–sphere quantities in equations (5.48)–(5.52), then it remains only to evaluate A_{ij}^* and $\eta_{ij}^{(0)}$ for all of the binary interactions to permit a calculation of the dense gas mixture viscosity. These latter quantities are just those required for the evaluation of the dilute–gas mixture viscosity and discussed in Chapter 4, so that the methods proposed there for their evaluation can be employed again here.

This description has made clear that in the procedure for the evaluation of the dense gas mixture viscosity, one needs the viscosity of each pure fluid as a function of density at the temperature for which the mixture property evaluation is required and two characteristics of each binary interaction in the limit of zero density. The procedure is automatically able to reproduce the properties of all the pure components in the mixture, so that use is made of the Enskog theory only to provide a reasonable basis for the interpolation between the properties of the pure components. It should be noted in conclusion that if the viscosity of a pure component is not available from experiment, it may itself be estimated by one of the procedures discussed in Section 5.3, preferably that which makes use of the concept of a temperature–independent excess viscosity.

Experimental data for the viscosity of multicomponent, or even binary, gas mixtures at elevated densities are rather scarce. However, whenever it has been possible to compare the results of calculations by this procedure with direct experimental data (Vesovic & Wakeham 1989a,b) the method described is able to predict the viscosity of a mixture to within a few percent over a very wide range of density and temperature.

5.5.1.2 Thermal conductivity

In the case of the thermal conductivity, as has been discussed in Chapter 4 and Section 5.2, it is necessary to account for the transport of internal energy as well as translational energy. For that reason the Thorne–Enskog equations in their original form for rigid, spherical molecules can be applied only to the translational component of the energy transport. Accordingly, before a method can be developed for the evaluation of the thermal conductivity of dense polyatomic gas mixtures, it is necessary to extend the Thorne–Enskog theory in a consistent fashion so as to treat internal energy transport. This extension was first carried out by Mason *et al.* (1978) by means of the assumption that the internal energy was transported by a diffusive mechanism, and that the diffusion coefficients for internal energy were the same as those for mass diffusion, so that the Thorne–Enskog theory can be used to describe their density dependence. This assumption implies that no account is taken explicitly of the processes of relaxation of internal energy to translational energy, so that the theory is equivalent to that leading to the Hirschfelder–Eucken equation for dilute–gas mixtures, as discussed in Section 4.3.

With this modification the thermal conductivity of an N–component dense–gas mixture can be written as the sum of two contributions (Mason *et al.* 1978; Kestin & Wakeham 1980; Vesovic & Wakeham 1991)

$$\lambda(\rho, T) = \lambda_{tr}(\rho, T) + \lambda_{int}(\rho, T). \tag{5.59}$$

Here the translational contribution is given by

$$\lambda_{tr} = - \frac{\begin{vmatrix} L_{11} & \cdots & L_{1N} & Y_1 \\ \vdots & \cdots & \vdots & \vdots \\ L_{N1} & \cdots & L_{NN} & Y_N \\ Y_1 & \cdots & Y_N & 0 \end{vmatrix}}{\begin{vmatrix} L_{11} & \cdots & L_{1N} \\ \vdots & \cdots & \vdots \\ L_{N1} & \cdots & L_{NN} \end{vmatrix}} + K_{mix} \tag{5.60}$$

where

$$Y_i = x_i \left[1 + \sum_{j=1}^{N} \frac{2m_i m_j}{(m_i + m_j)^2} x_j \gamma_{ij} \tilde{g}_{ij} \rho \right] \tag{5.61}$$

$$L_{ii} = \frac{x_i^2 \tilde{g}_{ii}}{\lambda_{tr,i}^{(0)}} + \sum_{j=1, j \neq i}^{N} \frac{x_i x_j \tilde{g}_{ij}}{2\lambda_{ij}^{(0)} A_{ij}^* (m_i + m_j)^2} \\ \times \left[\frac{15}{2} m_i^2 + \frac{25}{4} m_j^2 - 3m_j^2 B_{ij}^* + 4m_i m_j A_{ij}^* \right] \tag{5.62}$$

$$L_{ij}(i \neq j) = - \frac{x_i x_j m_i m_j \tilde{g}_{ij}}{2\lambda_{ij}^{(0)} A_{ij}^* (m_i + m_j)^2} \left[\frac{55}{4} - 3B_{ij}^* - 4A_{ij}^* \right] \tag{5.63}$$

$$K_{mix} = \left(\frac{16}{5\pi} \right) \frac{10}{9} \rho^2 \sum_{j=1}^{N} \sum_{i=1}^{N} \frac{x_i x_j m_i m_j}{(m_i + m_j)^2} \tilde{g}_{ij} \gamma_{ij}^2 \lambda_{ij}^{(0)} \tag{5.64}$$

and

$$\lambda_{\text{int}}(\rho, T) = \sum_{i=1}^{N} \left[\frac{\lambda_i^{(0)} - \lambda_{\text{tr},i}^{(0)}}{\tilde{g}_{ij}} \right] \left[1 + \sum_{j \neq i}^{N} \frac{x_j \lambda_{\text{tr},i}^{(0)} \tilde{g}_{ij} A_{ii}^*}{x_i \lambda_{ij}^{(0)} \tilde{g}_{ii} A_{ij}^*} \right]^{-1} \quad (5.65)$$

In these equations the pseudo–radial distribution functions for the real gas have been inserted in place of the radial distribution functions of hard spheres at contact, by analogy with the treatment above for the viscosity. Also, different mean free–path shortening parameters, γ_{ij}, have been introduced which are again related to the co–volume for molecules i and j. All other symbols have been previously defined, and it should be noted that $\lambda_{\text{tr},i}^{(0)}$ may be related to the viscosity of compound i by means of equation (4.44) of Chapter 4.

5.5.1.3 Estimation of the thermal conductivity

The method for predicting the thermal conductivity of a dense–gas mixture (Mason *et al.* 1978; Kestin & Wakeham 1980; Vesovic & Wakeham 1991) is analogous to that for the viscosity, so that only its main features need be described here. The pseudo–radial distribution function for the thermal conductivity of the pure species is obtained by solving equation (5.60) for $\tilde{g}_i$ at each temperature and the mixture molar density

$$\tilde{g}_i(\rho, T) = \frac{\beta_\lambda \left[\lambda_i - \rho \gamma_{ii} \lambda_{\text{tr},i}^{(0)} \right]}{2\rho^2 \gamma_{ii}^2 \lambda_{\text{tr},i}^{(0)}}$$

$$\pm \beta_\lambda \left[\left(\frac{\lambda_i - \rho \gamma_{ii} \lambda_{\text{tr},i}^{(0)}}{2\rho^2 \gamma_{ii}^2 \lambda_{\text{tr},i}^{(0)}} \right)^2 - \frac{\lambda_i^{(0)}}{\beta_\lambda \rho^2 \gamma_{ii}^2 \lambda_{\text{tr},i}^{(0)}} \right]^{1/2} \quad (5.66)$$

where

$$\frac{1}{\beta_\lambda} = \frac{1}{4} + \left(\frac{16}{5\pi} \right) \frac{5}{18} = 0.5329 \quad (5.67)$$

As for the viscosity, the solution for $\tilde{g}_i$ has two branches corresponding to the two roots of equation (5.66), $\tilde{g}_i^+$ and $\tilde{g}_i^-$. Using the same physical conditions as before, a smooth function $\tilde{g}_i$ can be obtained by a transition between the two solutions at a switchover density ρ^* obtained from the solution of the equation

$$\left[\frac{\partial \lambda_i(\rho, T)}{\partial \rho} \right]_T = \frac{\lambda_i(\rho, T)}{\rho} \quad (5.68)$$

while the parameter γ_{ii} for each pure species is calculated at each temperature from the equation

$$\frac{\lambda_i(\rho^*, T)}{\rho^* \gamma_{ii} \lambda_{\text{tr},i}^{(0)}} = 1 + \frac{2}{\sqrt{\beta_\lambda}} \left[\frac{\lambda_i^{(0)}(T)}{\lambda_{\text{tr},i}^{(0)}(T)} \right]^{1/2} \quad (5.69)$$

The functions $\tilde{g}_i$ derived from experimental thermal conductivity data and, indeed, the mean free–path parameters, are, naturally, different from those derived from viscosity data because of the inevitable failure of the Enskog theory to represent either property accurately. This fact inhibits the transposition of information on the density dependence of one property to the prediction of the second but is otherwise not significant in the prediction process.

In order to evaluate the thermal conductivity of a multicomponent mixture the functions $\tilde{g}_{ij}$ are generated from those for the pure gases by means of the combination rule of equation (5.57) while the mean free–path parameters γ_{ij} are generated by a combination rule analogous to that for the corresponding viscosity parameters so that

$$\gamma_{ij} = \frac{1}{8} \left(\gamma_{ii}^{1/3} + \gamma_{jj}^{1/3} \right)^3 \tag{5.70}$$

In order to calculate the thermal conductivity of a dense multicomponent gas mixture using the procedure outlined above, no information on the behavior of the gas mixture at elevated densities is required. However, the pure component thermal conductivity as a function of density must be available, together with three quantities characteristic of the zero–density state; namely A^*, B^* and $\lambda_{\mathrm{tr},ij}^{(0)}$. They can be easily obtained, for a large number of binary interactions, by the methods described in Chapter 4. Again, the procedure automatically reproduces the behavior of all of the pure components in the mixture and acts as an interpolatory formula between them. If the thermal conductivity of one of the pure components is not available as a function of density at the temperature of interest, it can be estimated by one of the methods described in Section 5.3, preferably that which makes use of the concept of a temperature–independent excess property.

Comparisons with the limited experimental data available at high density (Kestin & Wakeham 1980; Vesovic & Wakeham 1991) indicate that the method is capable of predicting the background thermal conductivity of a mixture to within ±4%, which is adequate for many practical purposes.

References

Chapman, S. & Cowling, T.G. (1952). *The Mathematical Theory of Non–Uniform Gases*, 2nd Ed. London: Cambridge University Press.

DiPippo, R., Dorfman, J.R., Kestin, J., Khalifa, H.E. & Mason, E.A. (1977). Composition dependence of the viscosity of dense gas mixtures. *Physica*, **86A**, 205–223.

Kestin, J. & Wakeham, W.A. (1980). Calculation of the influence of density on the thermal conductivity of gaseous mixtures. *Ber. Bunsenges. Phys. Chem.*, **84**, 762–769.

Mason, E.A., Khalifa, H.E., Kestin, J. & Dorfman, J.R. (1978). Composition dependence of the thermal conductivity of dense gas mixtures. *Physica*, **91A**, 377–392.

Sandler, S.I. & Fiszdon, J.K. (1979). On the viscosity and thermal conductivity of dense gases. *Physica*, **95A**, 602–608.

Vesovic, V. & Wakeham, W.A. (1989a). The prediction of the viscosity of dense gas mixtures. *Int. J. Thermophys.*, **10**, 125–135.

Vesovic, V. & Wakeham, W.A. (1989b). Prediction of the viscosity of fluid mixtures over wide ranges of temperature and pressure. *Chem. Eng. Sci.*, **44**, 2181–2189.

Vesovic, V. & Wakeham, W.A. (1991). Prediction of the thermal conductivity of fluid mixtures over wide ranges of temperature and pressure. *High Temp.–High Press.*, **23**, 179–190.

5.6 Fluid mixtures at high densities – liquid phase

J. H. DYMOND

The University, Glasgow, UK

and

M. J. ASSAEL

Aristotle University of Thessaloniki, Greece

In view of the success of the methods based on hard–sphere theories for the accurate correlation and prediction of transport properties of single–component dense fluids, it is worthwhile to consider the application of the hard–sphere model to dense fluid mixtures. The methods of Enskog were extended to mixtures by Thorne (see Chapman & Cowling 1952). The binary diffusion coefficient D_{12} for a smooth hard–sphere system is given by

$$[D_{12}]_{\text{SHS}} = \frac{3\,(k_{\text{B}}T)^{1/2}}{8n\sigma_{12}^2} \left(\frac{m_1 + m_2}{2\pi m_1 m_2}\right) \frac{1}{g_{12}\,(\sigma)} \tag{5.71}$$

where n is the total number density, $\sigma_{12} = (\sigma_1 + \sigma_2)/2$, for spheres of diameter σ, m_1, m_2 are the molecular masses and $g_{12}(\sigma)$ is the unlike radial distribution function at contact. This has to be corrected for correlated molecular motions using results from computer simulation studies $[D/D_{\text{E}}]_{\text{MD}}$ (Alder *et al.* 1974; Easteal & Woolf 1990) and a factor A_{12} introduced to account for effects of translational–rotational coupling in molecular fluids which behave as rough hard spheres rather than smooth hard spheres. Thus,

$$[D_{12}]_{\text{EXP}} = [D_{12}]_{\text{RHS}} = A_{12}\,[D_{12}]_{\text{SHS}}\,[D/D_{\text{E}}]_{\text{MD}} \tag{5.72}$$

This approach is widely used for the interpretation of binary diffusion data where there is just a trace of the second component (for example, Chen *et al.* 1982; Dymond & Woolf 1982; Erkey *et al.* 1989). The expression given for $g_{12}(\sigma)$ is based on computer calculations. Values for the coupling factors exhibit a linear variation with temperature which gives ease of interpretation and accurate calculation of tracer diffusion at other temperatures. However, it should be noted that there are uncertainties in the computed corrections, which means that the A_{12} are not uniquely defined, though at the present time there is no *a priori* method for calculating these factors.

For equimolar binary mixtures where the effects of molecular interactions are not strong, it was found (Easteal & Woolf 1984) that experimental interdiffusion coefficients (D_{ij}) were generally in close agreement with values computed for the hard–sphere mixture for the appropriate molecular mass and size ratio. The Enskog intradiffusion coefficients $[(D_i)_j]_{\text{E}}$ were calculated from

$$\frac{1}{[(D_i)_j]_{\text{E}}} = \frac{x_i}{g_{ii}\,(\sigma)\,[D_{ii}]_{\text{E}}} + \frac{x_j}{g_{ij}\,(\sigma)\,[D_{ij}]_{\text{E}}} \tag{5.73}$$

It is shown that interdiffusion coefficients can be predicted with some confidence on the basis of this model, but accurate computer simulations are required for corrections at other compositions.

Viscosity and thermal conductivity for dense hard–sphere mixtures can be calculated on the basis of Thorne's expressions (Kandiyoti & McLaughlin 1969), but a rigorous test of the applicability of this model for correlation of mixture data for real fluid mixtures awaits accurate computation of the corrections to these extended Enskog expressions. Following the success of their hard–sphere based method for correlation of transport properties of single–component fluids, Assael *et al.* (1992) proposed a pseudo one–fluid model for mixtures where there was negligible volume change on mixing. Molecular parameters V_0, the characteristic volume, and R_η and R_λ, the roughness factors, were given by the mole fraction average of the values for the pure components. Experimental viscosity and thermal conductivity data for alkane mixtures over wide temperature and pressure ranges were fitted generally to well within 6%. The outstanding advantage of this approach, which is described in detail in Chapter 10, is that there is no adjustable parameter, and no requirement for any experimental data for the pure components under the same conditions. Analysis of recent thermal conductivity measurements for benzene + 2,2,4–trimethylpentane up to 350 MPa (Mensah–Brown & Wakeham 1994) shows that the data for this system also can be predicted within 6% by this method. However, within this limit, there are systematic density–dependent differences, which suggests that the idea of a universal curve for reduced transport properties may have to be discarded for an exact data fit.

In spite of the simplicity of the hard–sphere model, this molecular approach has the greatest potential for the accurate correlation and prediction of dense liquid mixture transport properties at the present time.

However, in view of the present limitations of this method, semiempirical predictions such as those based on an extension of the principle of corresponding states, discussed in Chapter 12, and empirical estimations as described in Chapter 13 will continue to be widely used for their general applicability.

References

Alder, B.J., Alley, W.E. & Dymond, J.H. (1974). Studies in molecular dynamics XIV. Mass and size dependence of the binary diffusion coefficient. *J. Chem. Phys.*, **61**, 1415–1420.

Assael, M.J., Dymond, J.H., Papadaki, M. & Patterson, P.M. (1992). Correlation and prediction of dense fluid transport coefficients III. *n*–alkane mixtures. *Int J. Thermophys.*, **13**, 659–669.

Chapman, S. & Cowling, T.G. (1952). *The Mathematical Theory of Non–Uniform Gases*, New York: Cambridge University Press, Chap. 16.

Chen, S.–H., Davis, H.T. & Evans, D.F. (1982). Tracer diffusion in polyatomic liquids. *J. Chem. Phys.*, **77**, 2540–2544.

Dymond, J.H. & Woolf, L.A. (1982). Tracer diffusion of organic solutes in *n*–hexane at pressures up to 400 MPa. *J. Chem. Soc. Faraday Trans. 1*, **78**, 991–1000.

Easteal, A.J. & Woolf, L.A. (1984). Diffusion in mixtures of hard spheres at liquid densities: A comparison of molecular dynamics and experimental data in equimolar systems. *Chem. Phys.*, **88**, 101–111.

Easteal, A.J. & Woolf, L.A. (1990). Tracer diffusion in hard sphere liquids from molecular dynamics simulations. *Chem. Phys. Lett.*, **167**, 329–333.

Erkey, C., Rodden, J.B., Matthews, M.A. & Akgerman, A. (1989). Application of rough hard–sphere theory to diffusion in *n*–alkanes. *Int. J. Thermophys.*, **10**, 953–962.

Kandiyoti, R. & McLaughlin, E. (1969). Viscosity and thermal conductivity of dense hard sphere fluid mixtures. *Mol. Phys.*, **17**, 643–653.

Mensah–Brown, H. & Wakeham, W. A. (1994). Thermal conductivity of liquid mixtures of benzene and 2,2,4–trimethylpentane at pressures up to 350 MPa. *Int. J. Thermophys.*, **15**, 117–139.

6

The Critical Enhancements

J. V. SENGERS

University of Maryland, College Park, MD, USA,

and

Thermophysics Division, National Institute of Standards and Technology,
Gaithersburg, MD, USA

J. LUETTMER–STRATHMANN

University of Maryland, College Park, MD, USA

6.1 Introduction

Thermodynamic states of systems near a critical point are characterized by the presence of large fluctuations of the order parameter associated with the critical phase transition. For one–component fluids near the vapor–liquid critical point the order parameter can be identified with the fluctuating density ρ. That is, near the vapor–liquid critical point the local density becomes a function of the position $\mathbf{r}$. The density fluctuations have a pronounced effect on the behavior of both the thermodynamic and the transport properties of fluids in the critical region. Specifically, they lead to a strong enhancement of the thermal conductivity and to a weak enhancement of the viscosity of fluids in the critical region, as will be elucidated in this chapter.

To specify the spatial character of the density fluctuations, a correlation function $G(\mathbf{r})$ is introduced such that (Fisher 1964)

$$\rho^2 G(\mathbf{r}) = \langle \rho(\mathbf{r})\rho(0) \rangle - \rho^2 \tag{6.1}$$

where $\rho = \langle \rho(\mathbf{r}) \rangle$ is the average macroscopic density. For isotropic fluids the correlation function $G(\mathbf{r})$ is a function of the distance $r = |\mathbf{r}|$ only. Let T be the temperature, P the pressure, ρ the mass density and $\chi = (\partial \rho / \partial \mu)_T$, where μ is the chemical potential. The susceptibility χ, which is proportional to the isothermal compressibility, is related to the zeroth moment of $G(\mathbf{r})$ (Sengers & Levelt Sengers 1978)

$$k_B T \chi = \rho^2 \int d\mathbf{r}\, G(\mathbf{r}) \tag{6.2}$$

where k_B is the Boltzmann constant. The susceptibility χ diverges, and, consequently, the integral on the right–hand side of equation (6.2) diverges at the critical point. Thus the correlation function $G(\mathbf{r})$ becomes long range. The range ξ of this correlation function is defined by

$$\xi^2 = \int d\mathbf{r}\, r^2 G(\mathbf{r}) \bigg/ \int d\mathbf{r}\, G(\mathbf{r}) \tag{6.3}$$

Instead of $G(\mathbf{r})$, it is more convenient to consider its Fourier transform, the structure factor

$$\chi(\mathbf{k}) = \int d\mathbf{r}\, e^{i\mathbf{k}\cdot\mathbf{r}} G(\mathbf{r}) \tag{6.4}$$

For small values of the wave number k the structure factor is given by (Fisher 1964)

$$\chi(k) = \frac{\chi(0)}{1 + k^2 \xi^2} \tag{6.5}$$

where $\chi(0) = \chi$ is the thermodynamic susceptibility. In developing a theory for the effects of the critical fluctuations on the transport properties of one–component fluids in the critical region, occasionally results obtained for binary liquid mixtures near a consolute point are applied, that is, a critical point associated with the separation of the mixture into two liquid phases. For liquid mixtures near a consolute point the order parameter is to be identified with the concentration and χ with the osmotic susceptibility (Kumar *et al.* 1983).

6.2 Critical slowing down of the fluctuations

The transport properties are related to the dynamic behavior of the fluctuations. Hence, for a discussion of the transport properties we need to consider the dependence of the structure factor on the time t in accordance with (Forster 1975)

$$\chi(k, t) = \chi(k) e^{-Dk^2 t} \tag{6.6}$$

where $\mathcal{D}$ is the diffusivity associated with the fluctuations of the order parameter. For one–component fluids this diffusivity is to be identified with the thermal diffusivity (McIntyre & Sengers 1968)

$$a = \frac{\lambda}{\rho c_P} \tag{6.7}$$

where λ is the thermal conductivity and c_P the isobaric specific heat capacity. For a liquid mixture near a consolute point, $\mathcal{D}$ in equation (6.6) is to be identified with the mutual mass diffusivity D_{12} (Matos Lopes *et al.* 1992), which is related to the osmotic susceptibility χ as (Mistura 1972)

$$D_{12} = \frac{L}{\chi} \tag{6.8}$$

where L is a mass conductivity. It should be noted that c_P in equation (6.7) is related to the susceptibility χ by

$$\rho c_P = \rho c_V + T\rho^{-2}(\partial P/\partial T)_\rho^2 \chi \tag{6.9}$$

where c_V is the isochoric specific heat capacity. Since c_V is only weakly divergent, as discussed in Section 6.3, while $(\partial P/\partial T)_\rho$ remains finite at the critical point, it follows that c_P diverges like the susceptibility χ. As a consequence, the diffusivity $\mathcal{D}$ in equation (6.6) vanishes at the critical point. Thus the fluctuations near the critical point not only extend over long distances, that is, long compared to the range of the intermolecular interactions, but also decay very slowly with time. This phenomenon has been referred to as the critical slowing down of the fluctuations.

In the classical theory, the thermal conductivity and the viscosity of fluids are related to the microscopic molecular collisions, which are assumed to be effective at short ranges only. As a consequence, it had been assumed that transport coefficients like the thermal conductivity λ (or L) and the shear viscosity η are not affected by the long–range critical fluctuations. Thus, the dependence of these transport properties on temperature and density in the critical region was assumed to be qualitatively similar to that observed outside the critical region, and no divergent behavior of these transport properties was expected (Felderhof 1966). Thus in the classical theory the critical slowing down of the fluctuations was solely associated with the divergent behavior of the thermodynamic susceptibility χ, while the thermal conductivity λ in equation (6.7), or the mass conductivity L in equation (6.8), was expected to remain finite at the critical point (Van Hove 1954).

The classical theory of critical slowing down of the fluctuations became untenable when it was established experimentally that the thermal conductivity λ of fluids diverges at the vapor–liquid critical point (Sengers & Michels 1962; Michels & Sengers 1962). The nonclassical dynamic behavior of the critical fluctuations was subsequently confirmed by light–scattering experiments (Swinney & Henry 1973). Furthermore, it had been noticed from the beginning of the century that the viscosity of liquid mixtures exhibits a critical enhancement near a consolute point, as documented in earlier reviews (Sengers 1966, 1971). A first attempt to explain the critical behavior of the transport properties of fluids was made by Fixman (1962, 1964, 1967). His ideas were further developed by Kawasaki (1966) and Kadanoff & Swift (1968) into what has become known as the mode–coupling theory of dynamic phenomena in fluids (Kawasaki 1970, 1976; Pomeau & Résibois, 1975; Keyes 1977). Subsequently, the renormalization–group theory, originally developed to explain the critical behavior of equilibrium properties, was extended to deal with dynamic properties also (Siggia *et al.* 1976). The relationship between the mode–coupling theory and the renormalization–group theory of dynamic critical phenomena has been elucidated by Hohenberg & Halperin (1977) and by Gunton (1979).

6.3 Critical power laws for equilibrium properties

Before presenting the theoretical predictions for the critical behavior of the transport properties, first some aspects of the critical behavior of the equilibrium properties are reviewed. Specifically, the thermodynamic properties satisfy asymptotic power laws when the critical point is approached along specific paths. Let T_c, ρ_c and P_c be the temperature, density and pressure at the critical point. In particular, the dependence of some properties as a function of the reduced temperature $\Delta T^* = (T - T_c)/T_c$ is considered, when the critical temperature is approached along the critical isochore $\rho = \rho_c$ from above in the one–phase region. Along the critical isochore the isochoric specific heat capacity c_V diverges as

$$\rho c_V T_c / P_c = (A/\alpha)(\Delta T^*)^{-\alpha} \tag{6.10}$$

the susceptibility diverges as

$$\chi P_c / \rho_c^2 = \Gamma(\Delta T^*)^{-\gamma} \tag{6.11}$$

and the correlation length ξ diverges as

$$\xi = \xi_0 (\Delta T^*)^{-\nu} \tag{6.12}$$

where the exponent ν is related to the exponent α by $\nu = (2-\alpha)/3$, while the coefficients A, Γ and ξ_0 are system-dependent amplitudes.

Systems near a critical point are classified in terms of universality classes. The universality classes depend on the dimensionality of the system, on the number of components of the order parameter and whether the interactions between the constituent particles are short or long range. Systems within the same universality class have the same values of the critical exponents α, γ and ν. Fluids near the vapor–liquid critical point and liquid mixtures near the consolute point belong to the universality class of three–dimensional Ising–like systems with exponent values (Sengers & Levelt Sengers 1986; Liu & Fisher 1989).

$$\alpha = 0.110 \pm 0.003, \quad \gamma = 1.239 \pm 0.002, \quad \nu = 0.630 \pm 0.001 \tag{6.13}$$

Moreover, the amplitude ξ_0 of the power law (6.12) for ξ is related to the amplitude A of the power law (6.10) for c_V by

$$\xi_0 (A P_c / k_B T_c)^{1/3} = R_\xi \tag{6.14}$$

with

$$R_\xi = 0.27 \pm 0.01 \tag{6.15}$$

The expression (6.5) for the structure factor is strictly valid only for $k\xi \ll 1$. For $k\xi \gg 1$ it varies as $(k\xi)^{-\gamma/\nu}$ (Fisher 1964). The value of the exponent $\gamma/\nu = 1.97$ is close to 2, so that to a good approximation equation (6.5) can be used for all relevant values of $k\xi$ (Chang *et al.* 1979). Hence, this approximation will be adopted in the theory for the critical behavior of the transport properties.

It should be noted that the validity of the critical power laws (6.10)–(6.11) is restricted to a very small range of temperatures near the critical point (Levelt Sengers & Sengers 1981). Equations of state that incorporate these critical power laws but which remain valid in an appreciable range of densities and temperatures in the critical region have been developed (Sengers 1994).

6.4 Behavior of the transport properties in the near vicinity of the critical point

For a quantitative representation of the behavior of the transport properties in the critical region, the thermal conductivity λ and the viscosity η are decomposed as (Sengers & Keyes 1971; Sengers 1971, 1972):

$$\lambda = \bar{\lambda} + \Delta\lambda_c, \qquad \eta = \bar{\eta} + \Delta\eta_c \qquad (6.16)$$

where $\bar{\lambda}$ and $\bar{\eta}$ are background contributions and $\Delta\lambda_c$ and $\Delta\eta_c$ are enhancements due to the critical fluctuations. The background contributions $\bar{\lambda}$ and $\bar{\eta}$ are in turn decomposed as

$$\bar{\lambda} = \lambda^{(0)}(T) + \Delta\lambda(\rho, T), \qquad \bar{\eta} = \eta^{(0)}(T) + \Delta\eta(\rho, T) \qquad (6.17)$$

where $\lambda^{(0)}(T)$ and $\eta^{(0)}(T)$ are the thermal conductivity and the viscosity in the dilute-gas limit, $\rho \to 0$, while $\Delta\lambda(\rho, T)$ and $\Delta\eta(\rho, T)$ are commonly referred to as excess thermal conductivity and excess viscosity, respectively, which depend only weakly on temperature (Sengers & Keyes 1971; see Chapter 5). The decomposition is illustrated for the thermal conductivity in Figure 3.1 in Chapter 3 of this volume. Equations for the background transport coefficients $\bar{\lambda}$ and $\bar{\eta}$ are discussed in Chapters 4 and 5. In this chapter the attention is focused on the critical enhancements $\Delta\lambda_c$ and $\Delta\eta_c$ in equation (6.16). The decomposition of λ into $\bar{\lambda}$ and $\Delta\lambda_c$ implies a corresponding decomposition of the thermal diffusivity

$$a = \bar{a} + \Delta a_c \qquad (6.18)$$

with $\bar{a} = \bar{\lambda}/\rho c_P$.

The concept of universality classes, mentioned in the previous section, can be extended so as to be applicable to the characterization of the asymptotic critical behavior of dynamic properties (Hohenberg & Halperin 1977). Two systems belong to the same dynamic universality class when they have the same number and types of relevant hydrodynamic modes. Thus the asymptotical critical behavior of the mutual mass diffusivity D_{12} and of the viscosity η of liquid mixtures near a consolute point will be the same as that of the thermal diffusivity a and the viscosity η of one–component fluids near the vapor–liquid critical point (Sengers 1985). Hence, in analogy with equation (6.16) for liquid mixtures near a consolute point it can be written

$$L = \bar{L} + \Delta L_c, \qquad \eta = \bar{\eta} + \Delta\eta_c \qquad (6.19)$$

and in analogy with equation (6.18)

$$\mathcal{D} = \bar{\mathcal{D}} + \Delta\mathcal{D}_c \qquad (6.20)$$

with $\bar{\mathcal{D}} = \bar{L}/\chi$.

Near the critical point, $\Delta\mathcal{D}_c$ satisfies an equation of the form (Kawasaki 1976; Ferrell 1970; Hohenberg & Halperin 1977)

$$\Delta\mathcal{D}_c = \frac{R_D k_B T}{6\pi\eta\xi} \tag{6.21}$$

where R_D is a universal amplitude. For one–component fluids near the vapor–liquid critical point, $\Delta\mathcal{D}_c$ in equation (6.21) is to be identified with Δa_c, so that

$$\Delta\lambda_c = \frac{R_D k_B T}{6\pi\eta\xi}\rho c_P \tag{6.22}$$

Equation (6.21) may be interpreted as the diffusivity of a droplet–like cluster with radius ξ with a Stokes law friction coefficient $6\pi\eta\xi$.

The viscosity η diverges asymptotically as (Hohenberg & Halperin 1977)

$$\eta = \bar{\eta}(Q\xi)^z \tag{6.23}$$

where Q is a system–dependent amplitude and where z is a universal critical exponent. It should be noted that the critical enhancement of the viscosity is a multiplicative enhancement; that is, $\Delta\eta_c = \bar{\eta}[(Q\xi)^z - 1]$ is proportional to the background viscosity $\bar{\eta}$ (Sengers 1971; Ohta 1977).

The resolution of experiments in the near vicinity of the vapor–liquid critical point is hampered by the presence of gravitationally induced density gradients, and the resolution of optical experiments is hampered by multiple scattering and turbidity (Moldover *et al.* 1979). These problems can be avoided in experiments with liquid mixtures near a consolute point. Furthermore, the range of validity of the asymptotic equations (6.21) and (6.23) appears to be smaller in fluids near the vapor–liquid critical point than in liquid mixtures near a consolute point. Finally, while the background thermal diffusivity $\bar{a} = \bar{\lambda}/\rho c_P$ goes to zero faster than Δa_c at the critical point, the background contribution $\bar{\lambda}$ is not small enough for $\bar{a}$ ever to be negligible (Güttinger & Cannell 1980).

On the other hand, for certain binary liquids near the consolute point it appears that $\bar{\mathcal{D}} = \bar{L}/\chi$ does become vanishingly small, so that the measured diffusivity $\mathcal{D}$ can indeed be identified with $\Delta\mathcal{D}_c$ (Burstyn *et al.* 1980; Burstyn & Sengers 1982). Hence, the more accurate confirmations of the theoretical predictions for the asymptotic critical behavior of the transport properties have been obtained from experiments in liquid mixtures near consolute points. For the amplitude R_D in the Stokes–Einstein law (6.21) for the diffusivity one has found (Burstyn *et al.* 1980; Burstyn & Sengers 1982; Hamano *et al.* 1985, 1986; Matos Lopes *et al.* 1992)

$$R_D = 1.02 \pm 0.06 \tag{6.24}$$

and for the exponent z in the power law (6.23) for the viscosity (Berg & Moldover 1988;

Nieuwoudt & Sengers 1989; Krall *et al.* 1989)

$$z = 0.065 \pm 0.004 \tag{6.25}$$

As an example, Figure 6.1 shows the mutual diffusivity of a mixture of hexane and nitrobenzene at the critical composition as a function of $T - T_c$ as measured by Taylor dispersion (Matos Lopes *et al.* 1992) and by light scattering (Wu *et al.* 1988); the solid curve represents the Stokes–Einstein diffusion law (6.21) with $R_D = 1.04 \pm 0.06$. In Figure 6.2 a *log-log* plot of the viscosity ratio $\eta/\bar{\eta}$ of a mixture of 3–methylpentane and nitroethane at the critical composition is shown as a function of the correlation length ξ. The solid curve represents the power law $(Q\xi)^z$ with $Q = 1.4\,\mathrm{nm}^{-1}$ and $z = 0.063$ (Burstyn *et al.* 1983). Experiments in fluids near the vapor–liquid critical point are consistent with these results (Güttinger & Cannell 1980; Berg & Moldover 1990).

The diffusivity $\mathcal{D}$ in the vicinity of the critical point can be determined from Rayleigh scattering experiments (Chu 1974). However, such light–scattering experiments yield $\mathcal{D}$ at a finite value of the wave number k, while equation (6.21) applies only in the limit $k \to 0$, that is, for $k\xi \ll 1$. This condition is no longer satisfied in light–scattering experiments near the critical point. The dependence of $\Delta\mathcal{D}_c$ on k enters through the scaled variable $u = k\xi$ and is given by (Burstyn *et al.* 1983)

$$\frac{\Delta\mathcal{D}_c(k\xi)}{\Delta\mathcal{D}_c(0)} = K(k\xi)\left[1 + \left(\frac{k\xi}{2}\right)^2\right]^{z/2} \tag{6.26}$$

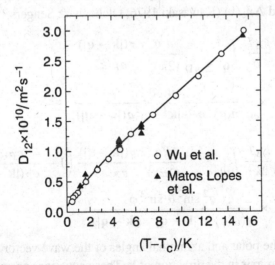

Fig. 6.1. Mutual diffusivity $\mathcal{D}$ of a mixture of hexane and nitrobenzene at the critical composition as a function of $T - T_c$. The solid curve represents equation (6.21) (Matos Lopes *et al.* 1992).

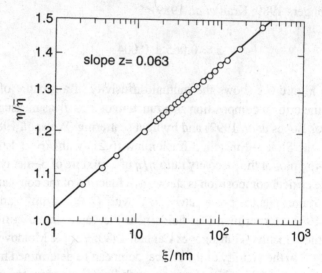

Fig. 6.2. *Log–log* plot of the viscosity ratio $\eta/\bar\eta$ of a mixture of 3–methylpentane and nitroethane at the critical composition as a function of the correlation length ξ. The solid curve
represents the power law $(Q\xi)^z$ with $z = 0.063$ (Burstyn *et al.* 1983).

where $K(u)$ is a so–called Kawasaki function

$$K(u) = \frac{3}{4u^2}\left[1 + u^2 + (u^3 - u^{-1})\arctan u\right] \tag{6.27}$$

such that $K(u) \to 1$ as $u \to 0$.

The mode–coupling theory of critical dynamics yields a set of coupled integral equations for $\Delta a_{\mathrm c}(k)$ and $\Delta\eta_{\mathrm c}(k)$ (Kawasaki 1976; Olchowy & Sengers 1988)

$$\Delta a_{\mathrm c}(k) = \frac{k_{\mathrm B}T}{\rho}\int_{|\mathbf q|\le q_{\mathrm D}}\frac{d^3 q}{(2\pi)^3}\frac{c_P(|\mathbf k - \mathbf q|)}{c_P(k)}$$

$$\times\frac{\sin^2\theta}{q^2\eta(q)/\rho + |\mathbf k - \mathbf q|^2 a(|\mathbf k - \mathbf q|)} \tag{6.28}$$

$$\Delta\eta_{\mathrm c}(k) = \frac{k_{\mathrm B}T}{2k^2}\int_{|\mathbf q|\le q_{\mathrm D}}\frac{d^3 q}{(2\pi)^3}\frac{c_P(|\mathbf k - \mathbf q|)}{c_P(q)}\left(1 - \frac{c_P(q)}{c_P(|\mathbf k - \mathbf q|)}\right)^2$$

$$\times\frac{q^2\sin^2\theta\,\sin^2\phi}{q^2 a(q) + |\mathbf k - \mathbf q|^2 a(|\mathbf k - \mathbf q|)} \tag{6.29}$$

where θ and ϕ are the polar and azimuthal angles of the wave vector $\mathbf q$ in a coordinate
system with the polar axis in the direction of $\mathbf k$. These equations account for the effects
of critical fluctuations with wave number $|\mathbf q|$ smaller than a maximum cutoff wave
number $q_{\mathrm D}$. For liquid mixtures $\Delta\mathcal{D}_{\mathrm c}(k)$ and $\Delta\eta_{\mathrm c}(k)$ are also given by equations (6.28)

and (6.29), provided that c_P is replaced everywhere by the osmotic susceptibility χ. Since near the vapor–liquid critical line $c_V \ll c_P$ in equation (6.9), it follows that the asymptotic critical behavior of $\Delta \mathcal{D}_c$ and $\Delta \eta_c$ in fluids near the consolute point will indeed be identical to that of Δa_c and $\Delta \eta_c$ in fluids near the vapor–liquid critical point. However, corrections to the asymptotic critical behavior will no longer be identical in the two types of systems.

Near the critical point $\Delta \mathcal{D}_c$ and $\Delta \eta_c$ not only depend on the wave number k, but also become functions of the frequency ω. For the diffusivity $\Delta \mathcal{D}_c$ this frequency dependence has been determined, but it is a very small effect (Burstyn & Sengers 1983). However, in the case of the viscosity, frequency effects can become appreciable (Perl & Ferrell 1972; Bhattacharjee & Ferrell 1983a; Nieuwoudt & Sengers 1987). In addition the viscosity may be affected by the magnitude of the shear rate (Oxtoby 1975; Nieuwoudt & Sengers 1989). In this chapter the frequency dependence of the transport coefficients is not considered, nor is a discussion of shear–induced critical effects included. However, actual viscosity measurements obtained near the critical point may have to be corrected for these effects (Nieuwoudt & Sengers 1989), unless they have been obtained at very low values of the frequency and of the shear rate (Berg & Moldover 1988). For the viscosity data shown in Figure 6.2 these effects turned out to be small.

To deduce the asymptotic critical behavior of Δa_c from equation (6.28), one introduces the following approximations. Since at the critical point $\xi \to \infty$, the integral (6.28) is evaluated for all wave numbers q from zero to infinity. Since $c_V \ll c_P$, c_P in equations (6.28) and (6.29) is everywhere replaced by the susceptibility χ, whose dependence on the wave number is represented by equation (6.5). Since the background contribution $\bar{\eta}$ to the viscosity is rather large, while the critical exponent z for $\Delta \eta_c$ is small, the dependence of η on the wave number q in equation (6.28) is in practice neglected and $\eta(q)$ is identified with the hydrodynamic viscosity η. Finally, since a goes to zero at the critical point, its contribution to the integrand in equation (6.28) is neglected. One then obtains (Kawasaki 1976)

$$\Delta a_c = \frac{k_B T}{6\pi \eta \xi} K(k\xi) \tag{6.30}$$

and in the limit $k\xi \to 0$ one recovers equation (6.21) with $R_D = 1$. However, equation (6.30) yields an incorrect dependence of Δa_c on $k\xi$ for large values of $k\xi$. Inclusion of the dependence of η on wave number and frequency yields the correction factor $[1 + (k\xi/2)^2]^{z/2}$ in equation (6.26), while the value of the amplitude R_D increases from unity to about 1.03 (Burstyn *et al.* 1983).

For the determination of the asymptotic critical behavior of $\Delta \eta_c$ from equation (6.29) one must retain a finite value of the cutoff wave number q_D, since the integral diverges otherwise (Perl & Ferrell 1972). Assuming an Ornstein–Zernike wave vector dependence for c_P in accordance with equation (6.5) in the integrand of equation (6.29), then in the limit $k \to 0$ (Bhattacharjee *et al.* 1981)

$$\bar{\eta} + \Delta\eta_c = \bar{\eta}\left[1 + \frac{8}{15\pi^2}\ln Q\xi\right] \tag{6.31}$$

with

$$Q = 2e^{-4/3}\frac{q_C q_D}{q_C + q_D} \tag{6.32}$$

For fluids near the vapor-liquid critical point

$$q_C \approx \frac{k_B T}{16\bar{\eta}\bar{\lambda}}\frac{\rho c_P}{\xi^2} \tag{6.33}$$

and for liquid mixtures near the consolute point

$$q_C \approx \frac{k_B T}{16\bar{\eta}\bar{L}}\frac{\chi}{\xi^2} \tag{6.34}$$

On the other hand, the renormalization–group theory of dynamic critical phenomena predicts that η asymptotically diverge as ξ^z (Hohenberg & Halperin 1977). The two predictions can be reconciled if one considers equation (6.31) as a first approximation to the power law (6.23), with $z = 8/15\pi^2 = 0.054$ as a first–order estimate for the exponent z. Until recently, the theoretical estimates for the exponent z appeared to be consistently lower than the experimental values $z \approx 0.065$ (Bhattacharjee & Ferrell 1983b). But this problem is now resolved, and the most recent theoretical estimate is $z = 0.063$ (Hao 1991).

6.5 Global behavior of the transport properties in the critical region

Equations (6.22) and (6.23) represent the singular behavior of the thermal conductivity and the viscosity asymptotically close to the critical point. Unfortunately, for fluids near the vapor–liquid critical point the range of validity of these asymptotic equations is very limited. In fact, there are very few experimental thermal–conductivity data sufficiently close to the critical point to satisfy this asymptotic behavior. On the other hand, a critical enhancement of the thermal conductivity is observed in a large range of densities and temperatures around the critical point. As an example, in Figure 6.3 the range of reduced densities ρ/ρ_c and reduced temperatures T/T_c is shown where the critical enhancement $\Delta\lambda_c$ for carbon dioxide yields a contribution to the total thermal conductivity of more than 1% (Vesovic *et al.* 1990). It is seen that a critical enhancement is present at densities up to twice the critical density and at temperatures up to more than 45% above the critical temperature, which for CO_2 corresponds to a temperature range of about 145°C. To deal with the actual critical enhancement observed, a treatment in terms of the asymptotic Stokes–Einstein law (6.22) is inadequate and a nonasymptotic model is needed for the effects of the critical fluctuations on the thermal conductivity, including a crossover from asymptotic singular behavior at the critical point to regular background behavior

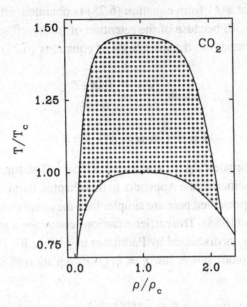

Fig. 6.3. Region in reduced temperature T/T_c and reduced density ρ/ρ_c where the ratio $\Delta\lambda_c/\lambda$ of CO_2 exceeds 1%.

far away from the critical point. A critical enhancement of the viscosity of fluids is observed in only a rather small range of densities and temperatures near the vapor–liquid critical point. But even for the viscosity the simple asymptotic power law (6.23) cannot account for the smooth transition between the divergence of the viscosity at the critical point and its regular background behavior away from the critical point (Basu & Sengers 1979). Thus one needs a global solution rather than an asymptotic solution of the mode–coupling equations (6.28) and (6.29). A preliminary solution for the viscosity was proposed by Bhattacharjee *et al.* (1981), and a more complete solution for both thermal conductivity and viscosity was obtained by Olchowy & Sengers (1988).

As a first step it is noted that a finite cutoff q_D should be retained in the evaluation of both mode–coupling integrals given by equations (6.28) and (6.29), since away from the critical point q_D^{-1} will no longer be negligible compared to the finite correlation length ξ. Here only a solution for the transport properties in the macroscopic hydrodynamic limit $k \to 0$ is needed. From the information presented in the previous section it appears that the effect of any wavenumber dependence of the viscosity on $\Delta a_c(k)$ through equation (6.28) becomes negligibly small in this limit. Hence, any wavenumber dependence of the viscosity η in equation (6.28) can be neglected. In the same vein, any wavenumber dependence of c_V in the relationship (6.9) between c_P and χ is neglected, since c_V has a weak singularity only. However, the thermal diffusivity a in the integrand of equation (6.28) can no longer be neglected. It is then possible to obtain an approximate solution of equations (6.28) and (6.29) by iteration (Olchowy 1989). That is, first a preliminary solution for $a(k)$ from equation (6.28) is obtained which differs from the asymptotic solution (6.30) because of the retention of the cutoff q_D. This preliminary solution is then substituted into the integrands of equations (6.28) and (6.29), and for $k \to 0$ it is found

$$\Delta a_c = \frac{\Delta \lambda_c}{\rho c_P} = \frac{R_D k_B T}{6\pi \eta \xi} \Omega \tag{6.35}$$

$$\eta = \bar{\eta} + \Delta \eta_c = \bar{\eta}(1 + zH) \tag{6.36}$$

with the first–order estimates $R_D = 1$ and $z = 8/15\pi^2$. The functions Ω and H are crossover functions specified in the Appendix to this chapter. It must be stressed that the equations for Ω and H presented here are simpler than the equations originally obtained by Olchowy & Sengers (1988). The earlier equations contained a matrix inversion that can be solved explicitly, as discussed by Perkins *et al.* (1991a,b).

To recover the asymptotic power law $\eta = \bar{\eta}(Q\xi)^z$, equation (6.36) is exponentiated to

$$\eta = \bar{\eta} \exp(zcH) \tag{6.37}$$

where the exponent z is identified with the most accurate theoretical value $z \approx 0.063$ currently available (Hao 1991). The constant c, given by $c^{-1} = 2 - (\alpha + \gamma)/2\nu \approx 0.93$ (Luettmer–Strathmann 1994), has up to now been approximated by unity.

To evaluate the crossover functions Ω and H, not only the specific heat capacities c_P and c_V are needed, but also the correlation length ξ. From equations (6.11) and (6.12) it follows that at $\rho = \rho_c$ asymptotically close to the critical temperature

$$\xi = \xi_0 \left(\Gamma^{-1} \chi^* \right)^{\nu/\gamma} \tag{6.38}$$

with $\chi^* = \chi P_c/\rho_c^2$. Reliable expressions for the nonasymptotic behavior of ξ are not available, and equation (6.38) has been used to estimate ξ as a function of ρ and T anywhere in the critical region. However, the crossover function H in equation (6.37) does not vanish when $q_D\xi = 1$, as it would be expected if ξ were the actual correlation length, but in the limit $q_D\xi \to 0$. Apparently ξ must be interpreted as the part of the correlation length that is associated with the long–range critical fluctuations only. To assure that ξ indeed becomes vanishingly small far away from the critical point, equation (6.38) is modified to (Olchowy & Sengers 1988)

$$\xi = \xi_0 \left(\Gamma^{-1} \Delta \chi^* \right)^{\nu/\gamma} \tag{6.39}$$

with

$$\Delta \chi^* = \chi^*(\rho, T) - \chi^*(\rho, T_r)T_r/T \tag{6.40}$$

where T_r is a reference temperature sufficiently far above T_c. The results are not sensitive to the value chosen for T_r; values ranging from $(3/2)T_c$ to $(5/2)T_c$ have been used in practice.

The critical enhancement Δa_c does not vanish in the limit $q_D\xi \to 0$. The reason is that there are also long–range dynamic contributions to the transport properties at dense fluid states far away from the critical point, commonly referred to as 'long–time–tail' contributions (Pomeau & Résibois 1975; Ernst & Dorfman 1972; Dorfman *et al.* 1994). Here such noncritical long–range contributions in the effective background contributions $\bar\lambda$ and $\bar\eta$ are to be included, so that the expression (6.35) for Δa_c vanishes far away from the critical point. This is accomplished by replacing equation (6.35) with

$$\Delta a_c = \frac{R_D k_B T}{6\pi \eta \xi} (\Omega - \Omega_0) \qquad (6.41)$$

where Ω_0 is an empirical subtraction term also included in the Appendix to this chapter. Slightly different forms for Ω_0 have also been used (Perkins *et al.* 1991a,b). Equation (6.41) implies for the critical enhancement of the thermal conductivity:

$$\Delta\lambda_c = \rho c_P \frac{R_D k_B T}{6\pi \eta \xi} (\Omega - \Omega_0) \qquad (6.42)$$

It should be noted that there are in principle also other mode–coupling effects, not included in equations (6.28) and (6.29), that may contribute to the transport properties in the critical region (Kawasaki 1976). However, the mode–coupling integrals (6.28) and (6.29) are the ones determining the dominant behavior of the transport properties in the critical region. Estimates of secondary effects from other mode–coupling contributions to the transport properties in the critical region are not yet available.

The theoretical expressions for the crossover behavior of the transport properties in the critical region have been used to represent experimental transport–property data for a variety of fluids including helium (Agosta *et al.* 1987), methane (Olchowy & Sengers 1989), carbon dioxide (Vesovic *et al.* 1990), ethane (Mostert *et al.* 1990; Hendl *et al.* 1994; Vesovic *et al.* 1994), oxygen (Laesecke *et al.* 1990), nitrogen (Perkins *et al.* 1991a), argon (Perkins *et al.* 1991b) and the refrigerant R134a (Krauss *et al.* 1993). In comparing the theory with experiment, q_D is the only adjustable parameter which, with one exception (Agosta *et al.* 1987), is taken to be the same for Δa_c or $\Delta\lambda_c$ and $\Delta\eta_c$. The comparison yields values for q_D^{-1} of a few Å, confirming that q_D^{-1} represents a microscopic cutoff distance. Examples of the application of the theory representing the transport properties of fluids in the critical region are presented in Chapter 14 of this book. In Figures 6.4 and 6.5 we show a comparison with experimental thermal–diffusivity data (Becker & Grigull 1978) and thermal–conductivity data (Michels & Sengers 1962) for CO_2, where the crossover equation for $\Delta\lambda_c$ is supplemented with suitable equations for $\bar\lambda$ and $\bar\eta$ (Vesovic *et al.* 1990). A comparison of the theory with viscosity data is difficult, because $\Delta\eta_c$ is very small and, therefore, sensitive to the choice made for the background viscosity $\bar\eta$. Because of inconsistencies in reported experimental viscosity data, it has not been possible to determine $\bar\eta$ from data outside

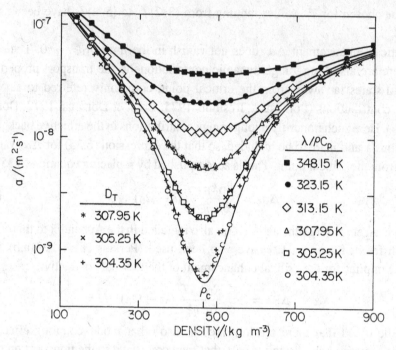

Fig. 6.4. The thermal diffusivity of CO_2 in the critical region. The symbols indicate experimental data, and the curves represent calculated values of $a = \bar{a} + \Delta a_c$ (Vesovic *et al.* 1990).

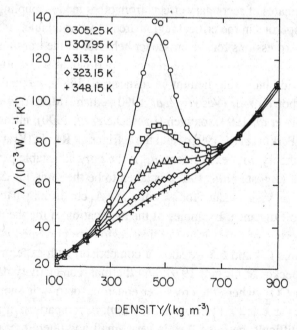

Fig. 6.5. The thermal conductivity of CO_2 in the critical region. The symbols indicate experimental data, and the curves represent calculated values of $\lambda = \bar{\lambda} + \Delta \lambda_c$ (Vesovic *et al.* 1990).

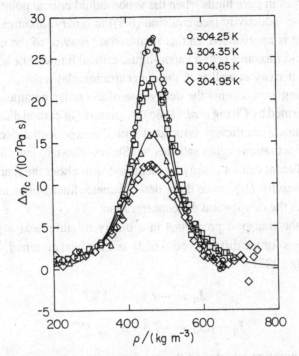

Fig. 6.6. Critical viscosity enhancement $\Delta\eta_c$ of CO_2 deduced from data obtained by Iwasaki and Takahashi (1981). The curves represent calculated values of $\Delta\eta_c$ (Vesovic *et al.* 1990).

the critical region with sufficient accuracy (Vesovic *et al.* 1990, 1994). In practice it has been necessary to represent $\bar{\eta}$ in equation (6.37) simultaneously by a function with adjustable parameters. If such a procedure is adopted, the results shown in Figure 6.6 for $\Delta\eta_c$ of CO_2 are obtained (Vesovic *et al.* 1990), when compared with experimental viscosity data reported by Iwasaki & Takahashi (1981).

The equation (6.42) for the critical enhancement $\Delta\lambda_c$ is based on the mode–coupling theory of the critical fluctuations and contains only one adjustable parameter q_D. Empirical equations for $\Delta\lambda_c$ containing more adjustable parameters have also been proposed (Sengers *et al.* 1984; Sengers 1985; Roder 1985; Roder *et al.* 1989; Perkins *et al.* 1991b). An example of such an empirical equation for $\Delta\lambda_c$ of argon is presented in Chapter 14 of this volume. Empirical equations with an adequate number of adjustable parameters can be used to represent sets of experimental thermal–conductivity data. However, they cannot be used to predict the thermal conductivity of fluids in the critical region from a limited data set.

6.6 Compressible mixtures near the critical line

In the preceding sections it was pointed out that pure fluids and binary fluid mixtures near the consolute point belong to the same universality class. Accordingly, the thermal conductivity diverges in pure fluids when the vapor–liquid critical point is approached, whereas the mass–conductivity (see equation (6.8)) in binary mixtures diverges when the consolute point is approached. In this section the behavior of the transport coefficients in binary fluid mixtures near a vapor–liquid critical line will be addressed where fluctuations in the density as well as in the concentration play a role.

In light–scattering experiments the decay rate of the order parameter is measured. Experiments performed by Chang *et al.* (1986) on mixtures of carbon dioxide and ethane show that the diffusion coefficient associated with the exponential decay (6.5) of the order–parameter fluctuations again satisfies the Stokes–Einstein law (6.21). However, the diffusion coefficient cannot simply be identified with either the thermal diffusivity *a* or the mass diffusivity D_{12}, since the order–parameter fluctuations now incorporate fluctuations in both the density and the concentration.

To investigate the transport properties in a binary mixture near a plait point, the Onsager expressions for the diffusion current $\mathbf{J_d}$ and the heat current $\mathbf{J_q}$ are considered (Landau & Lifshitz 1987)

$$\mathbf{J_d} = -\alpha \, \nabla\mu - \beta \, \nabla T \tag{6.43}$$

$$\mathbf{J_q} - \mu\mathbf{J_d} = -T\beta \, \nabla\mu - \gamma \, \nabla T \tag{6.44}$$

In contrast to the notation adopted in the previous section, $\mu = \mu_1 - \mu_2$ is now the difference of the specific chemical potentials of the two components, while α, β and γ are kinetic coefficients, not to be confused with the critical exponents introduced earlier. The coefficient α is related to the mass diffusivity D_{12} through

$$D_{12} = \frac{\alpha}{\rho} \left(\frac{\partial\mu}{\partial x} \right)_{T,P} \tag{6.45}$$

where ρ is the total mass density of the mixture and x is the mass fraction of component 1. A comparison with equation (6.8) shows that α is related to the mass conductivity $L = \alpha/\rho$. The coefficients β and γ are not directly measurable, but they are connected with experimentally accessible quantities. The thermal diffusion coefficient $k_T D_{12}$, where k_T is the thermal diffusion ratio, is related to α and β through

$$k_T D_{12} = \frac{T}{\rho} \left[\alpha \left(\frac{\partial\mu}{\partial T} \right)_{x,P} + \beta \right] \tag{6.46}$$

Finally, the thermal conductivity is defined by $\mathbf{J_q} = -\lambda\nabla T$, with $\mathbf{J_d} = 0$, and, therefore, it is given by

$$\lambda = \gamma - T\beta^2/\alpha \tag{6.47}$$

In analogy with equation (6.19) each of the coefficients introduced in equations (6.43) and (6.44) is written as a sum of a regular background contribution and a critical enhancement, that is, $\alpha = \bar{\alpha} + \Delta\alpha_c$, etc. Since the mode–coupling theory provides information about the critical part of diffusion coefficients, the diffusion coefficients associated with the transport coefficients β and γ are defined as

$$D_\beta = \frac{\beta}{\rho} \left(\frac{\partial T}{\partial x} \right)_{\mu, P} \tag{6.48}$$

$$D_\gamma = \frac{\gamma}{\rho T} \left(\frac{\partial T}{\partial s} \right)_{\mu, P} \tag{6.49}$$

where s is the specific entropy. The critical contributions to all these diffusion coefficients have been predicted (Mistura 1972) to follow the Stokes–Einstein diffusion law equation (6.21) in the asymptotic critical region. Therefore, the kinetic coefficients can be expressed as

$$\alpha = \frac{k_B T \rho}{6 \pi \eta \xi} \left(\frac{\partial x}{\partial \mu} \right)_{T, P} + \bar{\alpha} \tag{6.50}$$

$$\beta = \frac{k_B T \rho}{6 \pi \eta \xi} \left(\frac{\partial x}{\partial T} \right)_{\mu, P} + \bar{\beta} \tag{6.51}$$

$$\gamma = \frac{k_B T^2 \rho}{6 \pi \eta \xi} \left(\frac{\partial s}{\partial T} \right)_{\mu, P} + \bar{\gamma} \tag{6.52}$$

where the amplitude R_D has been approximated by unity. Since, according to Griffiths & Wheeler (1970), the derivative of a density variable with respect to a field variable diverges strongly, with an exponent γ, when two field variables are kept constant, all the kinetic coefficients are expected to diverge with an exponent $\gamma - \nu(1 + z) \approx \nu$. Next it is observed that the critical enhancements of α and β are not independent in the asymptotic region but are related through $\Delta \beta_c = -(\partial \mu / \partial T)_{x, P} \Delta \alpha_c$, where the derivative $(\partial \mu / \partial T)_{x, P}$ stays finite at the critical point (Anisimov & Kiselev 1992). Hence, the combination $\Delta \beta_c + (\partial \mu / \partial T)_{x, P} \Delta \alpha_c$ vanishes and the thermal diffusion coefficient $k_T D$ shows no critical enhancement. This result is confirmed by dynamic renormalization–group calculations (Siggia et al. 1976; Folk & Moser 1993).

Mistura (1972) went on to estimate the enhancement of the thermal conductivity by inserting the critical enhancements of the transport coefficients into equation (6.47) and concluded that the enhancement vanishes at the critical point. When Mostert & Sengers (1992) and Anisimov & Kiselev (1992) included the background effects, a qualitatively different result was obtained, namely, that the enhancement of the thermal conductivity is finite at the critical point.

To determine the asymptotic behavior of the thermal conductivity λ, equations (6.50)– (6.52) are inserted into equation (6.47) and the background is isolated by defining

$$\bar{\lambda} = \bar{\gamma} - T \bar{\beta}^2 / \bar{\alpha} \tag{6.53}$$

The critical contribution to λ is then found by expanding the remaining expression in powers of $\bar{\alpha}/\Delta\alpha_c$. To leading order, this means

$$\Delta\lambda_c = \frac{k_B T \rho}{6\pi\eta\xi} c_{P,x} + \left(\rho\bar{k}_T\bar{D}_{12}\right)^2 / T\bar{\alpha} \tag{6.54}$$

where $\bar{k}_T\bar{D}_{12}$ is the background value of the thermal diffusion coefficient (equation (6.46)) and $c_{P,x} = T(\partial s/\partial T)_{\mu,P} - T(\partial\mu/\partial T)_{x,P}^2(\partial x/\partial\mu)_{T,P}$ was applied. The second term in equation (6.54) goes to a constant at the critical point, whereas the first term vanishes, since the isobaric specific heat at constant concentration $c_{P,x}$ is expected to diverge only weakly (Griffiths & Wheeler 1970). Thus the critical enhancement of the thermal conductivity is finite at the critical point and determined by the background values of the thermal diffusion ratio and the mutual diffusion coefficient.

It should be pointed out here that the asymptotic description of the thermal conductivity is valid only extremely close to the critical point. Measurements on ^{3}He + ^{4}He (Cohen *et al.* 1982), methane + ethane (Friend & Roder 1985; Roder & Friend 1985) and CO_2 + ethane (Mostert *et al.* 1992) seem to indicate that the thermal conductivity exhibits a critical enhancement similar to that observed for pure fluids. In Figure 6.7, as an example the experimental thermal–conductivity results for CO_2 + ethane for a mole fraction of 25% CO_2 in the one–phase region close to the critical isochore are presented, which were obtained by Mostert (1991). To reconcile the experimental data with the asymptotic result of equation (6.54), again a crossover theory is needed. Thermophysical quantities in fluid mixtures near a plait point undergo two types of crossover as the

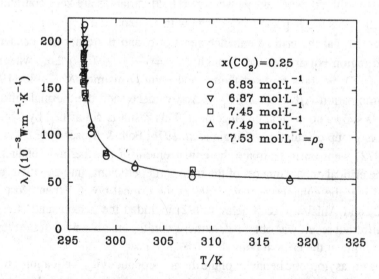

Fig. 6.7. The thermal conductivity in a mixture of ethane with CO_2 ($x_{CO_2} = 0.25$). The symbols indicate the experimental data by Mostert (1991), the solid line is obtained from a mode–coupling calculation (Luettmer–Strathmann & Sengers 1994).

critical point is approached: from classical behavior without critical contributions to pure–fluid–like critical enhancements, and then, when the critical point is approached even more closely, from pure–fluid–like critical behavior to mixture–like critical behavior. This phenomenon is well documented for the static properties of a fluid (Anisimov & Kiselev 1992; Jin *et al.* 1993) and is also predicted for transport properties, albeit with different crossover temperatures (Onuki 1985; Kiselev & Kulikov 1994). Investigations of the crossover of the thermal conductivity in mixtures of carbon dioxide and ethane (Luettmer–Strathmann 1994; Luettmer–Strathmann & Sengers 1994) indicate that for the data presented in Figure 6.7 the thermal conductivity behaves like that of a pure fluid for reduced temperatures up to 10^{-4} and is approaching its asymptotic value only for reduced temperatures smaller than 10^{-7}.

For the mass–diffusion coefficient D_{12}, insertion of (6.50) into equation (6.45) yields

$$D_{12} = \frac{k_B T \rho}{6 \pi \eta \xi} + \frac{\bar{\alpha}}{\rho} \left(\frac{\partial \mu}{\partial x} \right)_{T,P} \tag{6.55}$$

Hence, the behavior of the mass–diffusion coefficient of a mixture in the critical region is controlled by the behavior of $(\partial \mu / \partial x)_{T,P} = \chi^{-1}$. As pointed out in Section 6.1, in binary mixtures the osmotic susceptibility χ diverges near a consolute point and leads to the critical slowing down of the mass diffusion. Near a vapor–liquid critical point, on the other hand, where the thermodynamic properties undergo a crossover from pure–fluid–like behavior before they display their asymptotic mixture behavior (Jin *et al.* 1993), the osmotic susceptibility does not exhibit a critical behavior except at temperatures very close to the plait–point temperature (for the system mentioned above, the reduced temperature has to be smaller than 5×10^{-3}). Therefore, not too close to the critical point, the mutual diffusivity is dominated by its background value $\bar{\alpha}/(\rho \chi)$, and the critical slowing down that follows the Stokes–Einstein diffusion law is not seen in the mass diffusion coefficient.

Both renormalization–group as well as mode–coupling calculations predict a weak divergence of the viscosity, which has been observed by Wang & Meyer (1987) for mixtures of ^{3}He–^{4}He near the vapor–liquid critical line. The viscosity is found to behave just like that of a pure fluid; that is, the viscosity exhibits a weak divergence in accordance with equation (6.23).

Several theoretical attempts have been made recently to give a more complete description of the transport coefficients of mixtures in the critical region. Folk & Moser (1993) performed dynamic renormalization–group calculations for binary mixtures near plait points and obtained nonasymptotic expressions for the kinetic coefficients. Kiselev & Kulikov (1994) derived phenomenological crossover functions for the transport coefficients by factorizing the Kubo formulas for the transport coefficients α, β and γ. This approach is referred to as the decoupled–mode approximation (Ferrell 1970). Their calculations yield for the thermal conductivity the expected finite enhancement in the asymptotic critical region and also a smooth crossover to the background far away from the critical point. The solid curve in Figure 6.7 represents the first results of a crossover theory for the transport properties in binary fluid mixtures which were derived from the mode–coupling theory (Luettmer–Strathmann 1994; Luettmer–Strathmann & Sengers 1994). In the extreme critical region this crossover model reproduces the asymptotic laws described in the text, and far away from the critical point the model reduces to the background values of the transport properties.

Appendix to Chapter 6

Auxiliary functions

$$y_D = \arctan(q_D\xi),$$

$$y_\delta = \left\{\arctan(q_D\xi/(1 + q_D^2\xi^2)^{1/2}) - y_D\right\} \bigg/ \left(1 + q_D^2\xi^2\right)^{1/2}$$

$$y_\alpha = \rho k_B T \big/ 8\pi \bar{\eta}^2 \xi ,$$

$$y_\beta = \bar{\lambda} \big/ \bar{\eta}(c_P - c_V) ,$$

$$y_\gamma = c_V \big/ (c_P - c_V) ,$$

$$y_\eta = (y_\delta + y_\beta/y_\alpha) \big/ y_D ,$$

$$y_\nu = y_\gamma y_\delta \big/ y_D$$

$$F(x, y_D) = (1 - x^2)^{-1/2} \ln \left[\frac{1 + x + (1 - x^2)^{1/2} \tan(y_D/2)}{1 + x - (1 - x^2)^{1/2} \tan(y_D/2)}\right]$$

The crossover function $\Omega - \Omega_0$

$$\Omega(\{y_i\}) = \frac{2}{\pi} \frac{1}{1 + y_\gamma} \left[y_D - \sum_{i=1}^{4} \frac{(a_3 z_i^3 + a_2 z_i^2 + a_1 z_i + a_0)}{\prod_{j=1, j\neq i}^{4}(z_i - z_j)} F(z_i, y_D)\right]$$

$$\Omega_0 = \frac{2}{\pi} \frac{1 - \exp\left(-\left[(q_D\xi)^{-1} + (q_D\xi\rho_c/\rho)^2/3\right]^{-1}\right)}{1 + y_\alpha(y_D + y_\delta) + y_\beta(1 + y_\gamma)^{-1}}$$

$$\prod_{i=1}^{4}(z + z_i) = z^4 + b_3 z^3 + b_2 z^2 + b_1 z + b_0 = 0$$

$a_0 = y_\gamma^2 - y_\alpha y_\gamma y_\delta$	$b_0 = y_\alpha y_\gamma y_\delta$
$a_1 = y_\alpha y_\gamma y_D$	$b_1 = y_\alpha y_\gamma y_D$
$a_2 = y_\gamma - y_\beta - y_\alpha y_\delta$	$b_2 = y_\gamma + y_\beta + y_\alpha y_\delta$
$a_3 = y_\alpha y_D$	$b_3 = y_\alpha y_D$

The crossover function H

$$H(\{y_i\}) = h(\{y_i\}) + \sum_{i=1}^{3} \frac{(c_2 v_i^2 + c_1 v_i + c_0)}{\prod_{j=1, j \neq i}^{3} (v_i - v_j)} F(v_i, y_D)$$

$$h(\{y_i\}) = \left[3 y_\gamma y_\eta + 3 y_\eta/2 - y_\eta^3 - y_\nu\right] y_D + \left[y_\eta^2 - 2 y_\gamma - 5/4\right] \sin y_D$$
$$- y_\eta (\sin 2 y_D)/4 + (\sin 3 y_D)/12$$
$$+ \frac{[y_\gamma (1 + y_\gamma)]^{3/2}}{(y_\nu - y_\gamma y_\eta)} \arctan \left([y_\gamma/(1 + y_\gamma)]^{1/2} \tan y_D\right)$$

$$\prod_{i=1}^{3} (v + v_i) = v^3 + y_\eta v^2 + y_\gamma v + y_\nu = 0,$$

$$c_0 = y_\gamma y_\nu (y_\eta^2 - 3 y_\gamma - 2) + y_\nu^2 - y_\gamma y_\nu (1 + y_\gamma)^2/(y_\nu - y_\gamma y_\eta)$$

$$c_1 = (y_\nu - y_\gamma y_\eta)(y_\eta^2 - 3 y_\gamma - 2) + y_\gamma^2 (1 + y_\gamma)^2/(y_\nu - y_\gamma y_\eta),$$

$$c_2 = y_\eta^4 - 4 y_\gamma y_\eta^2 - 2 y_\eta^2 + 2 y_\eta y_\nu + 3 y_\gamma^2 + 4 y_\gamma + 1$$

Acknowledgments

The authors acknowledge valuable discussions with M. A. Anisimov, S. B. Kiselev, R. Mostert and W. A. Wakeham. The research is supported by the Division of Chemical Sciences of the Office of Basic Energy Science of the U.S. Department of Energy under Grant DE–FG05–88ER13902.

References

Agosta, C. C., Wang, S., Cohen, L. H., & Meyer, H. (1987). Transport properties of helium near the liquid–vapour critical point IV. The shear viscosity of ^{3}He and ^{4}He. *J. Low Temp. Phys.*, **67**, 237–289.

Anisimov, M. A. & Kiselev, S. B. (1992). Transport properties of critical dilute solutions. *Int. J. Thermophys.*, **13**, 873–893.

Basu, R. S. & Sengers, J. V. (1979). Viscosity of nitrogen near the critical point. *J. Heat Transfer, Trans. ASME*, **101**, 3, 575–578.

Becker, H. & Grigull, U. (1978). Messung der Temperatur- und der Wärmeleitfähigkeit von Kohlendioxid im kritischen Gebiet mittels holographischer Interferometrie. *Wärme–Stoffübertragung*, **11**, 9–28.

Berg, R. F. & Moldover, M. R. (1988). Critical exponent for the viscosity of four binary liquids. *J. Chem. Phys.*, **89**, 3694–3704.

Berg, R. F. & Moldover, M. R. (1990). Critical exponent for the viscosity of carbon dioxide and xenon. *J. Chem. Phys.*, **93**, 1926–1938.

Bhattacharjee, J. K. & Ferrell, R. A. (1983a). Frequency–dependent critical viscosity of a classical fluid. *Phys. Rev. A*, **27**, 1544–1555.

Bhattacharjee, J. K. & Ferrell, R. A. (1983b). Critical viscosity exponent for a classical fluid. *Phys. Rev. A*, **28**, 2363–2369.

Bhattacharjee, J. K., Ferrell, R. A., Basu, R. S. & Sengers, J. V. (1981). Crossover functions for the critical viscosity of a classical fluid. *Phys. Rev. A*, **24**, 1469–1475.

Burstyn, H. C. & Sengers, J. V. (1982). Decay rate of critical fluctuations in a binary liquid. *Phys. Rev. A*, **25**, 448–465.

Burstyn, H. C. & Sengers, J. V. (1983). Time dependence of critical concentration fluctuations in a binary liquid. *Phys. Rev. A*, **27**, 1071–1085.

Burstyn, H. C., Sengers, J. V., Bhattacharjee, J. K. & Ferrell, R. A. (1983). Dynamic scaling function for critical fluctuations in classical fluids. *Phys. Rev. A*, **28**, 1567–1578.

Burstyn, H. C., Sengers, J. V. & Esfandiari, P. (1980). Stokes–Einstein diffusion of critical fluctuations. *Phys. Rev. A*, **22**, 282–284.

Chang, R. F., Burstyn, H. C. & Sengers, J. V. (1979). Correlation function near the critical mixing point of a binary liquid. *Phys. Rev. A*, **19**, 866–882.

Chang, R. F., Doiron, T. & Pegg, I. L. (1986). Decay rate of critical fluctuations in ethane + carbon dioxide mixtures near the critical line including the critical azeotrope. *Int. J. Thermophys.*, **7**, 295–304.

Chu, B. (1974). *Laser Light Scattering*. New York: Academic.

Cohen, L. H., Dingus, M. L. & Meyer, H. (1982). Transport properties of helium near the liquid–vapour critical point. II. Thermal conductivity of a ^{3}He–^{4}He mixture. *J. Low Temp. Phys.*, **49**, 545–558.

Dorfman, J. R., Kirkpatrick, T. R. & Sengers, J. V. (1994). Generic long–range correlations in molecular fluids. *Ann. Rev. Phys. Chem.*, **45**, 213–239.

Ernst, M. H. & Dorfman, J. R. (1972). Nonanalytic dispersion relations for classical fluids II. The general fluid. *J. Stat. Phys.*, **12**, 311–359.

Felderhof, B. U. (1966). Onsager relations & the spectrum of critical opalescence. *J. Chem. Phys.*, **44**, 602–609.

Ferrell, R. A. (1970). Decoupled–mode dynamical scaling theory of the binary–liquid phase transitions. *Phys. Rev. Lett.*, **24**, 1169–1172.

Fisher, M. E. (1964). Correlation functions and the critical region of simple fluids. *J. Math. Phys.*, **5**, 944–962.

Fixman, M. (1962). Viscosity of critical mixtures. *J. Chem. Phys.*, **36**, 310–318.

Fixman, M. (1964). The critical region. *Adv. Chem. Phys.*, **6**, 175–228.

Fixman, M. (1967). Transport coefficients in the gas critical region. *J. Chem. Phys.*, **47**, 2808–2818.

Folk, R. & Moser, G. (1993). Critical dynamics near the liquid vapour transition in mixtures. *Europhys. Lett.*, **24**, 533–538.

Forster, D. (1975). *Hydrodynamic Fluctuations, Broken Symmetry and Correlation Functions*. Reading, MA: Benjamin.

Friend, D. G. & Roder, H. M. (1985). Thermal–conductivity enhancement near the liquid–vapour critical line of binary methane–ethane mixtures. *Phys. Rev. A*, **32**, 1941–1944.

Griffiths, R. B. & Wheeler, J. C. (1970). Critical points in multicomponent systems. *Phys. Rev. A*, **2**, 1047–1064.

Gunton, J. D. (1979). Mode–coupling theory in relation to the dynamical renormalisation–group method, in *Dynamical Critical Phenomena & Related Topics, Vol. 104 of Lecture Notes in Physics*, ed. C.P. Enz. pp. 1–13. Berlin: Springer.

Güttinger, H. & Cannell, D. S. (1980). Correlation range & Rayleigh linewidth near the critical point. *Phys. Rev. A*, **22**, 285–286.

Hamano, K., Kawazura, T, Koyama, T. & Kuwahara, N. (1985). Dynamics of concentration fluctuations for butylcellosolve in water. *J. Chem. Phys.*, **82**, 2718–2722.

Hamano, K., Teshigawara, S., Koyama, T. & Kuwahara, N. (1986). Dynamics of concentration fluctuations along the coexistence curve of a binary mixture. *Phys. Rev. A*, **33**, 485–489.

Hao, H. (1991). *Aspects of dynamic critical phenomena in a single component fluid*. Ph.D. Thesis. College Park, MD: University of Maryland.

Hendl, S., Millat, J., Vogel, E., Vesovic, V., Wakeham, W. A., Luettmer–Strathmann, J., Sengers, J. V. & Assael, M. J. (1994). The transport properties of ethane I. Viscosity. *Int. J. Thermophys.*, **15**, 1–32.

Hohenberg, P. C. & Halperin, B. I. (1977). Theory of dynamic critical phenomena. *Rev. Mod. Phys.*, **49**, 435–479.

Iwasaki, H & Takahashi, M. (1981). Viscosity of carbon dioxide & ethane. *J. Chem. Phys.*, **74** 1930–1943.

Jin, G. X., Tang, S. & Sengers, J. V. (1993). Global thermodynamic behaviour of fluid mixtures in the critical region. *Phys. Rev. E*, **47**, 388–402.

Kadanoff, L. P. & Swift, J. (1968). Transport coefficients near the liquid-gas critical point. *Phys. Rev.*, **166**, 89–101.

Kawasaki, K. (1966). Correlation–function approach to the transport coefficients near the critical point I. *Phys. Rev.*, **150**, 291–306.

Kawasaki, K. (1970). Kinetic equations and time correlation functions of critical fluctuations. *Ann. Phys.*, **61**, 1–56.

Kawasaki, K. (1976). Mode coupling & critical dynamics, in *Phase Transitions & Critical Phenomena*, eds. C. Domb & M. S. Green, pp. 165–403. New York: Academic Press.

Keyes, T. (1977). Principles of mode–mode coupling theory, in *Statistical Mechanics Part B: Time–Dependent Processes*, ed. B. J. Berne, pp. 259–309. New York: Plenum Press.

Kiselev, S. B. & Kulikov, V. D. (1994) Crossover behaviour of the transport coefficients of critical binary mixtures. *Int. J. Thermophys.*, **15**, 283–308.

Krall, A. H., Sengers, J. V. & Hamano, K. (1989). Viscosity of a phase–separating critical mixture. *Int. J. Thermophys.*, **10**, 309–319.

Krauss, R., Luettmer–Strathmann, J., Sengers, J. V. & Stephan, K. (1993). Transport properties of 1,1,1,2–tetrafluorethane (R134a). *Int. J. Thermophys.*, **14**, 951–988.

Kumar, A., Krishnamurthy, H. R. & Gopal, E. S. R. (1983). Equilibrium critical phenomena in binary liquid mixtures. *Phys. Rep.*, **98**, 57–143.

Laesecke, A., Krauss, R., Stephan, K. & Wagner, W. (1990). Transport properties of fluid oxygen. *J. Phys. Chem. Ref. Data*, **19**, 1089–1121.

Landau, L. D. & Lifshitz, E. M. (1987). *Fluid Mechanics*, Chapter VI. Oxford: Pergamon Press.

Levelt Sengers, J. M. H. & Sengers, J. V. (1981). How close is 'Close to the critical point,' in *Perspectives in Statistical Physics*, ed. H.J. Raveché, pp. 239–271. Amsterdam: North–Holland.

Liu, A. J. & Fisher, M. E. (1989). The three–dimensional Ising model revisited. *Physica A*, **156**, 35–76.

Luettmer–Strathmann, J. (1994). *Transport properties of fluids and fluid mixtures in the critical region*. Ph.D. Thesis. College Park, MD: University of Maryland.

Luettmer–Strathmann, J. & Sengers, J. V. (1994). Transport properties of fluid mixtures in the critical region. *Int. J. Thermophys.*, **15**, 1241–1249.

Matos Lopes, M. L. S., Nieto de Castro, C. A., Sengers, J. V. (1992). Mutual diffusivity of a mixture of *n*–hexane and nitrobenzene near its consolute point. *Int. J. Thermophys.*, **13**, 283–294.

McIntyre, D. & Sengers, J. V. (1968). Study of fluids by light scattering, in *Physics of Simple Liquids*, eds. H. N. V. Temperly, J. S. Rowlinson & G. S. Rushbrooke, pp. 447–505. Amsterdam: North–Holland.

Michels, A. & Sengers, J. V. (1962). The thermal conductivity of carbon dioxide in the critical region. Part II. *Physica*, **28**, 1216–1237.

Mistura, L. (1972). Transport coefficients near a critical point in multicomponent fluid systems. *Nuovo Cimento*, **12**, 35–42.

Moldover, M. R., Sengers, J. V., Gammon, R. W. & Hocken, R. J. (1979). Gravity effects in fluids near the gas–liquid critical point. *Rev. Mod. Phys.*, **51**, 79–99.

Mostert, R. (1991). *The thermal conductivity of ethane and of its mixtures with carbon dioxide in the critical region.* Ph.D. Thesis. Amsterdam, The Netherlands: Universiteit van Amsterdam.

Mostert, R. & Sengers, J. V. (1992). Asymptotic behaviour of the thermal conductivity of a binary fluid near the plait point. *Fluid Phase Equil.*, **75**, 235–244; **85**, 347 (1993).

Mostert, R., van den Berg, H. R., van der Gulik, P. S., & Sengers, J. V. (1990). The thermal conductivity of ethane in the critical region. *J. Chem. Phys.*, **92**, 5454–5462.

Mostert, R., van den Berg, H. R., van der Gulik, P. S., & Sengers, J. V. (1992). The thermal conductivity of carbon dioxide–ethane mixtures in the critical region. *High Temp.–High Press.*, **24**, 469–474.

Nieuwoudt, J. C. & Sengers, J. V. (1987). Frequency dependence of the transport properties of fluids near the critical point. *Physica A*, **147**, 368–386.

Nieuwoudt, J. C. & Sengers, J. V. (1989). A re–evaluation of the viscosity exponent for binary mixtures near the consolute point. *J. Chem. Phys.*, **90**, 457–482.

Ohta, T. (1977). Multiplicative renormalisation of the anomalous shear viscosity in classical liquids. *J. Phys. C*, **10**, 791–793.

Olchowy, G. A. (1989). *Crossover from singular to regular behaviour of the transport properties of fluids in the critical region.* Ph.D. Thesis. College Park, MD: University of Maryland.

Olchowy, G. A. & Sengers, J. V. (1988). Crossover from singular to regular behaviour of the transport properties of fluids in the critical region. *Phys. Rev. Lett.*, **61**, 15–18.

Olchowy G. A. & Sengers, J. V. (1989). A simplified representation for the thermal conductivity of fluids in the critical region. *Int. J. Thermophys.*, **10**, 417–426.

Onuki, A. (1985). Statics and dynamics in binary mixtures near the liquid–vapour critical line. *J. Low Temp. Phys.*, **61**, 101–139.

Oxtoby, D. W. (1975). Nonlinear effects in the shear viscosity of fluids near the critical point. *J. Chem. Phys.*, **62**, 1463–1468.

Perkins, R. A., Friend, D. G., Roder, H. M. & Nieto de Castro, C. A. (1991a). Thermal conductivity surface of argon: A fresh analysis. *Int. J. Thermophys.*, **12**, 965–983.

Perkins, R. A., Roder, H. M., Friend, D. G. & Nieto de Castro, C. A. (1991b). Thermal conductivity and heat capacity of fluid nitrogen. *Physica A*, **173**, 332–362.

Perl, R. & Ferrell, R. A. (1972). Decoupled mode theory of critical viscosity and diffusion in the binary–liquid phase transition. *Phys. Rev. A*, **6**, 2358–2369.

Pomeau, Y. & Résibois, P. (1975). Time dependent correlation functions and mode–coupling theories. *Phys. Rep.*, **19**, 63–139.

Roder, H. M. (1985). Thermal conductivity of methane for temperatures between 110 and 310 K with pressures up to 70 MPa. *Int. J. Thermophys.*, **6**, 119–142.

Roder, H. M. & Friend, D. G. (1985). Thermal conductivity of methane–ethane mixtures at temperatures between 140 and 330 K and pressures up to 70 MPa. *Int. J. Thermophys.*, **6**, 607–617.

Roder, H. M., Perkins, R. A. & Nieto de Castro, C. A. (1989). The thermal conductivity and heat capacity of gaseous argon. *Int. J. Thermophys.*, **10**, 1141–1164.

Sengers, J. V. (1966). Behaviour of viscosity and thermal conductivity of fluids near the critical point, in *Critical Phenomena, Proceedings of a Conference, NBS Miscellaneous*

Publ. 273., ed. M. S. Green & J. V. Sengers, pp. 165–178. Washington DC: US Gov't Printing Office.

Sengers, J. V. (1971). Transport properties of fluids near critical points, in *Critical Phenomena, Varenna Lectures, Course LI.*, eds. M. S. Green, pp. 445–507. New York: Academic Press.

Sengers, J. V. (1972). Transport processes near the critical point of gases and binary liquids in the hydrodynamic regime. *Ber. Bunsenges. Phys. Chem.*, **76**, 234–249.

Sengers, J. V. (1985). Transport properties of fluids near critical points. *Int. J. Thermophys.*, **6**, 203–232.

Sengers, J. V. (1994). Effects of critical fluctuations on the thermodynamic and transport properties of supercritical fluids, in *Supercritical Fluids: Fundamentals for Applications*, eds. E. Kiran & J. M. H. Levelt Sengers, pp. 231–271. Dordrecht: Kluwer.

Sengers, J. V. & Keyes, P. H. (1971). Scaling of the thermal conductivity near the gas–liquid critical point. *Phys. Rev. Lett.*, **26**, 70–73.

Sengers, J. V. & Levelt Sengers, J. M. H. (1978). Critical phenomena in classical fluids, in *Progress in Liquid Physics*, ed. C. A. Croxton, pp. 103–174. New York: Wiley.

Sengers, J. V. & Levelt Sengers, J. M. H. (1986). Thermodynamic behaviour of fluids near the critical point. *Ann. Rev. Phys. Chem.*, **37**, 189–222.

Sengers, J. V. & Michels, A (1962). The thermal conductivity of carbon dioxide in the critical region, in *Proceedings Second Symposium on Thermophysical Properties*, eds. J. F. Masi & D. H. Tsai, pp. 434–440. New York: American Society of Mechanical Engineers.

Sengers, J. V., Watson, J. T. R., Basu, R. S., Kamgar–Parsi, B. & Hendricks, R. C. (1984). Representative equations for the thermal conductivity of water substance. *J. Phys. Chem. Ref. Data*, **13**, 893–933.

Siggia, E. D., Halperin, B. I. & Hohenberg, P. C. (1976). Renormalisation group treatment of the critical dynamics of the binary–fluid and gas–liquid transitions. *Phys. Rev. B*, **13**, 2110–2122.

Swinney, H. L. & Henry, D. L. (1973). Dynamics of fluids near the critical point: Decay rate of the order–parameter fluctuations. *Phys. Rev. A*, **8**, 2586–2617.

Van Hove, L. (1954). Time–dependent correlations between spins and neutron scattering in ferromagnetic crystals. *Phys. Rev.*, **95**, 1374–1384.

Vesovic, V., Wakeham, W. A., Luettmer–Strathmann, J., Sengers, J. V., Millat, J., Vogel, E. & Assael, M. J. (1994). The transport properties of ethane II. Thermal conductivity. *Int. J. Thermophys.*, **15**, 33–66.

Vesovic, V., Wakeham, W. A., Olchowy, G. A., Sengers, J. V., Watson, J. T. R. & Millat, J. (1990). The transport properties of carbon dioxide. *J. Phys. Chem. Ref. Data*, **19**, 763–808.

Wang, S. & Meyer, H. (1987). Transport properties of helium near the liquid–vapour critical point V. The shear viscosity of ^{3}He–^{4}He mixtures. *J. Low Temp. Phys.*, **69**, 377–390.

Wu, G., Fiebig, M. & Leipertz, A. (1988). Messung des binären Diffusionskoeffizienten in einem Entmischungssystem mit Hilfe der Photonen–Korrelationsspektroskopie. *Wärme–Stoffübertragung*, **22**, 365–371.

Part three

DATA REPRESENTATION

7

Correlation Techniques*

D. G. FRIEND and R. A. PERKINS

Thermophysics Division
National Institute of Standards and Technology
Boulder, CO, USA

7.1 Introduction

The motivation for establishing accurate values of the various transport properties has been discussed in previous chapters. The current status of fundamental kinetic and statistical mechanical theory, computer simulation and experimental technique and data acquisition impose severe limits on the accuracies achievable in any description of a transport property surface for a real fluid. For instance, it will not be possible to determine viscosities to one part in 10^5 for the near future. Thus, the accuracies associated with primary standards for transport properties fall an order of magnitude below those associated with primary measurement standards for equilibrium thermodynamic properties. Fortunately, the technological applications of transport property information do not require extreme accuracies.

The most accurate description of a transport property surface can be determined from a thorough critical evaluation of abundant data, preferably obtained with a variety of techniques, using independently obtained samples, and from several laboratories. A correlation of such data, that is, a closed algebraic equation expressing the transport property of interest in terms of the independent variables defining the fluid state point, is often the most useful description. Alternative representations, such as lookup algorithms from tables of raw, or smoothed, experimental data or curves drawn through the data, may also have limited utility.

7.2 Functional representation of data sets

A general representation of any transport property correlation can be written as

$$X_{exp}(z_1, z_2, \ldots) = F(z_1, z_2, \ldots) + \Delta(z_1, z_2, \ldots) \tag{7.1}$$

The symbols in equation (7.1) are defined as follows. The experimental value of the transport property of interest is denoted X_{exp} (for example, η, λ). The state point of the

* Contribution of the National Institute of Standards and Technology, not subject to copyright.

system is represented by scalars z_i. For a pure fluid, it is represented by two independent thermodynamic variables. For a point on the coexistence surface, only one variable is needed, and for a mixture, the number of variables is determined from the Gibbs phase rule. Only Newtonian fluids are considered here.

The functional representation of the correlation is denoted by F and is often chosen to be a polynomial in the independent state variables. A general correlation, however, may be analytic or nonanalytic in its domain of definition and may include rational polynomials, algebraic or transcendental functions, Padé approximants or any combination of mathematical forms. The correlating function for the thermal conductivity is denoted by λ_{cor} and for the viscosity η_{cor}. The residuals Δ are defined by equation (7.1) and have a domain of definition of zero measure comprising the subset of $\{z_i\}$ containing points in the experimental data set. (X shares this limited domain of definition.) Although the goal of a correlation is usually considered to be the minimization of the residuals (in some sense, as discussed below), it cannot be overemphasized that this does not ensure that the correlating function F provides a good (or even fair) representation of the physical quantity being considered (denoted X in general or simply λ and η). The uncertainty of a correlation cannot be measured solely by a study of the residuals.

A correlation may be thought of primarily as a scheme to interpolate between data points (especially those separated by differences in but a single independent variable). If the experimental data set is sufficiently dense, a linear interpolation or a spline interpolation may be adequate or even preferable. However, this is seldom the case in practice. A good correlating equation provides the ability to generate a continuum of mutually consistent transport coefficients. In addition, a correlation can serve to compensate for random errors in experimental data. This may be true for simple correlations, with a minimum of fitting parameters, and for correlations appropriately guided by theoretical or empirical precepts. When a set of experimental data is overfit (described by a model that is more complex than the data and our knowledge of the theory warrants), uncertainty in the correlation may become amplified, and the ability to interpolate becomes compromised (Box *et al.* 1978). It is useful, when correlating experimental transport data, to take an extremely skeptical point of view and assume that

- it is impossible to determine the correct value of a transport property without measuring it at the state point of interest, and
- it is impossible to perform a perfect experimental determination of a transport coefficient at any state point.

The discussion above notwithstanding, a reasonable set of experimental data with a reasonable correlation will certainly provide adequate values of a transport property when interpolated between data points. Extrapolation beyond the limits in the independent state variables (and in the desired transport property) can introduce severe errors

(Box *et al*. 1978). Low–order polynomial fits may preserve the shape of the transport property surface in a fairly broad region beyond the range of the data. Whether this shape becomes essentially a linear extrapolation or preserves several orders of derivatives, the uncertainty associated with the extrapolation is difficult to ascertain. Also, uncertainties in experimental measurements often increase as the apparatus limits are approached. More complicated correlating equations often introduce anomalous behavior even if the state point is close to the experimental data. The recognition of anomalous behavior in an empirical correlation is a sufficient, but not a necessary, indication to conclude that the correlation is not adequate. A theoretically based correlation may provide a more useful extrapolation. However, even in this circumstance, the uncertainty of the calculated transport property will certainly become larger as the extrapolation is extended (Vesovic *et al*. 1990). The existence of new phenomena, such as the rise in the thermal conductivity associated with critical–region fluctuations, may not be recognized by any correlation outside its original range of development. Corresponding–states calculations can often be used to estimate the ability of a given correlation to extrapolate beyond the range of the data (Chapters 11 and 12).

The statistical measures of uncertainty associated with a correlation of experimental data are straightforward and are discussed briefly below (Draper & Smith 1980; Cornell 1981). In this section, some aspects of the uncertainty in experimental data and their relationship to correlations will be discussed. Again, the skeptical viewpoint asserts that experimental papers often underestimate the uncertainties associated with the accompanying data. This phenomenon is frequently exhibited as unexplained discrepancies among data generated by different groups and techniques. It is often useful to examine the historical evolution of data generated and reported within a given experimental laboratory to quantify the uncertainty. Data at different state points tabulated in an experimental paper do not represent independent observations: the internal consistency of these data does not place any bounds on their uncertainty.

It is unusual to find a paper which tabulates transport property experimental data and which presents a complete and believable uncertainty analysis (International Organization for Standardization 1993). The effects of impurities on the uncertainty to be associated with a pure fluid property (and of various demixing processes for mixture measurements) should be addressed. The uncertainties associated with the estimated corrections for the recognized systematic effects should be carefully accounted for by statistical or other means. The enumeration of possible systematic effects should be reassessed periodically in an ongoing experimental program: our understanding of the theory of instruments and measurement processes is constantly changing. The 'theory of the instrument' will never be complete, and there will never be a 'complete working equation for the instrument.' Evolution of instrument techniques and understanding may lead to the identification of some previously unknown systematic effects that were not considered in uncertainty estimates for measurements made 10 years ago. It is impossible to characterize unknown systematic effects in measurements, due to approximations and assumptions which are present in the instrument model, and experimenters may continue to underestimate the systematic uncertainty.

There are statistical, and perhaps systematic, uncertainties associated with the primary measured quantities used to define the thermodynamic state: these should be considered in the reporting of experimental uncertainties of the measured quantity and accounted for in the regressions used to determine a correlation. All the uncertainties are likely to have state point dependence, so that the assessment in the dilute gas regime will differ markedly from that in the critical region or in the compressed liquid (Perkins *et al.* 1991a). The uncertainty reported for a correlation should be a function of the fluid state point. The coverage factor (based on, perhaps, two standard deviations in a normal distribution) should be applied to the appropriate distribution associated with the combined standard uncertainty of the correlation.

The examination and comparison of data generated from independent laboratory experiments are important aspects in the establishment of the most useful correlations. Multiple data sets provide input to the scientific judgment needed to determine realistically quantitative standard uncertainties. The use of different experimental techniques and data from separate laboratories is needed to discover systematic biases which may be entrenched in an established experimental program.

It is usually not beneficial to generate a correlation by averaging between systematically offset data sets. When data sets provide results in different portions of the phase diagram, it is important to ensure that a physically reasonable correlation is maintained in the region of overlap or in any gap in the coverage. Care must be taken when selecting primary data – that is, those data used to determine the final fitting coefficients (Chapter 3). It is not inappropriate to ignore outliers when generating final parameters in a correlation, provided that these data are considered when establishing uncertainty estimates. In fact, it is not necessary to promote entire data sets to primacy or to relegate entire sets to a secondary status. This is because uncertainties are intrinsically state dependent and because coverage using a data set with a large uncertainty is usually preferable to establishing a correlation based on extrapolation.

Weighting schemes based on the state variables and the source of the data can be implemented objectively when accurate realistic combined standard uncertainties have been established for each point in an arbitrary number of data sets (Friend *et al.* 1989). Personal communications with researchers providing experimental data are often useful in establishing better estimates of uncertainty and resolving discrepancies. The imposition of corrections to a data set, in the absence of such communication, is usually not warranted. When discrepancies remain, it may be necessary to assume a uniform (or rectangular) probability distribution in the estimate of standard uncertainties. The use of smoothed data to generate a correlation is to be discouraged, although a treatment of uncertainties can still be developed to provide a useful correlation.

7.3 Selection of correlation variables

The state variables z_i in equation (7.1), chosen to parameterize the correlation, are arbitrary provided that the state point is uniquely defined. In most studies, the transport properties show the smoothest behavior when temperature T and density ρ are selected; see, however, Laesecke *et al.* (1990) for a discussion of formulations based on T and P. For mixtures, the composition variables x_i are usually added to this set, since they are most accessible to experimental control. When the variables are made dimensionless, the orders of magnitude of the coefficients in a polynomial fit become more similar (Friend *et al.* 1989). This can be useful in judging the selection and importance of terms in the correlation, but must be considered carefully, since neither the pure fluid viscosity nor thermal conductivity has a finite limit at the critical point.

Transport property measurements are normally reported in terms of the measured state variables of temperature, pressure and composition. Density is not usually measured at each state point but is instead obtained from an equation of state formulation. It is crucial to consider the uncertainty in the density which is calculated by the equation of state. As in the case for the transport property correlations, the uncertainty in the fluid density is not uniform over the entire $PVT(x)$ surface (Younglove 1982). The uncertainty in the fluid density in the critical region is almost certainly larger than it is in the dilute–gas limit or near the phase boundaries far from the critical point. Chapter 8 discusses equations of state and their importance in the analysis of transport properties.

Unfortunately, the discussion of density uncertainties and equation of state formulations cannot be limited entirely to the estimation of the density assigned to a given state point. The kinematic viscosity ($\nu = \eta/\rho$) is often measured (Diller & van der Gulik 1991; Kawata *et al.* 1991a,b; Nieuwoudt & Shankland 1991) and transformed by the original researchers to the reported viscosity using an equation of state. The correlation of such transformed viscosities incurs errors from the initial equation of state used to obtain the viscosity from the raw experimental data, and second from the equation of state (not necessarily the same equation of state) used to obtain the state density for the correlation. If the original kinematic viscosities are available, it is obviously more consistent to use the same equation of state to generate the viscosity from the kinematic viscosity as is used in the development of the correlation in order to reduce the uncertainty of the correlation. A more complex situation occurs when thermal diffusivity data ($a = \lambda/\rho C_P$), often from light–scattering measurements (Weber 1982; Shaumeyer *et al.* 1991), are used to generate thermal conductivity data. In this case, both the density *and* the isobaric heat capacity (C_P) are required from an equation of state. In general, the uncertainty of density from an equation of state is an order of magnitude less than the uncertainty in the isobaric heat capacity (Younglove 1982). Thus, thermal conductivity data derived from experimental thermal diffusivities have substantially greater uncertainty than the original thermal diffusivity data, and they may be biased by systematic errors in the equation of state. Once again, it is important to use the same equation of state to generate the thermal conductivity data as one uses to develop the correlation. Finally, instrumental corrections which define the working equations for an experimental apparatus often require thermodynamic quantities from an equation of state. But, since uncertainties in the corrections usually constitute a higher–order effect, the equation of state used for this purpose is not as critical (Assael *et al.* 1991).

In summary, the conventional choices of independent variables for the correlation – T, ρ (and x) – are often the most useful. When developing standard reference correlations for the transport properties, it is essential that a standard correlation for the thermodynamic surface be used and adequately described. The temperature scale used (currently ITS–90), fundamental constants and fluid–dependent parameters and their uncertainties must be specified in such a correlation.

7.4 Correlation development

The ultimate utility of a correlation is often determined at the outset when the form of the correlating function F is selected. The purpose for which the correlation is being developed must first be decided. The probability of success is based on the extent and quality of available data and the status of the available theory. The range of independent variables which define the fluid state of interest in this volume is extremely broad. It is bounded, in principle, by the solid–fluid phase boundaries, the ultradilute region (Knudsen regime, see Chapter 4) and the (pressure–dependent) temperature at which the molecular species in the system decompose or ionize. For this reason, the independent variables in a transport property correlation must be restricted. Typical global correlations include the vapor, liquid and supercritical fluid states from the triple–point temperature to a maximum temperature of perhaps 300 to 1000 K, with pressures ranging from the dilute–gas region to a maximum of perhaps 20 to 300 MPa (which may be temperature–dependent). The actual extremes of independent variables depend on the availability of data, the willingness of the correlator to extrapolate and the applications for which the correlation is intended. More restricted correlations are also common: there is theoretical and practical interest in correlations of the dilute gas, the liquid state, the saturation boundary and limited ranges of temperature or pressure that define an industrial process. These types of correlations are discussed briefly below.

To illustrate the development of a transport property correlation, discussion is focused here on the relevant example of binary mixtures of two new alternative refrigerants. A global correlation, based on theory where possible, is desired for the transport properties of mixtures of difluoromethane (R32) and pentafluoroethane (R125). This activity represents a portion of a project currently under way at National Institute of Standards and Technology (NIST). Viscosity data for R125 have been published (Diller & Peterson 1993), and the thermal conductivity data surface for R125 is shown in Figure 7.1. Primary data will be available for kinematic viscosity, thermal conductivity and thermal diffusivity from 180 to 400 K in the vapor, liquid and supercritical phases as well as

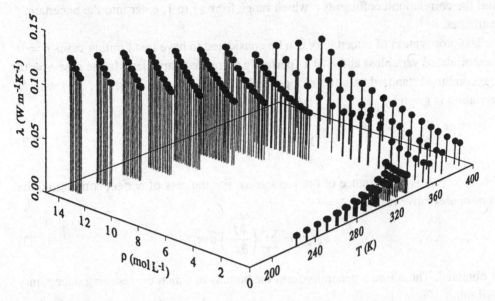

Fig. 7.1. The thermal conductivity data surface for pentafluoroethane (R125) in terms of temperature and density. The critical enhancement of the thermal conductivity is apparent in the isotherms closest to the critical point.

very near the saturated liquid and saturated vapor lines. These data will have a nominal uncertainty of less than 2% and extend to 70 MPa (or 35 MPa for viscosity). The mixture forms an azeotrope (Shankland & Lund 1990) with a composition that is nearly constant, and this composition is represented by one of the mixture samples. The second mixture composition is equimolar. Neither density nor heat capacity has been measured at each state point. Based on present knowledge of transport phenomena, as summarized in other chapters of this book, what are the options in the development of these transport property surfaces?

7.4.1 Reduction of experimental data

First, the reduction of the available raw data will be considered for use in the correlations. The kinematic viscosity ν must be reduced to yield estimates of the shear viscosity $(\eta = \rho\nu)$, and the thermal diffusivity data must be reduced to yield estimates of the thermal conductivity $(\lambda = a\rho C_p)$. The data reduction procedure produces additional uncertainty to the transport properties of interest. This can be assessed using the law of propagation of standard uncertainty (International Organization for Standardization 1993). One difficulty in estimating the standard uncertainty is in the quantification of the degree of correlation among the variables entering into the equations for the transport properties. Correlation between variables can be estimated from the covariance matrix, and the correlation coefficients r which range from -1 to 1, enter into the uncertainty estimates.

The propagation of uncertainty can be considered to have two limiting cases $r = 0$ (uncorrelated variables) and $r = 1$ (perfectly correlated variables). In the case $r = 0$, the combined standard variance of a general function $y = f(x_1, x_2, \ldots, x_N)$ with N variables is given by

$$u_c^2(y) = \sum_{i=1}^{N} \left(\frac{\partial f}{\partial x_i}\right)^2 u^2(x_i) \tag{7.2}$$

where $u^2(x_i)$ is the variance of the variable x_i. For the case of perfect correlation, the very conservative result,

$$u_c(y) = \sum_{i=1}^{N} \left(\frac{\partial f}{\partial x_i}\right) u(x_i) \tag{7.3}$$

is obtained. These limits generally form the bounds in which the resulting uncertainty will fall.

The kinematic viscosity is usually measured as a function of temperature and pressure. If these state variables are known with an uncertainty of 0.002% and an equation of state is used to obtain the fluid density, an uncertainty in density of 0.2% may be expected. Knowing the functional form of the equation of state, it is possible to use equations for the propagation of error to see how uncertainties in the measured temperature and pressure influence the calculated density. Except perhaps in the critical region, this influence is small and the uncertainty in the density is due to the uncertainty in the equation of state itself. The uncertainty in the shear viscosity is required, assuming that the kinematic viscosity data have an uncertainty of 2%. If the density and kinematic viscosity are not correlated ($r = 0$), then evaluating the combined standard uncertainty in $\eta = \rho v$ with equation (7.2) gives

$$u_c^2(\eta) = v^2 u^2(\rho) + \rho^2 u^2(v) \tag{7.4}$$

The relative uncertainty is obtained by dividing through by $(\eta = \rho v)^2$ and taking the square root to yield

$$\frac{u_c(\eta)}{\eta} = \sqrt{\left(\frac{u(\rho)}{\rho}\right)^2 + \left(\frac{u(v)}{v}\right)^2} \tag{7.5}$$

If the variables are perfectly correlated, equation (7.3) yields

$$u_c(\eta) = v u(\rho) + \rho u(v) \tag{7.6}$$

so that

$$\frac{u_c(\eta)}{\eta} = \frac{u(\rho)}{\rho} + \frac{u(v)}{v} \tag{7.7}$$

Equation (7.7) gives a very conservative outer bound for the estimate of uncertainty. In the absence of information to evaluate the correlation coefficients, it must be considered even though it may be too pessimistic. For the given example, with relative uncertainty values of 2% for kinematic viscosity and 0.2% for density, the relative combined uncertainty estimates for the shear viscosity are 2.01% for the uncorrelated case and 2.2% for the correlated case.

The combined uncertainty due to processing of the thermal diffusivity data can be estimated in a similar way. The expression which must be analyzed is $\lambda = a\rho C_P$, and both ρ and C_P are obtained from an equation of state. Following the analysis described for the viscosity above, the expression for the combined relative uncertainty for the uncorrelated case is

$$\frac{u_c(\lambda)}{\lambda} = \sqrt{\left(\frac{u(\rho)}{\rho}\right)^2 + \left(\frac{u(C_P)}{C_P}\right)^2 + \left(\frac{u(a)}{a}\right)^2} \qquad (7.8)$$

and for the case of perfect correlations, the combined relative uncertainty is given by

$$\frac{u_c(\lambda)}{\lambda} = \left(\frac{u(\rho)}{\rho}\right) + \left(\frac{u(C_P)}{C_P}\right) + \left(\frac{u(a)}{a}\right) \qquad (7.9)$$

It is assumed that the relative uncertainty in the thermal diffusivity measurement is 2% and the uncertainties in the density and heat capacity (from the equation of state) are 0.2% and 5% respectively. Substituting this information into equations (7.8) and (7.9), the estimated uncertainty in the processed thermal diffusivity data is 5.4% for the uncorrelated case and 7.2% for the correlated case.

Treatment of the data for the two binary mixtures is even more complicated and subject to greater uncertainty, since, in general, the density and heat capacities are not available at the exact compositions at which transport property data are available. For the R32/R125 mixture, the exact azeotropic composition and even the existence of a true azeotrope is not clear (Shankland & Lund 1990; Widiatmo *et al.* 1994). For this example it will be assumed that density, heat capacity, kinematic viscosity, thermal conductivity and thermal diffusivity data are all available for the fixed azeotropic composition and that this system forms a true azeotrope. Because the compositions of the vapor and liquid are equal for an azeotrope, demixing processes do not have to be considered as a significant contribution to the uncertainty for the azeotropic composition. For the equimolar mixture, additional uncertainty in the measurements must be acknowledged and quantified due to uncertain and changing composition. If the two–phase region is penetrated at any point in the measurements, preferential vaporization of one of the components in the mixture may be a factor. Furthermore, the composition of nonazeotropic mixtures will vary with the length of time the sample is held at these conditions. Thus, demixing processes are present for all nonazeotropic mixtures, and additional uncertainty is present in the composition at each state point. Demixing processes may be present whenever thermal or pressure gradients are present in the system (Hirschfelder *et al.* 1954). The flow necessary to fill the sample cell can result in composition gradients, since diffusion and relative mass flow depend on the species.

In general for mixtures, calculational models, such as extended corresponding–states (ECS) algorithms, must be used to provide the densities and isobaric heat capacities needed to reduce the available data to viscosity and thermal conductivity. Several implementations of the ECS model are available, including two computer packages from NIST. Mixture equations of state generally have a greater uncertainty associated with the calculated density and isobaric heat capacity than pure fluid equations of state. The exceptions include equations of state developed for specific compositions, such as the R32/R125 azeotrope, standard air or the thoroughly studied methane–ethane binary at certain fixed compositions (Haynes *et al.* 1985).

For the present discussion, it will be assumed that a fluid–specific equation of state for the azeotropic mixture is available which provides density with an uncertainty of 0.5% and isobaric heat capacity with an uncertainty of 7%. Thus, the uncertainty in viscosity obtained from kinematic viscosity degrades slightly from 2.06% (uncorrelated) to 2.5% (correlated), and the uncertainty in thermal conductivity obtained from thermal diffusivity data degrades from 7.29% (uncorrelated) to 9.5% (correlated). The region near the critical point which must be excluded from analysis is large in the case of mixtures, and the problem of determining the critical point at each mixture composition is encountered. For the equimolar mixture, an ECS model is assumed to be available to provide density within the order of 1% and isobaric heat capacity within 10%. The uncertainty in the viscosity obtained from kinematic viscosity is now from 2.24% (uncorrelated) to 3% (correlated), and the uncertainty in the thermal conductivity data from thermal diffusivity data is from 10.25% (uncorrelated) to 13% (correlated), for the equimolar fluid mixture. Of course, the reliability and consistency must be checked by comparison of the general corresponding–states predictions for ρ and C_p to available data and the values obtained from correlations for the pure fluids and the azeotrope. In conclusion, it can be very difficult to develop accurate correlations for the transport properties of fluid mixtures, unless the thermodynamic properties of the mixtures are also determined accurately.

7.4.2 Pure fluid surfaces

The transport property correlation of equation (7.1) can be written as a general polynomial in the independent variables:

$$F(z_1, z_2, \ldots) = \sum_{i,j,\ldots=0} a_{i,j,\ldots} z_1^{n_i} z_2^{m_j} \ldots \tag{7.10}$$

Although such a polynomial could be reasonably optimized and truncated with sufficient data and diligence, it is convenient to use a theoretical point of view to separate contributions from the various physical processes which determine the transport coefficients; the separation is useful in the development of the correlating function F. The dilute–gas region is discussed in Chapter 4. Formally, this contribution to F can be defined by

$$F^{(0)}(T,\dots) = \lim_{\rho \to 0} F(\rho, T, \dots)\bigg|_{T,\dots} \tag{7.11}$$

The first density coefficient is discussed in Section 5.2 and can be written

$$F^{(1)}(T,\dots) = \rho \lim_{\rho \to 0} \frac{\partial F(\rho, T, \dots)}{\partial \rho}\bigg|_{T,\dots} \tag{7.12}$$

From a theoretical viewpoint, the continued density expansion breaks down because of nonanalytic terms and from the more practical point of view, selected terms in the density series are statistically significant; the optimum set of terms, however, differs among different fluid systems. The first density correction will not be emphasized in this chapter (see Chapter 5), but rather these terms will be included in the excess transport properties as defined below. The transport coefficients in the critical region are dominated by effects associated with the long–range of density (or composition) fluctuations rather than by binary (or few particle) collisions; the thermal conductivity and viscosity of pure fluids are, in principle, divergent at the critical point (Chapter 6). An extended critical region, or crossover region, is defined by the competition between classical collisional processes and fluctuation–induced processes. Although it is not possible to separate these effects unambiguously, theoretical arguments and the empirical extrapolation of the critical divergence allow a contribution $\Delta F_c(\rho, T, \dots)$ to be examined. The remaining portion of the transport property surface, which can be dominant in the liquid, well away from the critical region, is denoted the 'excess contribution.' In studies of the critical region, the sum of the dilute gas and excess contributions, $F^{(0)} + \Delta F$, is called the background transport property. Instead of the general form of equation (7.10), then, a global correlation for transport properties is often written as

$$F(T, \rho, \dots) = F^{(0)}(T, \dots) + \Delta F(T, \rho, \dots) + \Delta F_c(T, \rho, \dots) \tag{7.13}$$

where F can be either λ or η. The terms in equation (7.13) are not truly independent from the point of view of a global correlation. However, the theoretical basis for this division is sufficiently strong to motivate their use. The excess and critical enhancement terms approach 0 in the limit of zero density. It is nearly impossible to separate unambiguously the excess function term from the critical enhancement term using experimental data (see Chapter 14).

If correlations of a spherical monatomic gas are considered, the dilute–gas effective collision cross sections or collision integrals, evaluated for a specific intermolecular potential, can be substituted directly for calculation of the dilute–gas term. From this point of view, it may not be appropriate to fit dilute–gas thermal conductivity and viscosity data independently, since kinetic theory indicates that they should be directly related for monatomic species (Hirschfelder *et al.* 1954; Chapter 4). Standard potentials, such as the Lennard–Jones 12–6 and the more general m–6–8, have been used to calculate the collision integrals (Klein *et al.* 1974). The use of more complex potentials for fluids such as noble gases is discussed in Section 14.1. In this method, the thermodynamic second virial coefficient can, in principle, be described using the same information.

Because of the internal degrees of freedom available in polyatomic species, energy from collisions can be stored as translational kinetic energy or as internal rotational, vibrational and electronic energy (Uribe *et al.* 1989). Thus, the well–developed solutions to the simple Boltzmann equation must be modified. These corrections to the Chapman–Enskog solutions are not straightforward to implement for correlations of such systems. Since both R125 and R32 are nonspherical polyatomic molecules, a more empirical description of the dilute–gas viscosity and thermal conductivity will be considered. Alternatively, effective Lennard–Jones energy (ϵ) and distance (σ) parameters can be found, the molecules can be treated as spheres and these standard collision integrals can be used to correlate the viscosity (Mourits & Rummens 1977). In this method the two Lennard–Jones parameters are the only adjustable parameters in the correlation for viscosity, but empirical adjustments for the internal degrees of freedom can be applied to the thermal conductivity (Uribe *et al.* 1989). There is a great deal of variety in the functional form of the dilute–gas viscosity and thermal conductivity terms. Some researchers prefer to retain the collision cross section and collision integral nomenclature, discussed in detail in Chapters 4 and 11. Others acknowledge the empiricism and fit the dilute–gas thermal conductivity data to a simple polynomial function in temperature or reduced temperature. These empirical functions for the dilute–gas thermal conductivity should not be used to extrapolate beyond the range of data used in their development.

The definitions of the dilute–gas viscosity and thermal conductivity from equation (7.11) make them impossible to measure directly. The defining limit of zero density ensures that both the excess and critical enhancement terms must approach 0. Because these transport properties cannot be measured at zero density, many researchers have simply reported data at 0.1 MPa (1 bar). When the vapor pressure is less than 0.1 MPa, data for saturated vapor are given as the dilute–gas value. This has led to much confusion in the literature, which is often not reversible, since the authors report only the temperature and not the pressure of the 'dilute–gas' measurement. In developing a correlation for the dilute gas, one could fit the entire surface, including the dilute–gas, excess and critical enhancement terms. The limit of equation (7.11) would then permit the extraction of the dilute–gas transport properties. Alternatively, one can extrapolate data along isotherms, in terms of density, to obtain dilute–gas values which are subsequently correlated only as a function of temperature. When extrapolating to zero density, care must be taken to ensure that the results are significant when using the lowest–order polynomial for the operation: usually a linear equation in density can be used at sufficiently low densities, and adding a quadratic term (and increasing the range in density) should not change the value of the zero density limit (Kestin *et al.* 1971). The defining extrapolation in density can be converted to an extrapolation of data with pressure as the independent variable only if the fluid behaves as an ideal gas in the region for which data are available. The process of extrapolating along an isotherm to zero pressure should be avoided, since some of the data may be at pressures, especially at low reduced temperature, at which the gas does not behave ideally.

If the terms of equation (7.13) are truly independent, then it may be appropriate to fit the dilute–gas transport properties to primary data at low density; the dilute–gas contribution calculated from the resultant correlation can then be subtracted from the experimental data at higher densities to study the remaining excess and critical enhancement terms. Since the dilute–gas term is a function of temperature alone (or a reduced temperature such as $T^* = T/T_c$ or $T^* = k_B T/\epsilon$) and both the excess and critical enhancement terms are zero in the limit of zero density, the dilute–gas term is expected to be mathematically independent of the excess and critical enhancement terms. However, if the dilute–gas term has some cross correlation with respect to density (because of its determination from experimental data at nonzero density), then it may not be proper to fit the dilute gas data independently. Any cross correlation which is present in the data is not necessarily due to the behavior of the transport property itself, but may be due to systematic effects. This uncertainty may be a function of the other independent variables such as density or composition.

The residuals defined in equation (7.1) describe the agreement between a correlation and the experimental measurement. For the present discussion, the uncertainty in the measurement itself must also be addressed. For example, measured thermal conductivity data contain both systematic $\Delta\lambda_{syst}$ and random errors $\Delta\lambda_{rand}$ and may be written

$$\lambda_{meas}(T', \rho') = \lambda(T, \rho) + \Delta\lambda_{syst}(T', \rho') + \Delta\lambda_{rand}(T', \rho') \qquad (7.14)$$

It cannot be overemphasized that the values of any measurands are unknowable (International Organization for Standardization 1993), and, for transport property measurement, there is also uncertainty in values of the independent variables. Measured values of the gas thermal conductivity in the dilute region obtained with transient hot–wire apparatus usually increase with decreasing density at pressures below about 1 MPa. This is contrary to the expected behavior and is due to uncertainty in heat–transfer corrections to the raw transient hot–wire data. These corrections increase dramatically at low densities, so that one observes increasing random uncertainty and systematic density dependence in the dilute–gas region (Assael *et al.* 1991). Since uncertainty due to random effects increases, and may not even be bounded in the limit of zero density with this type of apparatus, there is a practical difficulty in obtaining a completely independent dilute–gas term. A dilute–gas correlation using only transient hot–wire data at 0.101 325 MPa (1 atm) would be systematically high relative to the data at higher densities and would distort the surface fit for the excess and critical enhancement terms. The low–density points which have significant density–dependent uncertainty should be removed from the primary data set prior to the fit of the dilute–gas function, or the extreme uncertainty must be accounted for in the regression.

If an accurate correlation for the dilute–gas contribution is available, the dilute–gas function, evaluated at the temperature of each measurement, can be subtracted from the measurement itself. These values can then be used for subsequent fitting of the excess and critical enhancement functions. Separation of the excess and critical enhancement terms is not nearly as straightforward as the dilute–gas term, because both the excess and critical enhancement terms are functions of temperature and density and the distinction in terms of physical processes is not easily determined by experiment. Kinetic theory provides some guidance for the first term, the first density coefficient of equation (7.12), and predicts the weak temperature dependence in this term of the excess contribution. Enskog dense fluid theory can also provide guidance at higher, liquid–like densities (Chapter 5). It is the density region around the critical density where there is less theoretical knowledge concerning the background. The approximation of temperature independence (see Chapter 5) may be valid to moderate pressures and greatly simplifies the development of surface fits for transport properties. The critical enhancement to both viscosity and thermal conductivity is divergent at the gas–liquid critical point. For viscosity, it is significant only over a very limited region of temperature and density near the gas–liquid critical point, whereas the thermal conductivity critical enhancement is significant over a much wider region of temperature and density (Chapter 6). As with the dilute–gas term, it is necessary to find a region of the fluid PVT surface where the term of interest, in this case the critical enhancement, is insignificant in order to unambiguously separate the excess term from the critical enhancement term. Along the critical isochore of a fluid, the critical enhancement is insignificant only in the supercritical gas region. For thermal conductivity, it is necessary to have data above about twice the critical temperature for the critical enhancement to be insignificant (Chapter 6). This condition is often not physically possible, since many fluids, such as alternative refrigerants, have critical temperatures around 400 K, and these fluids will begin to decompose at temperatures below 800 K (Doerr *et al.* 1993). It is therefore necessary either to fit the excess and critical enhancement terms simultaneously or to iterate between fits of the two terms. A great deal of judgment is involved in this process (see Section 14.1).

Once good fits for the background properties are found, they can be evaluated at each state point and subtracted from the data to obtain data for the critical enhancement. The asymptotic form of the critical enhancement for transport properties has been understood for some time (Sengers & Keyes 1971; Sengers 1972; see Chapter 6). Correlations which make use of this theory, as well as empirical approximations, such as Gaussian–type functions (Roder 1982), have been used to describe the enhancement term. In recent years there have been significant advances in mode–coupling crossover theory for the transport properties of pure fluids. The approach of Olchowy & Sengers (1988) has been used to correlate the critical enhancement of several fluids. This theory is described in detail in Chapter 6 of this book and will be described only briefly here. The theory requires good estimates of the fluid density, compressibility and heat capacities for each state point. There are a number of scaling parameters which are fluid–specific but which can be obtained independently from equilibrium data. For transport properties, a cutoff wavenumber q_D^{-1} is required, which can be obtained in a regression of transport property data. In addition, a reference temperature T_{ref} above which the critical enhancement becomes negligible has been introduced. This parameter can be approximated by a constant of the order $2T_c$. This requires an accurate (or at least consistent) equation of state which is valid at such temperatures. This mode–coupling theory gives reasonable results both above and below the critical temperature, when flexible background correlations are regressed simultaneously or iteratively relative to the enhancement term (Perkins *et al.* 1991a,b; Chapter 14).

When the equation of state is not sufficiently accurate in the critical region or at high reduced temperatures ($2T_c$), an empirical representation is required. An empirical Gaussian function (centered at the critical density) has been extensively used for thermal conductivity correlations (Roder 1982; Roder 1985; Roder & Friend 1985; Perkins & Cieszkiewicz 1991). A common expression of this form is given by

$$\Delta \lambda_c = \left(C_1 + C_2 T' + C_3 T'^2 + \frac{C_4}{(T' - C_5)} \right) \exp(-C_6 |\rho - \rho_c|) \qquad (7.15)$$

where C_i are empirical parameters and the adjusted temperature is given by $T' = T + |T - T_c|$. In practice, C_3 can often be set equal to 0, and C_5 is near or at the critical temperature. This simple empirical correlation may not represent the critical enhancement below the critical temperature as well as the theoretically motivated crossover approach and requires 4–6 adjustable parameters. However, it has advantages if there are deficiencies in the available equation of state or if a computationally fast and simple representation is required. If reference equations of state are available for both R32 and R125, the crossover theory can be used to represent the critical enhancement.

The excess functions of equation (7.13) are not unique but depend strongly on both the functional form and the optimized parameters of the enhancement terms. The optimized parameters of both terms depend on the selected equation of state used to determine the density (especially in the critical region) as well as the compressibility and heat capacity, which are required in the theoretically based critical enhancement models. It is necessary to have data over an extremely wide range of temperature and pressure to resolve any temperature dependence of the transport property excess functions. Detailed examples of such pure fluid correlations are presented in Chapter 14.

7.4.3 Mixture treated as a pure fluid surface

Because the azeotropic mixture of R32/R125 is of interest to the refrigeration industry, a standard reference equation of state should become available for this specific composition (assumed independent of other state variables) which provides the density of the azeotrope with 0.2% uncertainty and the heat capacity and compressibilities with an uncertainty of 5%. Since a mixture with this composition can be treated as a pure fluid, the development of the correlation proceeds as in the previous section. To employ the equations for the critical enhancement, the mixture critical point must be determined. If there is sufficient industrial interest in this composition, the critical point will be experimentally determined. This same critical point will be used in the reference equation of state. For azeotropic mixtures, the behavior of the transport properties in the critical region is similar to that of pure fluids. For mixtures in general, it may not be possible to pass continuously from high density to low density along a supercritical isotherm near the critical point. This is because phase separation can occur at temperatures higher than the critical temperature, up to the maxcondentherm temperature; the density of the system at the maxcondentherm may differ significantly from the critical density for pressures near the critical pressure. Thus, it is generally more difficult to determine the background transport properties for mixtures compared to those of pure fluids. Although the general behavior of the critical enhancement of transport properties in the region of liquid–vapor equilibrium is similar for mixtures and pure fluids, the similarity does not extend to the critical point itself. Theory indicates that the transport properties of mixtures are finite at the critical point, while pure fluid transport properties are divergent (Chapter 6). Because the crossover to a saturated transport property enhancement is quite close to the critical point, this effect may have little consequence for most practical thermal conductivity correlations. The extensive study of the methane–ethane binary systems has allowed development of accurate thermodynamic surfaces for each of three compositions at which transport property data have also been obtained. Thus, the development of mixture thermal conductivities in a form analogous to the correlations for pure fluids has been reported for this system (Roder & Friend 1985).

7.4.4 General mixture surfaces

In general for mixtures, a more predictive rather than a validated algorithm must be used for obtaining the required densities in a transport–property correlation, and a more empirical transport–property surface may be appropriate. It is convenient to use corresponding–states methods to obtain the equilibrium properties in order to establish the correlation in terms of density, temperature and composition. An approach which gives accurate equilibrium properties for the pure fluid limits is desirable in establishing a general transport–property mixture surface. To use the critical enhancement expressions above, the mixture critical point needs to be calculated at any composition. These critical parameters can be used to determine reduced parameters in general mixture correlations. There are many approximations which can be used such as

$$\rho_c(x_a) = \rho_c^b + x_a(\rho_c^a - \rho_c^b)$$ (7.16)

for the density and

$$T_c(x_a) = T_c^b + \frac{x_a M_b}{(1 - x_a)M_a + x_a M_b}(T_c^a - T_c^b)$$ (7.17)

for the critical temperature. In these expressions, x_a is the mole fraction of component a, M_i is the molar mass of component i, ρ_c^i is the critical density of component i, and T_c^i is the critical temperature of component i. A more rigorous estimate of the mixture critical point locus can be obtained from scaled models such as the modified Leung–Griffiths approach (Rainwater 1991) if critical region data, including data for vapor–liquid equilibrium, are available. The general mixture model should be validated by comparing the general surface to the pure fluid surfaces for both R32 and R125 and the constant composition surfaces developed for the azeotrope transport properties.

In studying mixtures, the expressions for the dilute–gas solution of the Boltzmann equation are also available but require evaluation of additional collision cross sections compared to those for pure fluids (Chapter 4). This situation is again more complicated for the thermal conductivity of mixtures than for viscosity because of the internal degrees of freedom. The excess function can be made essentially independent of composition by scaling the variables in terms of mixture–dependent critical parameters (as in equations (7.16) and (7.17)) or in terms of other mixture–dependent scale factors. Empirical correlations developed in this manner for the general methane–ethane system, based on extensive data for the pure fluids and three mixture compositions, have been presented for both viscosity and thermal conductivity (Friend & Roder 1987; Friend 1990). In Figure 7.2a, selected viscosity data are presented at 310 K for five compositions of the methane–ethane system. In Figure 7.2b, a reduced excess viscosity (for data extending from 100 K to 320 K) is seen to lie essentially on a single curve for all five compositions. Similar results have been obtained for the thermal conductivity (Friend & Roder 1987), but the analysis is complicated, since the critical enhancement covers an extended region of the $PVT(x)$ surface.

Several empirical methods to extend the critical enhancement term to mixtures have been explored (see also Chapters 6 and 15). The parameters in the mode–coupling approach mentioned above have been made composition–dependent and the results are reasonable. Luettmer–Strathmann (1994) has recently reported a new mode–coupling solution which describes the critical enhancements to the transport properties of fluid mixtures. Corresponding states algorithms, based on the mode–coupling solutions, have also been used to describe the thermal conductivity of mixtures (Huber *et al.* 1992).

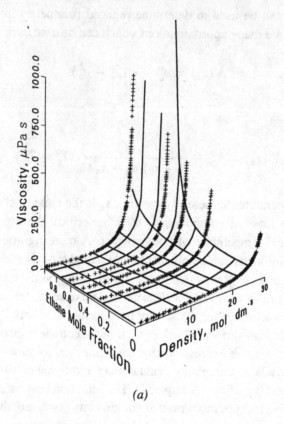

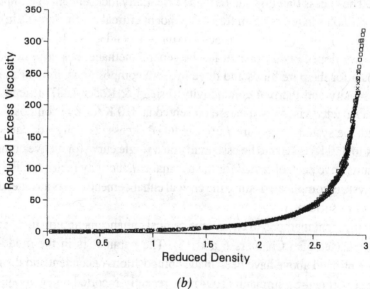

Fig. 7.2. (*a*) The viscosity data surface for mixtures of methane and ethane at temperatures near 310 K. (*b*) All of the excess viscosity data from 100 to 320 K are shown to lie on a single curve using reduced coordinates.

7.5 Restricted correlations

In the previous section, a hierarchy of techniques was examined to correlate transport–property surfaces of pure fluids and binary mixtures. It is often desirable to have correlations which cover a more limited region. Such correlations may provide higher accuracy in the restricted region than is practical in a global surface, may be designed to study particular theoretical questions, can be more easily generalized to additional fluid systems or may not require extensive transport property data and detailed and wide–range information on the equilibrium properties. To the extent that the structure and parameters of the pure fluid correlations discussed above are optimized to any finite set of data with limits of temperature and pressure, they can be considered to be restricted. More expansive correlations might allow choices of pure fluid identity or mixture composition among a class of fluids such as noble gases and alkanes. It is certainly feasible to develop such generalized correlations from an extensive set of data. Correlations which calculate properties only of the dilute gas, saturated vapor and saturated liquid are restricted in the sense of this section; such correlations reduce the number of independent variables (z_i of equation (7.1)) and thereby generate lines in the more general pure fluid transport property surface. Other restricted correlations maintain the dimensionality of the space of independent variables but may impose limits on their range; such restrictions as studies of the dilute–gas region, the critical region or the liquid phase region may have scientific or technical utility.

7.5.1 Dilute and moderately dense gas region

Developments in kinetic theory have led to theoretically based correlations of the transport properties of gases both in the dilute–gas limit and, more recently, at increasing densities. The forms of these limited surface correlations are given by

$$\eta = \eta^{(0)}(T^*, \sigma) + \eta^{(1)}(T^*, \sigma)\rho \qquad (7.18)$$

and

$$\lambda = \lambda^{(0)}(T^*, \sigma) + \lambda^{(1)}(T^*, \sigma)\rho \qquad (7.19)$$

where $T^* = k_B T/\epsilon$ and k_B is the Boltzmann constant, ϵ is the energy parameter of an intermolecular potential and σ is the distance parameter of the potential. It is apparent from these equations that the dilute–gas term and only the first term of the power series in density used in the more general transport property excess function have been included. Details of this correlation technique are provided in Chapters 4, 5 and 14. Dilute–gas terms obtained from evaluation of collision integrals for various intermolecular potentials combined with the first density coefficients of thermal conductivity and viscosity obtained for a Lennard–Jones (12–6) fluid have been shown to provide adequate descriptions of the surface for monatomic gases with no further parameters required from the transport property data. An extension of this approach using a semiempirical correction to account for internal degrees of freedom has also been reported for simple polyatomic gases (Nieto de Castro *et al.* 1990; see Section 5.2).

7.5.2 Liquid phase

Developments in liquid state theory and empirical studies of transport phenomena in liquids can be exploited by restricting the correlation to the region of saturated or compressed liquids away from the critical point (Brush 1962; Schwen & Puhl 1988). The basis of one such class of correlations is the recognition that liquid transport properties, and in particular viscosity, can be correlated well in terms of the difference between the molar volume and a compact packing volume which is fluid–specific and a weak function of temperature. The simplest form of this type of correlation is the fluidity (η^{-1}) versus molar volume $(V = \rho^{-1})$ approach described by Batschinski (1913) and Hildebrand (1977) for pure fluids based on experimental observations; these variables are linearly related to a very good approximation. Hard–sphere theories such as that proposed by Enskog and several modifications of this approach have also been used for liquids. Current work in this area has evolved significantly and is described in detail in Chapters 5 and 10.

7.5.3 Saturation boundary lines

The transport properties at saturation are often of great interest for industrial applications. Vapor–liquid compression cycles using steam or refrigerants are good examples. There is little or no theoretical basis for fitting these saturated vapor and liquid lines. For any subcritical isotherm, the saturated vapor and liquid densities represent the closest possible approach to the critical density. Hence along a subcritical isotherm, the critical enhancement contribution will be largest at the saturated vapor and liquid. It is often necessary to extrapolate to the phase boundary. It is extremely important to report both the temperature and pressure of transport property data even if the fluid is very near saturation. This allows adjustment of the data to saturation even if the accepted vapor pressure of a material changes.

For pure fluids, it is most common to represent the saturated vapor and saturated liquid transport properties as simple polynomial functions in temperature, although polynomials in density or pressure could also be used. Exponential expansions may be preferable in the case of viscosity (Brush 1962; Schwen & Puhl 1988). For mixtures, the analogous correlation of transport properties along dew curves or bubble curves can be similarly regressed. In the case of thermal conductivity, it is necessary to add a divergent term to account for the steep curvature due to critical enhancement as the critical point is approached. Thus, a reasonable form for a transport property, particularly thermal conductivity, along the saturation line is

$$X_{\text{sat}}(T) = a_1 + a_2 T + a_3 T^2 + \frac{a_4}{(T - T_c)} \tag{7.20}$$

where X_{sat} is the saturation transport property of interest. Chapter 6 provides the theoretical basis for the correct behavior in the critical limit. If the temperature range is sufficiently limited, a_3 can be set to 0. If the data are not close to the critical region, then a_4 can be set to 0. The general behavior of the critical enhancement of the transport property can be described by equation (7.20); however, the divergence of this expression at the critical temperature does not have the exponent, which has been determined from scaling theory studies.

Ancillary equations for the saturation transport properties must agree with surface correlations to within their mutual uncertainty. For thermal conductivity, this is an interesting situation, since the saturation values are divergent as the critical point is approached. Thus, evaluation of a transport–property surface, particularly a thermal conductivity surface, along the saturation boundary is a severe test for a surface fit of transport properties.

7.6 Summary

Theory is not presently capable of predicting the transport properties of either monatomic or polyatomic molecules over their entire thermodynamic surface with an accuracy which is generally sufficient for scientific or technological utility. The situation is even worse for the case of mixtures. Any correlation scheme for the entire thermodynamic surface is then a combination of as many theoretically based components as possible with enough empiricism to fill in the voids. Any valid correlation should be able to interpolate between the experimental data points which were used during its development – whether the correlation is theoretically based or entirely empirical. It should be realized that this may not be simple. For mixtures, the transport property at any composition can be considered to be an interpolation between the pure fluid limits. In the complex case of a global mixture correlation, it is certainly necessary to have a theoretical basis to predict transport properties for any given composition. The principal advantage of theoretically based terms is that they have a greater probability of being valid for extrapolations outside the region where experimental data are available. Examples of extrapolation include relatively simple cases such as use of the dilute gas term at low reduced temperatures where transport property data are not available due to the extremely low gas density, or extrapolation of the excess term to densities beyond the range of available experimental data. It is not appropriate to provide a rigid framework for the development of transport property surfaces, but instead we offer some ideas and warnings to allow the reader to use the tools described in this book. There are numerous equally valid approaches to correlation, and these will certainly evolve as the theoretical understanding of transport phenomena, the description of equilibrium properties, and the uncertainty and extent of experimental data are improved.

Acknowledgments

The authors acknowledge many valuable discussions with their colleagues at NIST. They gratefully acknowledge the long-term financial support of the NIST Standard Reference Data Program and the Division of Chemical Sciences, Office of Basic Energy Science, Office of Energy Research, U.S. Department of Energy.

References

Assael, M.J., Nieto de Castro, C.A., Roder, H.M. & Wakeham, W.A. (1991). Transient methods for thermal conductivity, in *Measurement of the Transport Properties of Fluids – Experimental Thermodynamics, Volume III*, eds. W.A. Wakeham, A. Nagashima & J.V. Sengers, pp. 164–94. Oxford: Blackwell Scientific Publications.

Batschinski, A. J. (1913). Untersuchungen über die innere Reibung der Flüssigkeiten I. *Z. Phys. Chem.*, **84**, 643–706.

Box, G.E.P., Hunter, W.G. & Hunter, J.S. (1978). *Statistics for Experimenters*. New York: Wiley.

Brush, S.G. (1962). Theories of liquid viscosity. *Chem. Rev.* **62**, 513–548.

Cornell, J. A. (1981). *Experiments with Mixtures: Designs, Models, and the Analysis of Mixture Data*. New York: Wiley.

Diller, D.E. & Peterson, S.M. (1993). Measurements of the viscosities of saturated and compressed fluid 1–chloro–1,2,2,2–tetrafluoroethane (R124) and pentafluoroethane (R125) at temperatures between 120 and 420 K. *Int. J. Thermophys.*, **14**, 55–66.

Diller, D.E. & van der Gulik, P.S. (1991). Vibrating viscometers, in *Measurement of the Transport Properties of Fluids – Experimental Thermodynamics, Volume III*, eds. W.A. Wakeham, A. Nagashima & J.V. Sengers, pp. 79–94. Oxford: Blackwell Scientific Publications.

Doerr, R. G., Lambert, D., Schafer, R. & Steinke, D. (1993). Stability studies of E245 fluoroether, $CF_3 - CH_2 - O - CHF_2$. *ASHRAE Trans.*, **99** (Part 2), 1137–1140.

Draper, N. & Smith, H. (1980). *Applied Regression Analysis*, 2nd Ed. New York: Wiley.

Friend, D.G., Ely, J.F. & Ingham, H. (1989). Thermophysical properties of methane. *J. Phys. Chem. Ref. Data*, **18**, 583–638.

Friend, D.G. & Roder, H.M., (1987). The thermal conductivity surface for mixtures of methane and ethane. *Int. J. Thermophys.*, **8**, 13–26.

Friend, D.G. (1990). Viscosity surface for mixtures of methane and ethane. *Cryogenics*, **30**, 105–112.

Haynes, W.M., McCarty, R.D., Eaton, B.E. & Holste, J.C. (1985). Isochoric (p, V(m), x, T) measurements on (methane + ethane) from 100 to 320 K at pressures to 35 MPa. *J. Chem. Thermodyn.*, **17**, 209–232.

Hildebrand, J.M. (1977). *Viscosity and Diffusivity*. New York: Wiley.

Hirschfelder, J.O., Curtiss, C.F. & Bird, R.B. (1954). *Molecular Theory of Gases and Liquids*. New York: Wiley.

Huber, M.L., Friend, D.G. & Ely, J.F. (1992). Prediction of thermal conductivity of refrigerants and refrigerant mixtures. *Fluid Phase Equil.*, **80**, 249–261.

International Organization for Standardization (1993). *Guide to the Expression of Uncertainty in Measurement*. Geneva, Switzerland: International Organization for Standardization.

Kawata, M., Kurase, K., Nagashima, A. & Yoshida, K. (1991a). Capillary viscometers, in *Measurement of the Transport Properties of Fluids – Experimental Thermodynamics, Volume III*, eds. W.A. Wakeham, A. Nagashima & J.V. Sengers, pp. 51–75. Oxford: Blackwell Scientific Publications.

Kawata, M., Kurase, K., Nagashima, A., Yoshida, K. & Isdale, J.D. (1991b). Falling–body viscometers, in *Measurement of the Transport Properties of Fluids – Experimental Thermodynamics, Volume III*, eds. W.A. Wakeham, A. Nagashima & J.V. Sengers, pp. 97–110. Oxford: Blackwell Scientific Publications.

Kestin, J., Paykoc, E. & Sengers, J.V. (1971). On the density expansion for viscosity in gases. *Physica*, **54**, 1–19.

Klein, M., Hanley, H.J.M., Smith, F.J. & Holland, P. (1974). *Tables of Collision Integrals and Second Virial Coefficients for the (m,6,8) Intermolecular Potential Function*, Natl. Stand. Ref. Data Ser. 47. Gaithersburg, MD: National Bureau of Standards.

Laesecke, A., Krauss, R., Stephan, K. & Wagner, W. (1990). Transport properties of fluid oxygen. *J. Phys. Chem. Ref. Data*, **19**, 1089–121.

Luettmer–Strathmann, J. (1994). *Transport Properties of Fluids and Fluid Mixtures in the Critical Region*. Ph.D. Thesis, University of Maryland, College Park, MD, USA.

Mourits, F.M. & Rummens, F.H.A. (1977). A critical evaluation of Lennard-Jones and Stockmayer potential parameters and of some correlational methods, *Can. J. Chem.*, **55**, 3007–3020.

Nieto de Castro, C.A., Friend, D.G., Perkins, R.A. & Rainwater, J.C. (1990). Thermal conductivity of a moderately dense gas. *Chem. Phys.*, **145**, 19–26.

Nieuwoudt, J.C. & Shankland, I.R. (1991). Oscillating–body viscometers, in *Measurement of the Transport Properties of Fluids – Experimental Thermodynamics Volume III*, eds. W.A. Wakeham, A. Nagashima & J.V. Sengers, pp. 9–48. Oxford: Blackwell Scientific Publications.

NIST Standard Reference Data (1992). *NIST Thermophysical Properties of Hydrocarbon Mixtures Database, SUPERTRAPP*, Version 1.04. Gaithersburg, MD: National Institute of Standards and Technology (U.S.).

NIST Standard Reference Data (1993). *NIST Mixture Property Database, NIST14*, Version 9.08. Gaithersburg, MD: National Institute of Standards and Technology (U.S.).

Olchowy, G.A. & Sengers, J.V. (1988). Crossover from singular to regular behavior of the transport properties of fluids in the critical region. *Phys. Rev. Lett.*, **61**, 15–18.

Perkins, R.A. & Cieszkiewicz, M.T. (1991). *Experimental Thermal Conductivity, Thermal Diffusivity, and Specific Heat Values for Mixtures of Nitrogen, Oxygen, and Argon*, NISTIR 3961. Washington, DC: National Institute of Standards and Technology (U.S.).

Perkins, R.A., Friend, D.G., Roder, H.M. & Nieto de Castro, C.A. (1991a). Thermal conductivity surface of argon: a fresh analysis. *Int. J. Thermophys.*, **12**, 965–984.

Perkins, R.A., Roder, H.M., Friend, D.G. & Nieto de Castro, C.A. (1991b). The thermal conductivity and heat capacity of fluid nitrogen, *Physica A* **173**, 332–362.

Rainwater, J.C. (1991). Vapour liquid equilibrium and the modified Leung–Griffiths model, in *Supercritical Fluid Technology*, eds. J.F. Ely & T.J. Bruno, pp. 57–162. Boca Raton, FL: CRC Press.

Roder, H.M. (1982). The thermal conductivity of oxygen. *J. Res. Nat. Bur. Stand. (U.S.)*, **87**, 279–310.

Roder, H.M. (1985). Thermal conductivity of methane for temperatures between 110 and 310 K with pressures to 70 MPa. *Int. J. Thermophys.*, **6**, 119–142.

Roder, H.M. & Friend, D.G. (1985). Thermal conductivity of methane–ethane mixtures at temperatures between 140 and 330 K and at pressures up to 70 MPa. *Int. J. Thermophys.*, **6**, 607–617.

Schwen, R. & Puhl, H. (1988). *Mathematische Darstellung der Viskosität-Temperatur-Abhängigkeit von Flüssigkeiten und Gasen*, Fortschritt–Berichte VDI Series 3, No. 166. Düsseldorf, Germany: VDI–Verlag GmbH.

Sengers, J.V. (1972). Transport processes near the critical point of gases and binary liquids in the hydrodynamic regime. *Ber. Bunsenges. Phys. Chem.*, **76**, 234–249.

Sengers, J.V. & Keyes, P.H. (1971). Scaling of the thermal conductivity near the gas–liquid critical point. *Phys. Rev. Lett.*, **26**, 70–73.

Shankland, I.R. & Lund, E.A.E. (1990). *Azeotrope–like compositions of pentafluoroethane and difluoromethane*. Washington, DC: U.S. Patent 4,978,467.

Shaumeyer, J.N., Gammon, R.W., Sengers, J.V. & Nagasaka, Y. (1991). Light scattering, in *Measurement of the Transport Properties of Fluids – Experimental Thermodynamics, Volume III*, eds. W.A. Wakeham, A. Nagashima & J.V. Sengers, pp. 197–225. Oxford: Blackwell Scientific Publications.

Uribe, F.J., Mason, E.A. & Kestin, J. (1989). A correlation scheme for the thermal conductivity of polyatomic gases at low density. *Physica A*, **156**, 467–491.

Vesovic, V., Wakeham, W.A., Olchowy, G.A., Sengers, J.V., Watson, J.T.R. & Millat, J. (1990). The transport properties of carbon dioxide. *J. Phys. Chem. Ref. Data*, **19**, 763–808.

Weber, L.A. (1982). Thermal conductivity of oxygen in the critical region. *Int. J. Thermophys.*, **3**, 117–135.

Widiatmo, J.V., Sato, H. & Watanabe, K. (1994). Saturated–liquid and bubble–point pressures of the binary system HFC–32 + HFC–125. *High Temp.–High Press.*, **25**, 677–683.

Younglove, B.A. (1982). Thermophysical properties of fluids. I. Argon, ethylene, parahydrogen, nitrogen, nitrogen trifluoride, and oxygen. *J. Phys. Chem. Ref. Data*, **11**, Suppl. 1.

8

Equations of State

K. M. de REUCK, R. J. B. CRAVEN,
and A. E. ELHASSAN

IUPAC Thermodynamic Tables Project Centre,
Imperial College, London, UK

8.1 Introduction

The purpose of this chapter, in a book about transport properties, is to give advice to
the reader on the best methods for converting the data, which are usually measured as
a function of P and T, to a function of ρ and T, which is the form required for the
correlating equations; and, in addition, to provide sources for values of the ideal–gas
isobaric heat capacities, which are also required for the transport–property calculations.
Both of these purposes can be fulfilled by calculations from a single equation of state
which has been fitted to the whole thermodynamic surface. Heat capacities of the real
fluid are required only for the calculation of the critical enhancement of the thermal
conductivity and viscosity, as described in Chapter 6; discussion of these properties in
this chapter will be restricted to Section 8.4.4.

An equation of state for a pure fluid relates the various equilibrium thermodynamic
properties to one another and will usually be largely an empirical function, although at
the limits it will approach theoretical values. In general, entirely theoretical equations
of state are unable to represent measured data to within their experimental accuracy.
If the accuracy of the calculated transport properties is to be as high as possible, it is
important that the most accurate equation of state be used for calculating the appropriate
densities.

8.2 Types of equation of state

The simplest form of equation of state is that for the ideal gas, $PV = RT$, which is
a combination of the laws of Boyle and Charles and is applicable to real gases at low
densities. From this equation two major types of equations of state have been developed.

The first follows from the improvements to the ideal–gas equation made by van der
Waals in 1873, when he included a parameter 'b' to account for the volume occupied
by the molecules and a term '(a/V^2)' to account for the intermolecular forces. In spite
of its simplicity, the van der Waals equation gives a good qualitative representation of
the behavior of a fluid, including the processes of condensation and evaporation, and

165

it does not lead to physically absurd results. This equation has been developed into a whole family of more than 100 'cubic' equations – so named because they are cubic in volume. They are easy to manipulate but are never able to reproduce the whole fluid surface to within its experimental accuracy; they are discussed in detail in Section 8.2.3.

The second line of development has occurred via the virial equation for gases, first suggested by Kamerlingh–Onnes (1901), where the compression factor $Z = P/(\rho RT)$ is expressed as a power series either in inverse volume or in pressure, which at the limit of zero pressure reduces to the ideal–gas equation. The virial coefficients themselves are functions of temperature only, and statistical thermodynamics has since shown that the second virial coefficient, B, arises from the interactions between pairs of molecules, and the third, C, from that between triplets. This equation is an infinite series which does not converge at higher densities and is most frequently used in a truncated form to represent the properties in the gas phase; these equations are discussed in Section 8.2.2. Benedict, Webb & Rubin (1940) extended the limits of applicability of the virial equation by adding an exponential term to a truncated virial equation, thus enabling both the gaseous and liquid phases to be represented. Further extensions of this form of equation have been (and are continuing to be) made, leading to the most accurate equations of the present day; they are able to represent all the equilibrium thermodynamic properties to within their experimental accuracy for a limited number of pure fluids. These accurate equations of state are discussed in Section 8.2.1.

8.2.1 Wide–range accurate equations of state

A wide–range equation of state will represent the thermodynamic properties, in principle, over the whole fluid range. In practice, the temperature range covered will usually be from the triple point to well beyond the critical value, and the pressure range will be between zero in the gas phase and the melting curve in the liquid phase. It will also represent the saturation properties, giving accurate values for the vapor pressure and the two saturated densities; all the single and two–phase properties will be thermodynamically consistent because they are calculated from a single equation. Properties calculated from the best of these equations will predict all the measured properties to within their experimental accuracies. There are fewer than 20 fluids for which such accurate and wide–range equations are available; a list of these is given in Section 8.3.

8.2.1.1 Helmholtz energy functions

The most common form of wide–range accurate equation being developed today is that of a dimensionless Helmholtz energy function, which can be written as

$$A(\delta, \tau)/RT = \alpha^{\mathrm{R}}(\delta, \tau) + \ln(\delta/\delta_{\mathrm{a}}) + \alpha^{\mathrm{id}}(\tau) \tag{8.1}$$

where $\delta = \rho/\rho^*$, $\tau = T^*/T$ and ρ^* and T^* are the reducing parameters: for some equations these reducing parameters will take the critical values, but as this is not always so, the user should be careful to use the values given by the author of the original paper. The term $\alpha^R(\delta, \tau)$ is the residual Helmholtz energy, which is the difference between the Helmholtz energy of the real fluid and that of the ideal gas at the point (δ, τ). The sum $\alpha^{id}(\tau) + \ln(\delta/\delta_a)$ is the Helmholtz energy of the ideal gas, the isobaric heat capacity of which is discussed in Section 8.4.5.

To obtain the density from equation (8.1) at a given pressure and temperature, only the function $\alpha^R(\delta, \tau)$ is required. This is related to the pressure by

$$P = \rho RT[1 + \delta(\partial \alpha^R/\partial \delta)_\tau] \tag{8.2}$$

The functional form of $\alpha^R(\delta, \tau)$ is usually

$$\alpha^R(\delta, \tau) = \sum_{i=1}^{M_1} N_i \delta^{r_i} \tau^{s_i} + \sum_{i=M_1+1}^{M_2} N_i \delta^{r_i} \tau^{s_i} \exp[-(q\delta)^{k_i}] + \cdots \tag{8.3}$$

where, for some fluids, there will be a few additional exponential terms of importance in the near–critical region. If the reducing parameters for ρ and T are the critical properties, then $q = 1$. Equation (8.3) may contain up to 50 coefficients, the numerical values of which will have been estimated by least–squares techniques. The equation is usually a linear function of the coefficients; all thermodynamic properties which are also linear functions of the coefficients such as $P\rho T$, ΔH and C_V as well as the Maxwell criterion for the saturation properties can be included in a linear fit. The data used to fit the equation are critically assessed, and preliminary fits are used to eliminate sets of measurements which deviate systematically from the bulk of the data. There are usually between 1500 and 3000 data points for which individual weights, inversely proportional to the variance, are estimated; then error propagation weights are calculated using preliminary fits to the data. Isobaric heat capacity, speed of sound and Joule–Thomson coefficient measurements can be included using nonlinear fitting techniques.

For the most accurate equations, as well as optimizing the values of the coefficients, the number of terms in the equation is also optimized using methods described by de Reuck & Armstrong (1979), Setzmann & Wagner (1989) and Ewers & Wagner (1982). Other wide–range equations are often fitted to equations with a fixed structure.

Numerical methods which can be used to solve equation (8.2) for density are given in Section 8.4. For some of these methods the partial derivative of the pressure with respect to density is needed, which is given by

$$(\partial P/\partial \rho)_T = RT[1 + 2\delta(\partial \alpha^R/\partial \delta)_\tau + \delta^2(\partial^2 \alpha^R/\partial \delta^2)_\tau] \tag{8.4}$$

8.2.1.2 *Pressure or compression factor functions*

Other wide–range equations of state are given in the form of either $P = f(\rho, T)$ or $Z = f(\rho, T)$, or either of these as functions of reduced density and temperature. In all these cases the isobaric heat capacity of the ideal gas will be given as a function of temperature in a separate equation, but many of the derived properties will have to be calculated by integration. For the real fluid, the dimensionless form of this type of equation is

$$P/(\rho RT) = 1 + \sum_{i=1}^{M_1} N_i \delta^{r_i} \tau^{s_i} + \exp[-(q\delta)^2] \sum_{i=M_1+1}^{M_2} N_i \delta^{r_i} \tau^{s_i} \qquad (8.5)$$

where, again, $q = 1$ if the critical properties are used for the reducing parameters. The derivative required for the numerical solution of the density is

$$(\partial P/\partial \rho)_T = RT \left[1 + \sum_{i=1}^{M_1} N_i (r_i + 1) \delta^{r_i} \tau^{s_i} + \right.$$

$$\left. \exp\left[-(q\delta)^2\right] \sum_{i=M_1+1}^{M_2} N_i (r_i + 1 - 2q\delta^2) \delta^{r_i} \tau^{s_i} \right] \qquad (8.6)$$

and suitable numerical methods are given in Section 8.4.

8.2.2 Limited–range accurate equations of state

Limited–range equations which also represent the experimental data to within their accuracy are mostly available for the gas phase. Virial equations which have been fitted to a wide temperature range are usually of the form

$$Z = \frac{PV}{RT} = 1 + \frac{B}{V} + \frac{C}{V^2} + \frac{D}{V^3} + \dots \qquad (8.7)$$

and given the pressure and temperature, the volume or density can be calculated iteratively by using one of the numerical methods described in Section 8.4. Since the left–hand side of the equation is dimensionless, the virial coefficients will have the dimensions of V, V^2, V^3 etc. The virial equation can also be given in terms of pressure instead of volume when

$$Z = PV/(RT) = 1 + B'P + C'P^2 + \dots \qquad (8.8)$$

This form is easier to solve for the required density or volume but is less reliable at the lowest temperatures. If a virial equation is used over a restricted range of the gas phase which includes the ideal gas region, it is usually truncated after the second or third term. When it is truncated after the third term, it becomes cubic in density and can be solved analytically, as described in Section 8.4.2.

Another less frequently used equation which can represent gas–phase properties is a reduced Gibbs function. This was successfully used by Craven *et al.* (1989) to correlate the gas–phase data for methanol.

8.2.3 Cubic equations of state

Cubic equations of state are only recommended for use when there are no other equations available which have been fitted to a wide range of accurate experimental data. In these circumstances it may be necessary for the reader to fit a cubic equation to whatever data are available, and since they contain few adjustable parameters these can be correlated with a limited amount of experimental data. Therefore we have included in this section some details about the various fitting methods.

Although the qualitative behavior of the van der Waals equation is good, quantitatively it cannot predict many of the saturation and single–phase properties for different fluids, including spherical nonpolar fluids such as argon. A major limitation is its inability to predict liquid–phase and saturation densities with reasonable accuracy, especially for polar fluids; at low densities the predictions are acceptable but they are too high at higher pressures, and Z_c is too high (0.375). The temperature dependence of the second virial coefficient given by $B = b - a/RT$ fails to produce a maximum for B at any temperature.

A major improvement on the van der Waals equation was made by Redlich & Kwong (1949) when they introduced a weak temperature dependence for the attractive term, resulting in the well–known equation

$$P = [RT/(V - b)] - [a/\{T^{1/2}V(V + b)\}] \tag{8.9}$$

This significantly improved the volumetric predictions, especially in the liquid region, although the two parameters remained as constants related to the critical properties. Also, the prediction of second virial coefficients for simple fluids was much improved. However, Z_c is still too high (0.333) for most fluids and the equation is not suitable for complex fluids with nonzero acentric factors, which leads to low values of Z_c of < 0.3.

A revival of interest in cubic equations in the sixties and seventies led to the introduction of two equations which have found great acceptance among engineers: the Soave–Redlich–Kwong (SRK) equation and the Peng–Robinson (PR) equation. The SRK equation (Soave 1972) has the form

$$P = [RT/(V - b)] - [a(T)/\{V(V + b)\}]$$

$$\text{where} \quad a(T) = 0.42747[1 + k(1 - T_r^{1/2})]^2 R^2 T_c^2/P_c$$

$$k = 0.48 + 1.574\omega - 0.176\omega^2$$

$$\text{and} \quad b = 0.08664RT_c/P_c \tag{8.10}$$

The parameter a is now temperature–dependent and has been fitted to the acentric factor ω for pure hydrocarbons, but the b parameter remains constant. The PR equation

(Peng & Robinson 1976) extends the denominator of the attractive term so that

$$P = [RT/(V - b)] - [a(T)/\{V(V + b) + b(V - b)\}]$$

where $a(T) = 0.45724[1 + k(1 - T_r^{1/2})]^2 R^2 T_c^2 / P_c$

$$k = 0.37464 + 1.54226\omega - 0.26992\omega^2$$

and $b = 0.0778 R T_c / P_c$ (8.11)

The introduction of the third parameter, ω, significantly improved the fit to the saturation properties of hydrocarbons for both equations. The SRK equation gives better predictions for the vapor pressure and saturated vapor volumes for alkanes from C_1 up to C_{10}. For saturated liquid densities, the compression factor in the liquid and densities above the critical temperature, the SRK equation gives better results for C_1 and C_2 only, while the PR equation is better for hydrocarbons higher than ethane (Yu *et al.* 1986). Thus the choice of the most suitable equation to use will depend on both the size of the molecule and the part of the surface to be considered.

Since the introduction of the SRK and PR equations, many other cubic equations of state have appeared with similar modifications to the a and b parameters. Of these, the Patel–Teja (PT) equation (Patel & Teja 1982) is one which is increasingly being used. The form of the PT equation is similar to the PR but introduces an extra parameter, c, and is written as

$$P = [RT/(V - b)] - [a(T)/\{V(V + b) + c(V - b)\}]$$

where $a(T) = \Omega_a \alpha(T) R^2 T_c^2 / P_c \quad b = \Omega_b R T_c / P_c$

$$c = \Omega_c R T_c / P_c$$

$$\Omega_a = 3Z_c^2 + 3(1 - 2Z_c)\Omega_b + \Omega_b^2 + 1 - 3Z_c$$

Ω_b equals the smallest positive root of

$$\Omega_b^3 + (2 - 3Z_c)\Omega_b^2 + 3Z_c^2\Omega_b - Z_c^3 = 0$$

$$\Omega_c = 1 - 3Z_c, \quad Z_c = 0.329032 - 0.076799\omega + 0.0211947\omega^2$$

$$\alpha(T) = \left[1 + k(1 - T_r^{1/2})\right]^2$$

and $k = 0.452413 + 1.30982\omega - 0.295937\omega^2$ (8.12)

Although this equation is more complicated than the SRK or the PR, it was designed to include the merits of both. Also, as $\omega \to 0$, that is, for small molecules, Z_c approaches the value of 0.333 given by the SRK equation, but for moderately large molecules with $\omega \approx 0.3$ the value of Z_c approaches the value 0.3074 given by the PR equation. The correlations for k and Z_c given above are adequate for representing the saturation properties for nonpolar fluids, but for polar fluids specific substance–dependent values

Table 8.1. *Values of coefficients of cubic equations in terms of their parameters.*

Equation	Coefficients			
	Ω_a	Ω_b	α	Ω_c
RK	0.4278	0.0867	$T_c^{1/2}$	0
SRK	0.42747	0.08664	$[1 + k_{SRK}^\dagger (1 - T_r^{1/2})]^2$	0
PR	0.45724	0.0778	$[1 + k_{PR}^\ddagger (1 - T_r^{1/2})]^2$	0
PT	★	★	$[1 + k_{PT}^* (1 - T_r^{1/2})]^2$	$1 - 3Z_c$

† See equation (8.10); ‡ see equation (8.11); * see equation (8.12).

for the parameters must be used, which have usually been found by fitting to vapor pressures and saturated liquid densities. Values for these parameters for more than 50 pure fluids, including alcohols and carboxylic acids, can be found in Georgeton *et al.* (1986).

All the above cubic equations contain only two or three adjustable parameters, and there is a choice of two methods for estimating them. The first is based on the corresponding–states principle where the a and b parameters are fitted to the critical constants P_c and T_c by constraining the respective equations to the critical point where $(\partial P/\partial V)_T = (\partial^2 P/\partial V^2)_T = 0$. The third parameter c used by the PT equation can be calculated from $Z_c = P_c V_c/(RT_c)$ using critical properties only or from a suitable correlation with the acentric factor ω as shown in equation (8.12) or taken from a compiled list such as that given by Georgeton *et al.* (1986). The general forms of these parameters then become

$$a = \alpha \Omega_a (R^2 T_c^2 / P_c), \quad b = \Omega_b (RT_c/P_c) \quad \text{and} \quad c = \Omega_c (RT_c/P_c) \qquad (8.13)$$

and the values of these coefficients for each of the cubic equations are listed in Table 8.1.

The second method is to fit the a and b parameters to vapor pressure and saturated liquid density data along the saturation curve, in which case they become temperature–dependent and can be correlated using a function such as the one proposed by Soave (1972). This requires the solution of the following three simultaneous equations in three unknowns:

$$P_\sigma = f(T, V_1, a, b), \qquad P_\sigma = f(T, V_g, a, b)$$

and

$$\int_{V_g}^{V_1} P(V) dV = P_\sigma (V_1 - V_g) \qquad (8.14)$$

The third equation represents the Maxwell equal–area constraint or, equivalently, the equality of the fugacities of the two equilibrium phases. By assigning experimental values to T, P_σ and V_1, the three unknown variables a, b and V_g can be found by solving equations (8.14) using an iterative technique such as Newton's method. Initial estimates must be given for the unknown variables.

In some applications, because of the weak dependence of the b parameter on temperature at low vapor pressures (Panagiotopoulos & Kumar 1985), it is held constant; also, some investigations (Trebble & Bishnoi 1986; Hnědkovský & Cibulka 1990) have shown that varying the b parameter with temperature may lead to thermodynamic inconsistencies in the compressed liquid region at elevated pressures and above the critical temperature, which result in the prediction of negative heat capacities and crossovers in enthalpy isotherms. In these cases equations (8.14) are solved for the unknowns a, V_1 and V_g by fitting to vapor pressure data alone.

8.2.4 Tait equation

The Tait equation is one of the simplest and yet more accurate empirical equations for fitting liquid density data of pure components over wide ranges of temperature and pressure. The general expression for this equation in terms of density along an isotherm is

$$(\rho - \rho_0)/\rho = C \log\left[(B + P)/(B + P_0)\right] \tag{8.15}$$

where C and B are adjustable parameters determined from fitting $P\rho$ data and ρ_0 is the density at a reference pressure P_0, usually taken as 0.1 MPa or the saturation pressure. The C parameter is fairly independent of temperature over a wide range of pressures, but the B parameter is usually correlated with a second–degree polynomial in reduced temperature such as

$$B = a + bT_r + cT_r^2 \tag{8.16}$$

The value of C ranges from 0.173 for 1–chlorobutane to 0.3146 for pentan–1,5–diol, but for all n–alkanes it can be taken as 0.2000. C varies over small temperature ranges for some fluids but can often be considered as constant except when the data extend to low temperatures (Dymond & Malhotra 1988). Values for both parameters for a number of substances over different ranges of temperature and pressure are given in Malhotra & Woolf (1991), Susnar *et al.* (1992) and Dymond *et al.* (1988).

An important advantage of the Tait equation is that when extrapolated to higher pressures and outside the experimental range, it continues to give reasonably accurate density results.

8.3 Selecting an equation of state

The most accurate equations of state are those where the coefficients have been estimated by multiproperty fitting of critically assessed and statistically weighted data and for which the form of the equation has also been optimized. Fluids for which this process has been completed are:

- Ammonia (Haar & Gallagher 1978)
- Carbon Dioxide (Angus *et al.* 1976)
- Ethene (Jacobsen *et al.* 1988)
- Fluorine (de Reuck 1990)
- Heavy water (Hill *et al.* 1982)
- Helium-4 (Angus *et al.* 1977)
- Methane (Setzmann & Wagner 1991)
- Methanol (de Reuck & Craven 1993)
- Nitrogen (Angus *et al.* 1979)
- Oxygen (Wagner & de Reuck 1987)
- Propene (Angus *et al.* 1980)
- R11 (Marx *et al.* 1992)
- R12 (Marx *et al.* 1992)
- R22 (Wagner *et al.* 1993)
- R113 (Marx *et al.* 1992)
- R134a (Tillner–Roth & Baehr 1994)
- Water (Saul & Wagner 1989)

Fluids for which the data have been fitted to an equation of fixed form, frequently a 32–term equation first fitted to nitrogen by Stewart & Jacobsen (Angus *et al.* 1979), or a 32–term equation with a different form first fitted to oxygen by Schmidt & Wagner (Wagner & de Reuck 1987), are:

- Air (Sytchev *et al.* 1978; Jacobsen *et al.* 1992)
- Argon (Stewart & Jacobsen 1989)
- Isobutane (Waxman & Gallagher 1983)
- Butane (Younglove & Ely 1987)
- Ethane (Friend *et al.* 1991)
- Hydrogen (McCarty 1974)
- Propane (Younglove & Ely 1987)
- R123 (Younglove 1993)

There are also equations for krypton (Juza & Sifner 1976; Polt & Maurer 1992), neon (Polt & Maurer 1992), sulphur dioxide (Polt & Maurer 1992) and xenon (Juza & Sifner 1977) and one for chlorine (Angus *et al.* 1985), where some of the data were predicted.

Useful sources for other wide–range equations of state are the *Journal of Physical and Chemical Reference Data* and Monographs, published by the U.S. National Bureau of Standards, now called the National Institute for Standards and Technology (NIST).

When using any of the above equations, or others found in the literature, care should be taken to note the limits on T, P given by the authors; extrapolation outside these is not recommended. If data are required only over a limited range of temperature and pressure in the gas phase, which includes the ideal–gas region, then the best choice will usually be a virial equation (8.7) or (8.8), truncated after the second or third term depending on the range required. Although it is possible to obtain approximate values for the virial coefficients from theoretical considerations, the most accurate equations will be those derived from fitting to experimental data over the range of validity of the truncated equations. Tsonopoulos & Heidman (1990), among others, have examined the validity range for virial equations truncated after the second and third terms, and found them to give good results for densities up to $0.25\rho_c$ and $0.5\rho_c$ respectively. For highly polar and associating fluids like hydrogen fluoride and the alcohols, the range of validity of such equations will usually be more restricted than the limits given here, and properties near the saturation curve may show anomalous behavior. Therefore where possible it is recommended that the deviations of all fitted data from the equation are examined and the validity limits reduced if necessary. If a choice is necessary between equations (8.7) and (8.8) truncated after the second term, then the pressure virial equation is superior at temperatures above the critical, and the density virial equation is better below the critical temperature. There are several methods for deriving truncated virial equations from experimental data. The first and preferred method is to fit all the critically assessed data with the correct weighting over the required density and temperature range. Usually only $P\rho T$ data are used, but this restriction is unnecessary, as other equilibrium properties may be included in the fitting. Another method is to determine values of the virial coefficients at a series of temperatures by plotting $(Z - 1)/\rho$ against ρ or $(Z - 1)/P$ against P along isotherms. The density or pressure virials are then fitted separately as a function of temperature; the disadvantages of this method have been discussed by Holleran (1970). A third method uses acoustic virial coefficients derived from speed of sound data. However, in order to convert these into $P\rho T$ virials, an expression for the intermolecular potential is often used (Ewing *et al.* 1987). Although speed of sound data can be very precise, the same is not true of the intermolecular potential. Also, since speed of sound data can be incorporated directly into the first method, use of the potential is unnecessary. Therefore, if a choice of virial equations is available for a particular fluid, preference should be given to one derived by simultaneous fitting.

There are very few limited–range virial equations for specific fluids in the literature. Values for the second virial coefficients of several gases at specified temperatures are given in Marsh *et al.* (1993) and Dymond & Smith (1980); if a correlating equation is required, it is recommended that these values are fitted to a polynomial in $1/T$.

Cubic equations will not be able to provide the accuracy of a wide–range equation of state nor that of the limited–range equations for their specified range and should therefore be used only when there are no other sources available. As shown in Section 8.2.3, a cubic equation will give better results if it has been fitted to experimental data in the region of interest rather than fitted to critical properties alone.

In the liquid phase the best choice may be a Tait equation, as discussed in Section 8.2.4.

8.4 Methods for calculating density

For all wide–range equations of state the calculation of density has to be carried out iteratively, although for some of the simpler equations this can be done directly.

8.4.1 Units and temperature scale conversion

For equations of state which are not in dimensionless form, care must be taken to adopt the units appropriate to the equation of state coefficients. This applies particularly to the numerical value used for the gas constant R.

For equations of state such as equation (8.1) or (8.5), each of which is a dimensionless function of reduced density and temperature, the user can in principle select the desired units by suitable choice of the reducing parameters ρ^* and T^*. While for ρ^* this is straightforward, over the years there have been several different agreed international temperature scales, and there are no simple conversion factors between them. In 1990 a change was made from the then current scale, known as IPTS–68, to the present approved scale ITS–90. Some accurate equations of state which are recommended in Section 8.3 are based on IPTS–68 and others on ITS–90; conversion from one scale to the other can be made using the equations given by Rusby (1991). For the temperature range $T_{68} = 73.15$ K to 903.89 K, which will cover most of the fluids of concern to readers of this book, the equation is

$$(T_{90} - T_{68})/\text{K} = \sum_{i=1}^{8} b_i [\{(T_{68}/\text{K}) - 273.15\}/630]^i \qquad (8.17)$$

with the coefficients b_i listed in Table 8.2. Conversions from earlier scales to that of ITS–90 are given in Goldberg & Weir (1992).

Table 8.2. *Numerical values of the coefficients b_i of equation (8.17).*

i	b_i	i	b_i
1	-0.148 759	5	-4.089 591
2	-0.267 408	6	-1.871 251
3	1.080 760	7	7.438 081
4	1.269 056	8	-3.536 296

For the highest accuracy all temperatures should be converted to the scale used for the equation of state before calculating any properties from it.

8.4.2 Single–phase regions

In the single phase, there will be only one physically correct solution for the density from an equation of state, although solution of the equation may produce several roots for a given pressure and temperature. In Section 8.5 numerical techniques for finding this density are given.

If a density virial equation, such as equation (8.7), is truncated after the second term it becomes quadratic in density and the physically correct root will correspond with the minimum Gibbs energy.

Cubic equations as well as the density virial equation truncated after the third term can be solved without iterative procedures. Following Abramowitz & Stegun (1972), a cubic equation of the form

$$\rho^3 + a_2\rho^2 + a_1\rho + a_0 = 0$$

will have roots

$$\rho_1 = (s_1 + s_2) - a_2/3$$
$$\rho_2 = (s_1 + s_2)/2 - a_2/3 + i\sqrt{3}(s_1 - s_2)/2$$
$$\text{and} \quad \rho_3 = (s_1 + s_2)/2 - a_2/3 - i\sqrt{3}(s_1 - s_2)/2$$
$$\text{where} \quad s_1 = [r + (q^3 + r^2)^{1/2}]^{1/3}$$
$$s_2 = [r - (q^3 + r^2)^{1/2}]^{1/3}$$
$$q = a_1/3 - a_2^2/9$$
$$\text{and} \quad r = (a_1a_2 - 3a_0)/6 - a_2^3/27 \tag{8.18}$$

If one root is real, this is the desired solution, but if all roots are real then the physically correct density is the root which corresponds with the minimum Gibbs energy.

For all other equations one of the numerical methods given in Section 8.5 can be used.

8.4.3 Saturation region

For the wide–range accurate equations of state the saturation region properties will be calculated by equating the Gibbs energies, so that along an isotherm,

$$A_l + P_\sigma/\rho_l = A_g + P_\sigma/\rho_g \tag{8.19}$$

In terms of the equation of state (8.1) this can be written as

$$P_\sigma/(\rho^* RT) = \delta_l\delta_g/(\delta_l - \delta_g)[\ln(\delta_l/\delta_g) + \alpha^R(\delta_l, \tau_\sigma) - \alpha^R(\delta_g, \tau_\sigma)] \tag{8.20}$$

Similarly, equation (8.2) can be used to give $P = P_\sigma$ when $\delta = \delta_1$ and also when $\delta = \delta_g$ at any given value of τ_σ, so that there are three equations with four unknowns: P_σ, τ_σ, δ_g and δ_1. If a value is assigned to either P_σ or τ_σ, then the other three can be calculated using an iterative technique such as those described in Section 8.5. Equation (8.20) is equivalent to the Maxwell equal–area principle for the P–V surface. Applying equation (8.19) to the pressure or compression factor functions, such as equation (8.5), requires integration and results in a more complicated expression. The expressions required for the Stewart & Jacobsen 32–term equation can be found in Angus *et al.* (1979), and for the 32–term Schmidt & Wagner equation they will be found in Wagner & de Reuck (1987).

An alternative method is to consider the A–V surface (Rowlinson & Swinton 1982), where the saturation pressure is given by the slope of the common tangent to the curve at the saturated volumes. The objective function, at constant temperature, is

$$\mathcal{F}(V_1, V_g) \equiv \left| \int_{V_1}^{V_g} A(V)dV \right| - \left| [A(V_1) + A(V_g)] \cdot \frac{(V_g - V_1)}{2} \right| \tag{8.21}$$

and at equilibrium $\mathcal{F}(V_1^\sigma, V_g^\sigma) = \max \mathcal{F}(V_1, V_g)$. This is called the Area method which is described in Elhassan *et al.* (1993) and was developed from the Area method for multicomponent vapor–liquid equilibrium by Eubank *et al.* (1992).

This equation can be solved numerically in three different ways. The first is by direct integration of equation (8.21) and is described in Section 8.5.2. The second method is to treat the objective function as an unconstrained optimization, which can be solved using any direct search technique such as Powell's method (1964). This approach is computationally faster than the direct integration but requires good initial estimates for the volumes. The third method is to differentiate equation (8.21) to give two nonlinear equations:

$$(\partial \mathcal{F}/\partial V_1)_{V_g} = [(V_g - V_1)/2](\partial A/\partial V_1)_{V_g} - [A(V_g) - A(V_1)]/2 = 0 \tag{8.22}$$

$$(\partial \mathcal{F}/\partial V_g)_{V_1} = [(V_1 - V_g)/2](\partial A/\partial V_g)_{V_1} - [A(V_1) - A(V_g)]/2 = 0 \tag{8.23}$$

which can then be solved simultaneously using the Newton–Raphson method. Again, this method is computationally fast but good initial estimates must be given if trivial solutions are to be avoided. An advantage of the Area method over the Maxwell principle, irrespective of the solution technique used, is that it reduces the number of independent variables to two, which removes the need to provide for or to calculate the saturation pressure during the computation.

For limited–range equations in the gas phase, the density of the saturated vapor may be found by inserting values for the saturated vapor pressure at the saturation temperature in the truncated virial equations or in the Gibbs equation (see Section 8.2.2) as appropriate and then proceeding as described for the single phase. In using this method care should be taken to use an accurate equation for the vapor pressure; accurate vapor pressure

measurements which have been fitted to a Wagner–type equation, such as those by
Ambrose (1986), will give good results for a wide variety of fluids.

8.4.4 Critical region

Values for the density, isothermal compressibility $\kappa = (1/\rho)(\partial \rho/\partial P)_T$ and isochoric
and isobaric heat capacities are needed to describe the observed enhancement of the
thermal conductivity and viscosity in the critical region. This phenomenon is adequately
discussed in Chapter 6, which also gives details for calculating these properties from
scaled equations of state. However, apart from a region very close to the critical point,
these properties may also be calculated with comparable accuracy and total thermodynamic consistency from the wide–range accurate equations of state described in
Section 8.2.1. Other forms of equations give results in the critical region which are too
inaccurate and so are not discussed here.

At a given pressure and temperature, the density must first be calculated before
the isothermal compressibility or the heat capacities can be obtained. The isothermal
compressibility may then be calculated directly from equation (8.3) or (8.5) by using
the derivatives given by equation (8.4) or (8.6). The isobaric heat capacity C_P at any
point is related to the isochoric heat capacity C_V by the expression

$$\frac{C_P(\delta, \tau)}{R} = \frac{C_V(\delta, \tau)}{R} + \frac{T(\partial P/\partial T)_\rho^2}{R\rho^2(\partial P/\partial \rho)_T} \tag{8.24}$$

where for the Helmholtz function

$$(\partial P/\partial T)_\rho = R\rho \left[1 + \delta(\partial \alpha^R/\partial \delta)_\tau - \delta\tau(\partial^2 \alpha^R/\partial \delta \partial \tau)\right] \tag{8.25}$$

and

$$C_V(\delta, \tau)/R = -\tau^2 \left[\left(d^2\alpha^{id}/d\tau^2\right) + \left(\partial^2 \alpha^R/\partial \tau^2\right)_\delta\right] \tag{8.26}$$

The quantity $-\tau^2 \left(d^2\alpha^{id}/d\tau^2\right)$ is related to the isobaric heat capacity of the ideal gas
C_P^{id} by

$$C_P^{id}/R - 1 = -\tau^2 \left(d^2\alpha^{id}/d\tau^2\right) \tag{8.27}$$

and the calculation of C_P and C_V therefore requires a knowledge of C_P^{id}, which is a
function of temperature only.

8.4.5 Heat capacity of the ideal gas

For monatomic fluids, C_P^{id} is simply $5R/2$; for more complex fluids, however, C_P^{id} is
best expressed by an equation of the form

$$\frac{C_P^{id}}{R} = \sum_{i=1}^{n} \frac{a_i u_i^2 \exp(u_i)}{[\exp(u_i) - 1]^2} + a_{n+1} \tag{8.28}$$

where the a_i are constants. For molecules possessing only rotational and n vibrational degrees of freedom, u_i is hv_i/kT, where v_i is the normal vibrational frequency of the ith degree of freedom. If values of v_i are known from spectroscopic data, equation (8.28) can be used to give C_P^{id} directly with good accuracy by setting a_{n+1} to 3.5 for linear or 4 for nonlinear molecules and $a_i=1$. Alternatively if values of C_P^{id} have been found by extrapolation to zero pressure of measurements of the isobaric heat capacity or speed of sound in the low–density gas, then these data can be fitted to an equation of the form of (8.28) by treating the a_i as adjustable parameters if approximate values of v_i are known. Alternatively both the a_i and v_i may be treated as adjustable parameters. This approach can also be adopted for molecules with more complex internal degrees of freedom such as internal rotation or for C_P^{id} data which have been derived from spectroscopy with corrections for anharmonicity. An equation of the form of (8.28) is recommended because it will give physically realistic values for C_P^{id} at all temperatures. Equations based on polynomials in temperature or reciprocal temperature will break down in the limit as $T \to 0$ or infinity. If such equations are used they should not be extrapolated outside their fitted range unless they have been compared with reliable data.

Useful sources for ideal–gas isobaric heat capacity equations are the *Journal of Physical and Chemical Reference Data* and *Monographs*, published by NIST. Data for the ideal gas can also be found in the *TRC Thermodynamic Tables – Hydrocarbons and Non–Hydrocarbons*, published by the Thermodynamic Research Center (TRC) at the Texas A&M University System, in Gurvich *et al.* (1978–1981) and Stull & Prophet (1971) and supplements to the *Joint-Army-Navy-Air-Force (JANAF) Tables* also published in the *Journal of Physical and Chemical Reference Data*.

8.5 Numerical methods

In this section two iterative schemes for calculating the density given values for the pressure and temperature are described: the bisection or interval–halving method, and the Newton–Raphson technique. These methods and others are described in more detail by Burden *et al.* (1978), who also give algorithms for these procedures.

8.5.1 Iterative techniques

Two of the most popular iterative techniques are the bisection and the Newton–Raphson methods. The former has the advantage that it always converges to a solution. However, it is linearly convergent, which can imply an excessive number of iterations; an intermediate solution may be closer to the correct answer than the final one. The Newton–Raphson technique is quadratically convergent and therefore requires fewer iterations than the bisection method. However, because of its small radius of convergence, it must be provided with good initial estimates for the density to avoid the problem of trivial solutions, especially around the critical region where the slope of the isotherm is close to zero.

The general expression for the Newton–Raphson method for an nth iteration is given as

$$\rho_{n+1} = \rho_n - f(\rho_n)/f'(\rho_n) \qquad (8.29)$$

where $f(\rho_n)$ represents the nonlinear equation for the density and $f'(\rho_n)$ its derivative. For any procedure some convergence criteria are necessary. Since the density can vary by several orders of magnitude over the gas and liquid phases, a test on the absolute density for convergence is unsuitable. Therefore, the following is recommended

$$| \, 1 - (\rho_{n-1}/\rho_n) \, | < \epsilon \qquad (8.30)$$

where ϵ is the desired accuracy.

For general purposes, the Newton–Raphson technique is recommended because of its simplicity and fast convergence. For calculations in the single phase well above the critical point, no convergence problems should be experienced, provided a suitable initial approximation for the density is given. At temperatures near but above the critical point, this method may fail and it will be necessary either to restrict the step size $f(\rho)/f'(\rho)$ or to use the bisection method. Below the critical temperature there is the problem of multiplicity of roots for noncubic equations, which will arise when the pressure lies between the spinodal values of the $P\rho$ isotherm. Here the initial density estimate will depend on whether the stable phase is vapor or liquid at the specified pressure and temperature. For the vapor phase an initial value of zero is recommended and for the liquid phase, a value of $3\rho_c$ or the triple–point liquid density, whichever is greater. As a precaution, the sign of $(\partial P/\partial \rho)_T$ should be checked to ensure that the final density does not correspond with physically unrealistic behavior. This precaution is sufficient in regions where the equation has only three real roots, but near the triple point, wide–range equations often exhibit multiple roots in the two–phase region. In this case, if doubt exists about the validity of the result the density should be checked against the appropriate saturation value at the experimental temperature. Techniques to find the saturation densities for a given experimental temperature are discussed in Section 8.4.3.

If the stable phase is unknown, two methods can be used. The first is to calculate the saturation pressure at the experimental temperature, as described in Section 8.4.3, and then compare it with the experimental value. If it is greater than P_σ, the stable phase is liquid; otherwise it is gas. The second method is to find all the roots of the equation and calculate the corresponding Gibbs energies. The stable phase corresponds with the density value, which gives the minimum Gibbs energy. This method has the advantage that only single phase calculations are necessary; they are computationally more efficient than those for the saturation curve. While the second method will give the thermodynamically correct answer in all cases, practical problems can arise when the wide–range equations exhibit multiple roots in the two–phase region; therefore, the first method is recommended.

8.5.2 Direct integration technique

This technique is applied solely to the Area method described in Section 8.4.3 for the solution of saturation properties. It is a powerful tool in the temperature range $0.9 < T_r < 0.999$, as it avoids the problem of initial estimates for either the pressure or the volumes. Its sole requirement is a fixed volume range, which is divided into a predetermined number of N nodes at which the Helmholtz function is evaluated. It was found from experience that a domain of volume $\{V \,|\, 0.4V_c, 7V_c\}$ and $N = 400$ ensure a reliable solution, in the temperature range given above, for most fluids – including polar fluids such as water and methanol.

The implementation of the technique is carried out in the following steps:

(i) Specify a value for the temperature, say at $T_r = 0.95$.

(ii) Evaluate the Helmholtz function at the N nodes. The first value is calculated at $V = 0.4V_c$, and the rest at volume increments of $(7V_c - 0.4V_c)/N$ up to the last value calculated at $V = 7V_c$. Values of the Helmholtz function and the corresponding volumes are then stored in computer memory.

(iii) Start evaluating and comparing the objective function, equation (8.21), two nodes at a time. The first node lies in the liquid phase and starts at $V = 0.4V_c$ and increases up to $V = V_c$. The second node lies in the vapor phase and starts just above the critical volume and varies up to $V = 7V_c$.

(iv) The two nodes at which the objective function reaches its maximum positive value correspond to the equilibrium volumes. The saturation pressure is then calculated from $P = -(\partial A/\partial V)_T$, which is evaluated at the vapor node.

The values given for the range and number of nodes are not fixed; they can be changed by the user according to the type of fluid and the particular isotherm.

8.6 Predicting volumetric properties of mixtures

There are no equations of state for mixtures with an accuracy comparable to that available for pure fluids. In practice, mixture properties are almost always predicted, using a variety of different equations of state and different mixing rules. This section lists a number of references to aid the reader when entering this very extensive literature. A general text, which discusses many different methods for the prediction of mixture properties, is to be found in the fourth edition of the volume by Reid, Prausnitz & Poling (1987).

8.6.1 Using the principle of corresponding states

The 'one–fluid' van der Waals model, based on that described by Rowlinson & Watson (1969), together with the shape factors defined by Leach *et al.* (1968), is frequently used to predict the properties of mixtures. Providing the binary interaction parameters are optimized using accurate measured data, this model will yield good values for mixture volumes. Either the user can optimize the binary interaction parameters for the particular fluids of concern or they can be taken from the literature, such as those listed by Mentzer, Greenkorn & Chao (1981) for mixtures of hydrocarbons on their own or mixed with carbon dioxide, hydrogen and hydrogen sulphide.

8.6.2 Using cubic equations of state

There is a very large literature related to the use of cubic equations of state to calculate $P - T - x$ surfaces of mixtures through applications of various mixing and combining rules to represent the parameters of these equations. Although the vapor–liquid equilibria (VLE) calculations by some of these cubic equations, with appropriate mixing and combining rules, produce results which are comparable in accuracy to the experimental data, the volumetric predictions for the mixtures are not, in general, satisfactory. Recently, Wong & Sandler (1992) have developed theoretical mixing rules that considerably improve the prediction of mixture volumetric properties.

8.6.3 Using BWR–type and virial–type equations of state

A useful set of computer programs for calculating the density of mixtures using the Benedict–Webb–Rubin (BWR) equation of state is given by Johnson & Colver (1968). Another computer program which improves the speed of calculation of the BWR–type equations is given by Plöcker & Knapp (1976). A virial–type equation, truncated after the fourth virial coefficient, has recently been used by Anderko & Pitzer (1991), together with mixing rules based on the rigorously known composition dependence of virial coefficients. This has been used to predict the volumetric properties of mixtures containing either methane or nitrogen with other nonpolar components such as the higher alkanes and benzene.

8.7 Acknowledgments

This work was supported by the Department of Trade and Industry of the United Kingdom.

References

Abramowitz, M. & Stegun, L.A. (eds.), (1972). *Handbook of Mathematical Functions with Formulas Graphs and Mathematical Tables.* NBS Applied Mathematics, Series 55, 10th Ed.

Ambrose, D. (1986). The correlation and estimation of vapour pressure IV. Observations on Wagner's method of fitting equations to vapour pressures. *J. Chem. Thermodyn.*, **18**, 45–51.

Anderko, A. & Pitzer, K. (1991). Equation of state for pure fluids and mixtures based on a truncated virial expansion. *AIChE J.*, **37**, 1379–1391.

Angus, S., Armstrong, B. & de Reuck, K. M. (1976). *International Thermodynamic Tables of the Fluid State–volume 3–Carbon Dioxide.* Oxford: Pergamon Press.

Angus, S., Armstrong, B. & de Reuck, K. M. (1977). *International Thermodynamic Tables of the Fluid State–volume 4–Helium.* Oxford: Pergamon Press.

Angus, S., Armstrong, B. & de Reuck, K. M. (1980). *International Thermodynamic Tables of the Fluid State–volume 7–Propylene.* Oxford: Pergamon Press.

Angus, S., Armstrong, B. & de Reuck, K. M. (1985). *International Thermodynamic Tables of the Fluid State–volume 8–Chlorine–Tentative Tables.* Oxford: Pergamon Press.

Angus, S., de Reuck, K. M. & Armstrong, B. (1979) *International Thermodynamic Tables of the Fluid State–volume 6–Nitrogen.* Oxford: Pergamon Press.

Benedict, M., Webb, G.B. & Rubin, L.C. (1940). An empirical equation for thermodynamic properties of light hydrocarbons and their mixtures I. Methane, ethane, propane and *n*–butane. *J. Chem. Phys.*, **8**, 334–345.

Burden, R.L., Faires, J.D. & Reynolds, A.C. (1978). *Numerical Analysis*, 2nd Ed. Boston: Prindle Weber and Schmidt.

Craven, R.J.B., de Reuck, K.M. & Wakeham, W.A. (1989). An equation of state for the gas phase of methanol. *Pure Appl. Chem.*, **61**, 1379–1386.

de Reuck, K. M. (1990). *International Thermodynamic Tables of the Fluid State –volume 11– Fluorine.* Oxford: Blackwell Scientific Publications.

de Reuck, K.M. & Armstrong, B.A. (1979). A method of correlation using a search procedure, based on a step–wise least–squares technique, and its application to an equation of state for propylene. *Cryogenics*, **19**, 505–512.

de Reuck, K. M. & Craven, R. J. B. (1993). *International Thermodynamic Tables of the Fluid State –volume 12– Methanol.* Oxford: Blackwell Scientific Publications.

Dymond, J. H. & Smith, E. B. (1980). *The Virial Coefficients of Pure Gases and Mixtures.* Oxford: Clarendon Press.

Dymond, J.H. & Malhotra, R. (1988). The Tait equation: 100 years on. *Int. J. Thermophys.*, **9**, 941–951.

Dymond, J.H., Malhotra, R. Isdale, J.D. & Glen, N.F. (1988). (p, ρ, T) of n–heptane, toluene, and oct–1–ene in the range 298 to 373 K and 0.1 to 400 MPa and representation by the Tait equation. *J. Chem. Thermodyn.*, **20**, 603–614.

Elhassan, A.E., Craven, R.J.B. & de Reuck, K.M. (1993). Using the area method to solve pure component vapour–liquid equilibrium problems. *The 1993 IChemE Research Event*, **2**, 744–746.

Eubank, P.T., Elhassan, A.E., Barrufet, M. & Whiting, W.B. (1992). Area method for prediction of fluid phase equilibria. *Ind. Eng. Chem. Res.*, **31**, 942–949.

Ewers, J. & Wagner, W. (1982). A method for optimizing the structure of equations of state and its application to an equation of state for oxygen. *Proc. 8th Symp. Thermophys. Prop.*, 78–87.

Ewing, M. E. B., Goodwin, A. R. H., McGlashan, M. L. & Trusler, J. P. M. (1987). Thermophysical properties of alkanes from speeds of sound determined using a spherical resonator I. Apparatus, acoustic model, and results for dimethylpropane. *J. Chem. Thermodyn.*, **19**, 721–739.

Friend, D. G., Ingham, H. & Ely, J. F. (1991). Thermophysical properties of ethane. *J. Phys. Chem. Ref. Data*, **20**, 275–347.

Georgeton, G.K., Smith, R.L. & Teja, A.S (1986). *Equations of State Theories and Applications*, eds. K.C. Chao & R.L. Robinson, Washington, D.C.: Amer. Chem. Soc.

Goldberg, R.N. and Weir, R.D. (1992). Conversion of temperatures and thermodynamic properties to the basis of the international temperature scale of 1990, *Pure Appl. Chem.*, **64**, 1545–1562.

Gurvich, L.V., Veits, I.V., Medvedev, V.A. *et al.* (1978–1981). *Thermodynamic Properties of Individual Substances, 1–3*. Moscow: Nauka.

Haar, L. & Gallagher, J.S. (1978). Thermodynamic properties of ammonia. *J. Phys. Chem. Ref. Data*, **7**, 635–792.

Hill, P. G., MacMillan, R. D. C. & Lee, V. (1982). A fundamental equation of state for heavy water. *J. Phys. Chem. Ref. Data*, **11**, 1–14.

Hnědkovský, L. & Cibulka, I. (1990). On a temperature dependence of the van der Waals volume parameter in cubic equations of state. *Fluid Phase Equil.*, **60**, 327–332.

Holleran, E. M. (1970). Accurate virial coefficients from P, V, T data. *J. Chem. Thermodyn.*, **2**, 779–786.

Jacobsen, R. T., Jahangiri, M., Stewart, R. B., McCarty, R. D., Levelt Sengers, J. M. H., White, H. J., Sengers, J. V. & Olchowy, G. A. (1988). *International Thermodynamic Tables of the Fluid State–volume 10–Ethylene*. Oxford: Blackwell Scientific Publications.

Jacobsen, R. T., Penoncello, S. G., Beyerlein, S. W., Clarke, W. P. & Lemmon, E. W. (1992). A thermodynamic property formulation for air. *Fluid Phase Equil.*, **79**, 113–124.

Johnson, D.W. & Colver, C.P. (1968). Mixture properties by computer. *Hydrocarbon Processing*, **47**, 79–84.

Juza, J. & Sifner, O. (1976). Modified equation of state and formulation of thermodynamic properties of krypton in a canonical form in the range from 120 to 423 K and 0 to 300 MPa. *Acta Tech. ČSAV*, **2**, 1–32.

Juza, J. & Sifner, O. (1977). Modified equation of state and formulation of thermodynamic properties of xenon in the range from 161.36 to 800 K and 0 to 350 MPa. *Acta Tech. ČSAV*, **22**, 1–32.

Kamerlingh–Onnes, H. (1901). Expressions of the equations for state of gases and liquids by means of series. *Commun. Phys. Lab. Leiden*, No. 71.

Leach, J. W., Chappelear, P. S. & Leland, T. W. (1968). Use of molecular shape factors in vapour–liquid equilibrium calculations with the corresponding states principle. *AIChE J.*, **14**, 568–576.

Malhotra, R. & Woolf, L.A. (1991) Thermodynamic properties of 2,2,2–trifluoroethanol. *Int. J. Thermophys.*, **12**, 397–407.

Marsh, K. N., Wilhoit, R. C. & Gammon, B. E. (1993). *TRC Thermodynamic Tables*. Thermodynamic Research Center, Texas A&M University System, Texas.

Marx, V., Pruß, A. & Wagner, W. (1992). Neue Zustandsgleichungen für R12, R22, R11 und R113. Beschreibung des thermodynamischen Zustandsverhaltens bei Temperaturen bis 525 K und Drücken bis 200 MPa. *Fortschr.–ber. VDI*, **19**, Nr. 57, 1–202.

McCarty, R. D. (1974). *A Modified Benedict–Webb–Rubin Equation of State for Parahydrogen*, NBSIR, 74–357.

Mentzer, R.A., Greenkorn, R.A. & Chao, K–C. (1981). Principle of corresponding states and vapor–liquid equilibria of molecular fluids, and their mixtures with light gases. *Ind. Eng. Chem. Process Des. Dev.*, **20**, 240–252.

Panagiotopoulos, A.Z. & Kumar, S.K. (1985). A generalized technique to obtain pure component parameters for two–parameter equations of state. *Fluid Phase Equil.*, **22**, 77–88.

Patel, N.C. & Teja, A.S. (1982). A new cubic equation of state for fluids and fluid mixtures. *Chem. Eng. Sci.*, **37**, 463–473.

Peng, D–Y. & Robinson, D.B. (1976). A new two–constant equation of state. *Ind. Eng. Chem. Fundam.*, **15**, 59–64.

Plöcker, U.J. & Knapp, H. (1976). Save time in computing density. *Hydrocarbon Processing*, 199–201.

Polt, A. & Maurer, G. (1992). The Bender equation of state for describing thermodynamic properties of krypton, neon, fluorine, sulfur dioxide and water over a wide range of state. *Fluid Phase Equil.*, **73**, 27–38.

Powell, M.J.D. (1964). An efficient method of finding the minimum of a function of several variables without calculating derivatives. *The Computer J.*, **7**, 155–162.

Redlich, O. & Kwong, J.N.S. (1949). On the thermodynamics of solutions. V. An equation of state. Fugacities of gaseous solutions. *Chem. Rev.*, **44**, 233–244.

Reid, R.C., Prausnitz, J.M. & Poling, B.E. (1987). *The Properties of Gases and Liquids*, 4th Ed. New York: McGraw–Hill.

Rowlinson, J.S. & Swinton, F.L. (1982). *Liquids and Liquid Mixtures*, 3rd Ed., p. 60. London: Butterworths.

Rowlinson, J. S. & Watson, I. D. (1969). The prediction of the thermodynamic properties of fluids and fluid mixtures I. The principle of corresponding states and its extensions. *Chem. Eng. Sci.*, **24**, 1565–1574.

Rusby, R. L. (1991). The conversion of thermal reference values to the ITS–90. *J. Chem. Thermodyn.*, **23**, 1153–1161.

Saul, A. & Wagner, W. (1989). A fundamental equation for water covering the range from the melting line to 1273 K at pressures up to 25 000 MPa. *J. Phys. Chem. Ref. Data*, **18**, 1537–1564.

Setzmann, U. & Wagner, W. (1989). A new method for optimizing the structure of thermodynamic correlation equations. *Int. J. Thermophys.*, **10**, 1103–1126.

Setzmann, U. & Wagner, W. (1991). A new equation of state and tables of thermodynamic properties for methane covering the range from the melting line to 625 K at pressures up to 1000 MPa. *J. Phys. Chem. Ref. Data*, **20**, 1061–1155.

Soave, G. (1972). Equilibrium constants from a modified Redlich–Kwong equation of state. *Chem. Eng. Sci.*, **27**, 1197–1203.

Stewart, R. B. & Jacobsen, R. T. (1989). Thermodynamic properties of argon from the triple point to 1200 K with pressures to 1000 MPa. *J. Phys. Chem. Ref. Data*, **18**, 639–798.

Stull, D.R. and Prophet, H. (1971). *JANAF Thermochemical Tables*, 2nd Ed. NSRDS–NBS37.

Susnar, S.S., Budziak, C.J. Hamza, H.A. & Neumann, A.W. (1992). Pressure dependence of the density of *n*–alkanes. *Int. J. Thermophys.*, **13**, 443–452.

Sytchev, V. V., Vasserman, A. A., Kozlov, A. D., Spiridonov, G. A. & Tsymarny, V. A. (1978). *Thermodynamic Properties of Air*. Washington: Hemisphere – Berlin: Springer–Verlag.

Tillner–Roth, R. & Baehr, H. D. (1994). An international standard formulation for the thermodynamic properties of 1,1,1,2–tetrafluoroethane (HFC–134a) for temperatures from 170 K to 455 K at pressures up to 70 MPa. *J. Phys. Chem. Ref. Data*, **23**, 657–729.

Trebble, M.A. & Bishnoi, P.R. (1986). Accuracy and consistency comparisons of ten cubic equations of state for polar and non–polar compounds. *Fluid Phase Equil.*, **29**, 465–474.

Tsonopoulos, C. & Heidman, J. L. (1990). From the virial to the cubic equation of state. *Fluid Phase Equil.*, **57**, 261–276.

Wagner, W. & de Reuck, K. M. (1987). *International Thermodynamic Tables of the Fluid State–volume 9–Oxygen*. Oxford: Blackwell Scientific Publications.

Wagner, W., Marx, V. and Pruß, A. (1993). A new equation of state for chlorodifluoromethane (R22) covering the entire fluid region from 116 K to 550 K at pressures up to 200 MPa. *Rev. Int. Froid.*, **16**, 373–389.

Waxman, M. & Gallagher, J. S. (1983). Thermodynamic properties of isobutane for temperatures from 250 to 600 K and pressures from 0.1 to 40 MPa *J. Chem. Eng. Data*, **28**, 224–241.

Wong, D.S.H. & Sandler, S.I. (1992). A theoretically correct mixing rule for cubic equations of state. *AIChEJ*, **38**, 671–680.

Younglove, B. A. (1993). Private communication to the IUPAC Thermodynamic Tables Project Centre.

Younglove, B. A. & Ely, J. F. (1987). Thermophysical properties of fluids II. Methane, ethane, propane, isobutane and normal butane. *J. Phys. Chem. Ref. Data*, **16**, 577–798.

Yu, J.M., Adachi, Y. & Lu, B.C.Y. (1986). *Equations of State Theories and Applications*, eds. K.C. Chao & R.L. Robinson. Washington, D.C.: Amer. Chem. Soc.

Part four

APPLICATION OF SELECTED METHODS

9

Computer Calculation

9.1 Introduction

Dynamic processes in atomic liquids are nowadays well understood in terms of kinetic theory and computer calculations. Transport coefficients and dynamic scattering functions of liquid argon, for example, are well reflected by kinetic theories and equilibrium (MD) as well as nonequilibrium molecular dynamics (NEMD) calculations using Lennard–Jones (LJ) pair potentials.

The situation for molecular liquids like carbon dioxide, benzene or cyclohexane, is not so satisfactory. Kinetic theories are practically lacking, and few MD studies have been performed. Nonetheless, during the last ten years our understanding of the dynamics in molecular liquids has been considerably improved by equilibrium MD, as shown in Section 9.2.

Section 9.3 reviews briefly the computer simulation of liquids by NEMD and outlines how the viscosity and thermal conductivity may be evaluated. Two examples, which demonstrate how NEMD algorithms are tools to understand transport phenomena better, are given: (i) a calculation of the non–Newtonian viscosity of a simple liquid, and (ii) the density dependence of the contribution to the thermal conductivity from internal degrees of freedom.

9.2 Equilibrium molecular dynamics

C. HOHEISEL

Ruhr–Universität Bochum, Germany

9.2.1 Atomic liquids

9.2.1.1 The equilibrium molecular dynamics method

Given a system of N interacting particles treatable by classical statistical mechanics, the N–particle trajectory may be computed by solving numerically the equations of motion.

Molecular dynamics computations for a system of $N(= 108)$ hard spheres were first performed by Alder and co–workers. In the case of hard spheres the motion of each particle is determined by the laws of elastic collisions. When a force on a particle i can be represented by the negative gradient of a given potential function, then the classical equations of motion may be written in the following form

$$\ddot{r}_i(t) = \frac{1}{m} F_i(r_1(t), \ldots, r_N(t)); \quad i = 1, N \qquad (9.1)$$

where F_i denotes the total force on particle i exerted by all the other particles in the N–particle system, r_i is the space vector coordinate, t the time and m is the particle mass (for simplicity, chosen to be equal for all the particles). Assuming furthermore that this force is composed of pair forces only, we can write

$$\ddot{r}_i(t) = -\frac{1}{m} \sum_{j=1, j\neq i}^{N} \frac{\partial}{\partial r_{ij}} \phi(r_{ij}(t)); \quad i = 1, N \qquad (9.2)$$

where $\phi(r_{ij})$ denotes the pair potential function depending on the separation $r_{ij} = r_i - r_j$ of two particles only.

Equations (9.2) represent a set of second–order differential equations coupled in time, which can be solved numerically on a large computer for $N \sim 1000$ and a certain small time period, if the pair potential is not too complicated. Here, we restrict ourselves to the application of the Lennard–Jones pair potential of the form

$$\phi^{LJ}(r) = 4\epsilon \left[\left(\frac{\sigma}{r}\right)^{12} - \left(\frac{\sigma}{r}\right)^6 \right] \qquad (9.3)$$

where ϵ is a characteristic energy parameter that denotes the potential minimum and σ the 'size parameter' of the particle interaction that indicates the zero crossing of potential function. For brevity, the subscripts in the notation of the pair separations are suppressed.

Of the various existing algorithms to solve equations (9.2) numerically, here the simplest one that is in current use will be discussed, the so–called Stoermer–Verlet algorithm. Equations (9.2) are of the general form

$$\ddot{x}(t) = b(x(t)) \qquad (9.4)$$

Writing (9.4) as a difference equation, it reads

$$x_{k+1} = 2x_k - x_{k-1} + b(x_k)(\Delta t)^2 \qquad (9.5)$$

where x_k denotes the value at time t and $x_{k\pm1}$ the values at time $t \pm \Delta t$. The last relation (9.5) yields directly the required formula for the numerical determination of the position of a particle i:

$$r_i(t + \Delta t) = 2r_i(t) - r_i(t - \Delta t) + \frac{1}{m} F_i(t)(\Delta t)^2 \qquad (9.6)$$

Thus, the Stoermer–Verlet algorithm does not need the velocity of a particle to compute the position. The velocity may, however, be computed via the central mean of the following form

$$\dot{r}_i(t) = \frac{1}{2\Delta t} \left[r_i(t + \Delta t) - r_i(t - \Delta t) \right] \tag{9.7}$$

There are, of course, a lot of other integration schemes, predominantly the predictor–corrector algorithms. Of these methods the Stoermer–Verlet algorithm has proved to be the best and simplest one.

Usually the computer calculation is started from a face–centered cubic (fcc) lattice. Then the total number of particles of the system amounts to 32, 108, 256, 500 etc. according to the relation $N = 4n^3$, where n denotes the number of elementary cells in one dimension. Even for a system of 6000 particles, today treatable by supercomputers, the size of the model cube is so small that surface effects would completely overwhelm the bulk properties. Consequently, so–called periodic boundary conditions are used. The original cube is assumed to be surrounded by replicas. Thus a particle leaving the basis box is replaced by the corresponding particle of one of the replicas. So the number of particles in the original cube is conserved. A further consequence of the periodic boundary conditions is that the particles near the boundaries of the original cube are allowed to interact with the particle replicas. However, in order to avoid interaction of the considered particle with its own images, an interaction range of a maximum of one half of the basic box length, $l/2$, is permitted. Therefore, each particle of the system can interact with only another particle or its replica. This so–called minimum image convention assures that only $N(N-1)$ interaction contributions appear for a potential range $l/2$. Smaller cutoff ranges of the potential resulting in less particle interactions can, however, be considered for the computation (Hansen & McDonald 1986; Hoheisel & Vogelsang 1988).

The structure of the computer program based on the Stoermer–Verlet integration algorithm may be as follows:

(i) starting from the fcc lattice, the positions of the $N = 32, 108, \ldots$ particles are known within a cube of volume V. Maxwell distribution of the momenta is set up according to the desired temperature with the constraint of zero total linear momentum. To start the Stoermer–Verlet algorithm, the particle positions at time $t - \Delta t$ are required rather than the velocities of the particles. These positions are estimated by exploiting expression (9.7),

$$r_i(t - \Delta t) = r_i(t) - \Delta t \, \dot{r}_i(t) \tag{9.8}$$

where Δt denotes the integration time step chosen to be of order $10^{-14} s$.

(ii) In order to apply the integration algorithm via relation (9.6), the total force acting on each particle of the N–particle system must be determined. These forces are evaluated through the $N(N-1)/2$ pair contributions arising from the negative gradients of the potential.

The corresponding part of the program requires by far the longest computation time being roughly proportional to N^2. Therefore, it is essential to optimize carefully this part of the computer program. Many schemes have been proposed to achieve this optimization. One of these is the use of neighbor tables (Hoheisel & Vogelsang 1988; Hoheisel 1993).

(iii) Evaluation of pair forces requires determination of all the pair separations between the atoms according to the minimum image convention.

(iv) Knowledge of the pair forces and the positions at times t and $t - \Delta t$ allows then the application of the Stoermer–Verlet scheme via equation (9.6) to obtain the particle positions at time $t + \Delta t$.

(v) After resetting all the coordinates by shifting their actual values to those of the preceding time step, the next integration step can be done by considering only the points (ii)–(v). The particle velocities are not needed for the algorithm. However, to check or to adjust the temperature of the system via the average value of the total kinetic energy, they can be evaluated by expression (9.7).

During the first few hundred time steps the desired temperature of the system may be adjusted by frequent rescaling of the velocities of the particles and subsequent reestimating of the position using equation (9.8). These adjustments at the initial stage of the computation lead usually to a final temperature which differs at most only by about 1% from the desired value. Total energy conservation is achieved to within a relative accuracy of 10^{-6}.

For Lennard–Jones–type potentials, which are of concern here, the forces between particles decay sufficiently quickly as a function of the separation that a long–range correction during the MD computation is superfluous. However, the systematic cutoff error in the average values of potential energy and the virial and related quantities must not be neglected. A reliable estimate of these long–range corrections may be obtained by the relevant integral relations containing the static pair correlation function $g(r)$. These expressions read for the potential energy and the compressibility factor as follows:

$$E_{\text{conf}} = \frac{N}{2}n \int_0^\infty \phi^{\text{LJ}}(r)g(r)4\pi r^2 dr \qquad (9.9)$$

$$\frac{PV}{Nk_{\text{B}}T}\bigg|_{\text{conf}} = -\frac{n}{6k_{\text{B}}T} \int_0^\infty r\frac{d}{dr}\phi^{\text{LJ}}g(r)4\pi r^2 dr \qquad (9.10)$$

where n denotes the number density, $\phi^{\text{LJ}}(r)$ the Lennard–Jones potential, k_{B} the Boltzmann constant and P the pressure. Substituting the cutoff radius for the lower integration limit and assuming that for larger distances than the cutoff radius $g(r)$ is unity to a very good approximation, the desired formulas for long–range corrections are found. Some problems arise near zero pressure states. Here, the correction value for the compressibility factor reaches the same order of magnitude as the computed value. In these instances more sophisticated long–range correction procedures have to be used. The $g(r)$ function determined by MD could, for example, be extended with the help of perturbation methods and then integrated numerically (Hoheisel 1990, 1993). Generally, average values are obtained as ensemble averages or time averages by exploiting the computed particle trajectories. Commonly, the so–called MD ensemble is used for average values of time–dependent properties. The MD ensemble is equivalent to the microscopic one under the additional constraint of constant total linear momentum. For MD calculations using other ensembles see Klein (1986).

9.2.1.2 Green–Kubo expressions and time correlation functions

Within the framework of linear response theory a phenomenological thermal transport coefficient L_{ij} can be shown to have the form of a Green–Kubo relation

$$L_{ij} = \frac{1}{k_B} \int_0^\tau < J_j(t) J_i(t+s) > ds \qquad (9.11)$$

where the integrand $< J_j(t) J_i(t+s) >$ is the equilibrium time correlation function of the relevant microscopic currents $J_j(t)$, $J_i(t+s)$ involved in the transport process. Of the time variables s and t the latter can be eliminated due to the stationarity condition valid for the correlation function. The total integration time τ must be long enough to ensure that the correlation function has decayed to zero.

The time correlation functions for the various thermal transport coefficients can be efficiently computed by MD with low statistical uncertainty. The results for transport coefficients are usually accurate to within 5–10%. One–particle time correlation functions are in general more accurate than collective functions due to the possible averaging over each single particle trajectory (Hansen & McDonald 1986; Hoheisel & Vogelsang 1988). So self–diffusion coefficients are, for instance, more accurately computable than mutual–diffusion coefficients.

In the following, the Green–Kubo formulas for self–diffusion, shear viscosity and thermal conductivity coefficients are compiled. A complete list of all the thermal transport coefficients of one– and two–component systems was given by Hoheisel & Vogelsang (1988).

Self–diffusion

$$D = \frac{1}{3} \int_0^\infty d\tau < \underline{v}(0) \underline{v}(\tau) > \qquad (9.12)$$

where D denotes the self–diffusion coefficient, $\underline{v}$ the single particle velocity and τ the time. The bracketed expression is the well–known velocity autocorrelation function of a particle.

Shear viscosity

$$\eta = \frac{1}{V k_B T} \int_0^\infty < \underline{\underline{J}}_p^{xy}(0) \underline{\underline{J}}_p^{xy}(t) > dt \qquad (9.13)$$

with

$$\underline{\underline{J}}_p = m \sum_{i=1}^N \underline{v}_i \underline{v}_i - \frac{1}{2} \sum_{i=1}^N \sum_{i \neq J}^N \underline{r}_{ij} \nabla \phi(r_{ij}) \qquad (9.14)$$

where η denotes the shear viscosity coefficient, V the volume, T the temperature and $\underline{\underline{J}}_p$ the stress tensor. The superscripts xy mean that only the nondiagonal elements of the stress tensor are used to build the correlation function. Equation (9.14) represents the microscopic form of the stress tensor, where the gradient of the pair potential $\phi(r_{ij})$ is formed with respect to the vector $\underline{r}_{ij}$. Obviously, the time correlation function in equation (9.13) may be determined by averaging over three contributions from the different coordinate sets, xy, xz and yz (Hoheisel & Vogelsang 1988; Hoheisel 1993).

Thermal conductivity

$$\lambda = \frac{1}{V k_B T^2} \int_0^\infty < \underline{J}_q^x(0) \, \underline{J}_q^x(t) > dt \qquad (9.15)$$

with

$$\underline{J}_q = \frac{m}{2} \sum_{i=1}^N v_i^2 \underline{v}_i - \frac{1}{2} \sum_{i=1}^N \sum_{i \neq j}^N \left[\underline{r}_{ij} \nabla \phi(r_{ij}) - \phi(r_{ij}) \, \underline{\underline{i}} \right] \underline{v}_i \qquad (9.16)$$

where λ denotes the thermal conductivity coefficient, $\underline{J}_q$ the heat flux and $\underline{\underline{i}}$ the unit tensor. The superscript x in formula (9.15) means an arbitrary component of the considered quantity. Note that the two terms in parentheses in equation (9.16) are to be evaluated as tensors (Hansen & McDonald 1986; Hoheisel & Vogelsang 1988; Hoheisel 1993).

Apparently, the heat flux has a complicated microscopic form, since the second part of the right–hand side of equation (9.16) contains also the particle velocities. So MD computation of $\underline{J}_q$ requires evaluation of pair velocities, which slows down the computation speed markedly (Hoheisel & Vogelsang 1988).

From equations (9.14) and (9.16) it becomes obvious that stress tensor and heat flux both involve kinetic and potential terms. Therefore, the correlation functions consist of kinetic–kinetic, kinetic–potential and potential–potential terms. For liquids, however, the potential–potential contribution is dominant (Hansen & McDonald 1986; Hoheisel 1993).

Examples of time correlation functions are shown in Figure 9.1, where the normalized autocorrelation functions for D, η and λ are displayed. The figure contains also the correlation function for the bulk viscosity, which is, however, not of concern in the

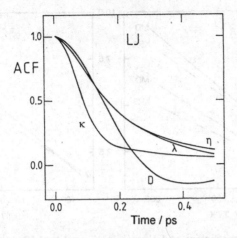

Fig. 9.1. Decay of the normalized Green–Kubo integrands for the self–diffusion coefficient D, the shear and bulk viscosity coefficients, η and κ, as well as the thermal conductivity coefficient λ for LJ argon near the triple point.

present report (see Hoheisel & Vogelsang (1988) and Hoheisel (1993) for more information). Figure 9.1 presents only the short–time behavior of these functions. For states near the triple point of a liquid, the long–time behavior is also important for the determination of the transport coefficient. Since the long–time behavior of the correlation functions plays a decisive role for molecular liquids, it will be discussed in more detail in Section 9.2.2.

9.2.1.3 Results

Two applications of numerically exact results of MD computations are considered. In the first case, MD results serve as reference data for theoretical approximations; in the second, they serve as 'computed' data for comparison with experiment.

Modern kinetic theory is able to predict the transport coefficients of the Lennard–Jones liquid (1–center Lennard–Jones interaction between particles) to a fairly good approximation (Karkheck 1986; Hoheisel 1993). The results of these theories have been compared in detail with the 'exact' MD computation results (Borgelt *et al.* 1990). Comparisons for self–diffusion, shear viscosity and thermal conductivity are presented in Figures 9.2–9.4.

Best agreement between MD and kinetic theory appears for lower densities, as expected. However, also for high densities the agreement is acceptable. On the whole, the kinetic theory values lie somewhat above the MD results, but the deviations do not exceed 30%.

For argon an extended comparison of measured and computed shear viscosities and thermal conductivities has recently been carried out (see, for instance, Hoheisel 1994). These results are reproduced in Figures 9.5–9.6. Apart from the values at the highest

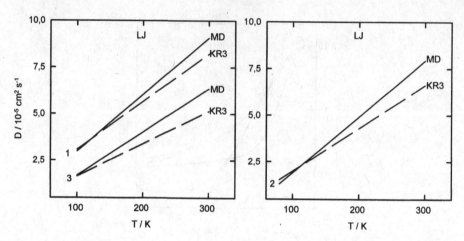

Fig. 9.2. Self–diffusion coefficient of LJ argon computed by MD and predicted by kinetic reference theory (version 3 = KR3). 1: 1.311 g cm^{-3}, 2: 1.414 g cm^{-3}, 3: 1.485 g cm^{-3}.

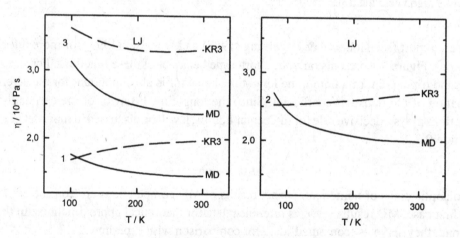

Fig. 9.3. As in Figure 9.2, but for shear viscosity coefficient.

density considered, the agreement between MD computation and experiment is excellent. The discrepancies at high densities and high temperatures may be smaller in reality, since the experimental values are somewhat uncertain at these high pressure states. It appears that the Lennard–Jones potential used for the MD is a very good representation of the interaction in liquid argon (Hansen & McDonald 1986; Hoheisel 1990, 1993).

9.2.2 Molecular liquids

9.2.2.1 The MD method of constraints

In this section, real molecules are considered as assemblies of rigid molecules composed of polyatomic interaction centers. There are several effective methods to treat rigid

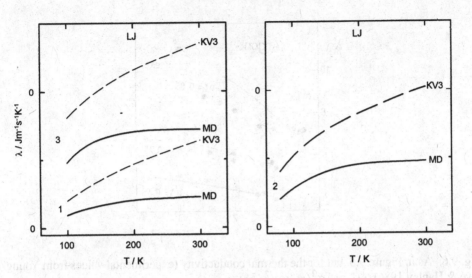

Fig. 9.4. As in Figure 9.2, but for thermal conductivity coefficient (KV3 denotes version 3 of a kinetic variational theory).

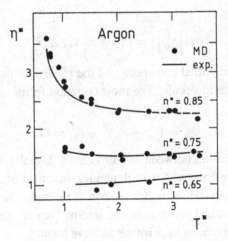

Fig. 9.5. Experimental and MD shear viscosities ($\eta^* = \eta\sigma^2/(m\epsilon)^{1/2}$) for liquid argon as a function of temperature and three densities. MD values are for the LJ liquid. Experimental values from Younglove & Hanley (1986).

molecules by molecular dynamics (see, for example, Hoheisel 1993). However, the constraints method has been used predominantly because of the important advantage that the MD program for particles needs to be modified only slightly to include molecules. So this method is described in detail here (Ciccotti & Ryckaert 1986; Hoheisel 1993).

A system of N molecules is considered, where each consists of n atoms which are represented by mass point interaction centers. The intra–molecular structure is given

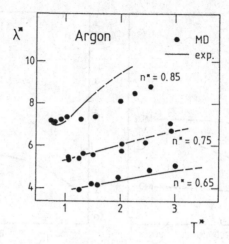

Fig. 9.6. As in Figure 9.5, but for the thermal conductivity (experimental values from Young-love & Hanley 1986). ($\lambda^* = \lambda\sigma^2(m/\epsilon)^{1/2}/k_B$).

by geometrical constraints. For each molecule, it is assumed that there are l holonomic constraints that relate the atoms. This means that $l = 3n - 5$ independent constraints must be imposed. Then

$$\sigma_k(\{r_\alpha^i\}) = 0; \quad k = 1, \ldots, l \tag{9.17}$$

where σ_k denotes the individual constraint and the r_α^i denote the position coordinates of the ith atom in the αth molecule. The most common forms of constraints are bond constraints:

$$\sigma_B = (r_\alpha^i - r_\alpha^j)^2 - d_{ij}^2 = 0 \tag{9.18}$$

where d_{ij} fixes the separation between atomic centers. Usually these bond constraints suffice to fix the molecule structure for the dynamics. In some instances, further vectorial linear constraints are necessary, as will be shown at the end of this section.

Apart from the geometrical constraints, the intermolecular interactions are assumed to be of atom–atom interactions in pairwise additive form

$$u_{\alpha\beta} = \sum_{i=1}^{n}\sum_{j=1}^{n}\phi_{ij}(r_{\alpha\beta}^{ij}); \quad \alpha \neq \beta \tag{9.19}$$

where ϕ_{ij} may denote the Lennard–Jones potential, which becomes effective only for atoms of different molecules.

The dynamics of a system of N molecules with l holonomic constraints per molecule can be described in Cartesian coordinates via the Lagrangian equations of the following form:

$$\frac{d}{dt}\frac{\partial\mathcal{L}}{\partial\dot{r}_\alpha^i} - \frac{\partial\mathcal{L}}{\partial r_\alpha^i} = G_\alpha^i; \quad i = l, n; \ \alpha = 1, N \tag{9.20}$$

where $\mathcal{L}$ denotes the Lagrange function and $\underline{G}_\alpha^i$ the constraint force on atom (i, α), which may be written as

$$\underline{G}_\alpha^i = -\sum_{k=1}^{l} \lambda_k^{(\alpha)} \underline{\nabla}_\alpha^i \sigma_k^{(\alpha)} \tag{9.21}$$

where $\lambda_k^{(\alpha)}$ denotes the Lagrangian multiplier associated with the $\sigma_k^{(\alpha)}$th constraint. The $3nN$ equations (9.20) together with equation (9.21) and the Nl geometrical constraints given by the following relations

$$\sigma_k^{(\alpha)}(\{\underline{r}_\alpha^i\}) = 0; \quad \alpha = 1, N; \quad k = 1, l \tag{9.22}$$

constitute a set of $(3n + l)N$ equations of motion for the unknown $\{\underline{r}_\alpha^i(t), \lambda_k^{(\alpha)}(t)\}$. Therefore, the trajectory of the molecular system is uniquely determined. It can be shown that equations (9.20) reduce to ordinary second–order differential equations which may be solved numerically. A numerical solution procedure would, however, induce numerical inaccuracies, and after several time steps the constraints could no longer be satisfied. This problem can be solved by the following special technique.

The explicit expression for $\lambda_k^{(\alpha)}$ defined in equation (9.21) may be derived by requiring that for all k the $\sigma_k^{(\alpha)}$ vanishes at all times. Using a certain numerical integration algorithm, the set of Lagrangian multipliers is no longer evaluated by equation (9.21) but by equations that contain free parameters $\{\gamma_k^{(\alpha)}\}$ corresponding to the $\{\lambda_k^{(\alpha)}\}$. These parameters are determined such that the atomic coordinates at the next time step in the integration scheme of the chosen algorithm do satisfy the constraints exactly.

For the Stoermer–Verlet integration algorithm

$$\underline{r}_i(t + h) = \underline{r}_i'(t + h) - \frac{h^2}{m} \sum_{k=1}^{l} \gamma_k \left(\frac{\partial \sigma_k}{\partial \underline{r}_i}\right)_{\{\underline{r}_i(t)\}} \tag{9.23}$$

where h denotes the time step and m the atomic mass of an interaction center. $\underline{r}_i'$ denotes the predicted position at time $(t + h)$ according to the ordinary force $\underline{F}_i(t)$ without the constraint force, which may be written as

$$\underline{r}_i'(t + h) = -\underline{r}_i(t - h) + 2\underline{r}_i(t) + \frac{h^2}{m} \underline{F}_i(t) \tag{9.24}$$

Substitution of equation (9.23) into the constraints conditions gives the following set of equations:

$$\sigma_k(\{\underline{r}_i(t + h)\}_{i=1,n}) = T_k(\{\gamma_{k'}\}_{k'=1,l}) = 0 \tag{9.25}$$

where $T_k(\{\gamma_{k'}\}_{k'=1,l})$ is a set of algebraic equations which in general can be solved by an iterative procedure (Klein 1986).

There exists, however, a simpler iterative solution of equations (9.25) that is based on local solutions for the individual constraints. A given constraint is considered out of the total number of l constraints, and it is assumed furthermore that it is in the Mth step of an iterative procedure for each constraint. Then the effect of the constraint forces may be given by the following expression:

$$r_j^{\text{new}} = r_j^{\text{old}} - \frac{h^2}{m} \gamma_k^{(M)} \frac{\partial \sigma_k}{\partial r_j}; \qquad j = 1, n_k \tag{9.26}$$

where r_j^{old} denotes the subset of n_k atomic positions involved in the particular constraint σ_k. The parameter $\gamma_k^{(M)}$ is thereby determined as the first-order solution of the scalar equation

$$\sigma_k(\{r_j^{\text{new}}\}) = 0 \tag{9.27}$$

The result is

$$\gamma_k^{(M)} = \frac{\sigma_k(\{r_j^{\text{old}}\})}{\frac{h^2}{m} \sum_{j=1}^{n_k} \left(\frac{\partial \sigma_k}{\partial r_j}\right)_{r^{\text{old}}} \left(\frac{\partial \sigma_k}{\partial r_j}\right)_{r(t)}} \tag{9.28}$$

If the iteration is started from an unconstrained position $r_i'(t+h)$, the following coordinates are obtained after completion of the Mth iterative step

$$r_i^{\text{new}} = r_i'(t+h) - \frac{h^2}{m} \sum_{k'=1}^{l} \beta_{k'}^{k,M} \left(\frac{\partial \sigma_{k'}}{\partial r_i}\right)_{r_i(t)} \tag{9.29}$$

with

$$\beta_{k'}^{k,M} = \sum_{M'=1}^{M} \gamma_{k'}^{(M')}; \quad k' \leq k \tag{9.30}$$

$$\beta_{k'}^{k,M} = \sum_{M'=1}^{M-1} \gamma_{k'}^{(M')}; \quad k' > k \tag{9.31}$$

The convergence of this iteration cycle is controlled by a certain parameter indicating the deviation of the $\sigma_k(\{r_j^{\text{old}}\})$ constraint from zero. In the final stage of the cycle the positions are given by

$$r_i^{\text{new}} = r_i'(t+h) - \frac{h^2}{m} \sum_{k=1}^{l} \beta_k^{l,M} \left(\frac{\partial \sigma_k}{\partial r_i}\right)_{r_i(t)} \tag{9.32}$$

The programming work consists essentially of coding expressions (9.28) and (9.32) in a suitable loop. For not too complex molecules, a few iterations suffice to fulfill the constraints conditions with a relative accuracy of 10^{-5}. This iterative method has been widely used to model molecular liquids. The great advantage over other methods is that the program structure of the ordinary MD method for particles can be transferred without any changes, and the additional part required for the iterative procedure to fulfill the individual constraints can be very efficiently coded even on vector machines. In general, the computation time necessary for the iterative process amounts to about 5–10% of the total computation time, when molecules of 2–10 centers are involved. However, it should be noted that a system of 108 six–center molecules consists of 648 'particles'!

Some final remarks on vector constraints may be in order: for linear and planar molecules, bond constraints are not satisfactory to guarantee the shape of the molecules during the described integration process. The reason for this is the absence of orthogonal forces that keep the molecular shape. In these instances, suitable primary atoms are chosen for the bond constraints, and the positions of the remaining secondary atoms are obtained simply by applying the vector constraints under the provision that the associated constraint forces are given. Vector constraints are of the form

$$\sigma = \underline{r}_{i\alpha} - \sum_{j=1}^{n} c_j \underline{r}_{j\alpha} = 0 \tag{9.33}$$

where the c_j are normalized to unity (Ciccotti & Ryckaert 1986).

9.2.2.2 Green–Kubo formulas and time correlation functions

The general form of Green–Kubo formulas has already been given in Section 9.1.1. However, for molecular liquids the microscopic formulation of the currents for self–diffusion, shear viscosity and thermal conductivity must be modified (Marechal & Ryckaert 1983; Hoheisel 1993).

Self–diffusion is characterized by the velocity autocorrelation function. Hence for a molecular system, the center of mass (c.o.m.) velocity of a molecule must be considered. So formula (9.12) is valid for the molecular liquid, when the atomic velocity is replaced by the c.o.m. velocity of a molecule.

More complex is the description of viscous flow and heat flux in molecular fluids due to the potential contributions coming from the anisotropic forces between the molecules, which govern the transport processes.

Focusing the discussion first on the stress tensor, which is the relevant dynamic variable for the viscosity, its microscopic form is expected to involve all the atomic pair forces constituting the pair forces of the molecules. A detailed consideration shows that projections of these forces on the difference vectors connecting the c.o.m.'s of the molecule pairs are needed. This leads to the following microscopic form of stress tensor for the pure molecular fluid (Marechal & Ryckaert 1983; Hoheisel 1993)

$$\underline{\underline{J}}_p = M \sum_{\alpha=1}^{N} \dot{\underline{R}}_\alpha \dot{\underline{R}}_\alpha - \frac{1}{2} \sum_{\alpha=1}^{N} \sum_{\beta \neq \alpha}^{N} \sum_{i=1}^{n} \sum_{j=1}^{n} \underline{R}_{\alpha\beta} \nabla \phi(r_{\alpha\beta}^{ij}) \tag{9.34}$$

where α, β number the molecules and i, j the atomic sites of a molecule. N denotes the total number of molecules in the system and n the number of interaction centers per molecule. M is the mass of a molecule, while $\phi(r_{\alpha\beta}^{ij})$ denotes the atomic pair potential depending on the separation of the centers. $\underline{R}_\alpha$ denotes the c.o.m. position vector of a molecule α and $\underline{R}_{\alpha\beta}$ the difference vector of the c.o.m. positions of a pair of molecules α, β. The summation in the potential part of the stress tensor runs essentially over the forces between centers projected onto the difference position vectors of molecular pairs. It is to be noted that for the molecular system the corresponding off–diagonal terms in the potential part of the stress tensor are not equal in general. Furthermore, it is necessary to know that using constraints MD for mimicking a molecular liquid, the c.o.m. velocities and the difference position vectors must be indirectly evaluated via the atomic quantities.

The microscopic form of the heat flux of a molecular system may be expressed accordingly:

$$\underline{J}_q = \frac{1}{2}M \sum_{\alpha=1}^{N} \dot{\underline{R}}_\alpha^2 \, \dot{\underline{R}}_\alpha - \frac{1}{2} \sum_{\alpha=1}^{N} \sum_{i=1}^{n} \dot{\underline{r}}_\alpha^i \sum_{\substack{\beta \neq \alpha}}^{N} \sum_{j=1}^{n} (\underline{R}_{\alpha\beta} \nabla \phi(r_{\alpha\beta}^{ij}) - \phi(r_{\alpha\beta}^{ij})\underline{i}) \qquad (9.35)$$

where $\dot{\underline{r}}_\alpha^i$ denotes the velocity of the ith atomic site of molecule α. As well as the stress tensor, the heat flux has a kinetic and a potential part. However, the potential part involves additionally the atomic velocities and the pair energies. This complicates the computation of the thermal conductivity via equation (9.16) enormously, particularly for molecules composed of several atomic centers.

A strict account of the kinetic part of the heat flux of molecules must allow for the rotational contribution (Hoheisel 1993). Denoting the kinetic part of $\underline{J}_q$ by $\underline{J}_q^{kin}$ it follows

$$\underline{J}_q^{kin} = \frac{1}{2}m \sum_{\alpha=1}^{N} \dot{\underline{R}}_\alpha \sum_{i=1}^{n} \dot{r}_\alpha^{i2} \qquad (9.36)$$

where m denotes the atomic mass assumed to be equal for each center. Use of the excellent approximation for $\underline{J}_q^{kin}$ shown in equation (9.35) simplifies the computation procedure. For liquids, the kinetic contribution to the whole current is very small, and therefore a rough approximation for $\underline{J}_q^{kin}$ would suffice.

Compared with the collective time correlation functions of atomic fluids the behavior of molecular time correlation functions is significantly different. This is illustrated by plots of the normalized correlation function for the shear viscosity of liquid model CF_4 in Figure 9.7. In the figure, the time correlation function for the four–center LJ liquid is plotted with that of the common Lennard–Jones liquid modeled to give roughly the same η value (Hoheisel 1993). All the LJ potential parameters used for the computations are listed in Table 9.1.

As is evident from the figure, the molecular time correlation function has a pronounced 'chair form' and decays initially much more quickly than the atomic function. Therefore, the long time branch of the molecular time correlation functions is difficult

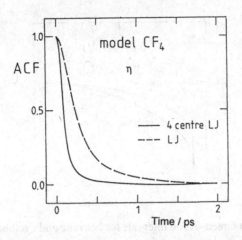

Fig. 9.7. Normalized time correlation functions of liquid CF_4 computed with an atomic and a molecular potential (see Table 9.1). 140 K; 1.632 g cm^{-3}.

Table 9.1. *Lennard–Jones potential parameters for CF$_4$.*

Number of centers	$\frac{\sigma}{nm}$	$\frac{\epsilon/k_B}{K}$	$\frac{d}{nm}$	Comment
1	0.415	175.0		Usual particle model
4	0.271	87.5	0.154[a]	4–site, rigid tetrahedral model without a central site; d denotes the distance between the c.o.m. and the atomic site.

[a] Experimental C–F bond length: 0.1322 nm.

to obtain due to its small amplitude. The extreme chair form of the Green–Kubo integrand for η of molecular liquids has an awkward consequence at triple–point conditions, where the integrand displays a tail of low amplitude up to several picoseconds. Heavy computations are required to determine the shear viscosity coefficient with reasonable accuracy. The tail of the Green–Kubo integrand contributes considerably to the integral value, that is, the η value. This is illustrated by plots of the integral value as a function of time for six–center LJ benzene and six–center LJ cyclohexane in Figure 9.8 (potential models are given in Table 9.5).

Apparently, the long time branch of the correlation functions contributes more than 40% to the total coefficient. In order to compute this slowly converging time integral with sufficient accuracy, systems of more than 32 molecules must be used. Modern MD calculations have revealed that the chair form of the time correlation function for η appears to depend particularly on the anisotropy of the molecules involved. Luo & Hoheisel (1991) have shown that the long–time behavior of the correlation function

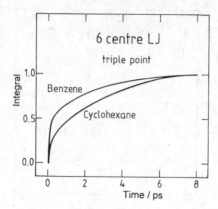

Fig. 9.8. Time dependent Green–Kubo integrals for benzene and cyclohexane near their triple point.

for the viscosity is due to long–lived orientational correlations occurring in liquids of significantly anisotropic molecules. This is confirmed by the lower–order reorientational correlation functions which show a similar long–time behavior (Luo & Hoheisel 1991). Since the reorientational correlation functions are one–molecule functions, they can be determined by short MD runs and may be conveniently exploited to estimate the tail behavior of the time correlation function for the viscosity.

9.2.2.3 Results

Liquid and fluid nitrogen is one of the best–studied molecular systems by both experiment and computer calculations (Vogelsang & Hoheisel 1987). Shear viscosity and thermal conductivity of liquid nitrogen have been calculated by MD using three different model potentials. The computations included a spherically symmetric LJ potential, a two–centre LJ potential and a two–centre LJ potential augmented by a quadrupole interaction term. Potential parameters are listed in Table 9.2.

Table 9.2. *Lennard–Jones potential parameters for the computation of liquid N_2: 1–center model; 2–center model; 2–center model augmented by a quadrupole interaction term.*

Sites	$\dfrac{\sigma}{nm}$	$\dfrac{\epsilon/k_B}{K}$	$\dfrac{d}{nm}$	$\dfrac{10^{26}Q}{esu}$
1	0.36360	101.6		
2	0.32932	36.5	0.1094	
3	0.33140	35.3	0.1101^a	1.52^b

[a] denotes bond length.
[b] Q is the quadrupole moment.

Table 9.3. *Computed and experimental transport coefficients of dense N_2.*

$\dfrac{\rho}{\mathrm{g\,cm^{-3}}}$	$\dfrac{T}{\mathrm{K}}$	$\dfrac{10^4\eta}{\mathrm{Pa\,s}}$	$\dfrac{10^3\lambda}{\mathrm{W\,m^{-1}\,K^{-1}}}$	Comment[a]
0.7003	100	0.825 ± 0.04	86.5 ± 4[b]	
		0.725 ± 0.03[b]	81.0 ± 4[b]	
		0.757 ± 0.03	85.3 ± 4	
		0.760 ± 0.05	100.0 ± 10	
0.7003	150	0.844 ± 0.04	95.1 ± 4	high pressure
		0.759 ± 0.04	95.1 ± 4	
		0.746 ± 0.04	95.1 ± 4	
		0.720 ± 0.05	100.0 ± 20	
0.7003	250	0.856 ± 0.05	101.9 ± 5	high pressure
		0.844 ± 0.05	103.8 ± 5	
		0.670 ± 0.05		
0.7003	400	0.865 ± 0.05	117.9 ± 5	high pressure
		0.874 ± 0.06	118.4 ± 5	
		0.600 ± 0.05		
0.8678	63	3.22 ± 0.3	141.0 ± 5	triple point
		2.02 ± 0.2		
		2.27 ± 0.2	140.5 ± 10	
		2.90 ± 0.1	156.0 ± 5	
0.8678	70	2.59 ± 0.1	144.1 ± 7	
		1.91 ± 0.1	142.8 ± 7	
		2.15 ± 0.1	142.3 ± 7	
		2.20 ± 0.1	149.5 ± 5	
0.8678	100	2.12 ± 0.1	146.1 ± 7	high pressure
		1.59 ± 0.1	157.5 ± 7	
		1.74 ± 0.1	164.0 ± 8	
		1.50 ± 0.2	180.0 ± 20	
0.8678	150	1.93 ± 0.1	168.0 ± 8	high pressure
		1.73 ± 0.1	161.5 ± 8	
		1.74 ± 0.1	167.8 ± 8	
		1.15 ± 0.2		
0.8678	250	1.69 ± 0.1	175.2 ± 8	high pressure
		1.50 ± 0.1	179.1 ± 8	
		1.70 ± 0.1	191.0 ± 10	
		0.90 ± 0.2		

[a] The rows given per state have the following order: result from 1-center LJ potential; result from 2-center LJ potential; result from 2-center LJ potential plus quadrupole interaction; experimental result. MD results were obtained by computations with 108 molecules.
[b] Values calculated with 32 molecules are: $\eta = 0.762$; $\lambda = 78.2$.

For nine thermodynamic states including the triple point of N_2, experimental and MD computed η and λ are compared in Table 9.3. All the considered potentials give quite good results in comparison with experiment. However, the molecular potentials show better agreement in general. It appears that the thermal conductivity is much less

Table 9.4. *Computed and experimental transport coefficients for liquid SF_6. (The 6–center Lennard–Jones potential model is given in Table 9.5.)*

$\frac{T}{K}$	$\frac{\rho}{g\,cm^{-3}}$	\multicolumn{3}{c}{$\frac{10^3\lambda}{W\,m^{-1}\,K^{-1}}$}	\multicolumn{3}{c}{$\frac{10^4\eta}{Pa\,s}$}
		Exp.[a]	MD[b]	MD[c]	Exp.[a]	MD[b]	MD[c]
250	1.500	49	28 ± 2		1.65	1.20 ± 0.05	
250	1.600	55	32 ± 2		2.15	1.40 ± 0.05	
250	1.700	65	41 ± 3		2.65	1.75 ± 0.10	
250	1.800	75	46 ± 3		3.65	3.25 ± 0.15	
223	1.848	86	54 ± 3	(62 ± 6)	4.85	3.00 ± 0.20	(3.05 ± 0.3)

[a] Taken from Ulybin & Makarushkin (1977).
[b] Results from runs with 32 molecules.
[c] Values in parentheses from runs with 108 molecules.

sensitive to the details of the potential function than the shear viscosity. The quadrupole interaction has only little effect on the transport coefficients, but improves the agreement with experimental data in a few instances.

In view of the fact that the two–center LJ potential has already proved to give thermo-dynamic and one–particle properties of N_2 to a very good approximation, the excellent representation of collective properties confirms the validity of this potential. One may say that the two–center LJ pair potential for dense N_2 plays the same role as the one–center LJ potential does for dense argon.

Much less investigated are simple dense molecular fluids which contain relatively large molecules of roughly globular shape. Recently, liquid SF_6 was studied with use of a six–center potential (Hoheisel 1993). For several dense states of SF_6, experimental shear viscosities and thermal conductivities are compared with computer data in Table 9.4. The comparison shows significant discrepancies. Apparently, the potential parameters optimized with respect to thermodynamic properties are insufficient for the description of dynamic properties. Hence, the interaction in liquid SF_6 seems to be more anisotropic than describable by a six–center LJ potential.

By contrast, for liquid benzene and cyclohexane, the six–center potential model works well both for thermodynamic and transport properties. Results for η and λ as determined by experiment and computer calculation are shown in Tables 9.6–9.7. For potential models, see Table 9.5. The tables show that in particular the strong increase of η on approaching the triple–point conditions is well predicted by the MD calculations. It appears that the employed six–center potential type can be regarded as well suited for liquid benzene and cyclohexane.

The transport properties of liquid butane have been studied by MD computations at least by four different groups (Marechal *et al.* 1987; Edberg *et al.* 1987). However, most of the calculations were only performed for the shear viscosity at the boiling

Table 9.5. *Lennard–Jones potential models for various substances.*

Sites	Substance	$\frac{\sigma}{nm}$	$\frac{\epsilon/k_B}{K}$	$\frac{d}{nm}$	Comment
6	SF_6	0.2700	60.0	0.1561	rhombic
6	C_6H_6	0.3350	95.5	0.1765	flat ring
6	C_6H_{12}	0.3970	78.0	0.1600	chair form
4	C_4H_{10}	0.3923	72.0	0.1530	bond angle 109.5°

Table 9.6. *Computed and experimental transport coefficients of liquid benzene.*

$\frac{T}{K}$	$\frac{\rho}{g\,cm^{-3}}$		$\frac{10^3\lambda}{W\,m^{-1}\,K^{-1}}$			$\frac{10^4\eta}{Pa\,s}$	
		Exp.[a]	MD[b]	MD[c]	Exp.[a]	MD[b]	MD[c]
353	0.8134				3.17	3.10 ± 0.3	(3.20 ± 0.2)
343	0.8247				3.50	3.55 ± 0.3	
333	0.8359	133	117 ± 15		3.89	4.00 ± 0.3	
323	0.8469	140	121 ± 15	(135 ± 10)	4.35	4.40 ± 0.3	(4.50 ± 0.2)
313	0.8577				4.91	5.10 ± 0.4	(5.00 ± 0.3)
303	0.8684	143	127 ± 15		5.60	5.35 ± 0.5	(5.70 ± 0.3)
293	0.8790	146.5	133 ± 15	(145 ± 10)	6.47	6.35 ± 0.6	(6.60 ± 0.4)
283	0.8895				7.57	6.80 ± 0.7	(7.30 ± 0.4)

[a] Taken from Landolt–Börnstein, Vol. 2, Part 5.
[b] Results from runs with 32 molecules.
[c] Values in parentheses from runs with 108 molecules.

Table 9.7. *Computed and experimental transport coefficients of liquid cyclohexane.*

$\frac{T}{K}$	$\frac{\rho}{g\,cm^{-3}}$		$\frac{10^3\lambda}{W\,m^{-1}\,K^{-1}}$		$\frac{10^4\eta}{Pa\,s}$
		Exp.[a]	MD	Exp.[a]	MD
293	0.7786	127	118 ± 15	9.97	7.5 ± 0.7[b]
			124 ± 10		10.0 ± 0.7[c]
					10.1 ± 0.5[d]
303	0.7693	125	112 ± 15	8.24	6.8 ± 0.6[b]
					8.6 ± 0.5[c]
313	0.7405	122	109 ± 15	7.02	6.2 ± 0.6[b]
			115 ± 10		7.3 ± 0.4[c]
343	0.7598	118	107 ± 15	5.26	5.0 ± 0.5[b]
					5.5 ± 0.3[c]

[a] Taken from Landolt–Börnstein, Vol. 2, Part 5.
[b] Results from runs with 32 molecules.
[c] Results from runs with 108 molecules.
[d] Results from runs with 256 molecules.

Table 9.8. *Shear viscosity coefficients of liquid butane calculated by MD and compared with experiment. (The 4–center Lennard–Jones potential used for the computations is given in Table 9.5.)*

$\frac{T}{K}$	$\frac{\rho}{g\,cm^{-3}}$	$\frac{10^4\eta}{Pa\,s}$	
		Exp.[a]	MD[b]
223	0.6519	3.5	3.1 ± 0.2
213	0.6616	4.0	3.5 ± 0.5
203	0.6712	4.6	4.3 ± 0.7
193	0.6807	5.3	4.8 ± 0.7
183	0.6900	6.3	5.6 ± 1.0

[a] Younglove & Ely (1987).
[b] Results from runs with 108 molecules.

point using the semiflexible four–site model potential. Unfortunately, the various MD results for the shear viscosity are based on somewhat different potential parameter sets, which complicates a fair comparison of the findings (see Hoheisel 1994 for a detailed discussion). Some of the computed shear viscosity values lie far off the experimental value (Marechal *et al.* 1987). Recently it was shown that a rigid four–center Lennard–Jones potential suffices to generate the thermodynamics and the transport coefficients in reasonable agreement with experiment (Luo & Hoheisel 1993). In Table 9.8, the shear viscosity coefficients computed by MD and obtained by experiment are listed for a number of states including a state in the neighborhood of the triple point.

The table demonstrates the acceptable agreement between experiment and calculation, even for the near triple–point state.

9.2.3 Conclusions

Only the transport properties of pure substances are considered here. However, also for binary mixture systems, there are a lot of interesting results for transport coefficients of atomic liquids and a few studies for molecular liquids. Obviously, for atomic liquids all the thermal transport coefficients, including the thermal diffusion coefficient, can be obtained by MD calculations with reasonable accuracy. For molecular mixtures, the presently available theoretical investigations are too rare to give a sufficient picture of the transport phenomena. To be more specific, even the sign of the thermal diffusion coefficient of a molecular liquid mixture of about equal masses of the component molecules is difficult to obtain by MD calculations.

References

Borgelt, P., Hoheisel, C. & Stell, G. (1990). Exact molecular dynamics and kinetic theory results for thermal transport coefficients of the Lennard–Jones argon fluid in a wide range of states. *Phys. Rev.*, **A42**, 789–794.

Ciccotti, G. & Ryckaert, J. P. (1986). Molecular dynamics simulation of rigid molecules. *Comput. Phys. Rep.*, **4**, 345–392.

Edberg, R., Morriss, G. P. & Evans, D. J. (1987). Rheology of *n*–alkanes by nonequilibrium molecular dynamics. *J. Chem. Phys.*, **86**, 4555–4570.

Hansen, J. P. & McDonald, I. R. (1986). *Theory of Simple Liquids*. New York: Academic Press.

Hoheisel, C. & Vogelsang, R. (1988). Thermal transport coefficients for one– and two–component liquids from time correlation functions computed by molecular dynamics. *Comput. Phys. Rep.*, **8**, 1–69.

Hoheisel, C. (1990). Memory functions and the calculation of dynamical properties of atomic liquids. *Comput. Phys. Rep.*, **12**, 29–66.

Hoheisel, C. (1993). *Theoretical Treatment of Liquids and Liquid Mixtures*. Amsterdam: Elsevier.

Hoheisel, C. (1994). Transport properties of molecular liquids. *Phys. Rep.*, **245**, 111–157.

Karkheck, J. (1986). Kinetic theory and ensembles of maximum entropy. *Kinam*, **A7**, 191–208.

Klein, M. (1986). *Molecular-Dynamics Simulation of Statistical–Mechanical Systems*, eds. G. Ciccotti & W. G. Hoover, Amsterdam: North–Holland.

Landolt–Börnstein, Zahlenwerte and Funktionen – Transportpläne, Band 2, Teil 5. Berlin, Göttingen, Heidelberg: Springer–Verlag.

Luo, H. & Hoheisel, C. (1991). Behaviour of collective time correlation functions in liquids composed of polyatomic molecules. *J. Chem. Phys.*, **94**, 8378–8383.

Luo, H. & Hoheisel, C. (1993). Thermodymamic and transport properties of *n*–butane computed by molecular dynamics using a rigid interaction model. *Phys. Rev.*, **E47**, 3956–3961.

Marechal, G. & Ryckaert, J. P. (1983). Atomic versus molecular description of transport properties in polyatomic fluids: *n*–Butane as an illustration. *Chem. Phys. Lett.*, **101**, 548–554.

Marechal, G., Ryckaert, J. P. & Bellemans, A. (1987). The shear viscosity of *n*–butane by equilibrium and nonequilibrium molecular dynamics. *Mol. Phys.*, **61**, 33–49.

Ulybin, S. A. & Makarushkin, V. I. (1977). Viscosity of sulfur hexafluoride at 230–800 K and up to 50 MPa. (Russ.) *Teplofiz. Vys. Temp.*, **15**, 1195–1201.

Vogelsang, R. & Hoheisel, C. (1987). Comparison of various potential models for the simulation of the pressure of liquid and fluid N_2. *Phys. Chem. Liq.*, **16**, 189–203.

Younglove, B. A. & Hanley, H. J. M. (1986). The viscosity and thermal conductivity coefficients of gaseous and liquid argon. *J. Phys. Chem. Ref. Data*, **15**, 1323–1337.

Younglove, B. A. & Ely, J. F. (1987). Thermophysical properties of fluids II. Methane, ethane, propane, isobutane and normal butane. *J. Phys. Chem. Ref. Data*, **16**, 577–798.

9.3 Transport properties and nonequilibrium molecular dynamics[a]

H. J. M. HANLEY

Thermophysics Division,
National Institute of Standards and Technology,
Boulder, CO, USA

and

D. J. EVANS

The Australian National University, Canberra, Australia

9.3.1 Introduction

A simulation is a tool to judge a statistical mechanical theory. Because a computer simulation generates essentially exact results for a given molecular potential model, and because the only approximations are numerical rather than physical, the calculations can be made more and more accurate by simulating larger and larger systems over longer time spans. A theory is assumed successful if it reproduces experiment, and unsuccessful if it does not. But this criterion is not complete for there may be no clear–cut way to tell if any discrepancies arise from inadequacies in the theory itself, or in the model potential function it employs. Hence, the assumptions of the theory are best assessed by comparing its predictions with those from a simulation modeled on the identical potential function. If the results disagree, the theory is inadequate, but one now has a basis to modify it.

A simulation is also a tool to calculate the thermophysical properties of a real fluid, and there are a number of circumstances for which computer simulations using reasonably accurate potentials provide the only reliable 'experimental' information. Estimating the properties of a fluid under extreme conditions of temperature and pressure is one example; calculating the properties of a system far from equilibrium is another. Clearly, then, the most useful simulation algorithm should both evaluate a property to within a reasonable accuracy, assuming the pair potential is known to a first approximation, and give an insight into the behavior of the system.

The first systematic attempts to calculate the viscosity and thermal conductivity of dense liquids evaluated the Green–Kubo (G–K) equations, and the Einstein relations for diffusion. The G–K expressions relate the numerical values of linear nonequilibrium transport coefficients to the relaxation of appropriate fluctuations in the equilibrium state (see Section 9.2). These studies of about 20 years ago were most important in that they gave a quantitative insight into the behavior of nonequilibrium systems and opened the door to the modern view of statistical mechanics (Hansen & MacDonald 1986; McQuarrie 1976; Egelstaff 1992). From the practical standpoint, however,

---

[a] Work carried out in part at the National Institute of Standards and Technology; not subject to copyright.

reliable calculations were limited to the diffusion coefficient (Hanley & Watts 1975). The computational resources available at that time were too limited to calculate the more industrially significant viscosity and thermal conductivity coefficients to within acceptable accuracy, largely because results depended strongly on the system size and the duration of the simulation. In fact, even the best of the early calculations for the simple Lennard–Jones fluid argon were accurate only to about 30% for the viscosity, and to a factor of 2 for the thermal conductivity (Hansen & MacDonald 1986). Simulations of the transport properties of complex fluids were out of the question. Today, of course, powerful computers are commonplace, and these particular criticisms of the Green–Kubo method no longer apply. Nonetheless, its initial drawbacks encouraged alternative simulation techniques to be developed, and nonequilibrium molecular dynamics (NEMD) has emerged as the most useful.

In NEMD, transport coefficients are calculated as an experimentalist would measure them – by time averaging the nonequilibrium steady–state ratio of a nonequilibrium flux induced by a thermodynamic force. For example, the shear viscosity coefficient, η, follows directly by evaluating the momentum flux in a system subjected to a known applied shear rate, γ. It turns out that statistical errors in NEMD simulations are not nearly as sensitive to the system size as the G–K counterparts, and calculations of the transport properties for simple liquids whose intermolecular interactions are approximated by an isotropic pair potential – the rare gases and low molecular weight nonpolar hydrocarbons – give results that are within experimental error (Evans & Hanley 1980). Calculations for more complex fluids, such as 'floppy' molecules with weak internal conformational energy barriers, or associated liquids which interact through strong many–body potentials, are not yet at this stage, but generally the viscosity and thermal conductivity coefficients can be evaluated to within 20%. Cummings & Evans (1992) report typical comparisons for alkanes, water and carbon dioxide in their recent survey. The difficulty to overcome is that sufficiently accurate molecular potential functions for complex fluids – if they exist – are so complicated that computer simulations using them are extremely expensive. Better agreement between simulation and experiment must follow as computer speeds and molecular potential functions continue to improve.

The strongest appeal of NEMD, however, is its scope and versatility. In particular, the technique allows one to compute the rich and varied nonlinear behavior of systems far from equilibrium. This asset has been used to great advantage to study the atomic basis of the rheological non–Newtonian behavior of liquids.* There was *no* practical technique to investigate the rheological properties of fluids from an atomistic viewpoint before NEMD. (Indeed the word 'rheology' was rarely mentioned in a statistical–mechanical context.)

---

* This alone is an advantage over the earlier G–K approach, which could calculate only the linear transport coefficients which describe transport processes arbitrarily close to equilibrium. Evans & Morriss (1990), however, have shown that the G–K relations are simple examples of more general nonlinear, fluctuation dissipation relations which are valid for nonequilibrium steady states.

Furthermore, NEMD enables the fluid microstructure to be studied in nonequilibrium steady states and to compare this structure with experiment (Hess & Hanley 1982). The nonequilibrium distributions of particle positions and momenta are reflected in the thermophysical properties of viscosity, dilatancy and normal pressure differences. In molecular fluids such as lubricants, nonlinear fluid behavior is brought about in part by shear induced changes in molecular conformation.[†]

9.3.2 Molecular dynamics

The starting point of a molecular dynamics simulation is to set up a model fluid of N particles of mass m by assigning initial values of the positions and peculiar momenta, $\mathbf{r}_i$ and $\mathbf{p}_i$ $(i = 1, N)$ in volume V, where the peculiar momenta are measured with respect to the streaming velocity of the fluid. The evolution of these variables with time is calculated from equations of motion such as

$$\frac{d\mathbf{r}_i}{dt} = \frac{\mathbf{p}_i}{m} \qquad\qquad \frac{d\mathbf{p}_i}{dt} = \mathbf{F}_i - \alpha\mathbf{p}_i \qquad (9.37)$$

where $\mathbf{F}_i$ is the total intermolecular force on particle i, due to the $N - 1$ other particles, usually written as a pairwise additive sum of forces computed from a pair potential, $\phi(\mathbf{r}_{ij})$, where

$$\mathbf{r}_{ij} = \mathbf{r}_j - \mathbf{r}_i, r_{ij} = \mathbf{r}_{ij}/|\mathbf{r}_{ij}| \qquad (9.38)$$

and

$$\mathbf{F}_i = -\sum_j \partial\phi(r_{ij})/\partial\mathbf{r}_i \qquad (9.39)$$

The equations of motion as written here contain a term with a thermostatting multiplier, α, to ensure that the kinetic temperature, $T = \sum_i p_i^2/3Nmk_B$, is a constant of motion; k_B is the Boltzmann constant. The multiplier is given by

$$\alpha = \frac{\sum\limits_{i-1}^{N} \mathbf{p}_i \bullet \mathbf{F}_i}{\sum\limits_{i-1}^{N} \mathbf{p}_i \bullet \mathbf{p}_i} \qquad (9.40)$$

Since the equations of motion (9.37) constrain T, they are generally closer to experiment than the traditional Newtonian equations (i.e., with $\alpha = 0$) which constrain the total energy. The particular functional form of the thermostat (9.37) and (9.40) is derived using Gauss's Principle of Least Constraint. Today, however, several thermostatting schemes have been proposed and the choice is one of convenience. (See Evans & Morriss 1990 and Allen & Tildesley 1990 for a detailed discussion of thermostats.)

[†] In the linear regime, changes in fluid microstructure are in principle calculable using G–K methods but, in practice, such calculations are so difficult that no one has yet attempted them.

A simulation can involve only a small number of particles compared with the Avogadro number of $\sim 10^{23}$, but one compensates by using periodic boundary conditions, where the unit cell is surrounded by an infinite array of its periodic images (Allen & Tildesley 1990) (see Figure 9.9). Provided the pair potential is short–ranged with respect to the length, $V^{1/3}$, the force on particle i is evaluated by considering the shortest distance between particles i and j, whether j is in the unit cell or is an image. Further, should a particle leave the cell in the course of time, it is replaced by its image from a neighboring cell.

9.3.3 Nonequilibrium molecular dynamics

The general NEMD algorithm (Evans & Morriss 1990) introduces a known but (usually) fictitious applied field, X, into the equations of motion and hence generates a dissipative thermodynamic flux, J. The dissipative flux is defined as the work performed on the system per unit time, $\dot{H}_0^{ad}$, by the applied field, X, where

$$\dot{H}_0^{ad} = -JX \tag{9.41}$$

This applied field, and its coupling to the system, must be chosen to be consistent with periodic boundary conditions which ensure that the simulation sample remains homogeneous. It is often convenient to consider the isothermal steady state, whence the linear transport coefficient, L, follows from the limiting constitutive relation

$$L = \lim_{X \to 0} \lim_{t \to \infty} \left(\frac{<J>}{X} \right) \tag{9.42}$$

If we design the coupling of the external field to the system in such a way that the dissipative flux is equal to one of the Navier–Stokes fluxes (such as the shear stress in planar Couette flow or the heat flux in thermal conductivity), it can be shown – provided the system satisfies a number of fairly simple conditions (Evans & Morriss 1990) – that the response is proportional to the Green–Kubo time integral for the corresponding Navier–Stokes transport coefficient. This means that the linear response of the system to the fictitious external field is exactly related to linear response of a real system to a real Navier–Stokes force, thereby enabling the calculation of the relevant transport coefficient.

9.3.3.1 Shear viscosity

Since the early 1970s a number of NEMD algorithms have been proposed to calculate the shear viscosity (Ashurst & Hoover 1972; Lees & Edwards 1972; Gosling *et al.* 1973; Ciccotti *et al.* 1979; Evans 1979). The early algorithms were *ad hoc*, and deterministic thermostats were unknown, but Hoover *et al.* (1980) derived their so–called DOLLS tensor algorithm for shear viscosity and proved it was exact in the linear limit. Hoover *et al.* (1982) and Evans (1982) independently, but simultaneously, proposed the first deterministic thermostat – the Gaussian thermostat of equation (9.40). In 1983 Evans (1983) and Evans & Morriss (1984) derived the equations of motion now known as the SLLOD equations and proved that they give an exact description of adiabatic planar Couette flow arbitrarily far from equilibrium. The SLLOD equations are the basis for most of the viscosity algorithms in use today. For a system undergoing planar Couette flow with strain rate, $\gamma = \partial u_x/\partial y$, the thermostatted SLLOD equations are

$$\dot{\mathbf{r}}_i = \frac{\mathbf{p}_i}{m} + \mathbf{i}\gamma y_i \qquad\qquad \dot{\mathbf{p}}_i = \mathbf{F}_i - \mathbf{i}\gamma p_{y_i} - \alpha\mathbf{p}_i \qquad (9.43)$$

where $\mathbf{i}$ is the x–unit vector. From these equations of motion it follows that

$$\dot{H}_0^{ad} = -P_{xy}\gamma V \qquad (9.44)$$

where P_{xy} is the xy element of the microscopic pressure tensor, $\mathbf{P}$, defined by,

$$\mathbf{P}V = \frac{\sum \mathbf{p}_i \mathbf{p}_i}{m} - \frac{1}{2}\sum \mathbf{r}_{ij}\mathbf{F}_{ij} \qquad (9.45)$$

As in equilibrium molecular dynamics, the equations of motion have to be solved for a system with periodic boundaries. For shear, the boundaries are modified to become the Lees–Edwards 'sliding brick' conditions (Lees & Edwards 1972), in which periodic images of the simulation cell above and below the unit cell are moved in opposite directions at a velocity determined by the imposed shear rate (see Fig. 9.9). The properties of the system follow from the appropriate time averages, $< .. >$, usually (but not necessarily) after the system has reached the steady state. Given, for example, a system at a number density, $n = N/V$, under an applied shear rate, the kinetic temperature is constrained with an appropriate thermostat. Different properties can then be evaluated, for example, the internal energy,

$$\langle H_0 \rangle = \left\langle \sum_i p_i^2/2m + \frac{1}{2}\sum_{i,j}\phi_{ij} \right\rangle \qquad (9.46)$$

the microscopic hydrostatic pressure, defined as

$$p(\gamma) = \frac{1}{3}tr(< \mathbf{P} >) \qquad (9.47)$$

and the viscosity using equation (9.42), in the form

$$< P_{xy} > = -\eta(\gamma)\gamma \qquad (9.48)$$

If the system is far from equilibrium, it will display non–Newtonian, rheological behavior characterized by normal pressure differences $[P_{xx} - P_{yy}]$ and $[1/2(P_{xx} + P_{yy}) - P_{zz}]$, where P_{ii} are the diagonal elements of $\mathbf{P}$, and by a shear–rate dependent pressure and viscosity coefficient.

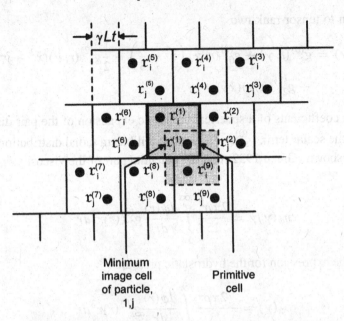

Fig. 9.9. Lees–Edwards periodic boundary conditions. At equilibrium, $\gamma = 0$, the cells form a time–independent square lattice. Under shear, the cells are displaced by $\gamma L t$, where L is the height of each cell. The minimum image convention dictates that particle 1, j (for example) interacts with particle 9, i.

9.3.3.2 Shear–induced microstructure

The pressure tensor can be written as the sum of kinetic and potential terms. The potential part arises from the direct intermolecular forces and is therefore a function of the local fluid microstructure. In a dense liquid the shear viscosity is overwhelmingly dominated by the potential contribution.*

In 1952 Green showed that the potential part of the pressure tensor, $\mathbf{P}_\phi$, can be written as an integral over the pair correlation function, $g(\mathbf{r}, \gamma)$, in a fluid where the intermolecular forces are a sum of pair forces. The pair correlation function is the ratio of the probability, per unit volume, of finding a particle at a position $\mathbf{r}$ from a given particle, to the number density of the fluid. At large distances there is no correlation between particle positions in a fluid and $\lim_{r \to \infty} g(\mathbf{r}, \gamma) = 1$.

For fluids at equilibrium the pair correlation function is independent of the orientation of the particle separation vector, $\hat{\mathbf{r}} = \mathbf{r}/|\mathbf{r}|$, and is dependent only on the separation distance, $r = |\mathbf{r}|$. For fluids out of equilibrium, however, the function could be anisotropic. However, for a given separation distance, r, the pair correlation function can be expanded in terms of spherical harmonics or Cartesian tensors (Green 1952; Hanley *et al.* 1987). For simplicity we write down the result for a fluid of spherical

---

* At the triple point of argon, about 98% of the viscosity arises from the potential contribution.

particles, taken to tensor rank two

$$g(\mathbf{r}, \gamma) = g_s^{(0)}(r, \gamma) + g_0^{(2)}(r, \gamma)\left(\hat{z}^2 - \frac{1}{3}\right) + \frac{1}{2}g_1^{(2)}(r, \gamma)(\hat{x}^2 - \hat{y}^2)$$
$$+ g_2^{(2)}(r, \gamma)(\hat{x}\hat{y}) + \dots \tag{9.49}$$

where $g_k^{(2)}$ are coefficients of a spherical harmonic expansion of the pair distribution. When $\gamma = 0$, the scalar term, $g_s^{(0)}(r, 0)$, is the equilibrium radial distribution function. Now it can be shown (Green 1952) that the potential part of the viscosity, η_ϕ, is given by

$$\eta_\phi(\gamma)\gamma = \frac{2\pi\rho^2}{15} \int\limits_0^\infty \frac{d\phi(r)}{dr} g_2^{(2)}(r)r^3 dr \tag{9.50}$$

analogous to the expression for the hydrostatic pressure

$$P_\phi(\gamma) = -\frac{2\pi\rho^2}{3} \int\limits_0^\infty \frac{d\phi(r)}{dr} g_s^{(0)}(r)r^3 dr \tag{9.51}$$

The potential contributions to the viscosity and to the thermodynamic variables can therefore be calculated if the terms of the expansion are known. It is straightforward in an NEMD simulation to calculate the ensemble averages of $g(\mathbf{r}, \gamma)$ and $g_k^{(i)}(r, \gamma)$ from the appropriate histograms.

The discussion on the distorted fluid microstructure demonstrates the utility of NEMD as a practical tool and as a means to get an insight into the physics of viscous phenomena. The shear–rate dependent non–Newtonian properties of a liquid can be calculated from the appropriate steady–state averages, but the mechanical transport properties can be obtained indirectly from the nonequilibrium pair distribution function using equation (9.49), and a more physical picture of the behavior of a nonequilibrium fluid is obtained. As an example, the phenomenological theory of Hanley et al. (1987, 1988) is outlined where the functionals $g_k^{(i)}(r)$ are evaluated from a relaxation time approximation. The confidence in the qualitative reliability of this theory is based on the comparisons of theoretical functionals with corresponding data from the NEMD simulations. It should be noted that real 'experimental' data which could in principle come from radiation scattering experiments would be far too imprecise to be of use in this context. Remarkably, the theory uses only the equilibrium radial distribution function and a single relaxation time, τ_0, to predict the harmonic expansion of $g(\mathbf{r}, \gamma)$, at least to the fourth–order spherical harmonic, as a function of shear rate. The theory is outlined here.

The pair correlation function is assumed to obey the Kirkwood–Smoluchowski kinetic equation (Hess 1980). In the case of Couette flow, the equation is

$$\frac{\partial g}{\partial t} + \gamma y \frac{\partial g}{\partial x} + \Omega(g) = 0 \tag{9.52}$$

Here Ω is a diffusion–like operator with the property that $\Omega(g_{eq}) = 0$. Equation (9.52) is then solved in a relaxation time approximation

$$\Omega\left[g_s^{(0)}(r, \gamma)\right] = \frac{g_s^{(0)}(r, \gamma) - g_s^{(0)}(r, \gamma = 0)}{\tau(\gamma)} \tag{9.53}$$

$$\Omega\left[g_k^{(i)}(r, \gamma)\right] = \frac{g_k^{(i)}(r, \gamma)}{\tau(\gamma)}, \qquad \forall i = 1, 2, 3, \ldots; k \leq i \tag{9.54}$$

The major approximations made here in going from equation (9.52) to equations (9.53)–(9.54) are that the same $\tau(g)$ applies for all values of r and for all allowed values of i, k. On inserting the expansion (9.49) into equation (9.52) and disregarding tensorial contributions above rank two, it can be shown that in the stationary state:
– to zero order in τ

$$g_s^{(0)}(r, \gamma) = g_s^{(0)}(r, 0) \tag{9.55}$$

– to first order

$$g_s^{(2)}(r, \gamma) = -\tau(\gamma)\gamma \frac{r d g_s^{(0)}(r, 0)}{dr} \tag{9.56}$$

– and to second order

$$g_s^{(0)}(r, \gamma) = g_s^{(0)}(r, 0) - \frac{1}{15}\tau(\gamma)\gamma\left[r\frac{d}{dr} + 3\right]g_2^{(2)}(r, \gamma) \tag{9.57}$$

$$g_1^{(2)}(r, \gamma) = \tau(\gamma)\gamma g_2^{(2)}(r, \gamma) \tag{9.58}$$

$$g_s^{(2)}(r, \gamma) = \frac{1}{7}\tau(\gamma)\gamma\left[r\frac{d}{dr} + \frac{3}{2}\right]g_2^{(2)}(r, \gamma) \tag{9.59}$$

There are also terms (not written here) with $i = 4$, to second order in τ. This expansion in τ is in fact an expansion about the Newtonian limit (that is, $\tau\gamma \to 0$).

The final approximation of the theory is due to Hess, invoking an argument based on the Kirkwood–Smoluchowski equation for the pressure tensor (Hess 1982) to yield

$$\tau(\gamma) = \tau_0\left[1 - \frac{1}{2}\sqrt{\pi\tau_0\gamma}\right] \tag{9.60}$$

where τ_0 is the relaxation time at zero shear. The viscosity and pressure can be calculated using equations (9.55)–(9.57) and (9.60), provided the relaxation time at zero shear is known and we have an estimate of the equilibrium radial distribution function. For simple molecules, the equilibrium radial distribution function is easily obtained from equilibrium computer simulations, and τ_0 can be estimated from the Maxwell equation

$$\tau_0 = \frac{\eta(\gamma = 0)}{G_\infty(\gamma = 0)} \tag{9.61}$$

where $G_\infty(\gamma = 0)$ is the infinite frequency shear modulus evaluated at equilibrium.

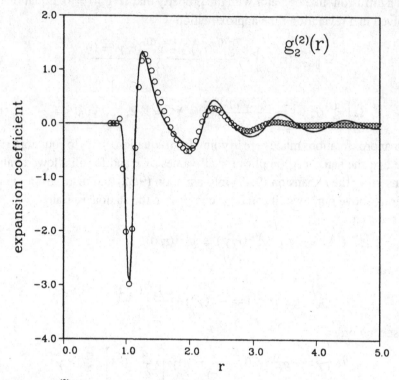

Fig. 9.10. Plot of $g_2^{(2)}$ obtained to first order in t using equation (9.56) for the soft sphere liquid at $\rho = 0.7$ and $\gamma = 1.0$. The curve is predicted from the relaxation time theory using equation (9.60) with $\tau_0 = 0.24$. The points are from the NEMD simulation of Hanley *et al.* (1987).

As an example we reproduce, without elaboration, some of the results reported in the paper of Hanley *et al.* (1988). The properties of a soft sphere liquid – a liquid whose particles interact with a $1/r^{12}$ potential in this case – undergoing Couette flow at a density equal to 7/8 of the freezing density were simulated. The radial distribution function was calculated by the procedure of Rogers & Young (1984), and the relaxation time $\tau_0 = 0.24$ was estimated from the Einstein frequency. Figures 9.10–9.12 display how the theory agrees with the NEMD. Since the curves are entirely predictive, the agreement is excellent for these reduced variables.

9.3.3.3 Thermal conductivity

The NEMD algorithm for computing the thermal conductivity of simple fluids was developed by Evans (1982). The driving force in the algorithm is a fictitious heat field vector, $\mathbf{F}_Q$, which gives rise to the heat flux $\mathbf{J}_Q$. The equations of motion for the system are

$$\dot{\mathbf{r}}_i = \frac{\mathbf{p}_i}{m} \tag{9.62}$$

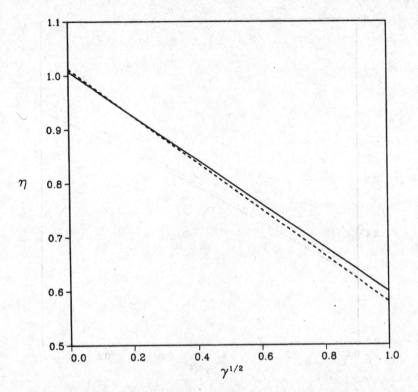

Fig. 9.11. The viscosity, η, of the soft sphere at a density of 0.7 calculated from equation (9.50) with equation (9.56), solid curve, plotted as a function of the shear rate. Note the $\gamma^{1/2}$ dependence. The dashed curve is the NEMD result.

$$\dot{\mathbf{p}}_i = \mathbf{F}_i + (e_i - \bar{e})\mathbf{F}_Q(t) - \frac{1}{2}\sum_{j=1}^{N}\mathbf{F}_{ij}\left(\mathbf{r}_{ij}\cdot\mathbf{F}_Q(t)\right)$$

$$+ \frac{1}{2N}\sum_{j,k}^{N}\mathbf{F}_{jk}\left(\mathbf{r}_{jk}\cdot\mathbf{F}_Q(t)\right) - \alpha\mathbf{p}_i \tag{9.63}$$

where e_i is the instantaneous energy of atom i, $\bar{e}$ is $(1/N)\sum_i e_i$ and α is the Gaussian thermostat multiplier, which becomes in this case

$$\alpha = \frac{1}{\sum\limits_{i=1}^{N} p_i^2}\left\{\sum_{i=1}^{N}\left[\mathbf{p}_i \bullet \left(\mathbf{F}_i + (e_i - \bar{e})\,\mathbf{F}_Q(t) - \frac{1}{2}\sum_{j=1}^{N}\mathbf{F}_{ij}\left(\mathbf{r}_{ij}\cdot\mathbf{F}_Q(t)\right)\right.\right.\right.$$

$$\left.\left.\left.+ \frac{1}{2N}\sum_{j,k}^{N}\mathbf{F}_{jk}\left(\mathbf{r}_{jk}\cdot\mathbf{F}_Q(t)\right)\right)\right]\right\} \tag{9.64}$$

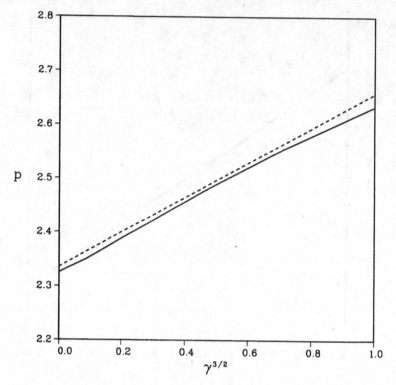

Fig. 9.12. The pressure, p, evaluated from equation (9.51) plotted versus the shear rate, solid curve. Note the approximate $\gamma^{3/2}$ dependence. The dashed curve is the NEMD result.

The heat flux is given by

$$\mathbf{J}_Q(t)V = \sum_i \frac{\mathbf{p}_i}{m} e_i - \frac{1}{2} \sum_{i,j\neq i} \mathbf{r}_{ij}\mathbf{F}_{ij} \cdot \frac{\mathbf{p}_i}{m} \tag{9.65}$$

It is straightforward to show that these equations of motion have the heat–flux vector as their dissipative flux (that is, $\dot{H}_0^{ad} = -\mathbf{J}_Q \bullet \mathbf{F}_Q$), and therefore the thermal conductivity can be calculated from the relation

$$\lim_{\mathbf{F}_Q \to 0} \lambda = \left\langle \frac{\mathbf{J}_Q \bullet \mathbf{F}_Q}{T F_Q^2} \right\rangle \tag{9.66}$$

Mathematically this corresponds to identifying (in the zero heat field limit) the heat field, $\mathbf{F}_Q$, with the negative logarithmic temperature gradient, $-\nabla \ln T$. However, as can be seen from the equations of motion, the system is homogeneous. The heat flow is generated in the absence of an actual temperature gradient. Its place is taken by the fictitious field, $\mathbf{F}_Q$. The fact that such a system is homogeneous greatly increases the efficiency of this algorithm over alternatives where an actual temperature gradient is employed.

Using this method, Evans calculated the thermal conductivity of argon as a Lennard–Jones fluid to higher precision than has been achieved experimentally. Consequently, if we use an accurate many–body potential for the inert gases (such potentials already exist), we should be able to calculate the thermal conductivity of the inert gases with higher accuracy than is possible experimentally at this time. Murad, Hanley & Evans (personal communication) have also simulated the conductivity of an inert gas in the critical region and found the critical point enhancement observed experimentally.

9.3.4 *The thermal conductivity of benzene; effect of internal degrees of freedom*

As an example of the practical utility of NEMD simulations of the thermal conductivity, a summary of a calculation by Ravi *et al.* (1992) is given. The object was to explore the contribution of internal degrees of freedom to the conductivity of a polyatomic molecule. This contribution has not been quantitatively demonstrated until very recently (Murad *et al.* 1991). In fact, it is usually assumed that this contribution is independent of number density and is given by the dilute–gas value at the corresponding temperature. In the paper of Ravi *et al.* (1992), heat flow is discussed for a model benzene–like liquid with a six–centered Lennard–Jones pair potential.

9.3.4.1 *Heat–flow algorithm for rigid polyatomic molecules*

The equations of motion for molecules that can be considered as classical rigid bodies are very similar to those above for atomic fluids. A difference is that the energy e_i must include the rotational kinetic energy contribution. The heat flux vector likewise contains a new contribution from the work performed by molecule i against the principal intermolecular torque about the center of mass of i due to molecule j so that

$$\mathbf{J}_Q V = \frac{1}{2} \sum_i \mathbf{v}_i \left[m\mathbf{v}_i^2 + \varpi_{pi} \bullet \mathbf{I}_p \bullet \varpi_{pi} + \sum_{i,j=1}^{N} \phi_{ij} \right]$$

$$+ \frac{1}{2} \sum_{i,j=1}^{N} \mathbf{r}_{ij} (\mathbf{v}_i \bullet \mathbf{F}_{ij} + \varpi_{pi} \bullet \Gamma_{pij}) \qquad (9.67)$$

In this equation, ϖ_{pi} and Γ_{pij} are the principal components of the angular velocity of molecule i about its center of mass and the torque on molecule i due to molecule j. $\mathbf{I}_p$ is the principal inertia tensor for the particle. The rotational contribution of the heat flux from equation (9.67), $\mathbf{J}_Q^r$, is

$$\mathbf{J}_Q^r V = \frac{1}{2} \sum_i \mathbf{v}_i (\varpi_{pi} \bullet \mathbf{I}_p \bullet \varpi_{pi}) + \frac{1}{2} \sum_{i,j} \mathbf{r}_{ij} (\varpi_{pi} \bullet \Gamma_{pij}) \qquad (9.68)$$

The terms on the right–hand side are defined as the rotational kinetic energy (rke) and the torque contributions, respectively, hence

$$J_Q^{rke} V = \frac{1}{2} \sum_i \mathbf{v}_i \left(\varpi_{pi} \bullet \mathbf{I}_p \bullet \varpi_{pi} \right) \tag{9.69}$$

and

$$J_Q^{torque} V = \frac{1}{2} \sum_{i,j} \mathbf{r}_{ij} \left(\varpi_{pi} \bullet \Gamma_{pij} \right) \tag{9.70}$$

The corresponding translational flux, J_Q^t, is defined by the remaining terms in equation (9.67). The decomposition of the heat flux, $(J_Q = J_Q^t + J_Q^r)$, implies a corresponding two–way decomposition of the thermal conductivity, thus

$$J_Q = -\lambda(1/T)\nabla T, \qquad J_Q^t = -\lambda^t(1/T)\nabla T, \qquad J_Q^r = -\lambda^r(1/T)\nabla T \tag{9.71}$$

and, finally, the total thermal conductivity is the sum of the contributions

$$\lambda = \lambda^t + \lambda^r \tag{9.72}$$

9.3.4.2 Thermal conductivity of benzene

Benzene was simulated with a six–centered Lennard–Jones potential

$$\phi_{12} = \sum_{i_1=1}^{6} \sum_{j_2=1}^{6} 4\epsilon \left[\left(\frac{\sigma}{r_{i_1 j_2}} \right)^{12} - \left(\frac{\sigma}{r_{i_1 j_2}} \right)^{6} \right] \tag{9.73}$$

where $\epsilon/k_B = 77$ K and $\sigma = 0.35$ nm are the site–site energy and length potential parameters selected from fits of the second virial coefficient and the lattice parameters of the crystal (Evans & Watts 1976). The temperature, density and conductivity are reduced by the potential parameters, ϵ, σ, and the interaction–site mass (that is, the mass of a CH group) in the usual way. The benzene critical temperature of 562.16 K is $T^* = 7.3$ in reduced units. Likewise, the critical density of 3.861 mol L^{-1} corresponds to $n^* = 0.1$. The sites were situated on a ring of radius 0.502 (0.1756 nm). The calculations were carried out for a 108– and 256–molecule system (648 or 1536 sites) with $|F_e| = 0.2$. The results are given in Table 9.9.

It is clear that the internal contributions are a strong function of density at a given temperature; moreover, their contribution is large at the densities considered, about 33% of the total. The ratio of the rotational to total conductivity is approximately constant.

Finally, it is of interest to see how well the benzene simulations represent experiment. It cannot be expected that there will be an exact match but, if a calculation is to have a practical value, it should give a sensible answer. To test this, the simulation is compared with thermal conductivity data at a reduced temperature of $T^* = 7.3$.

Table 9.9. *Thermal conductivity of model benzene.*

T^*	n^*	λ	λ^r	λ^{torque}
3.9	0.24	3.54 ± 0.05	1.21 ± 0.07	1.16 ± 0.06
	0.29	6.19 ± 0.22	2.11 ± 0.16	2.04 ± 0.14
6.0	0.29	6.90 ± 0.06	2.25 ± 0.09	2.16 ± 0.08
6.5	0.29	6.85 ± 0.16	2.28 ± 0.05	2.18 ± 0.04
7.3	0.15	1.33 ± 0.02	0.47 ± 0.02	0.35 ± 0.01
	0.17	1.76 ± 0.04	0.58 ± 0.04	0.47 ± 0.02
	0.20	2.42 ± 0.02	0.81 ± 0.01	0.69 ± 0.01
	0.24	4.05 ± 0.02	1.33 ± 0.02	1.21 ± 0.01
	0.29	6.90 ± 0.02	2.29 ± 0.02	2.18 ± 0.01
8.44	0.17	1.82 ± 0.02	0.57 ± 0.04	0.43 ± 0.03
	0.24	4.35 ± 0.09	1.30 ± 0.06	1.18 ± 0.05
	0.29	7.32 ± 0.12	2.38 ± 0.10	2.27 ± 0.09

First, however, the contribution of vibration to the thermal conductivity, which is excluded from the simulation, has to be estimated. Following Evans & Watts (1976), a dilute–gas vibrational contribution represented by a Eucken–type formula is considered

$$\lambda_{\text{vib}}^{(0)} = \lambda_{\text{tr}}^{(0)} + \frac{C_{\text{int,exp}}^{\text{id}}}{C_{\text{int,rig}}^{(\text{id})}} \lambda_{\text{rot}}^{(0)} \tag{9.74}$$

Here, $\lambda_{\text{tr}}^{(0)}$ and $\lambda_{\text{rot}}^{(0)}$ are the contributions of translational and rotational degrees of freedom to the thermal conductivity of the dilute gas evaluated from the six–site model. $C_{\text{int,exp}}^{\text{id}}$ is the experimental internal heat capacity at constant volume, whereas $C_{\text{int,rig}}^{(\text{id})}$ is the equipartition value obtained by assuming the benzene molecule is rigid, that is, $C_{\text{int,exp}}^{\text{id}} = C_V^{\text{id}} - (3/2)R$, where R is the universal gas constant. It turns out that $\lambda_{\text{vib}}^{(0)} = 1.04$ for $T^* = 7.3$.

Experimental data are represented by the 'SUPERTRAPP' procedure (STRAPP) of Ely & Hanley (1983; Ely & Huber 1990), which predicts the density, viscosity and thermal conductivity of pure species and mixtures (see Chapter 12 for details).

The data of Table 9.9 for $T^* = 7.3$, with a vibrational contribution added from equation (9.74), are compared with the STRAPP prediction in Figure 9.13. The simulation was carried out for the model fluid with the ring parameter $B = 0.1756$ nm; a value consistent with the Lennard–Jones σ and ϵ parameters. The vibrational contribution is a constant for the isotherm. Since the uncertainty in the STRAPP experimental data is about 10%, the agreement between the STRAPP curve and the simulated data is considered reasonable, and it could be improved by manipulating the potential parameters and the ring size.

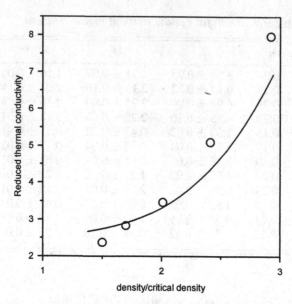

Fig. 9.13. Comparison between STRAPP (—) and simulation (O).

Acknowledgments

D. J. Evans thanks the staff of the Thermophysical Properties Division of NIST for their kind hospitality during his visit while this paper was being prepared. The contributions of M. L. Huber, J. C. Rainwater, S. Murad, P. T. Cummings and S. Hess, whose work is quoted here, are gratefully acknowledged. The work was supported in part by the U.S. Department of Energy, Office of Basic Energy Sciences.

References

Allen, M. P. & Tildesley, D. J. (1990). *Computer Simulation of Liquids.* Oxford: Clarendon.

Ashurst, W. T. & Hoover, W. G. (1972). Nonequilibrium molecular dynamics: Shear viscosity and thermal conductivity. *Bull. Amer. Phys. Soc.*, **17**, 1196.

Ciccotti, G., Jacucci, G. & McDonald, I. R. (1979). Thought experiments by molecular dynamics. *J. Stat. Phys.*, **21**, 1–22.

Cummings, P. & Evans, D. J. (1992). Review of: Nonequilibrium molecular dynamics approaches to transport properties and non–Newtonian fluid rheology. *Ind. Eng. Chem. Res.*, **31**, 1237–1252.

Egelstaff, P.A. (1992). *An Introduction to the Liquid State.* Oxford: Clarendon, 2nd Ed.

Ely, J. F. & Hanley, H. J. M. (1983). Prediction of transport properties 2. Thermal conductivity of pure fluids and fluid mixtures. *Ind. Eng. Chem. Fund.*, **22**, 90–97.

Ely, J. F. & Huber, M. L. (1990). NIST Standard Reference Database 4. Computer Program SUPERTRAPP, *NIST Thermophysical Properties of Hydrocarbon Mixtures*, Version 1.0.

Evans, D. J. (1979). The frequency dependent shear viscosity of methane. *Mol. Phys.*, **37**, 1745–1754.

Evans, D J. (1982). Homogeneous NEMD algorithm for thermal conductivity: Application of non–canonical linear response theory. *Phys. Lett.*, **91A**, 457–460.

Evans, D J. (1983). Computer 'experiment' for nonlinear thermodynamics of couette flow. *J. Chem. Phys.*, **78**, 3297–3330.

Evans, D. J. & Hanley, H. J. M. (1980). Computer simulation of an m–6–8 fluid under shear. *Physica*, **A 103**, 343–353.

Evans, D. J. & Morriss, G. P. (1984). Non–Newtonian molecular dynamics. *Comput. Phys. Rep.*, **1**, 297–343.

Evans, D. J. & Morriss, G. P. (1990). *Statistical Mechanics of Nonequilibrium Fluids.* London: Academic Press.

Evans, D. J. & Watts, R. O. (1976). A theoretical study of transport coefficients in benzene vapour, *Mol. Phys.*, **31**, 995–1015.

Green, H. S. (1952). *The Molecular Theory of Fluids.* Amsterdam: North–Holland.

Gosling, E. M., McDonald, I. R. & Singer, K. (1973). On the calculation by molecular dynamics of the shear viscosity of a simple fluid. *Mol. Phys.*, **26**, 1475–1484.

Hanley, H. J. M., Rainwater, J. C. & Hess, S. (1987). Shear–induced angular dependence of the liquid pair correlation function. *Phys. Rev. A*, **36**, 1795–1802.

Hanley, H. J. M., Rainwater, J. C. & Huber, M. L. (1988). Prediction of shear viscosity and non–Newtonian behaviour in the soft sphere liquid. *Int. J. Thermophys.*, **9**, 1041–1050.

Hanley, H. J. M. & Watts, R. O. (1975). The self–diffusion coefficient of liquid methane. *Mol. Phys.*, **29**, 1907–1917.

Hansen, J.–P. & McDonald, I. R. (1986). *Theory of Simple Liquids.* London: Academic Press, 2nd Ed.

Hess, S. (1980). Shear–flow–induced distortion of the pair–correlation function. *Phys. Rev. A*, **22**, 2844–2848.

Hess, S. (1982). Viscoelasticity of a simple liquid in the prefreezing regime. *Phys. Lett. A*, **90**, 293–296.

Hess, S. & Hanley, H. J. M. (1982). Distortion of the structure of a simple fluid, *Phys. Rev. A*, **25**, 1801–1804.

Hoover, W. G., Evans, D. J., Hickman, R. B., Ladd, A. J. C., Ashurst, W. T. & Moran, B. (1980). Lennard–Jones triple point bulk and shear viscosities. Green–Kubo theory, Hamiltonian mechanics and nonequilibrium molecular dynamics. *Phys. Rev. A*, **22**, 1690–1697.

Hoover, W. G., Ladd, A. J. C. & Moran, B. (1982). High strain rate flow studied via nonequilibrium molecular dynamics, *Phys. Rev. Lett.*, **48**, 1818–1820.

Lees, A. W. & Edwards, S. F. (1972). The computer study of transport processes under extreme conditions. *J. Phys.*, **C5**, 1921–1929.

McQuarrie, D. A. (1976). *Statistical Mechanics.* New York: Harper and Row.

Murad, S., Singh, D. P., Hanley, H. J. M. & Evans, D. J. (1991). Thermal conductivity of a model diatomic fluid, *Mol. Phys.*, **72**, 487–490.

Ravi, P., Murad, S., Hanley, H. J. M. & Evans, D.J. (1992). The thermal conductivity of polyatomic molecules: Benzene. *Fluid Phase Equil.*, **76**, 249–257.

Rogers, F. J. & Young, D. A. (1984). New thermodynamically consistent, integral equation for simple fluids. *Phys. Rev. A*, **30**, 999–1007.

10

Modified Hard–Spheres Scheme

J. H. DYMOND

The University, Glasgow, UK

M. J. ASSAEL

Aristotle University, Thessaloniki, Greece

10.1 Introduction

At the present time, the most successful correlations of dense–fluid transport properties are based upon consideration of the hard–sphere model. One reason for this is that, as discussed in Chapter 5, it is possible to calculate values from theory for this model at densities from the dilute–gas state up to solidification. Second, this is physically a reasonably realistic molecular model because the van der Waals model, which has been successfully applied to equilibrium properties of dense fluids, becomes equivalent to the hard–sphere model for transport properties.

The van der Waals picture of a fluid presented in 1873 (van der Waals 1873) is of an assembly of molecules having a hard–core repulsive–interaction energy and a weak very long–range attractive interaction as in Figure 10.1a. This has important consequences for transport properties, because the attractive interactions lead to a practically uniform attractive energy surface throughout the fluid and the molecules then move in straight lines between core collisions. Computer studies on hard–sphere and square–well fluids published by Einwohner & Alder (1968) clearly indicate that this description of molecular motion based on the van der Waals model is physically much more realistic for

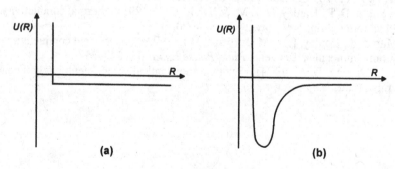

Fig. 10.1. Intermolecular pair potential curves, (a) van der Waals model, (b) realistic pair potential.

real dense fluids than descriptions based upon the activation–energy model (Glasstone *et al.* 1941) or quasi–brownian–motion approach (Davis *et al.* 1961).

For real molecules the pair–potential energy curve has the characteristic shape shown in Figure 10.1b. A real fluid can therefore be expected to behave as a van der Waals fluid only at high densities, in the region of critical density and above, when the average intermolecular separation corresponds to points just to the right of the minimum in the intermolecular potential curve. The range of the attractive potential energy is then long with respect to the average intermolecular spacing; providing the kinetic energy of the molecules is sufficiently high, that is, for temperatures in the region of the critical temperature and above, the molecular motion will approximate closely a succession of linear trajectories and hard–core collisions. Insofar as there are nonuniformities in the attractive potential surface, the effect of such deflections, or soft collisions, can be gained by comparing values for the transport coefficients calculated on the basis of the van der Waals theory with experiment.

One other difference between a real fluid and a van der Waals fluid is that the pair intermolecular–potential energy curve for real molecules, though steep, is not infinitely steep. Therefore, a very important consequence in applying the van der Waals model to real fluids is that the core size must be temperature–dependent, decreasing as the temperature increases to reflect the somewhat soft repulsive energy of real fluids. For transport properties of monatomic fluids, therefore, it is a reasonable approximation to represent the fluid as an assembly of hard spheres with temperature–dependent core sizes.

In the following section, expressions are given for the transport coefficients of dense assemblies of smooth hard–spheres and applied to the rare gases. The only adjustable parameter here is the core size. For molecular fluids, there is the possibility of translational–rotational coupling. The effects of this are discussed in Section 10.3 in terms of the rough hard–sphere model, which leads to the introduction of one additional parameter – the translational–rotational coupling factor – for each property, and the results are applied to methane. For dense nonspherical molecular fluids, it is assumed that the transport properties can also be related directly to the smooth hard–sphere values with proportionality factors ('roughness factors') which account for effects of nonspherical shape. The application of this method is described in 10.4 with reference to alkanes, aromatic hydrocarbons, alkanols and certain other compounds.

Basic to all these methods is the fact that it is the molar volume which is the important dependent variable, not the pressure. However, for engineering purposes it is necessary to know the transport properties for a given temperature and pressure, and therefore accurate knowledge of P, ρ, T behavior is required. The best representation is given by the Tait equation (Dymond & Malhotra 1987), as described in 10.5. By combination of these results with the equations given for the transport properties, it is now possible to calculate transport properties for pure fluids at any given temperature and pressure with an accuracy estimated to be generally better than 5%. The final section describes the extension of these procedures to the accurate correlation of transport properties of certain dense fluid mixtures.

10.2 Smooth hard–sphere fluid

A kinetic theory for transport coefficients of a dense hard–sphere system has been given by Enskog (1922). In a dense system, the collision rate is higher than in a dilute system because the diameter of the molecule is no longer negligible compared with the interparticle distance. The Enskog theory of diffusion assumes that the high–density system behaves exactly as a low–density system except that the collision frequency is increased by a factor of $g(\sigma)$, where $g(\sigma)$ is the radial distribution function at contact for spheres of diameter σ. The solution of the Boltzmann equation valid at low density is merely scaled in time to give the ratio of the diffusion coefficient D_E at high number density n relative to that at low density, superscript (0):

$$\frac{n D_E}{[n D]^{(0)}} = \frac{1}{g(\sigma)} \tag{10.1}$$

$g(\sigma)$ is obtained from computer simulation studies, given by the Carnahan and Starling equation (Carnahan & Starling 1969) as

$$g(\sigma) = \frac{1 - 0.5\xi}{(1 - \xi)^3} \tag{10.2}$$

where $\xi = b/4V$ for a molar volume V, and $b = 2\pi N_A \sigma^3/3$. $D^{(0)}$ is related to the number density $n^{(0)}$ at temperature T by the expression

$$[n D]^{(0)} = \frac{3}{8\pi\sigma^2} \left(\frac{\pi k_B T}{m}\right)^{1/2} \tag{10.3}$$

where m is the molecular mass and k_B is the Boltzmann constant.

For diffusion the particles themselves must move, but for viscosity and thermal conductivity there is the additional mechanism of collisional transfer whereby momentum and energy can be passed to another molecule upon collision. The Enskog theory for the viscosity η_E and the thermal conductivity λ_E in terms of the low–density coefficients accordingly contains additional terms

$$\frac{\eta_E}{\eta^{(0)}} = \frac{1}{g(\sigma)} + \frac{0.8b}{V} + 0.761 g(\sigma) \left(\frac{b}{V}\right)^2 \tag{10.4}$$

$$\frac{\lambda_E}{\lambda^{(0)}} = \frac{1}{g(\sigma)} + \frac{1.2b}{V} + 0.755 g(\sigma) \left(\frac{b}{V}\right)^2 \tag{10.5}$$

where the low–density coefficients are given to first–order approximation by

$$\eta^{(0)} = \frac{5}{16\pi\sigma^2} (\pi m k_B T)^{1/2} \tag{10.6}$$

$$\lambda^{(0)} = \frac{75}{64\pi\sigma^2} \left(\frac{\pi k_B^3 T}{m}\right)^{1/2} \tag{10.7}$$

In order to apply equations (10.1), (10.4) and (10.5) to the calculation of dense gas transport coefficients, it is necessary to assign a value to the core size. However, the Enskog theory is not exact, since it is based on the molecular chaos approximation. A sphere is considered as always colliding with other spheres approaching from random directions with random velocities from a Maxwell–Boltzmann distribution for the appropriate temperature. However, molecular dynamics calculations published by Alder & Wainwright (1963, 1967) have shown that there are correlated molecular motions in hard–sphere systems. At high densities, the principal correlation effect is back–scattering, whereby a sphere closely surrounded by a shell of spheres is most likely to have its velocity reversed on collision with its neighbors, which leads, in the case of diffusion, to a decreased diffusion coefficient. At intermediate densities, there is a different correlation effect associated with an unexpected persistence of velocities which leads to an enhanced diffusion coefficient. Alder *et al.* (1970) calculated the corrections to the Enskog theory for diffusion, viscosity and thermal conductivity for finite systems of 108 and 500 particles and showed that they were a function only of the ratio (V/V_0) where V_0, given by $N_A\sigma^3/2^{1/2}$, is the volume of close–packing of spheres. Erpenbeck & Wood (1991) confirmed these results for diffusion, extrapolated to an infinite–sized system. However, in the case of the viscosity and the thermal conductivity, the equivalent correction factors show larger uncertainties associated with the necessarily larger computing time. The realization that there were uncertainties in determining the core size (or the volume of close–packing of spheres) led to the proposition of a scheme by Dymond (1973) for comparing calculated and experimental transport coefficients without a prior estimation of the core size. According to this scheme, reduced quantities D^*, η^* and λ^* were introduced for diffusion, viscosity and thermal conductivity, as

$$D^* = \frac{nD}{[nD]^{(0)}} \left(\frac{V}{V_0}\right)^{2/3} \tag{10.8}$$

$$\eta^* = \frac{\eta}{\eta^{(0)}} \left(\frac{V}{V_0}\right)^{2/3} \tag{10.9}$$

$$\lambda^* = \frac{\lambda}{\lambda^{(0)}} \left(\frac{V}{V_0}\right)^{2/3} \tag{10.10}$$

These are all independent of the molecular diameter. From equations (10.1), (10.4) and (10.5) and the qualitative results of the molecular dynamics simulations discussed in the previous section, it is apparent that the above reduced coefficients are functions only of the ratio (V/V_0) – for example, equation (10.9) above can also be rewritten as $\eta^* = (\eta/\eta_E)_{MD}(\eta_E/\eta^{(0)})(V/V_0)^{2/3}$, where $(\eta/\eta_E)_{MD}$ is the computed correction from molecular dynamics simulations and $(\eta_E/\eta^{(0)})$ is given by equation (10.4).

These reduced coefficients can also be calculated from experimental data on the assumption that the real fluid is behaving like an assembly of hard spheres, since on substituting for the hard–sphere expressions (equations (10.3), (10.6) and (10.7)), the above expressions become

$$D^* = 5.030 \cdot 10^8 \left(\frac{M}{RT}\right)^{1/2} DV^{-1/3} = F_D \left(\frac{V}{V_0}\right) \tag{10.11}$$

$$\eta^* = 6.035 \cdot 10^8 \left(\frac{1}{MRT}\right)^{1/2} \eta V^{2/3} = F_\eta \left(\frac{V}{V_0}\right) \tag{10.12}$$

$$\lambda^* = 1.936 \cdot 10^7 \left(\frac{M}{RT}\right)^{1/2} \lambda V^{2/3} = F_\lambda \left(\frac{V}{V_0}\right) \tag{10.13}$$

where M is the molar mass, T the absolute temperature, R the gas constant, V the molar volume and all quantities are expressed in SI units. Although in the particular case of the diffusion coefficient the functional F_D can be calculated from the Enskog theory and the molecular dynamics simulation corrections (Alder *et al.* 1970; Erpenbeck & Wood 1991), in view of the uncertainties in the molecular dynamics results it was decided to use accurate experimental measurements of the three properties also in the determination of the three functionals. The determination of V_0 for a particular transport property, X, for a given compound with molecules behaving as smooth–hard spheres, is accomplished by curve–fitting. In principle, plots of $logX^*$ versus $logV$ from experiment at different temperatures produce parallel curves. A horizontal shift of each curve to produce coincidence with a reference curve will produce a ratio of the values of V_0 for the two particular temperatures for that compound. Absolute V_0 are obtained by superimposing all the curves on a given reference isotherm in this way and then adjusting this curve to give closest agreement with the computed hard–sphere curve of $logX^*$ versus $log(V/V_0)$ for the appropriate property. It should be noted that since V_0 is characteristic of the fluid, this procedure should be accomplished simultaneously for the three properties in order to produce the best values for V_0 and the universal curves.

10.2.1 Monatomic fluids

To apply the above scheme, accurate experimental measurements for the transport properties of the monatomic fluids were collected. In Table 10.1 the experimental measurements of diffusion, viscosity and thermal conductivity used for the correlation scheme are shown. This table also includes a note of the experimental method used, the quoted accuracy, the temperature range, the maximum pressure and the number of data sets. The data cover the range of compressed/gas and the liquid range but not the critical region, where there is an enhancement (Chapter 6) which cannot be accounted for in terms of this simple molecular model.

Table 10.1. *Comparison of calculated transport properties values with experimental values for the monatomic fluids.*

	Authors' references		Acc. /%	Temp. range /K	Max. press. /MPa	No. of data	Data < 10 /%	> 5 > 10 /%
							< 10	> 10
							/%	/%
	Self–Diffusion							
Ar	Cini–Castagnoli & Ricci (1960)	A	2.0	85	0.1	1	-	1
Kr	Carelli *et al.* (1973)	A	5.0	180-220	10	16	3	2
Xe	Peereboom *et al.* (1989)	B	1.0	240-340	140	26	1	-
						Total	43	4	3
	Viscosity							
Ne	Trappeniers *et al.* (1964)	C	2.5	290-350	180	21	-	-
	Vermesse & Vidal (1975)	A	0.5	298	610	16	-	-
Ar	Mostert *et al.* (1989)	C	2.5	174	470	25	-	-
	Trappeniers *et al.* (1980)	C	2.5	220-320	900	44	3	-
	Haynes (1973)	D	1.0	80-220	40	69	3	-
	Michels *et al.* (1954)	C	2.5	270-350	160	49	-	-
	Vermesse & Vidal (1973)	A	0.5	308	600	21	-	-
Kr	Trappeniers *et al.* (1965a)	C	1.0	290-400	210	42	1	-
Xe	Trappeniers *et al.* (1965b)	C	2.5	290-350	120	9	-	-
						Total	296	7	0
	Thermal Conductivity							
Ne	Tufeu (1992)	E	1.0	298	900	9	5	-
Ar	Calado *et al.* (1987)	F	0.5	100-130	10	65	-	-
	Roder *et al.* (1987)	F	1.0	110-140	70	214	17	-
	Michels *et al.* (1963)	G	2.0	270-350	170	16	5	-
	Tufeu (1992)	E	1.0	298	900	9	-	-
	Bailey & Kellner (1968)	E	2.0	90-300	50	119	15	-
Kr	Tufeu (1992)	E	1.0	298	700	6	-	-
Xe	Tufeu (1992)	E	1.0	298	400	2	-	-
						Total	440	42	0

A : Capillary;   B : NMR;   C : Vibrating Wire;
D : Torsional Crystal; E : Concentric Cylinders; F : Transient Hot-Wire;
G : Parallel Plate.

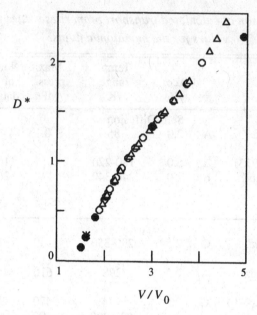

Fig. 10.2. Reduced self-diffusion, D^*, for monatomic fluids as a function of (V/V_0): ＊ Ar; △ Kr; ○ Xe; ● computed values for hard spheres.

Values for V_0 were derived for the rare gases by curve–fitting, as described above. The reduced transport coefficients can then be represented as a function of reduced volume, V/V_0, to give the universal curve characteristic for each property. Figures 10.2, 10.3 and 10.4 show the experimental results for diffusion, viscosity and thermal conductivity, together with the computed values for hard spheres. The density dependence for diffusion and viscosity for the rare gases mimics very closely the behavior found for a smooth hard–sphere system. Only for diffusion at $V/V_0 > 4$ does the difference become significant, but this is expected as a result of effects of the attraction forces in a real fluid. In the case of thermal conductivity, there is good agreement for $2 < V/V_0 < 4$, but experimental results at the highest densities show a less steep density dependence than that predicted by a finite–sized hard–sphere system (Dymond 1987). In view of the fact that data points for different rare gases fall on a single curve for each property, these were taken as the functionals $F_D(V/V_0)$, $F_\eta(V/V_0)$ and $F_\lambda(V/V_0)$ for $V/V_0 \to 5$.

For convenience, the universal curves are expressed as

$$\log D^* = \sum_{i=0}^{5} a_{Di} \left(\frac{V}{V_0}\right)^{-i} \tag{10.14}$$

$$\log \eta^* = \sum_{i=0}^{7} a_{\eta i} \left(\frac{V}{V_0}\right)^{-i} \tag{10.15}$$

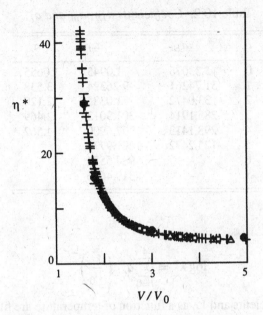

Fig. 10.3. Reduced viscosity, η^*, for monatomic fluids as a function of (V/V_0): □ Ne; + Ar; △ Kr; O Xe; ● computed values for hard spheres.

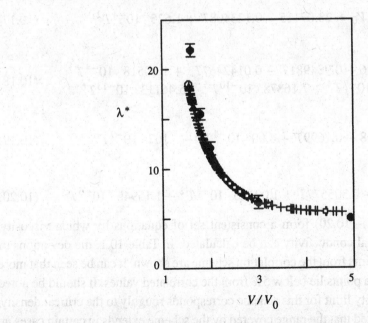

Fig. 10.4. Reduced thermal conductivity, λ^*, for monatomic fluids as a function of (V/V_0): □ Ne; + Ar; △ Kr; O Xe; ● computed values for hard spheres.

Table 10.2. Coefficients a_{Di}, $a_{\eta i}$ and $a_{\lambda i}$.

i	a_{Di}	$a_{\eta i}$	$a_{\lambda i}$
0	3.33076	1.0945	1.0655
1	-31.74261	-9.26324	-3.538
2	133.0472	71.0385	12.120
3	-285.1914	-301.9012	-12.469
4	298.1413	797.6900	4.562
5	-125.2472	-1221.9770	-
6	-	987.5574	-
7	-	-319.4636	-

and

$$\log \lambda^* = \sum_{i=0}^{4} a_{\lambda i} \left(\frac{V}{V_0} \right)^{-i} \tag{10.16}$$

Values for the coefficients and V_0 as a function of temperature are fitted by a regression analysis. The coefficients a_{Di}, $a_{\eta i}$ and $a_{\lambda i}$ are given in Table 10.2, and $V_0(/10^{-6}\text{m}^3 \text{ mol}^{-1})$ is given by

Ne:

$$V_0 = 28.07115 - 0.123408T + 1.625 \cdot 10^{-4}T^2 \tag{10.17}$$

Ar:

$$V_0 = -8.25916 + 0.984981T - 0.0142467T^2 + 1.00518 \cdot 10^{-4}T^3$$
$$-3.76457 \cdot 10^{-7}T^4 + 7.16878 \cdot 10^{-10}T^5 - 5.46113 \cdot 10^{-13}T^6 \tag{10.18}$$

Kr:

$$V_0 = 41.18 - 0.1699T + 4.00 \cdot 10^{-4}T^2 - 3.162 \cdot 10^{-7}T^3 \tag{10.19}$$

Xe:

$$V_0 = 57.6744 - 0.305577T + 1.00116 \cdot 10^{-3}T^2 - 1.15946 \cdot 10^{-6}T^3 \tag{10.20}$$

Equations (10.11)–(10.20) form a consistent set of equations by which viscosity, diffusion and thermal conductivity can be calculated. In Table 10.1, the deviations of the experimental points from the correlation scheme are shown. It can be seen that more than 93% of the data points lie below 5% from the correlated values. It should be noted that the lowest density limit for this scheme corresponds roughly to the critical density. It should also be noted that the range covered by the scheme extends in certain cases up to 9000 atmospheres pressure. The maximum pressure effects are shown for argon in the case of diffusion at 298 K, which decreases by a factor of about 5 over the pressure range considered. The viscosity at 174 K increases by a factor of 11, and the thermal conductivity at 298 K increases by a factor of 7. The accuracy of this scheme, which is estimated to be 5%, is thus most satisfactory.

It is of interest to note that the data fit using equations (10.14)–(10.16) covers the reduced volume (V/V_0) range down to 1.6 for diffusion, to 1.5 for viscosity and thermal conductivity – that is, for densities up to about 3 times the critical density. Although initial expectation was that the temperature should be greater than the critical temperature for the model to be valid, it is found that the data can be fitted at reduced temperatures T/T_c down to 0.6. It must be emphasized that this model can in no way account for the critical enhancement; experimental data showing this effect were therefore not considered.

10.3 Rough hard–sphere fluid

In the case of spherical molecules where transfer of rotational momentum can occur, Chandler (1975) showed that, to a good approximation, self–diffusion and viscosity for real fluids could be taken as equal to the rough hard–sphere values for densities greater than about twice the critical density. Furthermore, he showed that the rough hard–sphere coefficients were proportional to the smooth hard–sphere values. Thus

$$D \approx D_{RHS} \approx A \cdot D_{SHS} \qquad (10.21)$$

$$\eta \approx \eta_{RHS} \approx C \cdot \eta_{SHS} \qquad (10.22)$$

where A, C are the translational–rotational coupling factors. These are assumed to be constant for any given fluid, and independent of density and temperature. Chandler (1975) showed that coupling between translational and rotational motions is expected to lead to a decrease in diffusion, $0 \le A \le 1$, but to an increased viscosity, $C \ge 1$.

In the application of this rough hard–sphere theory for the interpretation of transport properties of dense pseudo–spherical molecules, it is assumed that equations (10.21) and (10.22) are exact. Reduced quantities for diffusion and viscosity, similar to those defined by equations (10.11) and (10.12), are given by

$$D^* = 5.030 \cdot 10^8 \left(\frac{M}{RT} \right)^{1/2} \frac{DV^{-1/3}}{A} \qquad (10.23)$$

$$\eta^* = 6.035 \cdot 10^8 \left(\frac{1}{MRT} \right)^{1/2} \frac{\eta V^{2/3}}{C} \qquad (10.24)$$

The simplest pseudo–spherical molecule is methane, for which extensive accurate transport property data are available. A summary of the experimental measurements is given in Table 10.3. Values for the coupling factor for the given property and for V_0 at any temperature were derived from fitting experimental isotherms to the universal curves described by equations (10.14) and (10.15). V_0 was determined from the horizontal shift, as for monatomic fluids, and the coupling factor was derived from the vertical displacement.

Table 10.3. *Comparison of calculated self–diffusion and viscosity values with experimental values for methane.*

Authors' References	Acc. /%	Temp. Range /K	Max. Press. /MPa	No. of Data	Data < 10 /%	> 5 > 10 /%
		Self–Diffusion				
Harris & Trappeniers (1980)	A	110-160	210	30	3	-
Harris (1978)	A	220-330	160	31	-	-
			Total	61	3	0
		Viscosity				
Diller (1980)	B	2.0	100-300	33	112	4	-
van der Gulik et al. (1988)	C	0.5	298	1000	37	-	-
van der Gulik et al. (1991)	C	0.5	273	1000	124	24	4
			Total	273	28	4

A: NMR;   B : Oscillating Quartz;   C : Vibrating Wire.

An excellent fit was obtained, with a coupling factor of 1.0 for viscosity and 0.905 for diffusion. Similar results are also obtained by Harris (1992). Values for V_0 are given in the following section, equation (10.28). It is important to note that this correlation procedure is applicable at densities down to twice the critical density for diffusion and down to 1.1 times the critical density for viscosity. The maximum pressure effects are shown at 140 K for diffusion, which decreases by a factor of 4 over the pressure range considered. The viscosity at 273 K increases by a factor of 20. Agreement between experiment and calculated values is generally well within 5%. In the few cases where there are larger differences, uncertainties in the density are a contributing factor.

10.4 Nonspherical fluids

In view of the successful application of the rough hard–sphere theory for the correlation of diffusion and viscosity for dense fluid methane, a similar form of data representation has been considered for dense nonspherical molecular fluids. Instead of the translational–rotational coupling factors, a roughness factor R_X ($X = D, \eta, \lambda$) has been introduced for each property to account for effects of nonspherical shape. It is assumed that these are independent of temperature and density. Thus, reduced quantities D^* and η^* similar to those defined by equations (10.11) and (10.12) can be redefined as

$$D^* = 5.030 \cdot 10^8 \left(\frac{M}{RT}\right)^{1/2} \frac{DV^{-1/3}}{R_D} \qquad (10.25)$$

$$\eta^* = 6.035 \cdot 10^8 \left(\frac{1}{MRT}\right)^{1/2} \frac{\eta V^{2/3}}{R_\eta} \tag{10.26}$$

A similar relationship is assumed for the thermal conductivity

$$\lambda^* = 1.936 \cdot 10^7 \left(\frac{M}{RT}\right)^{1/2} \frac{\lambda V^{2/3}}{R_\lambda} \tag{10.27}$$

The determination of V_0 and the R_X factor at any temperature for the particular transport property for a given compound is accomplished by the curve–fitting procedure outlined in Section 10.3, where the R_X factor is obtained from the vertical displacement and V_0 from the horizontal shift. This procedure was found to work successfully in many different systems of fluids, such as alkanes, aromatic hydrocarbons, alkan-1-ols, refrigerants and others that will be described in the following sections.

10.4.1 Alkanes

In a recent paper Assael *et al.* (1992a) applied the aforementioned scheme to a large number of experimental measurements of self–diffusion, viscosity and thermal conductivity of unbranched alkanes from methane to hexadecane. The data cover the range of compressed gas and the liquid range but not the critical region. Typical temperature range for the higher alkanes was 270 to 370 K, and the pressure extended in some cases up to 500 MPa. Since then, new accurate experimental measurements were performed by Greiner–Schmid *et al.* (1991), Harris *et al.* (1993) and Oliveira (1991). The new measurements, since not included in the original derivation of the equations, serve as a test of the predictive power of the scheme.

According to the scheme discussed and the universal curves – equations (10.14)–(10.16) – the horizontal shifts of the isotherms for every fluid produced the values of V_0, while the vertical shifts produced the value of the R_X factor. In the particular case of the alkanes it was also possible to further correlate the V_0 values and the R_X factors as a function of the alkane carbon number, C_n. It should, however, be pointed out that for V_0, it was found that the first five members of the series behaved slightly differently from the rest. Thus a separate fit of V_0 for the first four alkanes and one for pentane were necessary to maintain the accuracy.

The resulting equations for $V_0(/10^{-6}\text{m}^3 \text{ mol}^{-1})$ as a function of the absolute temperature, T, and the alkane carbon number, C_n, also shown in Figure 10.5, are:
CH_4-C_4H_{10} :

$$\begin{aligned}
V_0 &= 45.822 - 6.1867T^{0.5} + 0.36879T - 0.007273T^{1.5} \\
&+ C_n(2.17871T^{0.5} - 0.185198T + 0.00400369T^{1.5}) \\
&+ C_n^2(6.95148 - 52.6436T^{-0.5}) + C_n^3(-7.801897 \\
&+ 42.24493T^{-0.5} + 0.4476523T^{0.5} - 0.009573512T)
\end{aligned} \tag{10.28}$$

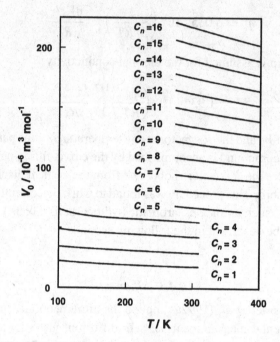

Fig. 10.5. V_0 as a function of temperature for alkanes.

C_5H_{12} :
$$V_0 = 81.1713 - 0.046169T \tag{10.29}$$

C_6H_{14}-$C_{16}H_{34}$:

$$
\begin{aligned}
V_0 = {} & 117.874 + 0.15(-1)^{C_n} - 0.25275T \\
& + 0.000548T^2 - 4.2464 \cdot 10^{-7}T^3 \\
& + (C_n - 6)(1.27 - 0.0009T)(13.27 + 0.025C_n)
\end{aligned} \tag{10.30}
$$

Similarly, the equations for the R_X factors as a function of the alkane carbon number C_n, are

$$R_D = 0.7445 + 0.2357C_n^{0.5} - 0.075C_n \tag{10.31}$$

$$R_\eta = 0.995 - 0.0008944C_n + 0.005427C_n^2 \tag{10.32}$$

$$
\begin{aligned}
R_\lambda = {} & -18.8416C_n^{-1.5} + 41.461C_n^{-1} - 30.15C_n^{-0.5} \\
& + 8.6907 + 1.3371 \cdot 10^{-3}C_n^{2.5}
\end{aligned} \tag{10.33}
$$

The dependence of R_D, R_η and R_λ on the length of the alkane chain is shown in Figure 10.6. The variation is smooth, except for ethane thermal conductivity, where the optimum value for R_λ is somewhat higher than expected. The universal equations (10.14)–(10.16), the reduced coefficients equations (10.25)–(10.27) and the aforementioned equations for V_0 and the R_X factors (10.28)–(10.33), form a consistent set by

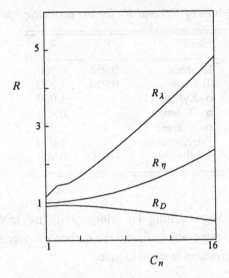

Fig. 10.6. Dependence of the R_X factors on alkane carbon number C_n.

which the transport properties of alkanes can be correlated and predicted in the ranges where no experimental measurements exist. As described in Assael et al. (1992a) and including the previously mentioned new data, in the case of self–diffusion about 10% of the points deviate between 5 and 10%, but this in many cases is attributed to inaccurate data. In the case of viscosity, where 1204 experimental points were used, only 4% of the points lie between 5 and 10%, while less than 1% of the experimental measurements lie above 10%. Considering that in some extreme cases the measurements extend to pressures up to 10,000 atmospheres, and that uncertainties in the density affect the calculated viscosity, the agreement is considered very satisfactory. Finally, in the case of the thermal conductivity measurements, from the 960 data examined only 1.5% of them deviate more than 5%.

10.4.2 *Aromatic hydrocarbons*

The correlation scheme has been applied to seven simple aromatic hydrocarbons: benzene, toluene, *o*–, *m*–, and *p*–xylene, mesitylene and ethylbenzene (Assael *et al.* 1992d). In addition to the large number of data used in the development of the correlation, new measurements can now be included (Kaiser *et al.* 1991; Krall *et al.* 1992; Harris *et al.* 1993; Yamada *et al.* 1993). The temperature and pressure ranges of the data used in this analysis are typically from 250 to 400 K and from 0.1 to 400 MPa. Transport coefficient measurements up to high pressures are available for most of these compounds. The roughness factors R_X were found to be temperature–independent, as in the case of alkanes. Slight adjustments were made for each compound, so that the R_X factors have a constant value and V_0 is property–independent and decreases smoothly with

Table 10.4. *Values of R_D and R_η for the aromatic hydrocarbons.*

Fluid	R_D	R_η
Benzene	0.958	0.960
Toluene	0.920	1.030
o–Xylene	-	1.050
m–Xylene	-	0.950
p–Xylene	-	1.033
Ethylbenzene	-	1.010
Mesitylene	0.910	0.900

increasing temperature. The resulting V_0 values ($/10^{-6} m^3$ mol^{-1}) were fitted as a function of the absolute temperature, T, and the carbon number, C_n, of the compound using a least–squares regression method to give

$$V_0 = -3324.7C_n^{-2} + 529.47C_n^{-1}$$
$$+T(9.48786C_n^{-2} - 8.55176 \cdot 10^{-2}C_n + 6.03463 \cdot 10^{-3}C_n^2) \qquad (10.34)$$
$$+T^2(-1.5797 \cdot 10^{-3} + 3.9901 \cdot 10^{-4}C_n - 2.2309 \cdot 10^{-5}C_n^2)$$

In Table 10.4 the values for R_D and R_η obtained are shown. No definite trend was observed in the R_η values in going from benzene to mesitylene. There were also insufficient R_D data to predict values for any other compound.

However, the R_λ values were found to vary linearly with carbon number as

$$R_\lambda = 1.528 + 0.212(C_n - 6) \qquad (10.35)$$

For self–diffusion, 13% of the 150 experimental measurements deviate by more than 5% from the correlation, but much of this is attributable to experimental uncertainty. However, in the case of viscosity (845 experimental measurements) and thermal conductivity measurements (477 experimental measurements), the situation is excellent, with only 3% of viscosity measurements deviating more than 5% and no deviations above 5% for the thermal conductivity.

10.4.3 Alkan–1–ols

In the case of the alkan–1–ols, experimental measurements of self–diffusion, viscosity and thermal conductivity for methanol to decan–1–ol (Assael *et al.* 1994b) were used to calculate the values for V_0 and the R_X factors. However, in this case, parameters R_η and R_D were found to be weak functions of temperature, especially for the first few alkanols of the series, while only parameter R_λ was found to be still a constant for a liquid. Thus, these parameters had to be optimized together with the characteristic molar volumes.

Table 10.5. *Coefficients of equation (10.36).*

i	b_i	c_i	d_i
0	-28.842	486.505	34.6986
1	-2759.29	-478.064	-3.944
2	–	217.9562	0.169911
3	–	-47.901	$-3.681 \cdot 10^{-3}$
4	–	4.139	$3.94017 \cdot 10^{-5}$
5	–	–	$-1.6598 \cdot 10^{-7}$

The equation obtained for the characteristic molar volume, V_0, as a function of the absolute temperature, T, and the number of carbon atoms, C_n, in the alkanol molecule is

$$V_0 = \sum_{i=0}^{1} b_i T^{-i/2} + \sum_{i=0}^{4} c_i C_n^{(i+2)/2} + \sum_{i=0}^{5} d_i (C_n T)^{(i+1)/2} \qquad (10.36)$$

The coefficients b_i, c_i and d_i are shown in Table 10.5. For the higher alkanols, V_0 values decrease steadily as the temperature is increased, as expected when a hard–sphere model is applied to real molecules for which the repulsive part of the interaction energy curve is soft, and not infinitely steep. For the first few members of the series, V_0 shows remarkably little temperature dependence. The optimized equations obtained for the R_X factors, in the temperature range where alkan–1–ol measurements exist, are

a) Viscosity for $CH_3OH - C_5H_{11}OH$:

$$R_\eta = \sum_{i=0}^{2} e_i T^i \qquad (10.37)$$

for $C_6H_{13}OH - C_{10}H_{21}OH$:

$$R_\eta = 38.22 - 16.071 C_n + 2.353 C_n^2 - 0.1088 C_n^3 \qquad (10.38)$$

b) Self–diffusion for $CH_3OH - C_3H_7OH$:

$$R_D = \sum_{i=0}^{2} f_i T^i \qquad (10.39)$$

c) Thermal conductivity for $CH_3OH - C_6H_{13}OH$:

$$R_\lambda = 1.493 - 0.09139 C_n + 0.02804 C_n^2 \qquad (10.40)$$

The coefficients e_i and f_i are shown in Table 10.6. From the above equations it can be seen that R_η and R_D for the lower alkanols of the series are a function of temperature. R_η decreases with a rise in temperature and R_D increases as the temperature is increased. This is in keeping with the fact that these lower alkanols are associated liquids due to

Table 10.6. *Coefficients of equations (10.37)–(10.39).*

Alkanol	R_η			R_D		
	e_0	$\dfrac{e_1}{K^{-1}}$	$\dfrac{10^4 e_2}{K^{-2}}$	f_0	$\dfrac{10^4 f_1}{K^{-1}}$	$\dfrac{10^6 f_2}{K^{-2}}$
CH_3OH	41.15	-0.2175	3.057	-2.747	177.1	-25.98
C_2H_5OH	49.74	-0.2505	3.325	-0.255	7.150	2.368
C_3H_7OH	119.69	-0.6770	9.817	-0.360	-0.088	5.965
C_4H_9OH	55.07	-0.2950	4.110			
$C_5H_{11}OH$	46.11	-0.2580	3.802			

hydrogen bonding, where the extent of hydrogen bonding falls off as the temperature is raised. R_η and R_D should therefore exhibit a temperature dependence. It was found previously for the n–alkane series that R_η increased with increase in carbon number, a trend that was to be expected. Here a similar pattern is found with the highest–temperature R_η values for butanol and pentanol and the temperature–independent values for hexan–1–ol to nonan–1–ol. The value for decan–1–ol is low, but where measurements on only one property are available there is the possibility of different pairs of values for R_η and V_0 fitting the data almost equally well. There is a need for more, accurate data on the transport properties for higher members of this series.

A direct comparison of the available experimental data for the viscosity, self–diffusion and thermal conductivity with the calculated values, with V_0 and R_X described by equations (10.36)–(10.40), is presented by Assael *et al.* (1994b). In the case of viscosity only 25 data points (4.7%) out of 533 were found to show deviations larger than 5%. From the 127 experimental measurements of self–diffusion employed, 11 of them (8.7%) show deviations larger than 5%, while from the 107 thermal conductivity measurements, only 5 (4.7%) show deviations of more than 5%. The overall average absolute deviations of the experimental viscosity, self–diffusion and thermal conductivity measurements from those calculated by the correlation is 2.4, 2.6 and 2.0% respectively. Considering the fact that in some cases experimental measurements were at pressures up to 400 MPa, and that part of the difference arises from uncertainties in the density, this agreement is considered very satisfactory.

10.4.4 Refrigerants

As a further test of the applicability of the scheme, it was applied to refrigerants. Unfortunately, the accuracy of the experimental viscosity and thermal conductivity measurements is much lower for these compounds than for those considered earlier. Therefore, as an example, a selected number of experimental measurements of viscosity and thermal conductivity for R22, R123, R134a and R152a refrigerants were considered

Table 10.7. *Values of R_η, R_λ and coefficients of equation (10.41).*

Compound	R_η	R_λ	g_0	$\frac{g_1}{K^{-1}}$	$\frac{10^4 g_2}{K^{-2}}$	$\frac{10^6 g_3}{K^{-3}}$
R22	1.110	1.570	76.0767	-0.387740	12.7082	-1.4708
R123	1.500	1.795	54.4228	0.083439	-2.04092	-0.048762
R134a	1.020	1.550	50.9210	-0.025435	.142126	0
R152a	0.700	1.230	-18.3310	0.777820	-32.427	4.3860

(Assael *et al.* 1993), covering a temperature range of 250 to 360 K and pressures from saturation pressure up to 30 MPa.

The measurements were used to calculate the temperature dependence of the characteristic molar volume, $V_0(/10^{-6} m^3 mol^{-1})$ as

$$V_0 = \sum_{i=0}^{3} g_i T^i \tag{10.41}$$

and the values of the constants R_η and R_λ for each liquid. The results are given in Table 10.7. In the case of the viscosity, from a total of 463 experimental points considered, 20 (4.3%) deviated more than 5%, while only 3 (0.65%) deviated more than 10%. In the case of the thermal conductivity from a total of 559 experimental points considered, 55 points (9.8%) deviated more than 5%, while only 10 points (1.8%) deviated more than 10%. Considering the large problems encountered in the measurement of the viscosity and the thermal conductivity of these refrigerants whereby differences as high as 5% between results reported by reputable laboratories are not uncommon, as well as the fact that the scheme is very sensitive to the values of the density of the liquid, which is itself not accurately known in some cases, we consider the scheme to be potentially very useful here also. There is an obvious need for more accurate data for these, and other, refrigerants.

10.4.5 Other compounds

To conclude the application of the scheme to single liquids, and in addition to the groups of liquids considered above, some simple molecular fluids were examined by Assael *et al.* (1992b). These are CS_2, C_6H_{12}, CCl_4, CH_3CN and CH_3Cl, for which self–diffusion, viscosity and thermal conductivity measurements were all available to provide a more critical test of the correlation. As before, the data are used to calculate the temperature dependence of the characteristic molar volumes, $V_0(/10^{-6} m^3 mol^{-1})$, as

$$V_0 = \sum_{i=0}^{2} h_i T^i \tag{10.42}$$

and the values of the constants R_D, R_η and R_λ for each liquid. Typical temperature and

Table 10.8. *Values of* R_D, R_η, R_λ *and coefficients of equation (10.42).*

Compound	R_D	R_η	R_λ	h_0	$\dfrac{h_1}{K^{-1}}$	$\dfrac{10^5 h_2}{K^{-2}}$
CS_2	0.85	1.27	1.67	40.366	-0.020017	0
C_6H_{12}	1.15	0.93	1.35	83.489	-0.010676	-3.2796
CCl_4	0.84	1.07	1.57	89.860	-0.125050	14.4440
CH_3CN	0.68	1.51	–	38.208	-0.029402	0.7903
$CHCl_3$	0.79	1.30	–	65.444	-0.080932	905.02

pressure ranges examined were from 260 to 360 K and from 0.1 up to 400 MPa. The values of the R_X constants and the coefficients h_i are shown in Table 10.8. In the case of self–diffusion only 6 points (3.2%) from the 149 considered were found to deviate more than 5%. In the case of viscosity, from the 207 points considered 14 points (7%) deviated more than 5%, while from the 156 thermal conductivity points, none deviated more than 5%.

10.5 The density dependence

In the above sections, it has been shown that transport properties for dense fluids can be successfully correlated on the basis of universal curves in terms of the variables temperature and density. However, transport properties are required for given conditions of temperature and pressure, and hence it is essential to have an accurate representation for density. As discussed in Section 8.2.4, the modified Tait equation provides the simplest, yet most accurate, representation of dense fluid density data. It is usually expressed in the form

$$\frac{\rho - \rho_0}{\rho} = C \log\left(\frac{B + P}{B + P_0}\right) \tag{10.43}$$

where ρ and ρ_0 are the liquid densities at the corresponding pressures, P and P_0, and B and C are parameters. In most cases P_0 is taken to be one atmosphere pressure (0.101 MPa) and ρ_0 the corresponding atmospheric–pressure density. Many investigators have examined the dependence of the C parameter on temperature. It has generally been found that this parameter is either a constant independent of the temperature or is a weak function of the temperature. In their work, Dymond & Malhotra (1987) showed that for the unbranched alkanes examined, C could be taken as a constant, equal to 0.2000. Parameter B definitely varies with temperature, and this has been represented by many investigators by a variety of equations, usually in terms of the critical temperature, T_c, and the critical pressure, P_c. Dymond & Malhotra (1987) showed that, for the temperature range up to 0.66 times the critical temperature, a simple quadratic equation for B as a function of the reduced temperature gave very good results. Since the work of Dymond & Malhotra (1987), new experimental data for alkane densities have become available. In particular, new measurements of the density of pentane and low–temperature measurements of heptane and octane have enabled a reexamination of the equation for the parameter B (Assael *et al.* 1994a). Their revised equation for B for the alkanes, as a function of the reduced temperature, $T_R = T/T_c$, is: for C_2H_6 to $C_{16}H_{34}$:

$$B = 331.2083 - 713.86T_R + 401.61T_R^2 - D \qquad (10.44)$$

where for C_2H_6 to C_7H_{16}: $\qquad\qquad D = 0 \qquad\qquad\qquad (10.45)$

$\quad$ for C_8H_{18} to $C_{16}H_{34}$: $\quad D = 0.8(C_n - 7) \qquad\qquad (10.46)$

and where for CH_4: $\quad B = 175.8 - 314.57T_R + 134.3T_R^2 \qquad (10.47)$

The following points ought to be mentioned in relation to the above two equations. From the equation for methane it is apparent that parameter B for methane follows a dependence in temperature different from the rest of the alkanes. This is not unusual in the case of the first fluid of a homologous series. In relation to the first equation, it is interesting that from ethane to heptane the parameter D is zero. The temperature and pressure ranges covered by the above equation are from approximately 250 to 400 K and up to 400 MPa (for the higher alkanes). The average deviation of the experimental densities from the calculated values is 0.10%. Densities at atmospheric pressure as a function of temperature were correlated by Cibulka (1993) and are given by Assael *et al.* (1994a).

In the case of aromatic hydrocarbons, benzene, toluene, xylenes and ethyl benzene, Dymond & Malhotra (1987) showed that B can be represented as

$$B = 494 - 1110T_R + 672T_R^2 - (C_n - 6) \qquad (10.48)$$

In this case, however, the optimized value for the constant C in equation (10.43) was 0.216. The temperature and pressure ranges to which these equations apply are similar to those for the alkanes.

In the case of the alkan–1–ols, Assael *et al.* (1994b) represented the temperature dependence of B for methanol to decan–1–ol by the expression

$$B = 520.23 - 1240T_R + 827T_R^2 - F \qquad (10.49)$$

where for CH_3OH: $\qquad\qquad F = 11.8 \qquad\qquad\qquad (10.50)$

and for C_2H_5OH to $C_{10}H_{21}OH$: $\quad F = 0.015C_n(1 + 11.5C_n) \qquad (10.51)$

The optimum value of C in equation (10.43) for the alkanols was found equal to 0.2000, and atmospheric pressure densities were also given (Assael *et al.* 1994a). The range of temperatures and pressures considered was 270 to 370 K and up to 100 MPa, while the average deviation of the fit was 0.05%. A further advantage of representing fluid densities by the modified Tait equation is that, by smoothing the experimental density data, an improved fit is obtained between the experimental transport properties and values calculated by the correlation scheme described above. This is especially noticeable at high densities, where small differences in density have a significant effect on the calculated transport properties (Dymond 1987).

10.6 Mixtures

The predictive power of the scheme described earlier for transport–property calculation has been clearly demonstrated in the case of mixtures of alkanes (Assael *et al.* 1992c). To apply the scheme, the mixture density was calculated from the densities of the pure components, assuming there was no change of volume during mixing. The V_0 values and the R_X factors were calculated as a massic fraction average of the pure component values. Hence the mixture viscosity and thermal conductivity were calculated without any mixture measurements and with no additional parameters. In the case of viscosity, 67 points (5.2%) out of 1234 binary alkane and 50 ternary and quaternary alkane mixtures were found to deviate more than 5%, and 24 points (1.8%) more than 10% (the latter occurring in mixtures with components having widely different carbon numbers). In the case of the thermal conductivity, for which 76 experimental data were available, no points were found to deviate more than 5%. The temperature and pressure ranges covered were similar to those for the pure alkanes. Almost all combinations of pure components were covered. Considering that no mixture data were used for this prediction, the predictive power of the scheme is considered very satisfactory.

10.7 Conclusions

Dense fluid transport property data are successfully correlated by a scheme which is based on a consideration of smooth hard–sphere transport theory. For monatomic fluids, only one adjustable parameter, the close–packed volume, is required for a simultaneous fit of isothermal self–diffusion, viscosity and thermal conductivity data. This parameter decreases in value smoothly as the temperature is raised, as expected for real fluids. Diffusion and viscosity data for methane, a typical pseudo–spherical molecular fluid, are satisfactorily reproduced with one additional temperature–independent parameter, the translational–rotational coupling factor, for each property. On the assumption that transport properties for dense nonspherical molecular fluids are also directly proportional to smooth hard–sphere values, self–diffusion, viscosity and thermal conductivity data for unbranched alkanes, aromatic hydrocarbons, alkan-1-ols, certain refrigerants and other simple fluids are very satisfactorily fitted. From the temperature and carbon number dependency of the characteristic volume and the carbon number dependency of the proportionality (roughness) factors, transport properties can be accurately predicted for other members of these homologous series, and for other conditions of temperature and density. Furthermore, by incorporating the modified Tait equation for density into the scheme it is now possible to calculate transport properties for dense fluids at any given temperature and pressure with an accuracy estimated to be better than ±5%. This has been realized in a computer package TransP, which covers the pure fluids and mixtures discussed in this chapter. Extension of the correlation scheme to mixtures with simple combining rules for the characteristic volume and roughness parameters results in a wholly predictive method, since there are now no adjustable parameters. Viscosities and thermal conductivities calculated for multicomponent liquid *n*–alkane mixtures at given temperatures and pressures within a wide range are in very good agreement with experiment.

Acknowledgments

The authors particularly wish to thank Dr. P. M. Patterson, who significantly contributed to this work, and Ms. S. K. Polimatidou for her contribution in the sections on alkan–1–ols and refrigerants. It is a pleasure to acknowledge the continued support of NATO through their International Scientific Exchange Program.

References

Alder, B. J., Gass, D. M. & Wainwright, T. E. (1970). Studies in molecular dynamics VIII. The transport coefficients for a hard–sphere fluid. *J. Chem. Phys.*, **53**, 3813–3826.

Alder, B. J. & Wainwright, T. E. (1963). *The Many–Body Problem*, ed. J.K. Percus. New York: Interscience Publications, Inc.

Alder, B. J. & Wainwright, T. E. (1967). Velocity autocorrelations for hard spheres. *Phys. Rev. Lett.*, **18**, 988–990.

Assael, M. J., Dymond, J. H., Papadaki, M. & Patterson, P. M. (1992a). Correlation and prediction of dense fluid transport coefficients I. *n*–Alkanes. *Int. J. Thermophys.*, **13**, 269–281.

Assael, M. J., Dymond, J. H., Papadaki, M. & Patterson, P.M. (1992b). Correlation and prediction of dense fluid transport coefficients II. Simple molecular fluids. *Fluid Phase Equil.*, **75**, 245–255.

Assael, M. J., Dymond, J. H., Papadaki, M. & Patterson, P.M. (1992c). Correlation and prediction of dense fluid transport coefficients III. *n*–Alkane mixtures. *Int. J. Thermophys.*, **13**, 659–669.

Assael, M. J., Dymond, J. H., & Patterson P.M. (1992d). Correlation and prediction of dense fluid transport coefficients V. Aromatic hydrocarbons. *Int. J. Thermophys.*, **13**, 895–905.

Assael, M.J., Karagiannidis, E. & Polimatidou, S.K. (1993). Measurements of the thermal conductivity of R22, R123, R134a and R152a. *High Temp.–High Press.*, **25**, 259–267.

Assael, M. J., Dymond, J. H. & Exadaktilou, D. (1994a). An improved representation for *n*–alkane liquid densities. *Int. J. Thermophys.*, **15**, 155–164.

Assael, M. J., Dymond, J. H. & Polimatidou, S.K. (1994b). Correlation and prediction of dense fluid transport coefficients VI. *n*–Alcohols. *Int. J. Thermophys.*, **15**, 189–201.

Bailey, B. J. & Kellner, K. (1968). The thermal conductivity of liquid and gaseous argon. *Physica* **39**, 444–462.

Calado, J. C. G., Mardolcar, U. V., Nieto de Castro, C. A., Roder, H. M. & Wakeham, W. A. (1987). The thermal conductivity of liquid argon. *Physica*, **143A**, 314–325.

Carelli, P., Modena, I. & Ricci, F. P. (1973). Self–diffusion in krypton at intermediate density. *Phys. Rev.*, **A7**, 298–303.

Carnahan, N. F. & Starling, K. E. (1969). Equation of state for non–attractive rigid spheres. *J. Chem. Phys.*, **51**, 635–636.

Chandler, D. (1975). Rough hard sphere theory of the self–diffusion constant for molecular liquids. *J. Chem. Phys.*, **62**, 1358–1363.

Cibulka, I. (1993). Saturated liquid densities of 1–alkanols from C_1 to C_{10} and *n*–alkanes from C_5 to C_{16}: A critical evaluation of experimental data. *Fluid Phase Equil.*, **89**, 1–18.

Cini–Castagnoli, G. & Ricci, F. P. (1960). Self–diffusion in liquid argon. *J. Chem. Phys.*, **32**, 19–20.

Davis, H. T., Rice, S. A. & Sengers, J. V. (1961). On the kinetic theory of dense fluids IX. The fluid of rigid spheres with a square–well attraction. *J. Chem. Phys.*, **35**, 2210–2233.

Diller, D. E. (1980). Measurements of the viscosity of compressed gaseous and liquid methane. *Physica*, **104A**, 417–426.

Dymond, J. H. (1973). Transport properties in dense fluids. *Proc. 6th Symp. Thermophys. Prop.*, New York: ASME, pp. 143–157.

Dymond, J. H. (1987). Corrections to the Enskog theory for viscosity and thermal conductivity. *Physica*, **144B**, 267–276.

Dymond, J. H. & Malhotra, R. (1987). Densities of *n*–alkanes and their mixtures at elevated pressure. *Int. J. Thermophys.*, **8**, 541–555.

Einwohner, T. & Alder, B. J. (1968). Molecular dynamics VI. Free–path distributions and collision rates for hard–sphere and square–well molecules. *J. Chem. Phys.*, **49**, 1458–1473.

Enskog, D. (1922). Kinetische Theorie der Wärmeleitung, Reibung und Selbst–diffusion in gewissen verdichteten Gasen und Flüssigkeiten. *Kungl. Svenska. Vet.–Ak. Handl.*, **63**, No. 4.

Erpenbeck, J. J. & Wood, W. W. (1991). Self–diffusion coefficient for the hard sphere fluid. *Phys. Rev. A*, **43**, 4254–4261.

Glasstone, S., Laidler, K. J. & Eyring, H. (1941). *The Theory of Rate Processes*. New York: McGraw–Hill.

Greiner–Schmid, A., Wappman, S., Has, M. & Luedemann, H. D. (1991). Self–diffusion in the compressed fluid lower alkanes: Methane, ethane and propane. *J. Chem. Phys.*, **94**, 5643–5649.

Harris, K. R. (1978). The density dependence of the self–diffusion coefficient of methane at -50, 25 and 50°C. *Physica*, **94A**, 448–464.

Harris, K. R. (1992). The self–diffusion coefficient and viscosity of the hard–sphere fluid revised: A comparison with experimental data for xenon, methane, ethene and trichloromethane. *Mol. Phys.*, **77**, 1153–1167.

Harris, K. R., Alexander, J .J., Goscinska, T., Malhotra, R., Woolf, L. A. & Dymond, J. H. (1993). Temperature and density dependence of the self–diffusion coefficients of liquid *n*–octane and toluene. *Mol. Phys.*, **78**, 235–248.

Harris, K. R. & Trappeniers, N. J. (1980). The density dependence of the self–diffusion coefficient of liquid methane. *Physica*, **104A**, 262–280.

Haynes, W. M. (1973). Viscosity of gaseous and liquid argon. *Physica*, **67**, 440–470.

Kaiser, B., Laesecke, A. & Stelbrink, M. (1991). Measurements of the viscosity of liquid toluene in the temperature range 218–378 K. *Int. J. Thermophys.*, **12**, 289–306.

Krall, A. H., Sengers, J. V. & Kestin J. (1992), Viscosity of liquid toluene at temperatures from 25 to 150°C and at pressures up to 30 MPa. *J. Chem. Eng. Data*, **37**, 349–355.

Michels, A., Botzen, A. & Schuurman, W. (1954). Viscosity of argon at pressures up to 2000 atmospheres. *Physica*, **20**, 1141–1148.

Michels, A., Sengers, J. V. & van der Kleindert, L. J. M. (1963). Thermal conductivity of argon at elevated densities. *Physica*, **29**, 149–160.

Mostert, R., van der Gulik, P. S. & van den Berg, H. R. (1989). Comment on the experimental viscosity of argon at high densities. *Physica*, **156A**, 921–923.

Oliveira, C. M. B. P. (1991). *Viscosity of liquid hydrocarbons at high pressures*. Ph.D. Thesis, Imperial College, London.

Peereboom, P. W. E., Luigjes, H. & Prins, K. O. (1989). An NMR spin–echo study of self–diffusion in xenon. *Physica*, **156A**, 260–276.

Roder, H. M., Nieto de Castro, C. A. & Mardolcar, U. V. (1987). The thermal conductivity of liquid argon for temperatures between 110 and 140 K with pressures to 70 MPa. *Int. J. Thermophys.*, **8**, 521–540.

Trappeniers, N. J., Botzen, A., van den Berg, H. R. & van Oosten, J. (1964). The viscosity of neon between 25°C and 75°C at pressures up to 1800 atmospheres. Corresponding states for the viscosity of the noble gases up to high pressures. *Physica*, **30**, 985–996.

Trappeniers, N. J., Botzen, A., van Oosten, J. & van den Berg, H. R. (1965a). The viscosity of krypton between 25°C and 75°C and at pressures up to 2000 atm. *Physica*, **31**, 945–952.

Trappeniers, N. J., Botzen, A., ten Seldam, C. A., van den Berg, H. R. & van Oosten, J. (1965b). Corresponding states for the viscosity of noble gases up to high densities. *Physica*, **31**, 1681–1691.

Trappeniers, N. J., van der Gulik, P. S. & van den Hooff, H. (1980). The viscosity of argon at very high pressure, up to the melting line. *Chem. Phys. Lett.*, **70**, 438–443.

Tufeu, R. (1992). Private communication.

van der Gulik, P. S., Mostert, R. & van den Berg, H. R. (1988). The viscosity of methane at 25°C up to 10 kbar. *Physica*, **151A**, 153–166.

van der Gulik, P. S., Mostert, R. & van den Berg, H. R. (1991). The viscosity of methane at 273 K up to 1 GPa. Presented at 11th Symp. Thermophys. Prop., Boulder, USA.

van der Waals, J.D. (1873). *On the continuity of the gaseous and the liquid state*. Dissertation, Leiden.

Vermesse, J. & Vidal, D. (1973). Mesure du coefficient de viscosité de l' argon à haute pression. *C. R. Acad. Sc. Paris*, **277B**, 191–193.

Vermesse, J. & Vidal, D. (1975). Mesure du coefficient de viscosité du néon à haute pression. *C. R. Acad. Sc. Paris*, **280B**, 749–751.

Yamada, T., Yaguchi, T., Nagasakai, Y. & Nagashima, A. (1993). Thermal conductivity of toluene in the temperature range 193–453 K. *High Temp.–High Press.*, **25**, 513–518.

11

The Corresponding–States Principle: Dilute Gases

E. A. MASON

Brown University, Providence, R. I., USA

F. J. URIBE

Universidad Autónoma Metropolitana, Iztapalapa, México

11.1 The corresponding–states principle – historical background

The principle of corresponding states provides correlations and predictive power that can save untold hours of labor in the laboratory.

From the present point of view, the principle – with all its far–flung applications – seems little more than fairly straightforward dimensional analysis applied to some rather basic theory. Thus it is a bit surprising to realize that such a simple, marvelous principle has had a rather bumpy history going back over 100 years, and that only in about the last 20 years has it been taken up again as a subject of serious study by chemists and physicists, although it was used extensively by engineers for a much longer time. In particular, it has now come to have an important influence on the correlation and prediction of transport properties, in addition to the much older applications to equilibrium properties.

Why has it taken so long? It would be gratifying to believe that it is because modern researchers are so much smarter than those old nineteenth–century scientists, but a little historical perspective suggests that the real reason is that now much more is known than before. In particular, more is known about transport theory and, especially, much more is known about intermolecular forces. This knowledge has provided both the insight and the quantitative details that are the main basis for Chapters 11 and 12.

Lack of knowledge about intermolecular forces first led the principle of corresponding states into too close an association with approximate equations of state, especially the van der Waals equation. Much later, especially in connection with transport properties, such lack led to too close an association with approximate models of the intermolecular potential, especially the Lennard–Jones (12–6) potential model. In both cases a sort of guilt by association followed, which kept many people from realizing that the principle was both more general and capable of much greater accuracy than seemed to be implied by these associations.

A little historical background helps to put matters in perspective. The principle of corresponding states springs from the work of J. D. van der Waals on the equation of state, published in his thesis (1873). (A recently published English translation of this thesis includes an excellent historical review by J. S. Rowlinson in the form of an introductory essay.) The van der Waals equation of state now appears in undergraduate textbooks, and relates the pressure, temperature and molar volume of a gas to each other through a simple cubic equation. This equation contains two constants characteristic of the particular fluid, universally known as a and b. The constant a corrects for the attractive forces between the molecules, and the constant b corrects for the volume excluded by the size of the molecules themselves. This is quite a remarkable equation: not only does it reduce to the perfect–gas equation at low densities, but it also predicts condensation, and a critical point at a pressure of $P_c = a/27b^2$ and a temperature of $RT_c = 8a/27b$.

Unfortunately, the van der Waals equation of state is quantitatively inaccurate (see Chapter 8) and was essentially abandoned by physicists and chemists, who considered it only a clever collage of insightful, but semiempirical, approximations. Its status was considerably enhanced, at least among theoreticians, by the work of Kac *et al.* (1963), who showed in what sense it could be considered an exact theoretical result for a definite, although physically unrealistic, model of hard spheres with weak long–range attractive forces. But this theoretical insight did not improve its accuracy.

The most useful and accurate by–product of the van der Waals equation of state was the principle of corresponding states. It may now seem obvious that the foregoing procedure can be reversed in order to find the constants a and b in terms of critical constants, instead of *vice versa*, but van der Waals did not take this decisive step until 1880. After a little algebra, the equation of state could then be written in a universal dimensionless form

$$P_r = f(V_r, T_r) \tag{11.1}$$

where $P_r = P/P_c$, $V_r = V/V_c$ and $T_r = T/T_c$ are dimensionless pressure, volume and temperature respectively. This transformation can in fact be carried out for any two–constant equation of state. Without specification of the function $f(V_r, T_r)$, other than that it would be the same for all ordinary fluids, the result is very general and much more accurate than the van der Waals equation itself, which gives a specific (and inaccurate) expression for the function. The function $f(V_r, T_r)$ could be determined experimentally once and for all by a compilation of measurements on various convenient gases. Thus the $P - V - T$ properties of all normal fluids could be scaled onto a universal dimensionless $P_r - V_r - T_r$ surface with just the critical constants characteristic of each individual fluid. A dramatic early application of this result was to predict the unknown critical constants of a gas by scaling some measurements of its $P - V$ isotherms to $f(V_r, T_r)$. Such prediction of the critical constants of hydrogen and helium played a significant role in the efforts to liquify the last two of the so–called 'permanent gases,' finally achieved for hydrogen by Dewar in 1898 and for helium by Kamerlingh Onnes in 1908.

It was immediately realized by Kamerlingh Onnes that the principle of corresponding states must be a macroscopic reflection of an underlying molecular principle of similarity, but he could do nothing quantitative with this insight because statistical mechanics was still in a rudimentary state and knowledge of intermolecular forces was virtually nil. The precise connection between corresponding states and intermolecular forces was not made until 1938–39, independently by de Boer and Michels and by Pitzer, and later extended and elaborated by Guggenheim. The gist of their arguments, here somewhat extended, is as follows.

The potential energy of all the molecules is assumed to be the sum over all pairs of the potential $U(r, \Omega)$, where r is the separation and Ω denotes the relative orientation angles of two molecules. The pair potential can always be written in a dimensionless form as

$$U(r, \Omega) = \epsilon \, \Phi(r/\sigma, \ \Omega, \ \alpha_1, \ \alpha_2, \ \ldots) \tag{11.2}$$

where ϵ is an energy parameter (often chosen to represent the depth of the potential well), σ is a range or distance parameter (often chosen to be the value of r for which $U = 0$), and $\alpha_1, \ \alpha_2, \ \ldots$ are dimensionless parameters characterizing the shape (rather than the scale) of the potential. Examples of such parameters are $C_6/\epsilon\sigma^6$, where C_6 is the r^{-6} London dispersion coefficient, and $\mu_d^2/\epsilon\sigma^3$, where μ_d is the molecular dipole moment. After this expression is substituted into the expression for the statistical–mechanical partition function, only algebra is required to show that the equation of state can also be written in a dimensionless form

$$P^* = F(V^*, \ T^*, \ \alpha_1, \ \alpha_2, \ \ldots) \tag{11.3}$$

where $P^* = P\sigma^3/\epsilon$, $V^* = V/\sigma^3$ and $T^* = k_B T/\epsilon$ are made dimensionless with molecular constants rather than critical constants. The function $F(\ldots)$ of course is not specified unless the functional form of the potential $\Phi(\ldots)$ is known, but it can always be determined from $P - V - T$ measurements. Thus the original insight of van der Waals was given a firm basis in statistical mechanics and intermolecular forces. In fact, the set of dimensionless parameters α_i can be extended to include quantum effects and deviations from pairwise additivity of the potential.

So much for historical matters. The current situation is in practice rather simpler than suggested by the function $F(\ldots)$. For practical reasons it is easier to use P_c and T_c than σ and ϵ, and it turns out that the whole set of parameters α_i can often be condensed into a single parameter. This parameter is now usually based on the slope of the vapor–pressure curve and is called the Pitzer acentric factor ω (Schreiber & Pitzer 1989). Values of P_c, T_c and ω have been tabulated for a large number of substances.

Turning now to transport properties, it is obvious that it is possible to extend the principle of corresponding states from the equation of state to the transport coefficients. This step was soon taken by engineers, using critical constants (Hirschfelder *et al.* 1964). For example, a dimensionless reduced viscosity could be defined as

$\eta_{\rm r} = \eta(RT_{\rm c})^{1/6}/M^{1/2}P_{\rm c}^{2/3}$, where M is the molar mass, and $\eta_{\rm r}$ would be assumed to be a universal function of $P_{\rm r}$ and $T_{\rm r}$. This procedure was eventually put on a molecular basis for dense fluids by Helfand & Rice (1960) using expressions for the transport coefficients in terms of autocorrelation functions, which play a role in transport theory analogous to the role of the partition function in equilibrium theory.

The current refinement of this procedure for dense fluids is described in Chapter 12. Scale factors near unity are used to adjust $P_{\rm r}$, $V_{\rm r}$ and $T_{\rm r}$ of a substance of interest so as to secure agreement of its reduced equation of state with that of a reference fluid, and then the same scale factors are used to predict its transport coefficients from the reduced transport coefficients of the reference fluid. The accuracy can be especially good if the reference fluid is similar to the fluid of interest. The scale factors are usually related empirically to the acentric factor ω. The method also gives reasonable results for dilute gases.

Special interest attaches to the transport coefficients of dilute gases because only pairwise interactions are relevant, and a very detailed and elaborate connection with the intermolecular forces is furnished by the classical kinetic theory of gases (see Chapter 4). However, to go from $U(r)$ to transport coefficients requires extensive numerical integrations, and in 1948 a great effort was required to carry out this numerical work for the Lennard–Jones (12–6) potential on desk calculators. The results, as put forth by Hirschfelder and co–workers, were enthusiastically embraced by engineers, who were eager to be able to make reasonable predictions of gas transport coefficients in the absence of measurements. The resulting efforts involved scale parameters ϵ and σ, and really amounted to the use of a principle of corresponding states, but closely tied to the (12–6) potential model. As later work began to show that the (12–6) potential was inadequate to do justice to a wide range of experimental data, the enterprise fell into disrepute, much as had happened earlier for equilibrium properties because of too close an association with the van der Waals equation of state.

In 1972 Kestin *et al.* (1972a,b) assembled a large body of consistent, accurate results on low–density gases, especially the viscosity, and decided to try taking the principle of corresponding states very seriously, at least for noble gases. They carefully avoided using potential models and based their work exclusively on general statistical–mechanical theory and accurate experimental measurements. To the surprise of nearly everyone, a remarkable accuracy was achieved, an improvement of nearly one order of magnitude, and all the low–density thermodynamic and transport properties of the noble gases and their mixtures could be correlated with just two adjustable–scale parameters. This advance was due almost entirely to improvements in the accuracy and range of experimental data, but at about the same time there were important advances occurring in the quantitative knowledge of intermolecular forces. This knowledge could be used as the basis for extensions and improvements to the two–parameter correlation. The results of the interplay between these two advances are described in this chapter.

In retrospect, the general principle of corresponding states has proved to be much better than it had often been thought in the past. It has a firm basis in statistical mechanics and kinetic theory, and has great range and accuracy if care is taken not to contaminate it with oversimplified models.

11.2 Introduction to dilute gases and gaseous mixtures

The restriction to low–density gases as defined in Chapter 4 trades breadth of application for depth of understanding. One gain is a detailed knowledge of how transport coefficients depend on intermolecular forces and other molecular properties. This knowledge allows information from sources other than transport coefficients to be factored in, and results in correlations of transport coefficients over a very large temperature range, often beyond the range of direct measurement. Another important gain is that multicomponent mixtures are easily included because many–body forces play no role. The loss, of course, is the ability to predict how transport coefficients vary with pressure or density.

The theoretical basis, naturally, is the kinetic theory as summarized in Chapter 4. The explicit formulas of the Chapman–Enskog theory given there are all that is needed to formulate a principle of corresponding states, and were the basis for the two–parameter correlation for the noble gases given in 1972 (Kestin *et al.* 1972a,b), and for the extended correlation for the noble gases (Najafi *et al.* 1983).

It is a virtue of the principle of corresponding states that most of the details given in several textbooks and summarized in Chapter 4 are not needed. None at all are needed for the two–parameter correlation, and only a few for the extended correlation.

Extension of the results to molecular gases involves, at least in principle, a reformulation of the Boltzmann equation and the Chapman–Enskog solutions to take account of noncentral intermolecular forces and inelastic molecular collisions. When this is done, the general forms of the expressions for the transport coefficients do not change, with the notable exception of the thermal conductivity, but the relations between the transport coefficients and the intermolecular forces and inelastic collisions become very complicated. No wide–ranging principle of corresponding states is thus apparently possible, but kinetic–theory calculations with models of nonspherical molecular interactions (Monchick & Mason 1961; Smith *et al.* 1967) suggested that a limited principle of corresponding states might still hold for viscosity and diffusion.

The exploration of the possibility of applying a corresponding–states correlation to molecular gases by Kestin *et al.* (1977) led to the surprising empirical result that viscosities could be correlated very well by the same two–parameter scheme as was developed for the noble gases, at least over the temperature range of their measurements, 298–673 K. Not only were no new parameters required, but the measurements fell very accurately on the same reduced curve that fitted the viscosities of the noble gases. What feature or features of the interactions could account for such apparently universal behavior? At the suggestion of J. Kestin, a direct inversion of the reduced viscosity correlation curve was carried out to find the effective reduced potential (Boushehri *et al.* 1978). This inversion showed that only the repulsive wall of the effective potential was involved in the temperature range covered by the correlation. It is not surprising that this rather featureless section of the potential could be fitted with only two parameters. However, this result suggests that any attempt to extend the correlation to lower temperatures is likely to fail, because an adjustment of potential–energy scale parameters that makes the repulsive walls agree is likely to ruin any agreement of the attractive wells and tails of the potentials, which dominate the low–temperature transport coefficients. But extension of the correlation to higher temperatures might still be possible.

The net results of the foregoing considerations are as follows for real systems:

- The two–parameter corresponding–states correlation holds very accurately for the noble gases and their mixtures from about $T^* = k_B T/\epsilon = 1$ to 25, where ϵ is the potential well–depth parameter, with an uncertainty of about 0.5% for the viscosity and thermal conductivity, and about 2% for the diffusion coefficient (Kestin *et al.* 1972a,b; Maitland *et al.* 1987).
- The second virial coefficients of the noble gases are also successfully correlated with the same two parameters.
- The viscosities and diffusion coefficients of molecular gases are also correlated in that temperature range to about the same accuracy with two parameters (Kestin *et al.* 1977; Maitland *et al.* 1987), but the thermal conductivities are not because the molecular internal degrees of freedom contribute directly to the transport of energy.
- It is noteworthy that this correlation fails for the second virial coefficients of molecular gases, showing that the second virial coefficient is sensitive to different portions of the potential than are the viscosity and diffusion coefficient in the same temperature range; in particular, the second virial coefficient is more sensitive to the well and tail regions of the potential, and to the nonspherical components of the potential, than is the viscosity.
- The extended corresponding–states correlation for the noble gases and their mixtures utilizes essentially the same two parameters as the two–parameter correlation, but adds three new ones that characterize the shape of the potential — one for the long–range potential tail and two for the short–range repulsive wall. Values of these three new parameters are independently available from theory and from other types of experiments. As a result, the temperature range of the correlation is extended to cover from (nominally) absolute zero to the onset of ionization. Theory also gives the asymptotic behavior of the transport coefficients at low and at high temperatures, which is a considerable help in devising mathematical expressions to fit the results. The accuracy of the extended correlation for the viscosity and thermal conductivity of the noble gases is about 0.5%, the same as for the two–parameter correlation but over a much greater temperature range. The accuracy of the diffusion coefficient

correlation is improved to about 1%, for reasons discussed below, and the thermal diffusion factor is now included, with an uncertainty of about 5%.

- For molecular gases the extended principle of corresponding states holds only in a limited form, for the reasons discussed above. In particular, it holds only in the temperature range in which the repulsive wall of an effective potential is dominant, and so fails at low temperatures ($k_B T/\epsilon < 1$); this eliminates the need for one of the three additional parameters introduced by the extended principle for the noble gases. However, the extension to higher temperatures is successful. The second virial coefficients can also be included, provided that corrections for specific contributions of nonspherical components of the potential are made. The overall accuracy of the correlation for molecular gases is estimated to be somewhat less than for the noble gases, but viscosity, diffusion and thermal diffusion are all included.

- The thermal conductivity of molecular gases requires special treatment because of the direct role played by the molecular internal degrees of freedom, especially the molecular rotation (see Chapter 4). If for simplicity the case is considered where only molecular rotation is important, the difficult quantity is the coefficient for the diffusion of rotational energy. There are only few direct measurements of this quantity (burdened with a high experimental uncertainty) which confirm that D_{rot} is often only approximately equal to the mass diffusion coefficient. In most cases the only hope for a correlation scheme is to work with the rotational–energy diffusion coefficient as obtained from measured thermal conductivities, provided it has some reasonable behavior. Such a scheme has proved moderately useful for evaluating and smoothing measurements but lacks predictive power. Predictive power was attained only when relations were found between the diffusion coefficient for rotational energy and the collision number for rotational relaxation by Sandler (1968) and more generally by Uribe *et al.* (1989). On this basis the principle of corresponding states can be applied to the thermal conductivity of simple molecular gases to produce a correlation over the temperature range from $T^* = 1$ to a nominal upper limit of 3000 K, with an uncertainty of a few percent.

- The thermal conductivity of mixtures of polyatomic gases requires even further consideration because the theoretical formulas are very complicated, and contain many interaction terms involving inelastic cross sections and relaxation times, which are known only poorly, if at all (see Chapter 4). Nevertheless, a careful study of simplifying approximations and major sources of errors produced a predictive algorithm with an accuracy of the order of 2% (Uribe *et al.* 1991). While not as good as the best experimental precision of about 0.5%, the result is very useful because of the practical impossibility of carrying out direct measurements on all mixtures of possible interest.

A selection of detailed results follows, including explicit formulas, skeleton tables, and references to the literature. However, because of space limitations some specific details on some of the less important mixture properties have been omitted, for which the reader must consult the references given.

11.3 Two–parameter corresponding–states principle

There are two reasons for discussing the two–parameter correlation, even though it has been superseded to a large extent by correlations based on the extended principle of corresponding states. First, the two–parameter principle exhibits the major features involved in a correlation, without some of the complexity introduced by the extended principle. Second, and more important, the two–parameter correlation is available in its more restricted temperature range for a number of molecular gases for which an extended correlation has not yet been developed. However, an explicit discussion of mixtures is omitted, since these are included under the extended principle of corresponding states.

The theoretical basis for the corresponding–states principle is as follows. Although molecules interact with a pair potential $U(r, \Omega)$ that depends on relative orientation angles Ω as well as on separation r, the formal extension of the Boltzmann equation and its subsequent solution by the methods of Chapman and Enskog show that the expression for the viscosity η and the so–called self–diffusion coefficient D have the same general form as for central potentials (Taxman 1958; Wang Chang *et al.* 1964)*

$$\eta = \frac{5}{16} \frac{(\pi m k_B T)^{1/2} f_\eta}{\langle \Omega^{(2,2)} \rangle} \tag{11.4}$$

$$D = \frac{3 k_B T}{8 P} \left(\frac{\pi k_B T}{m} \right)^{1/2} \frac{f_D}{\langle \Omega^{(1,1)} \rangle} \tag{11.5}$$

Here m is the molecular mass, k_B is Boltzmann's constant, T is the temperature, P is the pressure, and $\langle \Omega^{(2,2)} \rangle$ and $\langle \Omega^{(1,1)} \rangle$ are effective temperature–dependent cross sections or collision integrals that average over all elastic and inelastic molecular collisions governed by $U(r, \Omega)$, and which depend on all the parameters that characterize the magnitude and shape of $U(r, \Omega)$. These collision integrals are here defined so that they would each be equal to $\pi \sigma^2$ for collisions of hard spheres of diameter σ. The dimensionless quantities f_η and f_D are correction factors that follow from higher approximations made in solving the kinetic–theory equations; they are nearly unity and depend on ratios of various collision integrals.

The potential $U(r, \Omega)$ can always be written in the dimensionless form (11.2). If inelastic collisions are unimportant, then the details of molecular internal energy states are also unimportant, and it requires only dimensional analysis to show that

$$\langle \Omega^{(2,2)} \rangle = \pi \sigma^2 \Omega^{(2,2)*} (T^*, \alpha_1, \alpha_2, \ldots) \tag{11.6}$$

---

* In order to be consistent with the original papers, here the notation including Ω integrals is retained. Chapter 11 is complementary to Chapter 4 in the sense that equivalent kinetic theory relationships are expressed in terms of macroscopic quantities rather than effective cross sections.

where $\Omega^{(2,2)*}$ is a dimensionless function of the reduced temperature $T^* = k_B T/\epsilon$ and all the shape and anisotropy parameters $\alpha_1, \alpha_2 \ldots$ of the potential. A similar result holds for $\langle \Omega^{(1,1)} \rangle$ and for the correction factors f_η and f_D, except that the latter involve only ratios of collision integrals and hence do not depend on the scale parameter σ. It should be noted that no parameters pertaining to molecular internal energies appear in these collision integrals, because of the presumed unimportance of inelastic collisions for η and D. A similar result does not hold for the thermal conductivity, except for noble gases.

The specific assumption of the two–parameter correlation can be divided into two parts:

(i) the functions $\Omega^{(2,2)*}$, $\Omega^{(1,1)*}$, f_η and f_D have the same mathematical form for all the molecules considered, and

(ii) the parameters $\alpha_1, \alpha_2, \ldots$ are the same for all the molecules considered.

The first part is required in order to have any principle of corresponding states at all, and the second part allows a two–parameter correlation to be made with just σ and ϵ. Although these assumptions might seem reasonable for the noble gases, which are quite similar in structure, it seems quite surprising that they could hold very well for polyatomic molecules. There are essentially two reasons for the success:

- First, the part of $U(r, \Omega)$ that dominates the transport coefficients in the range $1 \leq T^* \leq 25$ is the relatively featureless repulsive wall, as was mentioned previously.
- Second, theoretical arguments and numerical calculations with simple models of $U(r, \Omega)$ indicate that η and D are only weakly affected by the anisotropy of the potential.

But whatever the underlying reasons may be, the important empirical result is that it is possible to scale measured values of $\langle \Omega^{(2,2)} \rangle / f_\eta = (5/16)(\pi m k_B T)^{1/2}/\eta$ as a function of temperature for different substances onto a single curve with just the two scale parameters σ and ϵ. The most recent formulation of the two–parameter correlation by Maitland *et al.* (1987) yields the following results

$$\frac{\Omega^{(2,2)*}}{f_\eta} = \exp\Big[0.46649 - 0.57015 \ln T^*$$
$$+ 0.19164(\ln T^*)^2 - 0.03708(\ln T^*)^3$$
$$+ 0.00241(\ln T^*)^4\Big] \quad (1 \leq T^* \leq 25) \tag{11.7}$$

$$\frac{\Omega^{(1,1)*}}{f_D} = \exp\Big[0.348 - 0.459 \ln T^* + 0.095 (\ln T^*)^2$$
$$- 0.010(\ln T^*)^3\Big] \quad (1 \leq T^* \leq 25) \tag{11.8}$$

where $\Omega^{(2,2)*}$ is defined by equation (11.6).

Six comments on the foregoing results should be noted:

(i) The form of equations (11.7) and (11.8) for the reduced collision integrals are purely empirical, and extrapolation outside of the specified T^* range is unreliable. In this connection, the original upper limit for $\Omega^{(2,2)*}$ was $T^* = 90$, but this result was based entirely on helium data and was not universal (Najafi *et al.* 1983).

(ii) Although D is often called the self–diffusion coefficient, it is actually measured by the diffusion of tagged molecules, usually isotopes, through the parent gas. True self–diffusion is unphysical.

(iii) The resulting uncertainty is about 0.5% for $\Omega^{(2,2)*}$ and about 2% for $\Omega^{(1,1)*}$. Within this uncertainty the value of f_D is essentially unity.

(iv) The formula for $\Omega^{(1,1)*}$ can also be used to calculate mutual diffusion coefficients D_{12}, which differ from D only by a mass factor within the uncertainty implied by f_D, provided that σ_{12} and ϵ_{12} are known. This point is not detailed here because mixtures are discussed later.

(v) These expressions can also be used to calculate the viscosity of a multicomponent mixture; the formulas are given later in the section on the extended principle. The only additional quantity needed is the ratio A^* (Maitland *et al.* 1987),

$$A^* = \Omega^{(2,2)*}/\Omega^{(1,1)*} \tag{11.9}$$

$$= \exp\left[0.1281 - 0.1108\ln T^* \right.$$
$$+ 0.0962\,(\ln T^*)^2 - 0.027\,(\ln T^*)^3$$
$$\left. + 0.0024\,(\ln T^*)^4\right] \quad (1 \leq T^* \leq 25) \tag{11.10}$$

(vi) Absolute values of the scale factors σ and ϵ cannot be determined from the principle of corresponding states alone, but only values relative to some chosen reference substance. In this case the reference was argon, with scale factors

$$\sigma_{\mathrm{Ar}} = 0.3350 \text{ nm} \qquad \epsilon_{\mathrm{Ar}}/k_{\mathrm{B}} = 141.6 \text{ K} \tag{11.11}$$

where σ is the value of r such that $U(\sigma) = 0$ and ϵ is the potential well depth. These values are believed to be close to the actual parameters for the argon pair potential, and differ by only a trivial amount from the values used with the extended principle of corresponding states.

Relative values of $\sigma/\sigma_{\mathrm{Ar}}$ and $\epsilon/\epsilon_{\mathrm{Ar}}$ are given in Table 11.1 for substances for which only a two–parameter correlation has been carried out.

11.4 Extended corresponding–states principle

In the decade following 1972 a number of experimental and theoretical advances were made in the knowledge of intermolecular potentials, especially for the noble–gas interactions, which allowed tests to be made of the Kestin–Ro–Wakeham assumption that

Table 11.1. *Scaling factors for the two–parameter correlation (Maitland et al. 1987).*

Gas	σ/σ_{Ar}	ϵ/ϵ_{Ar}
C_3H_8	1.4901	1.8964
C_4H_{10}	1.6496	2.0168
i-C_4H_{10}	1.6803	1.8423
CCl_3F	1.7186	1.8885
$CHClF_2$	1.3871	2.0005
C_3H_8–N_2	1.2774	1.1777
C_3H_8–CO_2	1.2844	1.7278
C_3H_8–CH_4	1.2853	1.5624
C_3H_8–C_2H_6	1.3795	1.8944
C_3H_8–C_4H_{10}	1.5555	2.0012
C_4H_{10}–N_2	1.3488	1.2428
C_4H_{10}–CO_2	1.3458	1.8436
C_4H_{10}–CH_4	1.3285	1.7929
C_4H_{10}–C_2H_6	1.4658	1.8957
C_4H_{10} – i-C_4H_{10}	1.6466	2.0018

all the shape and anisotropy parameters $\alpha_1, \alpha_2, \ldots$ were the same for a very large class of substances. It turns out that they are not all the same, and knowledge of how they differ can be used to formulate an extended principle of corresponding states. Because the results for the noble gases are much more extensive than those for molecular gases, it is useful to discuss them first.

11.4.1 Noble gases

By about 1980 the interaction potentials for the noble gases were rather accurately known (Aziz 1984; Scoles 1980). The specific advances that led to this situation have been discussed elsewhere (Najafi *et al.* 1983; Mason 1983) and can be briefly summarized as follows:

(i) Development of numerical methods for direct inversion of measured transport coefficients and second virial coefficients to find the potential, without any explicit assumption about its functional form.

(ii) Measurements of the scattering of noble gases by noble gases in the thermal energy range.

(iii) Accurate values of the coefficients of the long–range dispersion energy through a combination of quantum theory with dielectric and optical data.

(iv) Accurate information on the repulsive walls of the potentials from a synthesis of theoretical calculations and high–energy scattering measurements.

(v) Determination of vibrational levels in noble–gas dimers from their vacuum ultraviolet absorption spectra.

As a result of these advances, it could be seen directly that the potentials for the noble gases did not scale perfectly with only two parameters. That is, plots of U/ϵ versus r/r_m, where r_m is the position of the potential minimum, did not fall on a single curve. The curves corresponded rather closely around the minimum, but diverged from each other in both the tail and the repulsive–wall regions. In short, the shape parameters $\alpha_1, \alpha_2 \ldots$ of the potential were *not* the same for all the noble gases. The lack of correspondence in the potential tails would mean that the two–parameter principle of corresponding states would fail at low temperatures, and the lack of correspondence in the repulsive walls would mean that the principle would fail at high temperatures. Thus the range of validity of the two–parameter correlation could be rather precisely specified, and turned out to be about $1 \leq T^* \leq 25$, as given in equations (11.7) and (11.8).

To extend the corresponding–states correlation to lower and higher temperatures it is necessary to know what specific new parameters are needed, and this in turn requires some knowledge of the long–range and short–range behavior of the potential.

The long–range attractive part of the potential has the asymptotic form

$$U(r) = -\frac{C_6}{r^6} - \frac{C_8}{r^8} - \frac{C_{10}}{r^{10}} - \cdots \tag{11.12}$$

in which the dispersion coefficients $C_6, C_8, C_{10} \ldots$ are known independently. If a two–parameter principle of corresponding states held for the transport coefficients to very low temperatures, then the dimensionless parameters $C_6^* = C_6/\epsilon\sigma^6$, $C_8^* = C_8/\epsilon\sigma^8$, etc. would be the same for all the noble gases. The values of C_6^* show systematic variations, and so C_6^* must be included as an additional parameter in an extended principle of corresponding states. The values of C_8^* and C_{10}^* are probably also not universal, but their variations fall within their rather large uncertainties. In addition, the effects of C_8^* and C_{10}^* on the transport coefficients are small enough that these parameters can safely be assumed to have universal values. Thus the extension of the corresponding–states correlation to low temperatures requires one new parameter, C_6^*, in addition to ϵ and σ.

The short–range repulsive part of the potential can be represented by

$$U(r) = U_0 \exp(-r/\rho) \tag{11.13}$$

where U_0 and ρ (not to be confused with density) are energy and range parameters that are known independently. Since there are only two parameters in equation (11.13), a two–parameter correlation of transport coefficients must hold at high temperatures. Unfortunately, this is a different correlation than that given by ϵ and σ, as shown by the fact that the reduced parameters $U_0^* = U_0/\epsilon$ and $\rho^* = \rho/\sigma$ are not the same for all the noble gases. Thus the extension of the corresponding–states correlation to high temperatures requires two new parameters, U_0^* and ρ^*.

The extended principle of corresponding states developed by Najafi *et al.* (1983) is thus based on the following form for the potential

$$U(r) = \epsilon \Phi(r/\sigma, C_6^*, U_0^*, \rho^*) \tag{11.14}$$

which leads to reduced collision integrals of the form $\Omega^{(l,s)*}(T^*, C_6^*, U_0^*, \rho^*)$. In principle, there should be an additional parameter to allow for the fact that collisions at low energies do not follow classical mechanics. The resulting quantum deviations are usually described by a reduced de Broglie wavelength known as the de Boer parameter,

$$\Lambda^* \equiv h/\sigma(m\epsilon)^{1/2} \tag{11.15}$$

where h is the Planck constant and m is the atomic mass. Quantum effects at low temperatures are quite large for second virial coefficients, but are much less important for transport coefficients. No systematic calculations of quantum deviations of transport coefficients as a function of Λ^* and T^* have been carried out for realistic interaction potentials, and all the correlations presented here correspond to the classical limit. However, Najafi *et al.* (1983) have used available quantum–mechanical calculations for the (12–6) potential to estimate the regions in a (Λ^*, T^*) plane for which the classical–mechanical approximation is accurate to better than 1 or 2%.

In order to extend the corresponding–states correlations to lower and higher temperatures, it is important to know the asymptotic forms of the $\Omega^{(l,s)*}$ that are dictated by the limiting forms of $U(r)$ given in equations (11.12) and (11.13). The low–temperature asymptotic form can be found, from dimensional analysis and an argument based on 'capture' cross sections due to orbiting collisions, to be

$$\Omega^{(l,s)*}(T^*, C_6^*) \rightarrow A_0^{(l,s)}(C_6^*/T^*)^{1/3}\left[1 + \xi^{(l,s)}(C_8^*/C_6^*)(T^*/C_6^*)^{1/3} + \cdots\right] \tag{11.16}$$

where $A_0^{(l,s)}$ and $\xi^{(l,s)}$ are numerical constants of order unity, which can be found by quadrature.

The high–temperature asymptotic form can be found from dimensional analysis and numerical calculations to be

$$\Omega^{(l,s)*}(T^*, U_0^*, \rho^*) \rightarrow A_\infty^{(l,s)}(\rho^*)^2 \left[\ln(U_0^*/T^*)\right]^2 [1 + \cdots] \tag{11.17}$$

where the $A_\infty^{(l,s)}$ are numerical constants of order unity.

For practical purposes the correlations based on the extended principle of corresponding states are divided into three regions. In the middle temperature range, $1.2 \le T^* \le 10$, the two–parameter representation involving σ and ϵ works well. In the low–temperature range, $0 \le T^* \le 1.2$, the third parameter C_6^* is included with σ and ϵ and the results fitted with formulas based on the asymptotic form given by equation (11.16). In the high–temperature range, $T^* \ge 10$, the parameters U_0^* and ρ^* are included with σ and ϵ and the results fitted with formulas based on the asymptotic form given by equation (11.17). The formulas in the three ranges are arranged so that the reduced collision

Table 11.2. *Molar mass and scaling parameters for pure gases (Boushehri et al. 1987; Kestin et al. 1984).*

Gas	$\dfrac{M}{\text{g mol}^{-1}}$	$\dfrac{\sigma}{\text{nm}}$	$\dfrac{\epsilon/k_B}{\text{K}}$	$\rho^* = \dfrac{\rho}{\sigma}$	$U_0^* = \dfrac{U_0}{\epsilon}$
N_2	28.0134	0.3652	98.4	0.1080	5.308 (4)[a]
O_2	31.9988	0.3407	121.1	0.0745	1.322 (6)
NO	30.0061	0.3474	125.0	0.0883	2.145 (5)
CO	28.0106	0.3652	98.4	0.1080	5.308 (4)
CO_2	44.0100	0.3769	245.3	0.0720	2.800 (6)
N_2O	44.0128	0.3703	266.8	0.0730	2.600 (6)
CH_4	16.0430	0.3721	161.4	0.0698	3.066 (6)
CF_4	88.0048	0.4579	156.5	0.0200	1.460 (19)
SF_6	146.054	0.5252	207.7	0.0500	4.067 (8)
C_2H_4	28.0542	0.4071	244.3		
C_2H_6	30.0701	0.4371	241.9		
He	4.0026	0.2610	10.40	0.0797	8.50 (5)
Ne	20.179	0.2755	42.00	0.0784	11.09 (5)
Ar	39.948	0.3350	141.5	0.0836	5.117 (5)
Kr	83.80	0.3571	197.8	0.0831	4.491 (5)
Xe	131.29	0.3885	274.0	0.0854	3.898 (5)

[a] In all the tables values in parentheses are the powers of 10 by which the entries are multiplied, unless it is explicitly mentioned that this is not the case.

integrals $\Omega^{(l,s)*}$ are continuous, and have continuous first and second derivatives at the matching points of $T^* = 1.2$ and $T^* = 10$.

As in the case of the two–parameter correlation, only relative values of the scale parameters σ and ϵ can be determined, and some reference substance must be chosen. Argon was again chosen as reference, with

$$\sigma_{Ar} = 0.3350 \text{ nm} \qquad \epsilon_{Ar}/k_B = 141.5 \text{ K} \qquad (11.18)$$

which are essentially the same as used for the two–parameter correlation.

The formulas for the transport coefficients of the noble gases and their mixtures are given below, including the thermal conductivity, and values of the parameters are given in Tables 11.2–11.5. Details of the fitting procedures have been given by Najafi *et al.* (1983), and extensive tables and deviation plots have been prepared by Kestin *et al.* (1984). The latter include mixtures and second virial coefficients as well. The complicated formulas for the thermal diffusion factor, a quantity of secondary importance, are omitted here in order to conserve space; they can be found in the preceding two references.

Table 11.3. *Low–temperature scaling parameters (dispersion coefficients)* $C_6^* = C_6/\epsilon\sigma^6$ *for noble gases (Kestin et al. 1984).*

	He	Ne	Ar	Kr	Xe
He	3.09	2.940	2.681	2.498	2.346
Ne		2.594	2.429	2.424	2.204
Ar			2.210	2.426	2.053
Kr				2.164	2.051
Xe					2.162

Table 11.4. *Coefficients of the low–temperature formulas for* $\Omega^{(2,2)*}$ *and* $\Omega^{(1,1)*}$ *for noble gases (Kestin et al. 1984).*

i	a_{i1}	a_{i2}	b_{i1}	b_{i2}
3	-1.20407	-0.195866	10.0161	-10.5395
4	-9.86374	20.2221	-40.0394	46.0048
5	16.6295	-31.3613	44.3202	-53.0817
6	-6.73805	12.6611	-15.2912	18.8125

Table 11.5. *Coefficients of the high–temperature formulas for* $\Omega^{(2,2)*}$ *and* $\Omega^{(1,1)*}$ *(Kestin et al. 1984).*

i	a_{i1}	a_{i2}	a_{i3}	a_{i4}
2	-33.0838	20.0862	72.1059	8.27648
3	101.571	56.4472	286.393	17.7610
4	-87.7036	46.3130	277.146	19.0573

i	b_{i1}	b_{i2}	b_{i3}	b_{i4}
2	-267.00	201.570	174.672	7.36916
4	26700.	-19.2265 (3)	-27.6938 (3)	-3.29559 (3)
6	-8.90 (5)	6.31013 (5)	10.2266 (5)	2.33033 (5)

Four comments on these results may be noted:

(i) Readers wishing typical numerical results in order to check their computer codes should consult the compilation of Kestin *et al.* (1984), or possibly the skeleton tables for N_2 and CO_2 presented in Table 11.9.

(ii) A detailed estimate of the accuracy of the correlation for the transport coefficients of the noble gases and their mixtures in the temperature range 50–1000 K was given by Kestin *et al.* (1984). The uncertainty of the viscosity (η) is about 0.3%, that of the thermal conductivity (λ) about 0.7%, that of the mass diffusion coefficients (D and D_{12}) about 1% and that of the isotopic thermal diffusion factor about 5%.

(iii) The improvements in accuracy over the two–parameter correlation depend on information on $U(r)$ and do not follow directly from the principle of corresponding states itself. In particular, the collision integral $\Omega^{(1,1)*}$ for diffusion was not determined from direct measurements, but rather by inverting $\Omega^{(2,2)*}$ from measured viscosities to give $U(r)$ from which $\Omega^{(1,1)*}$ was then calculated. The reason for this indirect procedure is that measurements of viscosity are usually much more accurate than measurements of diffusion.

(iv) As an aside, there have been further improvements in knowledge of $U(r)$ for closed–shell atoms and ions, subsequent to the formulation of the extended principle of corresponding states. Use of a theoretical damping function to extend the range of validity of the dispersion energy given by equation (11.12) to small values of r shows that there is a relation between the parameters (U_0, ρ) and the parameters (ϵ, σ) (Koutselos *et al.* 1990). Thus the extended principle of corresponding states can be regarded as involving only three basic parameters rather than five: U_0, ρ and C_6, or ϵ, σ and C_6. In addition, improvements in the theoretical correlation of the dispersion coefficients have been made by Koutselos & Mason (1986). However, none of these developments affects either the use or the accuracy of the correlation presented here.

11.4.2 Summary of formulas for noble gases

11.4.2.1 General formulas

Pure substances:

$$\eta = \frac{5}{16} \left(\frac{mk_BT}{\pi} \right)^{1/2} \frac{f_\eta}{\sigma^2 \Omega^{(2,2)*}} \tag{11.19}$$

$$\lambda = \frac{75k_B}{64} \left(\frac{k_BT}{\pi m} \right)^{1/2} \frac{f_\lambda}{\sigma^2 \Omega^{(2,2)*}} \tag{11.20}$$

$$D = \frac{3}{8n} \left(\frac{k_BT}{\pi m} \right)^{1/2} \frac{f_D}{\sigma^2 \Omega^{(1,1)*}} \tag{11.21}$$

where $n = P/k_BT$ is the number density. The correction factors are given by

$$f_\eta = 1 + \frac{3}{196} (8E^* - 7)^2 \tag{11.22}$$

$$f_\lambda = 1 + \frac{1}{42} (8E^* - 7)^2 \tag{11.23}$$

$$f_D = 1 + \frac{(6C^* - 5)^2}{8(2A^* + 5)} \tag{11.24}$$

The dimensionless auxiliary functions A^*, C^* and E^* are defined by

$$A^* = \Omega^{(2,2)*}/\Omega^{(1,1)*} \tag{11.25}$$

$$C^* = \Omega^{(1,2)*}/\Omega^{(1,1)*} \tag{11.26}$$

$$E^* = \Omega^{(2,3)*}/\Omega^{(2,2)*} \tag{11.27}$$

They are all weak functions of T^* that are unity for hard spheres; they can be found from $\Omega^{(1,1)*}$, $\Omega^{(2,2)*}$ and the following recursion formula:

$$\Omega^{(l,s+1)*} = \Omega^{(l,s)*} + \frac{T^*}{s+2}\frac{d\Omega^{(l,s)*}}{dT^*} \tag{11.28}$$

Mixtures:

The simplest mixture property is the binary diffusion coefficient D_{12},

$$D_{12} = \frac{3}{8n}\left(\frac{k_B T}{\pi m_{12}}\right)^{1/2}\frac{f_{12}}{\sigma_{12}^2 \Omega_{12}^{(1,1)*}} \tag{11.29}$$

where $m_{12} = 2m_1 m_2/(m_1 + m_2)$ is twice the reduced mass, and $\Omega^{(1,1)*}$ is a function of the reduced temperature $T^* = k_B T/\epsilon_{12}$. This formula reduces to that for D when species 1 and 2 are mechanically similar but (in principle) are distinguishable. The correction f_{12} depends on the mixture composition, and can be approximated by the semiempirical formula

$$f_{12} \approx 1 + 1.3(6C_{12}^* - 5)^2\frac{ax_1}{1 + bx_1} \tag{11.30}$$

where x_1 is the mole fraction of the heavier component $(m_1 > m_2)$, and

$$a = \frac{2^{1/2}}{8[1 + 1.8(m_2/m_1)]^2}\frac{\Omega_{12}^{(1,1)*}}{\Omega_{22}^{(2,2)*}} \tag{11.31}$$

$$b = 10a[1 + 1.8(m_2/m_1) + 3(m_2/m_1)^2] - 1 \tag{11.32}$$

Diffusion in a multicomponent mixture is completely described by all the binary D_{ij} in the mixture, at least in a first approximation where all the f_{ij} are the same (Monchick *et al.* 1966).

The viscosity of a multicomponent mixture involves only quantities characterizing the binary interactions – A_{ij}^* and interaction viscosities η_{ij},

$$\eta_{ij} = \frac{5}{16}\left(\frac{m_{ij}k_B T}{\pi}\right)^{1/2}\frac{1}{\sigma_{ij}^2 \Omega_{ij}^{(2,2)*}} \tag{11.33}$$

where $m_{ij} = 2m_i m_j/(m_i + m_j)$ and η_{ij} is the viscosity of a hypothetical gas whose molecules have mass m_{ij} and whose interactions are described by the parameters σ_{ij}, ϵ_{ij}, $(C_6)_{ij}$, $(U_0)_{ij}$, ρ_{ij}. The viscosity of a mixture of ν components, η_{mix}, is given by

$$\eta_{\text{mix}} = -\frac{\begin{vmatrix} H_{11} & H_{12} & \cdots & H_{1\nu} & x_1 \\ H_{21} & H_{22} & \cdots & H_{2\nu} & x_2 \\ \vdots & \vdots & & \vdots & \vdots \\ H_{\nu 1} & H_{\nu 2} & \cdots & H_{\nu\nu} & x_\nu \\ x_1 & x_2 & \cdots & x_\nu & 0 \end{vmatrix}}{\begin{vmatrix} H_{11} & H_{12} & \cdots & H_{1\nu} \\ H_{21} & H_{22} & \cdots & H_{2\nu} \\ \vdots & \vdots & & \vdots \\ H_{\nu 1} & H_{\nu 2} & \cdots & H_{\nu\nu} \end{vmatrix}} \tag{11.34}$$

where

$$H_{ii} = \frac{x_i^2}{\eta_i} + \sum_{k=1, k\neq i}^{\nu} \frac{2x_i x_k}{\eta_{ik}} \frac{m_i m_k}{(m_i + m_k)^2} \left(\frac{5}{3A_{ik}^*} + \frac{m_k}{m_i} \right) \tag{11.35}$$

$$H_{ij} = -\frac{2x_i x_j}{\eta_{ij}} \frac{m_i m_j}{(m_i + m_j)^2} \left(\frac{5}{3A_{ij}^*} - 1 \right), \quad (i \neq j) \tag{11.36}$$

Note that the pure–component viscosities η_i are calculated with the correction factors f_η, but the interaction viscosities η_{ij} omit any f_η. This apparently slightly inconsistent procedure compensates for the error involved in the use of only a first–order kinetic–theory expression for η_{mix} (Najafi *et al.* 1983).

The thermal conductivity of a multicomponent mixture of noble gases also involves only binary–interaction quantities, A_{ij}^*, B_{ij}^*, λ_{ij} and f_{ij}, where B_{ij}^* is a ratio similar to A_{ij}^* and defined as

$$B_{ij}^* = \left[5\Omega_{ij}^{(1,2)*} - 4\Omega_{ij}^{(1,3)*} \right] / \Omega_{ij}^{(1,1)*} \tag{11.37}$$

The interaction thermal conductivity λ_{ij} is related to the interaction viscosity η_{ij},

$$\lambda_{ij} = \frac{15}{4} \frac{k_B}{m_{ij}} \eta_{ij} \tag{11.38}$$

The correction factors f_{ij} are included later to compensate for the error involved in the use of only a first–order kinetic–theory formula for λ_{mix}; they are the same as those in the formulas for the binary diffusion coefficients, with one small modification. The x_i in the formula for f_{ij} is interpreted to mean

$$x_i \rightarrow \frac{x_i}{x_i + x_j} \tag{11.39}$$

where i refers to the heavier component of the i–j pair (Najafi *et al.* 1983). The thermal conductivity of a mixture of ν components, λ_{mix}, is given by

$$\lambda_{\text{mix}} = -\frac{\begin{vmatrix} L_{11} & L_{12} & \cdots & L_{1\nu} & x_1 \\ L_{21} & L_{22} & \cdots & L_{2\nu} & x_2 \\ \vdots & \vdots & & \vdots & \vdots \\ L_{\nu 1} & L_{\nu 2} & \cdots & L_{\nu\nu} & x_\nu \\ x_1 & x_2 & \cdots & x_\nu & 0 \end{vmatrix}}{\begin{vmatrix} L_{11} & L_{12} & \cdots & L_{1\nu} \\ L_{21} & L_{22} & \cdots & L_{2\nu} \\ \vdots & \vdots & & \vdots \\ L_{\nu 1} & L_{\nu 2} & \cdots & L_{\nu\nu} \end{vmatrix}} \qquad (11.40)$$

where

$$L_{ii} = \frac{x_i^2}{\lambda_i}$$

$$+ \sum_{k=1, k \neq i}^{\nu} \frac{x_i x_k}{2\lambda_{ik} f_{ik}} \frac{\left[(15/2)m_i^2 + (25/4)m_k^2 - 3m_k^2 B_{ik}^* + 4m_i m_k A_{ik}^*\right]}{(m_i + m_k)^2 A_{ik}^*}$$

$$(11.41)$$

$$L_{ij} = -\frac{x_i x_j}{2\lambda_{ij} f_{ij}} \frac{m_i m_j}{(m_i + m_j)^2 A_{ij}^*} \left(55/4 - 3B_{ij}^* - 4A_{ij}^*\right), \quad (i \neq j) \qquad (11.42)$$

11.4.2.2 Collision integrals

$0 \leq T^* \leq 1.2$:

$$\Omega^{(2,2)*} = 1.1943(C_6/T^*)^{1/3}\left[1 + a_1(T^*)^{1/3} + a_2(T^*)^{2/3} + a_3(T^*) \right.$$
$$\left. + a_4(T^*)^{4/3} + a_5(T^*)^{5/3} + a_6(T^*)^2\right] \qquad (11.43)$$

$$\Omega^{(1,1)*} = 1.1874(C_6/T^*)^{1/3}\left[1 + b_1(T^*)^{1/3} + b_2(T^*)^{2/3} + b_3(T^*) \right.$$
$$\left. + b_4(T^*)^{4/3} + b_5(T^*)^{5/3} + b_6(T^*)^2\right] \qquad (11.44)$$

where

$$a_1 = 0.18, a_2 = 0, b_1 = 0, b_2 = 0$$

and the other coefficients are given in general by:

$$a_i = a_{i1} + a_{i2}(C_6^*)^{-1/3}, \quad i = 3, 4, 5, 6$$

$$b_i = b_{i1} + b_{i2}(C_6^*)^{-1/3}, \quad i = 3, 4, 5, 6$$

Numerical values for them are given in Table 11.4.

$1.2 \leq T^* \leq 10$:

$$\Omega^{(2,2)*} = \exp\left[0.46641 - 0.56991\,(\ln T^*) + 0.19591\,(\ln T^*)^2 \right.$$
$$\left. - 0.03879\,(\ln T^*)^3 + 0.00259\,(\ln T^*)^4 \right] \tag{11.45}$$

$$\Omega^{(1,1)*} = \exp\left[0.357588 - 0.472513\,(\ln T^*) + 0.0700902\,(\ln T^*)^2 \right.$$
$$\left. + 0.016574\,(\ln T^*)^3 - 0.00592022\,(\ln T^*)^4 \right] \tag{11.46}$$

$T^* \geq 10$:

$$\Omega^{(2,2)*} = (\rho^*)^2 \alpha^2 \left[1.04 + a_2(\ln T^*)^{-2} + a_3(\ln T^*)^{-3} + a_4(\ln T^*)^{-4} \right] \tag{11.47}$$

$$\Omega^{(1,1)*} = (\rho^*)^2 \alpha^2 \left[0.89 + b_2(T^*)^{-2} + b_4(T^*)^{-4} + b_6(T^*)^{-6} \right] \tag{11.48}$$

where $\alpha = \ln(U_0^*/T^*)$ and the coefficients a_i, b_i have the general form:

$$a_i = a_{i1} + (-1)^i (\alpha_{10}\rho^*)^{-2}[a_{i2} + (a_{i3}/\alpha_{10}) + (a_{i4}/\alpha_{10})^2], \qquad i = 2, 3, 4.$$

$$b_i = b_{i1} + (\alpha_{10}\rho^*)^{-2}[b_{i2} + (b_{i3}/\alpha_{10}) + (b_{i4}/\alpha_{10})^2], \qquad i = 2, 4, 6.$$

where $\alpha_{10} = \ln(U_0^*/10)$. Numerical values for the coefficients are given in Table 11.5.

These formulas apply to all interactions; it is necessary to use only the appropriate parameters ϵ_{ij}, σ_{ij}, $(C_6)_{ij}$, $(U_0^*)_{ij}$ and ρ_{ij}^* for the i-j interaction.

11.4.3 Polyatomic gases

As already mentioned, the extension of the principle of corresponding states is more limited for polyatomic gases than is the case for noble gases. This is caused in part by lack of detailed knowledge of interaction potentials and inelastic collisions and in part by a less extensive body of accurate experimental data on transport coefficients. Nevertheless, an extended corresponding-states correlation has been carried out for $T^* > 1$ for the 11 gases N_2, O_2, NO, CO, N_2O, CO_2, CH_4, CF_4, SF_6, C_2H_4 and C_2H_6 (Boushehri *et al.* 1987).

Except for the thermal conductivity, the general formulas for the transport coefficients are the same as for the noble gases. Formulas for the collision integrals are given below, and values of the parameters are given in Table 11.2, with argon as the reference substance according to equation (11.18). Details of the fitting procedures, and extensive tables and deviation plots, have been prepared by Boushehri *et al.* (1987), who also include second virial coefficients. Here only skeleton tables of transport coefficients for

N_2 and CO_2 are given as an aid in checking computer codes. Two comments on these results may be noted:

(i) The expressions used for $\Omega^{(2,2)*}$ are the same as for the noble gases, but $\Omega^{(1,1)*}$ has been adjusted slightly (less than 3%) to produce better agreement with the few accurate measurements of D.

(ii) Boushehri *et al.* (1987) have made detailed estimates of the accuracy of the correlation for polyatomic gases. The results are variable, and range from 0.3 to 1.0% for η, 1 to 5% for D and possibly as much as 25% for the thermal diffusion factor. The main advantage of the extended correlation over the two–parameter correlation is not greater accuracy, but rather the extended temperature range.

Mixtures involving the above 11 polyatomic gases with each other, and with the 5 noble gases have been correlated by Bzowski *et al.* (1990). Because of the lack of accurate experimental measurements on so many possible mixtures, it was necessary to predict the unlike interaction parameters ϵ_{ij}, σ_{ij}, $(C_6)_{ij}$, $(U_0)_{ij}$ and ρ_{ij} from the like parameters $\epsilon_{ii}, \sigma_{ii}, \ldots$ and $\epsilon_{jj}, \sigma_{jj}, \ldots$ by means of combination rules. Since such combination rules are not completely accurate in all cases, there was some loss of accuracy in the prediction of mixture properties. In compensation, the calculation of mixture properties now operates in an entirely predictive mode and does not depend on the existence of any accurate measurements on mixtures. The combination rules used and their theoretical bases are discussed by Bzowski *et al.* (1990).

Except for the thermal conductivity, the general formulas for the mixtures are the same as those already given for the noble gases. Values of the parameters are included in Tables 11.6 and 11.7. Details of the calculations, a selection of numerical tables, and extensive deviation plots have been given by Bzowski *et al.* (1990), who also include second virial coefficients. Here again only skeleton tables of transport coefficients for an equimolar $N_2 + CO_2$ mixture are given as an aid in checking computer codes. The general level of accuracy for the predicted transport coefficients is about 1% for η_{mix}, about 5% for D_{12} and roughly 25% for the thermal diffusion factor. These accuracies are only slightly worse than those of the pure polyatomic gases.

For more precise estimates of accuracy for particular polyatomic gases, it is necessary to examine the deviation plots given by Boushehri *et al.* (1987) and Bzowski *et al.* (1990).

11.4.4 Collision integrals for polyatomic gases

$1 \leq T^* \leq 10$:

$\Omega^{(2,2)*}$ same as for the noble gases.

$$\Omega^{(1,1)*} = \exp\Big[0.295402 - 0.510069 \ln T^* + 0.189395(\ln T^*)^2$$
$$- 0.045427(\ln T^*)^3 + 0.0037928(\ln T^*)^4\Big] \qquad (11.49)$$

$T^* \geq 10$:

$\Omega^{(2,2)*}$ and $\Omega^{(1,1)*}$ same as equations (11.47) and (11.48) for the noble gases.

Table 11.6. *Scaling parameters σ_{ij} (nm) (above the diagonal) and ϵ_{ij}/k_B (K) (below the diagonal) for mixtures of polyatomic and monatomic gases (Bzowski et al. 1990; Kestin et al. 1984).*

	N_2	O_2	NO	CO	CO_2	N_2O	CH_4
N_2		0.3529	0.3562	0.3652	0.3711	0.3676	0.3687
O_2	108.8		0.3441	0.3529	0.3596	0.3561	0.3569
NO	110.8	122.9		0.3562	0.3627	0.3592	0.3601
CO	98.4	108.7	110.8		0.3711	0.3676	0.3687
CO_2	155.0	168.8	172.9	155.2		0.3736	0.3745
N_2O	161.4	175.9	180.3	161.9	255.2		0.3710
CH_4	125.0	136.8	140.1	125.4	198.5	207.9	
CF_4	118.0	127.0	131.1	118.5	193.0	201.4	155.4
SF_6	121.2	127.6	132.7	121.5	204.1	210.4	161.6
C_2H_4	148.6	159.7	164.8	149.2	241.0	251.4	194.8
C_2H_6	146.8	157.4	162.5	147.3	238.9	248.7	192.5
He	23.42	27.94	27.56	23.23	34.14	35.82	27.78
Ne	52.97	61.77	61.24	52.41	70.70	80.37	62.64
Ar	117.7	130.9	132.9	117.6	182.2	190.2	148.1
Kr	139.4	152.8	156.2	139.7	219.6	229.5	178.6
Xe	159.3	171.7	177.0	160.0	256.3	268.0	208.2

	CF_4	SF_6	C_2H_4	C_2H_6	He	Ne	Ar
N_2	0.4132	0.4521	0.3878	0.4032	0.3243	0.3253	0.3499
O_2	0.4016	0.4409	0.3765	0.3919	0.3096	0.3118	0.3378
NO	0.4045	0.4435	0.3794	0.3948	0.3139	0.3157	0.3412
CO	0.4132	0.4520	0.3878	0.4032	0.3244	0.3253	0.3499
CO_2	0.4172	0.4536	0.3923	0.4074	0.3339	0.3343	0.3568
N_2O	0.4138	0.4503	0.3890	0.4042	0.3297	0.3304	0.3534
CH_4	0.4156	0.4530	0.3904	0.4057	0.3302	0.3307	0.3540
CF_4		0.4932	0.4323	0.4474	0.3814	0.3790	0.3986
SF_6	172.6		0.4672	0.4820	0.4298	0.4236	0.4377
C_2H_4	195.9	215.0		0.4221	0.3553	0.3537	0.3736
C_2H_6	194.8	216.1	242.9		0.3717	0.3697	0.3890
He	22.31	19.24	28.72	27.81		0.2691	0.3084
Ne	52.69	48.18	67.05	65.49	19.44		0.3119
Ar	137.4	137.4	172.6	170.0	30.01	64.17	
Kr	171.0	176.7	214.0	211.4	31.05	67.32	165.8
Xe	207.2	223.6	258.3	256.4	29.77	67.25	182.6

	Kr	Xe
N_2	0.3610	0.3778
O_2	0.3492	0.3665
NO	0.3524	0.3695
CO	0.3610	0.3778
CO_2	0.3671	0.3829
N_2O	0.3637	0.3796
CH_4	0.3645	0.3808
CF_4	0.4079	0.4228

	Kr	Xe
SF_6	0.4453	0.4580
C_2H_4	0.3830	0.3978
C_2H_6	0.3982	0.4128
He	0.3267	0.3533
Ne	0.3264	0.3488
Ar	0.3464	0.3660
Kr		0.3753
Xe	225.4	

Table 11.7. *Scaling parameters* $\rho_{ij}^* = \rho_{ij}/\sigma_{ij}$ *(above the diagonal) and* $(U_0^*)_{ij} = (U_0)_{ij}/\epsilon_{ij}$ *(below the diagonal) for mixtures (Bzowski et al. 1990; Kestin et al. 1984).*

	N_2	O_2	NO	CO	CO_2	N_2O
N_2		0.0918	0.0984	0.1080	0.0897	0.0904
O_2	1.790(5)		0.0815	0.0918	0.0730	0.0736
NO	9.563(4)	4.877(5)		0.0984	0.0797	0.0803
CO	5.311(4)	1.792(5)	9.566(4)		0.0897	0.0904
CO_2	2.420(5)	2.011(6)	7.094(5)	2.418(6)		0.0725
N_2O	2.322(5)	1.937(6)	6.812(5)	2.314(5)	2.705(6)	
CH_4	2.490(5)	2.070(6)	7.256(5)	2.482(5)	2.946(6)	2.827(6)
CF_4	2.060(7)	3.455(9)	2.797(8)	2.052(7)	4.520(9)	4.460(9)
SF_6	2.031(6)	3.039(7)	8.273(6)	2.026(6)	3.586(7)	3.513(7)
He	2.547(5)	1.546(6)	6.099(5)	2.568(5)	3.016(6)	2.880(6)
Ne	2.048(5)	1.464(6)	5.418(5)	2.070(5)	2.929(6)	2.584(6)
Ar	1.303(5)	8.057(5)	3.238(5)	1.304(5)	1.202(6)	1.155(6)
Kr	1.252(5)	7.062(5)	3.092(5)	1.250(5)	1.074(6)	1.030(6)
Xe	1.297(5)	7.346(5)	3.105(5)	1.291(5)	9.577(5)	9.182(5)

	CH_4	CF_4	SF_6	He	Ne	Ar
N_2	0.0887	0.0588	0.0727	0.0929	0.0938	0.0964
O_2	0.0719	0.0430	0.0586	0.0746	0.0753	0.0790
NO	0.0786	0.0492	0.0642	0.0820	0.0828	0.0860
CO	0.0887	0.0588	0.0727	0.0929	0.0938	0.0964
CO_2	0.0709	0.0435	0.0589	0.0718	0.0729	0.0773
N_2O	0.0714	0.0437	0.0592	0.0725	0.0736	0.0779
CH_4		0.0423	0.0576	0.0708	0.0719	0.0762
CF_4	5.669(9)		0.0359	0.0393	0.0406	0.0466
SF_6	4.060(7)	2.160(11)		0.0548	0.0565	0.0620
He	2.954(6)	9.329(9)	7.590(7)		0.0788	0.0791
Ne	2.689(6)	9.689(9)	5.904(7)	10.60(5)		0.0795
Ar	1.231(6)	9.765(8)	1.606(7)	9.740(5)	9.235(5)	
Kr	1.106(6)	6.522(8)	1.255(7)	10.89(5)	9.929(5)	4.849(5)
Xe	9.978(5)	3.322(8)	9.047(6)	13.37(5)	11.20(5)	4.878(5)

	Kr	Xe		Kr	Xe
N_2	0.0957	0.0961	CF_4	0.0476	0.0501
O_2	0.0788	0.0799	SF_6	0.0628	0.0649
NO	0.0856	0.0864	He	0.0772	0.0764
CO	0.0957	0.0961	Ne	0.0786	0.0785
CO_2	0.0774	0.0788	Ar	0.0833	0.0835
N_2O	0.0780	0.0793	Kr		0.0837
CH_4	0.0763	0.0777	Xe	4.337(5)	

11.5 Thermal conductivity

The internal degrees of freedom of polyatomic molecules contribute in a direct and specific way to the thermal conductivity and cannot be simply absorbed into an effective potential or reduced collision integral, as was the case for viscosity and diffusion. Nevertheless, it is still possible to pursue a strategy of a correlation based on a principle of corresponding states. The advantage of such a procedure is that it has predictive power and encompasses a variety of different gases. The disadvantage is that the correlation may be a little less accurate than correlations that merely fit the available data for each individual gas. Here, pure gases and mixtures are discussed separately, because the theoretical results for mixtures are so complicated that special approximations are necessary.

11.5.1 Pure Gases

The present correlation is based on a version of kinetic theory in which each molecular degree of freedom contributes separately to the thermal conductivity (usually referred to as Wang Chang–Uhlenbeck–de Boer (WCUB) theory) (Taxman 1958; Wang Chang *et al.* 1964; Mason & Monchick 1962; Monchick *et al.* 1965). Other procedures are also possible that are mathematically equivalent but make different special approximations; these are discussed briefly below. The thermal conductivity λ is written as a sum of contributions from translational, rotational, vibrational and electronic degrees of freedom

$$\lambda = \lambda_{tr} + \lambda_{rot} + \lambda_{vib} + \lambda_{elec} \tag{11.50}$$

To a first (rather crude) approximation these contributions are independent – the translational contribution is like that of a monatomic gas, and the other contributions correspond to the transport of molecular internal energy by a diffusion mechanism. Approximations of this sort actually predate the more elaborate kinetic–theory treatments based on an extended Boltzmann–like equation, and are often accurate to about a 10% level, with the notable exception of strongly polar gases, which have anomalously low thermal conductivities. In this approximation λ can be calculated from η, D and the specific heat of the gas.

This simple approximation is inaccurate for two main reasons:

(i) The separate contributions to λ are not independent. Interactions among the various degrees of freedom result in correction terms to these contributions.

(ii) Each internal degree of freedom requires its own diffusion coefficient, which is not necessarily well approximated by the mass diffusion coefficient D. An outstanding example is the diffusion coefficient for rotational energy of polar molecules. This coefficient appears anomalously low compared to D because a quantum of rotational energy can resonantly hop from one molecule to another in a distant encounter, thereby converting a glancing collision into an apparent nearly head–on collision.

Despite these complications, a reasonable correlation can be carried out for most small polyatomic molecules because of the following simplifications:

(i) Only the translational–rotation interaction is important for simple molecules, because it takes only a few collisions to effect rotational energy transfer.

(ii) Vibrational energy relaxation requires hundreds or thousands of collisions. The vibrational degrees of freedom can therefore be considered to behave independently, and it is a good approximation to take $D_{vib} \approx D$.

(iii) Electronic energy rarely plays any role, the most notable exception being NO. But even for NO the electronic degrees of freedom make only a small contribution to the specific heat, and the electronic relaxation time is nearly ten times longer than the rotational relaxation time. It is therefore a reasonable approximation to take the electronic degrees of freedom as independent, and to take $D_{elec} \approx D$.

The correlation problem thus reduces to the determination of the translational–rotation interaction and the diffusion coefficient for rotational energy. The interaction can be described by a collision number for rotational relaxation, ζ_{rot}, which can be found from independent measurements such as sound absorption (Lambert 1977). The temperature dependence of ζ_{rot} is reasonably well–determined from theory (Parker 1959; Brau & Jonkman 1970), and depends on T^* and one new parameter, the limiting high–temperature value ζ_{rot}^{∞}. However, this theory is valid only for the case of a single rotational degree of freedom.

The behavior of D_{rot} has not been found from independent measurements, and is calculated from ζ_{rot} (Uribe et al. 1989). It is this relation between D_{rot} and ζ_{rot} that makes a corresponding–states correlation possible.

There is also a small correction, usually less than 1%, due to the partial alignment of nonspherical molecules caused by collisions in the presence of a temperature gradient. This phenomenon is usually called spin polarization (see Chapter 4). This correction was included for the sake of completeness, even though its magnitude is probably within the overall uncertainty in the correlation (Uribe et al. 1989).

There is also a theoretical correction to D_{rot} to allow for resonant exchange of rotational energy caused by long–range dipole (Mason & Monchick 1962) and quadrupole (Nyeland et al. 1972) interactions. These corrections are very small for all the molecules considered here.

An extended corresponding–states correlation has been carried out for $T^* > 1$ for the nine gases N_2, O_2, NO, CO, N_2O, CO_2, CH_4, CF_4 and SF_6 (Uribe et al. 1990). These are the same gases as those for which the correlations for viscosity, diffusion and thermal diffusion were carried out by Boushehri et al. (1987), with the omission of

Table 11.8. *Parameters for the calculation of D_{rot}^{ex} and λ (Uribe et al. 1990).*

Gas	ζ_{rot}^{∞}	$10^3 C_{spin}$	T_{cross}^*	$\dfrac{\mu}{10^{-18}esu}$	$\dfrac{|\theta|}{10^{-26}esu}$	$\dfrac{\theta_{rot}}{K}$
N_2	29.5	5.7	6.70	0	1.4	2.88
O_2	28.0	6.4	7.3	0	0.39	2.07
NO	24.0	5.9	9.22	0.153	1.8	2.45
CO	22.2	6.1	10.48	0.112	2.5	2.77
CO_2	32.0	6.3	5.94	0	4.3	0.561
N_2O	36.2	6.3	4.99	0.167	3.0	0.603
CH_4	61.5	1.2	2.55	0	0	
CF_4	24.9	2.2	8.68	0	0	
SF_6	12.6	1.4	28.57	0	0	

C_2H_4 and C_2H_6. Here only skeleton tables for N_2 and CO_2 are given. A revision for the values of the constants ζ_{rot}^{∞}, C_{spin} and T_{cross} for O_2 is also included.

The formulas needed for the calculation of λ according to the extended principle of corresponding states are given below, and values of the necessary parameters are given in Table 11.8.

Five comments on these results may be noted:

(i) The correlation extends to a nominal upper limit of 3000 K. Uribe *et al.* (1990) have made detailed estimates of the uncertainty of the correlation. As was the case for η and D, the results are somewhat variable: about 1.5% in the range 300–500 K, deteriorating to about 3% at lower and higher temperatures, for all but O_2 and NO; the estimates are increased for O_2 and NO to 3% in the range 300–500 K, rising to 5% at lower and higher temperatures.

(ii) The correlation allows λ to be calculated from known quantities plus one new parameter, ζ_{rot}^{∞}. A single accurate measurement of λ for a gas not specifically included in the present correlation can determine ζ_{rot}^{∞} and permit the prediction of λ over a wide temperature range.

(iii) The correlation does not directly apply to complicated polyatomic molecules that have more than one easily excited internal degree of freedom. Asymmetric nonlinear molecules, for example, may have two or three different rotational degrees of freedom. Another example of an additional degree of freedom is hindered internal rotation about the C–C bond in C_2H_6.

(iv) Other correlations of a similar nature have been made, in which a separate correlation is made for each individual system rather than one general correlation. This procedure usually yields somewhat greater accuracy in the region where reliable measurements exist, but is less reliable for extrapolation and prediction. Millat *et al.* (1988) performed such correlations for N_2, CO, CO_2, CH_4 and CF_4 in the temperature range 300–1000 K, and Millat & Wakeham (1989) carried out a more extended correlation for N_2 (see Chapter 14) and CO. Direct comparison shows good agreement with the general correlation of Uribe *et al.* (1990).

(v) A simple and useful correlation for λ through the Prandtl number was developed by van den Oord & Korving (1988) (see Chapter 4), who essentially rearranged the kinetic–theory results described here into a form in which a rather complicated term turned out to be small enough to be neglected. Their result in terms of macroscopic quantities is

$$Pr \equiv \frac{C_P}{R} \left(\frac{\eta k_B}{\lambda m} \right) \approx \frac{2}{3} \left(1 + \frac{4}{3\pi \zeta_{rot}} \frac{C_{rot}}{R} \right) \tag{11.51}$$

where C_P is the molar heat capacity at constant pressure and C_{rot} the rotational contribution to the molar heat capacity at constant volume. The accuracy is somewhat less than the present correlation (except for SF_6, which is quite inaccurate), but it is much simpler to use.

11.5.2 Summary of formulas for the thermal conductivity of polyatomic gases

$$\lambda = \lambda_{tr} + \lambda_{rot} + \lambda_{vib} + \lambda_{elec}$$

$$\frac{m \lambda_{vib}}{\eta k_B} = \frac{\rho D_{vib}}{\eta} \frac{c_{vib}}{k_B} \approx \frac{\rho D}{\eta} \frac{c_{vib}}{k_B} = \frac{6}{5} A^* \frac{c_{vib}}{k_B} \tag{11.52}$$

$$\frac{m \lambda_{elec}}{\eta k_B} = \frac{\rho D_{elec}}{\eta} \frac{c_{elec}}{k_B} \approx \frac{\rho D}{\eta} \frac{c_{elec}}{k_B} = \frac{6}{5} A^* \frac{c_{elec}}{k_B} \tag{11.53}$$

$$\frac{m \lambda_{tr}}{\eta k_B} = \frac{5}{2} \left(\frac{3}{2} - \Delta_{rot} \right) (1 + \Delta_{spin}) \tag{11.54}$$

$$\frac{m \lambda_{rot}}{\eta k_B} = \frac{\rho D_{rot}^{ex}}{\eta} \left(\frac{c_{rot}}{k_B} + \Delta_{rot} \right) (1 + \Delta_{spin}) \tag{11.55}$$

where D_{rot}^{ex} is the diffusion coefficient for rotational energy with a correction for resonant energy exchange included. The correction terms Δ_{rot} and Δ_{spin} are

$$\Delta_{rot} = \frac{\frac{2}{\pi \zeta_{rot}} \frac{c_{rot}}{k_B} \left(\frac{5}{2} - \frac{\rho D_{rot}}{\eta} \right)}{1 + \frac{2}{\pi \zeta_{rot}} \left(\frac{5}{3} \frac{c_{rot}}{k_B} + \frac{\rho D_{rot}}{\eta} \right)} \tag{11.56}$$

$$\Delta_{spin} = \frac{C_{spin} \left(\frac{5}{2} + \frac{c_{rot}}{k_B} \right) \frac{\rho D_{rot}}{\eta}}{\left(1 + \frac{8}{15\pi \zeta_{rot}} \frac{c_{rot}}{k_B} \right) \frac{\rho D_{rot}}{\eta} + \frac{3}{5} \frac{c_{rot}}{k_B}} \tag{11.57}$$

where C_{spin} is a dimensionless constant of order $10^{-3} - 10^{-2}$. The temperature dependence of ζ_{rot} is given by

$$\frac{\zeta_{rot}^{\infty}}{\zeta_{rot}} = 1 + \frac{\pi^{3/2}}{2(T^*)^{1/2}} + \left(2 + \frac{\pi^2}{4}\right)\frac{1}{T^*} + \frac{\pi^{3/2}}{(T^*)^{3/2}} \tag{11.58}$$

The correlation for D_{rot} in terms of ζ_{rot} is given in two parts: a low–temperature part (Uribe *et al.* 1989), which switches to the high–temperature result of Sandler (1968) at a reduced temperature T^*_{cross},

$T^* \le T^*_{cross}$:

$$\frac{\rho D_{rot}}{\eta} = \left(\zeta_{rot}/\left(\zeta_{rot}^{\infty}\right)^{3/4}\right)(1.122 + 4.552/T^*) \tag{11.59}$$

$T^* \ge T^*_{cross}$:

$$\frac{\rho D_{rot}}{\eta} = (\rho D/\eta)\left(1 + 0.27/\zeta_{rot} - 0.44/\zeta_{rot}^2 - 0.90/\zeta_{rot}^3\right) \tag{11.60}$$

(Note that in equation (28) of Uribe *et al.* (1989) and equation (C5b) of Uribe *et al.* (1990) there is a misprint: ζ_{rot}^{∞} should be replaced by ζ_{rot}.) Numerical values of the constants T^*_{cross}, ζ_{rot}^{∞} and C_{spin} are given in Table 11.8.

The resonant energy exchange corrections to D_{rot} depend on the molecular dipole and quadrupole moments and on the molecular moment of inertia

$$\frac{D_{rot}}{D_{rot}^{ex}} = 1 + \Delta_{ex}^{\mu\mu} + \Delta_{ex}^{\mu\theta} + \Delta_{ex}^{\theta\theta} \tag{11.61}$$

$$\Delta_{ex}^{\mu\mu} = 0.44\, g^{\mu\mu}\left(\frac{3\pi^2}{2}\right)\left(\frac{\pi}{2}\right)^{1/2}\frac{\mu^2}{\hbar}\frac{\eta}{k_B T}\frac{\rho D_{rot}}{\eta}\left(\frac{\theta_{rot}}{T}\right)^{3/2} \tag{11.62}$$

$$\Delta_{ex}^{\mu\theta} = 0.51\, g^{\mu\theta}\left(\frac{56\pi^2}{45}\right)\left(\frac{3}{5}\right)^{1/2}\left(\frac{\pi^2}{6}\right)^{1/3}\left(\frac{\mu\,|\theta|}{\hbar}\right)^{2/3}$$

$$* \frac{\eta}{k_B T}\frac{\rho D_{rot}}{\eta}\left(\frac{\theta_{rot}}{T}\right)^{3/2} \tag{11.63}$$

$$\Delta_{ex}^{\theta\theta} = 1.31\, g^{\theta\theta}\left(\frac{7}{2}\pi^{3/2}\right)\Gamma\left(\frac{7}{4}\right)\left(\frac{\theta^2}{\hbar}\right)^{1/2}\left(\frac{k_B T}{m}\right)^{1/4}$$

$$* \frac{\eta}{k_B T}\frac{\rho D_{rot}}{\eta}\left(\frac{\theta_{rot}}{T}\right)^{3/2} \tag{11.64}$$

where μ is the molecular dipole moment θ is the quadrupole moment and $\theta_{rot} = \hbar^2/2k_B I$ is the characteristic rotational temperature, in which I is the molecular moment of inertia. The quantities $g^{\mu\mu}$, $g^{\mu\theta}$ and $g^{\theta\theta}$ are dimensionless factors of order unity that correct for the replacement of quantum–mechanical summations by integrations. Their high–temperature asymptotic forms are

$$g^{\mu\mu} = \exp\left(-2\theta_{\text{rot}}/3T\right)\left[1 - \left(\theta_{\text{rot}}/3T\right) + \ldots\right] \tag{11.65}$$

$$g^{\mu\theta} = \exp\left(-17\theta_{\text{rot}}/12T\right)\left[1 - \left(5\theta_{\text{rot}}/6T\right) + \ldots\right] \tag{11.66}$$

$$g^{\theta\theta} = \exp\left(-13\theta_{\text{rot}}/6T\right)\left[1 - \left(4\theta_{\text{rot}}/3T\right) + \ldots\right] \tag{11.67}$$

which are sufficiently accurate for most purposes. More accurate results for low T have been tabulated by Nyeland *et al.* (1972).

The above results are accurate from $T^* = 1$ to a nominal upper limit of 3000 K (Uribe *et al.* 1989, 1990). For the nine gases correlated by Uribe *et al.* (1990), the term Δ_{spin} amounts to a maximum correction of 1.5% and for all practical purposes is temperature–independent. The Δ_{ex} terms give less than 1% total correction to λ for these systems, but are much larger for strongly polar molecules with low moments of inertia, such as HCl, NH_3 and H_2O.

11.5.3 Mixtures

The kinetic theory expressions for the thermal conductivity of mixtures containing polyatomic molecules are very complicated and contain many essentially unknown quantities involving inelastic collisions and relaxation times (Monchick *et al.* 1965; see Chapter 4). Direct use of these formulas is essentially hopeless. However, by a careful application of some simplifying approximations and analysis of the major sources of errors, it was possible to obtain a fairly simple formula for predicting the composition dependence of λ_{mix} with an uncertainty of the order of 2% (Uribe *et al.* 1991).

The most crucial simplifying assumption is to use the kinetic–theory formulas for λ_{mix} as interpolation formulas to predict the composition dependence. In particular, this assumption means using accurate values for the pure–component thermal conductivities and viscosities in the formulas (not theoretical first approximations), and dropping all the explicit inelastic–collision terms. The resulting simplified expression is usually called the Hirschfelder–Eucken formula. The next important simplifying assumption is to replace all the ratios of diffusion coefficients for internal energy that appear in the Hirschfelder–Eucken formula by ratios of the corresponding mass diffusion coefficients. The ultimate justification for these simplifications is that a great deal of testing and numerical experimentation has shown that the inelastic–collision terms do not change the shape of λ_{mix} *versus* composition very much; they only move it up or down. Thus the Hirschfelder–Eucken formula gives good results if it is forced to the end points corresponding to the pure components.

Two secondary considerations are necessary in order to obtain accurate results for λ_{mix}. The first is a kinetic–theory convergence error that occurs when the masses of the components are very different. This has nothing to do with internal degrees of

Table 11.9. *Transport coefficients of N_2, CO_2 and an equimolar N_2+CO_2 mixture. The heat capacities used in this table for CO_2 are from the work by Woolley (1954).*

$\frac{T}{K}$	$\frac{\eta}{10^{-6}Pa\,s}$			$\frac{\lambda}{mW\,m^{-1}K^{-1}}$		
	N_2	CO_2	N_2+CO_2	N_2	CO_2	N_2+CO_2
100	6.71			8.63		
250	15.57	12.50	14.06	22.10	12.76	16.88
300	17.96	15.08	16.59	25.89	16.79	20.86
400	22.26	19.82	21.16	32.88	25.07	28.65
500	26.09	24.11	25.25	39.52	33.31	36.22
600	29.59	28.05	28.99	46.09	41.37	43.64
700	32.85	31.71	32.46	52.49	49.17	50.82
800	35.93	35.13	35.71	58.62	56.80	57.79
1000	41.65	41.44	41.71	70.61	71.10	71.04
2000	66.41	66.88	66.53	123.87	128.61	126.36
3300	95.20	93.03	93.46	184.06	186.21	184.66

$\frac{T}{K}$	$\frac{D(1atm)}{cm^2s^{-1}}$		
	N_2	CO_2	N_2+CO_2
100	0.0279		
250	0.1553	0.0829	0.1181
300	0.2138	0.1187	0.1652
400	0.3510	0.2050	0.2767
500	0.5129	0.3087	0.4089
600	0.6976	0.4280	0.5599
700	0.9038	0.5616	0.7285
800	1.1305	0.7086	0.9136
1000	1.6428	1.0400	1.3307
2000	5.3072	3.3491	4.2551
3300	12.6716	7.7430	9.9690

freedom and occurs even with mixtures of noble gases. The correction for this error is the appearance of the composition–dependent f_{ij} factors in the formulas for noble–gas λ_{mix}, namely equations (11.41) and (11.42). Finally, useful prediction of λ_{mix} requires accurate values for input data – in particular, all the λ_i, η_i and D for the pure components, the D_{ij} for all the gas pairs in the mixture and the values of A_{ij}^*, B_{ij}^* and C_{ij}^* for all the pairs. These are available from the correlations already discussed.

Uribe *et al.* (1991) surveyed all the modern measurements of λ_{mix}, obtained by the refined transient hot–wire technique, in order to assess the accuracy of their prediction scheme. The deviations were not particularly systematic, indicating that no single large effect remains unaccounted for. Improvements in the accuracy of calculated λ_{mix} will therefore probably be difficult to achieve, requiring improved treatment of many effects at once.

11.5.4 Formulas for mixtures

$$\lambda_{\text{mix}} = \lambda_{\text{mix}}(\text{mon}) + \sum_{i=1}^{\nu} [\lambda_i - \lambda_i(\text{mon})] \left[1 + \sum_{j=1, j \neq i}^{\nu} \frac{x_j D_{ii}}{x_i D_{ij}} \right]^{-1} \tag{11.68}$$

in which the λ_i are the experimental or accurately calculated thermal conductivities of the pure components, the D_{ii} are accurate self–diffusion coefficients of the pure components and the D_{ij} are accurate binary diffusion coefficients. The $\lambda_i(\text{mon})$ are hypothetical thermal conductivities that would be measured if the components behaved like monatomic gases, and are calculated from accurate values of the pure–component viscosities,

$$\lambda_i(\text{mon}) = \frac{15}{4} \frac{k_B}{m_i} \eta_i \tag{11.69}$$

The expression for $\lambda_{\text{mix}}(\text{mon})$ is the same as that given for the noble gases by equations (11.40)–(11.42) with use of $\lambda_i(\text{mon})$ and $\lambda_{ij}(\text{mon})$,

$$\lambda_{ij}(\text{mon}) = \frac{15}{4} \frac{k_B}{m_{ij}} \eta_{ij} \tag{11.70}$$

11.6 Conclusions

The principle of corresponding states, extended on the basis of kinetic theory and some knowledge of intermolecular forces, allows the correlation of the transport coefficients of low–density gases and their mixtures over a very wide temperature range. The ability to operate in a predictive mode is especially valuable for mixtures of any number of components, for which accurate experimental data are extremely scanty.

Although the collection of formulas presented here may appear forbiddingly complicated at first glance, they are actually quite easy to use, and the expressions are easy to program on a computer.

Acknowledgment

Work financed in part by CONACYT (0651–E9110, México).

References

Aziz, R. A. (1984). Interatomic potentials for rare–gases: Pure and mixed interactions, in *Springer Series in Chemical Physics, Vol. 34. Inert Gases: Potentials, Dynamics and Energy Transfer in Doped Crystals*, ed. M. L. Klein, pp. 5–86. Berlin: Springer–Verlag.

Boushehri, A., Bzowski, J., Kestin, J. & Mason, E. A. (1987). Equilibrium and transport properties of eleven polyatomic gases at low density. *J. Phys. Chem. Ref. Data*, **19**, 445–466.

Boushehri, A., Viehland, L. A. & Mason, E. A. (1978). On the extended principle of corresponding states and the pair interaction potential. *Physica*, **91A**, 424–436.

Brau, C. A. & Jonkman, R. M. (1970). Classical theory of rotational relaxation in diatomic gases. *J. Chem. Phys.*, **52**, 477–484.

Bzowski, J., Kestin, J., Mason, E. A. & Uribe, F. J. (1990). Equilibrium and transport properties of gas mixtures at low density: Eleven polyatomic gases and five noble gases. *J. Phys. Chem. Ref. Data*, **19**, 1179–1232.

Helfand, E. & Rice, S.A. (1960). Principle of corresponding states for transport properties. *J. Chem. Phys.*, **32**, 1642–1644.

Hirschfelder, J. O., Curtiss, C. F. & Bird, R. B. (1964). *Molecular Theory of Gases and Liquids*. 2nd printing, Section 9.1. New York: Wiley.

Kac, M., Uhlenbeck, G. E. & Hemmer, P. C. (1963). On the van der Waals theory of the vapor–liquid equilibrium I. Discussion of a one–dimensional model. *J. Math. Phys.*, **4**, 216–228.

Kestin, J., Ro, S. T. & Wakeham, W. A. (1972a). An extended law of corresponding states for the equilibrium and transport properties of the noble gases. *Physica*, **58**, 165–211.

Kestin, J., Ro, S. T. & Wakeham, W. A. (1972b). Viscosity of the noble gases in the temperature range 24–700°C. *J. Chem. Phys.*, **56**, 4119–4124.

Kestin, J., Khalifa, H. E., Ro, S. T. & Wakeham, W. A. (1977). The viscosity and diffusion coefficients of eighteen binary gaseous systems. *Physica*, **88A**, 242–260.

Kestin, J., Knierim, K., Mason, E. A., Najafi, B., Ro, S. T. & Waldman, M. (1984). Equilibrium and transport properties of the noble gases and their mixtures at low density. *J. Phys. Chem. Ref. Data*, **13**, 229–303.

Koutselos, A. D. & Mason, E. A. (1986). Correlation and prediction of dispersion coefficients for isoelectronic systems. *J. Chem. Phys.*, **85**, 2154–2160.

Koutselos, A. D., Mason, E. A. & Viehland, L. A. (1990). Interaction universality and scaling laws for interaction potentials between closed–shell atoms and ions. *J. Chem. Phys.*, **93**, 7125–7136.

Lambert, J. D. (1977). *Vibrational and Rotational Relaxation in Gases*. Oxford: Clarendon.

Maitland, G. C., Rigby, M., Smith, E. B. & Wakeham, W. A. (1987). *Intermolecular Forces: Their Origin and Determination*. Oxford: Clarendon, Appendix 3.

Mason, E. A. (1983). Extended principle of corresponding states and intermolecular forces. *Rev. Port. Quim.*, **25**, 1–10.

Mason, E. A. & Monchick, L. (1962). Heat conductivity of polyatomic and polar gases. *J. Chem. Phys.*, **36**, 1622–1639.

Millat, J., Vesovic, V. & Wakeham, W. A. (1988). On the validity of the simplified expression for the thermal conductivity of Thijsse et al. *Physica*, **148A**, 153–164.

Millat, J. & Wakeham, W. A. (1989). The thermal conductivity of nitrogen and carbon monoxide in the limit of zero density. *J. Phys. Chem. Ref. Data*, **18**, 565–581.

Monchick, L. & Mason, E. A. (1961). Transport properties of polar gases. *J. Chem. Phys.*, **35**, 1676–1697.

Monchick, L., Munn, R. J. & Mason, E. A. (1966). Thermal diffusion in polyatomic gases: A generalized Stefan–Maxwell diffusion equation. *J. Chem. Phys.*, **45**, 3051–3058; erratum, **48**, 3344.

Monchick, L., Pereira, A. N. G. & Mason, E. A. (1965). Heat conductivity of polyatomic and polar gases and gas mixtures. *J. Chem. Phys.*, **42**, 3241–3256.

Najafi, B., Mason, E. A. & Kestin, J. (1983). Improved corresponding states principle for the noble gases. *Physica*, **119A**, 387–440.

Nyeland, C., Mason, E. A. & Monchick, L. (1972). Thermal conductivity and resonant multipole interactions. *J. Chem. Phys.*, **56**, 6180–6192. (The formulae for $\Delta_{ex}^{\mu\theta}$ and $\Delta_{ex}^{\theta\theta}$ given in this paper are too large by factors of $2^{2/3}$ and 2, respectively, but the numerical results are correct.)

Parker, J. G. (1959). Rotational and vibrational relaxation in diatomic gases. *Phys. Fluids*, **2**, 449–462.

Sandler, S. I. (1968). Thermal conductivity of polyatomic gases. *Phys. Fluids*, **11**, 2549–2555.

Schreiber, D. R. & Pitzer, K. S. (1989). Equations of state in the acentric factor system. *Fluid Phase Equil.*, **46**, 113–130, and references therein.

Scoles, G. (1980). Two–body, spherical, atom–atom, and atom–molecule interaction energies. *Ann. Rev. Phys. Chem.*, **31**, 81–96.

Smith, F. J., Munn, R. J. & Mason, E. A. (1967). Transport properties of quadrupolar gases. *J. Chem. Phys.*, **46**, 317–321.

Taxman, N. (1958). Classical theory of transport phenomena in dilute polyatomic gases. *Phys. Rev.*, **110**, 1235–1239.

Uribe, F. J., Mason, E. A. & Kestin, J. (1989). A correlation scheme for the thermal conductivity of polyatomic gases at low density. *Physica*, **156A**, 467–491.

Uribe, F. J., Mason, E. A. & Kestin, J. (1990). Thermal conductivity of nine polyatomic gases at low density. *J. Phys. Chem. Ref. Data*, **19**, 1123–1136.

Uribe, F. J., Mason, E. A. & Kestin, J. (1991). Composition dependence of the thermal conductivity of low–density polyatomic gas mixtures. *Int. J. Thermophys.*, **12**, 43–51.

van den Oord, R. J. & Korving, J. (1988). The thermal conductivity of polyatomic molecules. *J. Chem. Phys.*, **89**, 4333–4338.

van der Waals, J. D. (1873). *On the Continuity of the Gaseous and Liquid States*. Dissertation, Leiden; translation edited with an introductory essay by J. S. Rowlinson (1988), in *Studies in Statistical Mechanics*, **14**, Amsterdam: North–Holland.

Wang Chang, C. S., Uhlenbeck, G. E. & de Boer, J. (1964). The heat conductivity and viscosity of polyatomic gases. *Studies in Statistical Mechanics, Vol. 2*, eds. J. de Boer & G. E. Uhlenbeck, pp. 241–268. Amsterdam: North–Holland.

Woolley, H. W. (1954). Thermodynamic functions for carbon dioxide in the ideal gas state. *J. Res. Nat. Bur. Stand. (U.S.)*, **52**, 289–292.

12

The Corresponding–States Principle: Dense Fluids

M. L. HUBER and H. J. M. HANLEY

Thermophysics Division,
National Institute of Standards and Technology,
Boulder, CO, USA

12.1 Introduction

The power and versatility of corresponding–states (CS) methods as a prediction tool has been pointed out by Mason and Uribe in Chapter 11 of this volume. Here, however, the strong point of corresponding–states principles is stressed: that methods based on the principle are theoretically based and predictive, rather than empirical and correlative. Thus, while CS cannot always reproduce a set of data within its experimental accuracy, as can an empirical correlation, it should be able to represent data to a reasonable degree but, more important, do what a correlation cannot do – estimate the properties beyond the range of existing data. In this chapter a particular corresponding–states method is reviewed that can predict the viscosity and thermal conductivity of pure fluids and their mixtures over the entire phase range from the dilute gas to the dense liquid with a minimum number of parameters. The method was proposed several years ago by Hanley (1976) and also by Mo & Gubbins (1974, 1976). It led to a computer program known as TRAPP ('TRAnsport Properties Prediction' (Ely & Hanley 1981a)). The method is also the basis for two NIST Standard Reference Databases – NIST Standard Reference Database 4 (SUPERTRAPP; Ely & Huber 1990) and NIST Standard Reference Database 14 (DDMIX; Friend 1992). Here, the original TRAPP procedure will be discussed as well as some more recent modifications to it. The performance of the model for viscosity and thermal conductivity prediction will also be examined for selected pure fluids and mixtures.

12.2 The TRAPP corresponding–states method

In the TRAPP procedure (Ely & Hanley 1981b, 1983), the viscosity η of a fluid at density ρ, temperature T (and for a mixture, of composition $\{x_i\}$) is equated to the viscosity of a hypothetical pure fluid. (The hypothetical pure fluid is obtained through the use of mixing and combining rules and is discussed later.) The hypothetical pure

283

fluid is then related to the viscosity of a reference fluid at a corresponding–state point (ρ_0, T_0). Thus,

$$\eta_{mix}(\rho, T, \{x_i\}) \equiv \eta_x(\rho, T) \equiv \eta_0(\rho_0, T_0) F_\eta \qquad (12.1)$$

where

$$F_\eta = \left(\frac{M_x}{M_0}\right)^{1/2} f_x^{1/2} h_x^{-2/3} \qquad (12.2)$$

Here the subscript 'x' refers to the fluid of interest (pure fluid or mixture), the subscript '0' refers to the reference fluid and M is the molecular mass. The state points T_0 and ρ_0 are defined by the ratios

$$T_0 = \frac{T}{f_x} \quad \text{and} \quad \rho_0 = \rho h_x \qquad (12.3)$$

where f_x and h_x are functions of the critical parameters and, in the case of TRAPP, the acentric factor. In the limit of two–parameter corresponding states between two pure fluids x and 0, the functions reduce to the ratios of the critical constants

$$f_x = \frac{T_{c,x}}{T_{c,0}} \quad \text{and} \quad h_x = \frac{\rho_{c,0}}{\rho_{c,x}} \qquad (12.4)$$

where the subscript 'c' denotes the critical value. Expressions for the thermal conductivity λ are similar, except that λ is separated into two contributions: λ' arising from the transfer of energy from translational effects, and λ'' from the internal degrees of freedom. Then,

$$\lambda_{mix} = \lambda'_{mix} + \lambda''_{mix} \qquad (12.5)$$

For λ' an expression analogous to the viscosity is used,

$$\lambda'_{mix}(\rho, T, \{x_i\}) \equiv \lambda'_x(\rho, T) \equiv \lambda'_0(\rho_0, T_0) F_\lambda \qquad (12.6)$$

where

$$F_\lambda = \left(\frac{M_0}{M_x}\right)^{1/2} f_x^{1/2} h_x^{-2/3} \qquad (12.7)$$

For a pure fluid, λ'' is assumed independent of density and is given in the TRAPP program by the modified Eucken correction for a polyatomic gas (Hirschfelder *et al.* 1954)

$$\frac{\lambda''_x M_x}{\eta_x^{(0)}} = 1.32 \left(C_P^{id} - \frac{5R}{2}\right) \qquad (12.8)$$

where $\eta_x^{(0)}$ is the dilute–gas viscosity of fluid x (which can be estimated from kinetic theory), C_P^{id} is the ideal–gas heat capacity and R is the universal gas constant. For a mixture, λ''_{mix} is estimated from the empirical mixing and combining rules (Ely & Hanley 1983)

$$\lambda''_{mix} = \sum_{i=1}^{n} \sum_{j=1}^{n} x_i x_j \lambda''_{ij} \tag{12.9}$$

$$\lambda''_{ij} = \frac{2\lambda''_i \lambda''_j}{\lambda''_i + \lambda''_j} \tag{12.10}$$

12.3 Extended corresponding states

The range of applicability of corresponding states in general, and of equations (12.1) and (12.6) in particular, can be broadened considerably by introducing the concept of extended corresponding states (Rowlinson & Watson 1969; Leland & Chappelear 1968). In extended corresponding states, the two–parameter corresponding–states formalism is maintained, except that the equivalent substance reducing ratios f_x and h_x become

$$f_x = \frac{T_{c,x}}{T_{c,0}} \theta_x \left(\rho_{r,x}, T_{r,x}, \omega_x \right) \tag{12.11}$$

$$h_x = \frac{\rho_{c,0}}{\rho_{c,x}} \phi_x \left(\rho_{r,x}, T_{r,x}, \omega_x \right) \tag{12.12}$$

where θ_x and ϕ_x are shape factors which are functions of the reduced temperature $T_r = T/T_c$, reduced density $\rho_r = \rho/\rho_c$ and the acentric factor ω. In principle, the shape factors may be determined exactly (Ely 1990) for a pure fluid with respect to a reference fluid by a simultaneous solution of the conformal equations (Rowlinson & Watson 1969)

$$A_x^{conf}(V, T) = f_x A_0^{conf}(V/h_x, T/f_x) - RT \ln(h_x) \tag{12.13}$$

$$Z_x(V, T) = Z_0(V_x/h_x, T_x/f_x) \tag{12.14}$$

where A is the dimensionless Helmholtz energy, $Z = P/\rho RT$ is the compressibility factor and the superscript 'conf' indicates the configurational contribution.

12.3.1 Mixing rules

Representation of mixtures is based on the one-fluid corresponding–states principle. Two steps are required to represent a mixture: the mixture is first characterized as a hypothetical pure fluid and then the resulting hypothetical pure fluid is related to the reference fluid using equations (12.1) and (12.6). The mixing rules are

$$h_x = \sum_{i=1}^{n} \sum_{j=1}^{n} x_i x_j h_{ij} \tag{12.15}$$

and

$$f_x h_x = \sum_{i=1}^{n} \sum_{j=1}^{n} x_i x_j f_{ij} h_{ij} \tag{12.16}$$

where

$$f_{ij} = \sqrt{f_i f_j} \left(1 - k_{ij} \right) \tag{12.17}$$

$$h_{ij} = \left(h_i^{1/3} + h_j^{1/3} \right)^3 \left(\frac{1 - l_{ij}}{8} \right) \tag{12.18}$$

In these equations x_i is the concentration of component i in the mixture, n is the total number of components in the mixture, f_x and h_x are the equivalent substance reducing ratios for the mixture and k_{ij} and l_{ij} are the binary interaction parameters that can be nonzero when $i \neq j$. For transport properties, additional mixing formulas are used for the mass. For the viscosity the expression reads (Ely & Hanley 1981b)

$$M_{x,\eta}^{1/2} = \left[\sum_{i=1}^{n} \sum_{j=1}^{n} x_i x_j M_{ij,\eta}^{1/2} f_{ij}^{1/2} h_{ij}^{4/3} \right] f_x^{-1/2} h_x^{-4/3} \tag{12.19}$$

$$M_{ij,\eta} = \frac{2 M_i M_j}{(M_i + M_j)} \tag{12.20}$$

and for the thermal conductivity (Ely & Hanley 1983)

$$M_{x,\lambda}^{-1/2} = \left[\sum_{i=1}^{n} \sum_{j=1}^{n} x_i x_j M_{ij,\lambda}^{-1/2} f_{ij}^{1/2} h_{ij}^{-4/3} \right] f_x^{-1/2} h_x^{4/3} \tag{12.21}$$

$$M_{ij,\lambda}^{-1} = \frac{1}{2} \left(M_i^{-1} + M_j^{-1} \right) \tag{12.22}$$

12.3.2 Implementation

In order to apply the method, the following are required:

(i) an equation of state for the reference fluid,
(ii) correlations for the viscosity and thermal conductivity of the reference fluid and
(iii) the critical parameters, acentric factor and molecular mass of the fluid of interest, or for each component of the mixture of interest.

The number of components in the mixture can be unlimited. Furthermore, if the necessary critical parameters and acentric factors are unknown, it is straightforward to incorporate a procedure to estimate them from the normal boiling point and specific gravity (Baltatu 1982; Baltatu *et al.* 1985). A structural method such as that of Lydersen (Reid *et al.* 1987) could also be used to estimate the critical properties needed in TRAPP. The early work on TRAPP used a 32–term modified Benedict–Webb–Rubin equation of state for methane (Ely & Hanley 1981a) for the reference fluid. Shape factors were found from a generalized expression, developed for pure normal paraffins C_1-C_{15} (Leach *et al.* 1968; Fisher & Leland 1970). The results were generally within 10% for viscosity and thermal conductivity of hydrocarbon mixtures (Ely & Hanley 1981a,b, 1983), although the viscosity predictions of mixtures containing cycloalkanes or highly branched alkanes were not always satisfactory. The reader is referred to the original papers (Ely & Hanley 1981a,b, 1983) for a complete description of the original TRAPP method.

12.3.3 Modifications to TRAPP

The current version of TRAPP has the following features:

(i) The reference fluid equation of state is the 32–term MBWR equation given by Younglove & Ely (1987) for propane (which alleviates problems encountered earlier when using methane at very low reduced temperatures).

(ii) The transport correlations for the propane reference fluid are those of Younglove & Ely (1987).

(iii) A simpler, density–independent correlation for shape factors using the propane reference fluid was developed by Erickson & Ely (1993) and is incorporated into TRAPP,

$$\theta_i \left(T_{r,i}, \rho_{r,i}, \omega_i \right) = 1 + (\omega_i - \omega_0) (\alpha_1 + \alpha_2 \ln(T_{r,i})) \qquad (12.23)$$

$$\phi_i \left(T_{r,i}, \rho_{r,i}, \omega_i \right) = \left(\frac{Z_{0,c}}{Z_{i,c}} \right) [1 + (\omega_i - \omega_0) (\beta_1 + \beta_2 \ln(T_{r,i}))] \qquad (12.24)$$

where the values of the coefficients are: $\alpha_1 = 0.5202976$, $\alpha_2 = -0.7498189$, $\beta_1 = 0.1435971$ and $\beta_2 = -0.2821562$.

(iv) An Enskog correction $\Delta\eta^{\text{Enskog}}$ has been included to account for size difference effects in the prediction procedure for mixture viscosity (Ely 1981). Equation (12.1) becomes

$$\eta_{\text{mix}} \left(\rho, T, x_\alpha \right) = \Delta\eta^{\text{Enskog}} + \eta_0 \left(\rho_0, T_0 \right) F_\eta \qquad (12.25).$$

(v) A thermal conductivity correction factor also improved the results. Equation (12.6) was modified to

$$\lambda'_x \left(\rho, T \right) = \lambda'_0 \left(\rho_0, T_0 \right) F_\lambda X_\lambda \qquad (12.26)$$

where

$$X_\lambda = \frac{1 + 2.186634 \left(\omega_j - \omega_0 \right)}{1 - 0.5050059 \left(\omega_j - \omega_0 \right)} \qquad (12.27)$$

The correlation for X_λ was developed by Ely (1990) using data from hydrocarbon mixtures.

(vi) For hydrocarbon mixtures the binary interaction parameters k_{ij} were determined using a generalization presented by Nishiumi *et al.* (1988), and l_{ij} set equal to 1. In practice, k_{ij} can also be set to 1 without causing significant errors.

12.4 Semiempirical modification: SUPERTRAPP

As will be shown in the next section, the modified TRAPP is very successful for the majority of fluids tested; the exception is that the viscosity prediction for cyclic compounds and some highly branched alkanes is not satisfactory. To correct this and slightly improve the overall representation of the data, a further modification is proposed. It is semiempirical and requires thermophysical property data as input. If these data are not available, however, the modified procedure will revert to TRAPP. The modifications are implemented in a computer program SUPERTRAPP (Ely & Huber 1990) and are as follows:

(i) The equilibrium shape factors θ and ϕ are determined by incorporating either information about the $P - V - T$ surface (Ely 1990) or the saturation boundary (Huber & Ely 1993) of the fluid of interest. The saturation boundary matching technique uses

$$P_i^{\text{sat}}(T) = P_0^{\text{sat}}\left(\frac{T}{f_i}\right)\frac{f_i}{h_i} \qquad (12.28)$$

$$\rho_i^{\text{sat}}(T) = \rho_0^{\text{sat}}\left(\frac{T}{f_i}\right)h_i \qquad (12.29)$$

A correlation for the vapor pressure and saturated liquid density is used with equations (12.28) and (12.29) to solve numerically for f and h.

(ii) In order to improve the viscosity prediction for cycloalkanes and highly branched alkanes, the concept of mass shape factors is introduced (Ely & Magee 1989)

$$F_\eta = f_x^{1/2}h_x^{-2/3}g_x^{1/2} \qquad (12.30)$$

where the additional factor g is defined by

$$g_i^{1/2} = \frac{\eta_i(\rho_i, T_i^*) - \eta_i^*(T_i)}{[\eta_0(\rho_0, T_0) - \eta_i^*(T_0)_i]f_i^{1/2}h_i^{-2/3}} \qquad (12.31)$$

$$g_{ij} = \frac{2}{(1/g_i + 1/g_j)} \qquad (12.32)$$

$$f_x^{1/2}h_x^{-4/3}g_x^{1/2} = \sum\sum x_ix_j f_{ij}^{1/2}h_{ij}^{-4/3}g_{ij}^{1/2} \qquad (12.33)$$

where η_i is the viscosity of pure component i at the indicated state point. When viscosity data are unavailable, the program sets the mass shape factor g in equation (12.30) to 1 and the SUPERTRAPP model defaults to the TRAPP model.

(iii) The procedure to evaluate thermal conductivity is unaltered, except that the equilibrium shape factors are those computed from equations (12.28) and (12.29) rather than equations (12.23) and (12.24).

12.5 Results

A full discussion on the original TRAPP equations plus extensive comparisons of the method with pure fluid and mixture data were given in the papers of Ely & Hanley (1981a,b, 1983) and will not be repeated here. Table 12.1 gives a comparison of experimental viscosity data for selected pure fluids with the modified TRAPP model and the SUPERTRAPP model, in terms of the average absolute percent deviation AAD and the average percentage error BIAS. For all fluids except the cyclic compounds and

Table 12.1. *Sample results for pure fluid viscosity.*

System & References	N	T /K (ρ/mol L^{-1})	TRAPP		SUPERTRAPP	
			AAD	BIAS	AAD	BIAS
ethane: Eakin *et al.* 1962; Abe *et al.* 1978; Diller & Saber 1981; Iwasaki & Takahashi 1981.	447	95-511 (0.02-22)	5.0	4.2	2.0	0.9
butane: Sage *et al.* 1939; Dolan *et al.* 1963; Abe *et al.* 1978; Diller & Van Poolen 1985.	268	130-468 (0.03-13)	2.3	-0.1	2.9	-0.9
isobutane: Sage *et al.* 1939; Giller & Drickamer 1949; Diller & Van Poolen 1985.	249	115-444 (0.03-13)	24	-23	3.8	-0.1
ethene: Kestin *et al.* 1977; Golubev 1970.	70	297-477 (0.03-18)	3.0	0.2	2.9	-0.4
cyclohexane: Geist & Cannon 1946; Jonas *et al.* 1980; Dymond & Young 1980.	35	283-393 (8.1-10)	49	-49	12	-7.7
benzene: Geist & Cannon 1946; Heiks & Orban 1956; Medani & Hasan 1977.	22	293-523 (7.2-11)	5.1	-2.4	1.3	0.1

Table 12.2. *Sample results for dense fluid mixture viscosity.*

System & References	N	T/K (ρ/mol L^{-1})	TRAPP AAD	TRAPP BIAS	SUPERTRAPP AAD	SUPERTRAPP BIAS
methane + propane: Giddings *et al.* 1966; Huang *et al.* 1967.	416	123-411 (0.03-24)	3.0	-0.2	3.4	-2.3
octane + hexadecane Wakefield *et al.* 1987.	21	318-338 (3.3-6)	1.9	1.9	1.2	-0.7
methane + nitrogen: Diller 1982.	298	100-300 (0.69-29)	5.0	3.3	5.0	-0.05
toluene + heptane: Mussche & Verhoeye 1975.	21	293-298 (6.8-9.4)	4.6	4.6	1.8	1.3
cyclohexane + hexane: Ridgeway & Butler 1967; Isdale *et al.* 1979; Wei & Rowley 1984.	127	298-373 (0.07-10)	24	-23	8.0	-2.7
hexane + octane + hexadecane: Dymond & Young 1980.	10	288-378 (3.6-5.2)	1.9	1.7	2.0	-0.7

highly branched alkanes, the viscosity prediction of the TRAPP model is good, within about 5%. The SUPERTRAPP model's use of the 'mass shape factors' described in equations (12.30)–(12.33) improves the estimates for these fluids. Table 12.2 gives sample results for the viscosity prediction of some dense fluid mixtures. Calculations for both models were made using binary interaction parameters obtained from the generalized method of Nishiumi *et al.* (1988). Again, except for mixtures containing cyclocompounds or highly branched alkanes, the TRAPP model is within about 5% of the experimental values. Table 12.3 presents corresponding comparisons for the thermal conductivity of pure fluids. The TRAPP model gives estimates within 10%, generally within the experimental error. Results for mixtures are given in Table 12.4. For mixtures, an assignment of 5–20% on the accuracy of the mixture data is reasonable, and the model gives estimates within this range. Other improvements to the basic extended corresponding–states method continue to be made. Monnery *et al.* (1991) developed new methods for obtaining shape factors. Hwang & Whiting (1987) extended the model to polar fluids using an association parameter and a viscosity acentric factor. The model has also been adapted for polar and nonpolar refrigerants and their mixtures (Huber & Ely 1992; Huber *et al.* 1992), high–temperature aqueous solutions (Levelt–Sengers & Gallagher 1990), coal liquids and petroleum fractions (Baltatu 1982; Baltatu *et al.* 1985).

Table 12.3. *Sample results for pure fluid thermal conductivity.*

System & References	N	T /K (ρ/mol L^{-1})	TRAPP AAD	BIAS
ethane: Keyes 1954; Prasad *et al.* 1984; Roder 1985; Leng & Comings 1957; Carmichael *et al.* 1963.	1047	112-600 (0.00-21.7)	9.0	-5.7
butane: Carmichael & Sage 1964; Kandiyoti *et al.* 1972; Nieto de Castro *et al.* 1983.	259	148-601 (0.03-12.5)	8.9	-7.4
isobutane: Nieuwoudt *et al. 1987*.	247	293-642 (0.03-11.3)	6.9	-3.3
ethene: Keyes 1954; Kolomiets 1974.	171	180-480 (0.00-21.1)	10.1	-6.0
cyclohexane: Li *et al.* 1984.	35	309-361 (8.9-9.8)	3.9	3.9
benzene: Horrocks & McLaughlin 1963; Venart 1965; Li *et al.* 1984.	77	273-361 (10.2-12.2)	1.8	1.0

Table 12.4. *Sample results for dense fluid mixture thermal conductivity.*

System & References	N	T /K	TRAPP	
methane + butane: Carmichael *et al.* 1968.	24	278-378 (7.8-13.3)	4.6	3.3
octane + heptadecane: Mukhamedzyanov *et al.* 1964.	6	300-392 (3.6-5.1)	4.4	-4.4
toluene + heptane: Parkinson 1974.	8	273-273 (6.9-9.4)	4.2	4.2
cyclopentane + heptane: Parkinson 1974.	8	273-273 (6.9-11.0)	1.0	-0.4
2,2,4-trimethylpentane + benzene + cyclohexane: Rowley & Gubler 1988; Rowley *et al.* 1988.	28	298-313 (6.0-11.0)	4.3	4.3

12.6 Conclusions

An extended corresponding–states procedure to estimate the viscosity and thermal conductivity of pure fluids and fluid mixtures has been discussed. The core of the method is the procedure TRAPP. It has been shown that, excepting mixtures containing cycloalkanes and highly branched alkanes, the viscosity and thermal conductivity are satisfactory, generally within 5%. The significant advantages of this model are that it is predictive and that the number of mixture components is unlimited in principle. For many engineering applications, however, only a correlation or interpolation of the transport coefficients is required. In this case, the requirement that TRAPP should be strictly predictive can be relaxed, and sensible use of available thermophysical data can be made, resulting in the procedure called SUPERTRAPP. SUPERTRAPP is recommended for the estimation of viscosity and thermal conductivity of hydrocarbons and their mixtures when a semiempirical model is required, whereas TRAPP is recommended when an entirely predictive model is acceptable.

Acknowledgments

The support of the Division of Chemical Sciences, Office of Basic Energy Sciences, Office of Energy Research, U.S. Department of Energy and the Standard Reference Data Program of the National Institute of Standards and Technology is acknowledged.

References

Abe, Y., Kestin, J., Khalifa, H. E. & Wakeham, W. A. (1978). The viscosity and diffusion coefficients of the mixtures of four light hydrocarbon gases. *Physica*, **93A**, 155–170.

Baltatu, M. (1982). Prediction of the liquid viscosity for petroleum fractions. *Ind. Eng. Chem. Process Des. Dev.*, **21**, 192–195.

Baltatu, Monica E., Ely, James, F. Hanley, H. J. M., Graboski, M. S., Perkins, R. A. & Sloan, E. D. (1985). Thermal conductivity of coal–derived liquids and petroleum fractions. *Ind. Eng. Process Des. Dev.*, **24**, 325–332.

Carmichael, L. T., Berry, V. M. & Sage, B. H. (1963). Thermal conductivity of fluids. Ethane. *J. Chem. Eng. Data*, **8**, 281–285.

Carmichael, L. T., Jacobs, J. & Sage, B. H. (1968). Thermal conductivity of fluids. A mixture of methane and n-butane. *J. Chem. Eng. Data*, **13**, 489–495.

Carmichael, L. T. & Sage, B. H. (1964). Thermal conductivity of fluids. n-butane. *J. Chem. Eng. Data*, **9**, 511–515.

Diller, D. E. (1982). Measurements of the viscosity of compressed gaseous and liquid nitrogen + methane mixtures. *Int. J. Thermophys.*, **3**, 237–249.

Diller, D. E. & Saber, J. M. (1981). Measurements of the viscosity of compressed and liquid ethane. *Physica*, **108A**, 143–152.

Diller, D. E. & Van Poolen, L. J. (1985). Measurements of the viscosities of saturated and compressed liquid normal butane and isobutane. *Int. J. Thermophys.*, **6**, 43–62.

Dolan, J. P., Starling, K. E., Lee, A. L., Eakin, B. E. & Ellington, R. T. (1963). Liquid, gas and dense fluid viscosity of n-butane. *J. Chem. Eng. Data*, **8**, 396–399.

Dymond, J. H. & Young, K. J. (1980). Transport properties of nonelectrolyte liquid mixtures I. Viscosity coefficients for n-alkane mixtures at saturation pressure from 283 to 378 K. *Int. J. Thermophys.*, **1**, 331–344.

Eakin, B. E., Starling, K. E., Dolan, J. P. & Ellington, R. T. (1962). Liquid, gas and dense fluid viscosity of ethane. *J. Chem. Eng. Data*, **7**, 33–36.

Ely, J. F. (1981). An Enskog correction for size and mass difference effects in mixture viscosity prediction. *J. Res. Nat. Bur. Stand. (U.S.)*, **86**, 597–604.

Ely, J. F. (1990). A predictive, exact shape factor extended corresponding states model for mixtures. *Adv. Cryog. Eng.*, **35**, 1511–1520.

Ely, J. F. & Hanley, H. J. M. (1981a). A computer program for the prediction of viscosity and thermal conductivity in hydrocarbon mixtures. *Nat. Bur. Stand. (U.S.) Tech. Note 1039*.

Ely, James F. & Hanley, H. J. M. (1981b). Prediction of transport properties 1. Viscosity of fluids and mixtures. *Ind. Eng. Chem. Fundam.*, **20**, 323–332.

Ely, J. F. & Hanley, H. J. M. (1983). Prediction of transport properties. 2. Thermal conductivity of pure fluids and fluid mixtures. *Ind. Eng. Chem. Fund.*, **22**, 90–97.

Ely, J. F. & Huber, M. L. (1990). NIST Standard Reference Database 4, Computer program SUPERTRAPP, *NIST Thermophysical Properties of Hydrocarbon Mixtures*, Version 1.0.

Ely, J. F. & Magee, J. W. (1989). Experimental measurement and prediction of thermophysical property data of carbon dioxide rich mixtures. *Proceedings of the 68th GPA Annual Convention*, 89–98.

Erickson, D. & Ely, J. (1993). Personal communication.

Fisher, G. D. & Leland, T. W. (1970). Corresponding states principle using shape factors. *Ind. Eng. Chem. Fundam.*, **9**, 537–544.

Friend, D.G. (1992). NIST Standard Reference Database 14, *NIST Mixture Property Database*, Version 9.08 (DDMIX).

Geist, J. M. & Cannon, M. R. (1946). Viscosities of pure hydrocarbons. *Ind. Eng. Chem.*, **18**, 611–613.

Giddings, J. G., Kao, J. T. & Kobayashi, R. (1966). Development of a high-pressure capillary–tube viscometer and its application to methane, propane and their mixtures in the gaseous and liquid regions. *J. Chem. Phys.*, **45**, 578–586.

Giller, E. B. & Drickamer, H. G. (1949). Viscosity of normal paraffins near the freezing point. *Ind. Eng. Chem.*, **41**, 2067–2069.

Golubev, I. F. (1970). *Viscosity of gases and gas mixtures*. Israel Program for Scientific Translations Ltd.

Hanley, H. J. M. (1976). Prediction of the viscosity and thermal conductivity coefficients of mixtures. *Cryogenics*, **16**, 643–651.

Heiks, J. R. & Orban, E. (1956). Liquid viscosities at elevated temperatures and pressures: Viscosity of benzene from 90 K to its critical temperature. *J. Phys. Chem.*, **60**, 1025–1027.

Hirschfelder, J. O., Curtiss, C. F. & Bird, R. B. (1954). *Molecular Theory of Gases and Liquids*. New York: Wiley.

Horrocks, J. K. & McLaughlin, E. (1963). Non-steady-state measurements of the thermal conductivities of liquid polyphenols. *Proc. Roy. Soc., London*, **273A**, 259.

Huang, E. T. S., Swift, G. W. & Kurata, F. (1967). Viscosities and densities of methane-propane mixtures at low temperature and high pressures. *AIChE J.*, **13**, 846–850.

Huber, M. L. & Ely, J. F. (1992). Prediction of viscosity of refrigerants and refrigerant mixtures. *Fluid Phase Equil.*, **80**, 239–248.

Huber, M. L. & Ely, J. F. (1993). A predictive extended corresponding states model for pure and mixed refrigerants including an equation of state for R134a. *Int. J. Refrig.*, **7**, 18–31.

Huber, M. L., Ely, J. F. & Friend, D. G. (1992). Prediction of thermal conductivity of refrigerants and refrigerant mixtures. *Fluid Phase Equil.*, **80**, 249–261.

Hwang, M. J. & Whiting, W. B. (1987). A corresponding states treatment for the viscosity of polar fluids. *Ind. Eng. Chem. Res.*, **26**, 1758–1766.

Isdale, J. D., Dymond, J. D. & Brawn, T. A. (1979). Viscosity and density of n-hexane cyclohexane mixtures between 25 and 100°C up to 500 MPa. *High Temp.-High Press.*, **11**, 571–580.

Iwasaki, H. & Takahashi, M. (1981). Viscosity of carbon dioxide and ethane. *J. Chem. Phys.*, **74**, 1930–1943.

Jonas, J., Hasha, D. & Huang, S. G. (1980). Density effects on transport properties in liquid cyclohexane. *J. Phys. Chem.*, **84**, 109–112.

Kandiyoti, R., McLaughlin, E. & Pittman, J. F. T. (1972). Liquid state thermal conductivity of n-paraffin hydrocarbons. *J. Chem. Soc. Faraday Trans.*, **68**, 860–866.

Kestin, J., Khalifa, H. E. & Wakeham, W. A. (1977). The viscosity of five gaseous hydrocarbons. *J. Chem. Phys.*, **66**, 1132–1134.

Keyes, F. G. (1954). Thermal conductivity of gases. *Trans. ASME*, **75**, 809–816.

Kolomiets, A. Y. (1974). Experimental study of the thermal conductivity of ethylene. *Heat Transfer Soviet Research (USA)*, **6**, 42–48.

Leach, J. W., Chappelear, P. S. & Leland, T. W. (1968). Use of molecular shape factors in vapor-liquid equilibrium calculations with the corresponding states principle. *AIChE J.*, **14**, 568–576.

Leland, T. W. & Chappelear, P. S. (1968). The corresponding states principle. *Ind. Eng. Chem.*, **60**, 15–43.

Leng, D. E. & Comings, E. W. (1957). Thermal conductivity of propane. *Ind. Eng. Chem.*, **49**, 2042–2045.

Levelt Sengers, J. M. H. & Gallagher, J. S. (1990). Generalized corresponding states and high-temperature aqueous solutions. *J. Phys. Chem.*, **94**, 7913–7922.

Li, S. F. Y., Maitland, G. C. & Wakeham, W. A. (1984). Thermal conductivity of benzene and cyclohexane in the temperature range 36-90 °C at pressures up to 0.33 GPa. *Int. J. Thermophys.*, **5**, 351–365.

Medani, M. S. & Hasan, M. A. (1977). Viscosity of organic liquids at elevated temperatures and the corresponding vapour pressures. *Can. J. Chem. Eng.*, **55**, 203–209.

Mo, K.C. & Gubbins, K. E. (1974). Molecular principle of corresponding states for viscosity and thermal conductivity of fluid mixtures. *Chem. Eng. Commun.*, **1**, 281–290.

Mo, K.C. & Gubbins, K. E. (1976). Conformal solution theory for viscosity and thermal conductivity of mixtures. *Mol. Phys.*, **31**, 825–847.

Monnery, W. D., Mehrotra, A. K. & Svrcek, W. Y. (1991). Modified shape factors for improved viscosity predictions using corresponding states. *Can. J. Chem. Eng.*, **69**, 1213–1219.

Mukhamedzyanov, G. Kh., Usmanov, A. G. & Tarzimanov, A. A. (1964). *Izv. Vyssh. Ucheb. Zaved Neft i Gaz*, **7**, 70–74.

Mussche, M. J. & Verhoeye, L. A. (1975). Viscosity of ten binary and one ternary mixtures. *J. Chem. Eng. Data*, **20**, 46–50.

Nieto de Castro, C. A., Tufeu, R. & Le Neindre, B. (1983). Thermal conductivity measurement of n-butane over wide temperature and pressure ranges. *Int. J. Thermophys.*, **4**, 11–33.

Nieuwoudt, J. C., Le Neindre, B., Tufeu, R. & Sengers, J. V. (1987). Transport properties of isobutane. *J. Chem. Eng. Data*, **32**, 1–8.

Nishiumi, H., Arai, T. & Takeuchi, K. (1988). Generalization of the binary interaction parameters of the Peng-Robinson equation of state by component family. *Fluid Phase Equil.*, **42**, 43–62.

Parkinson, W. J. (1974). *Thermal conductivity of binary liquid mixtures*. Ph.D. Thesis, University of Southern California.

Prasad, R. C., Mani, N. & Venart, J. E. S. (1984). Thermal conductivity of methane. *Int. J. Thermophys.*, **5**, 265–279.

Reid, R. C., Prausnitz, J. M. & Poling, B. E. (1987). *The Properties of Gases and Liquids*, 4th Ed. New York: McGraw-Hill.

Ridgeway, K. & Butler, A. (1967). Some physical properties of the ternary system benzene-cyclohexane-n-hexane. *J. Chem. Eng. Data*, **12**, 509–517.

Roder, H. M. (1985). Thermal conductivity of ethane at temperatures between 110 and 325 K and pressures to 70 MPa. *High Temp.–High Press.*, **17**, 453–460.

Rowley, R. L. & Gubler, V. (1988). Thermal conductivities in seven ternary liquid mixtures at 40°C and 1 atm. *J. Chem. Eng. Data*, **33**, 5–8.

Rowley, R. L., White, G. L. & Chiu, M. (1988). Ternary liquid mixture thermal conductivities. *Chem. Eng. Sci.*, **43**, 361–371.

Rowlinson, J. S. & Watson, I. D. (1969). The prediction of the thermodynamic properties of fluids and fluid mixtures I. The principle of corresponding states and its extensions. *Chem. Eng. Sci.*, **24**, 1565–1574.

Sage, B. H., Yale, W. D. & Lacey, W. N. (1939). Effect of pressure on viscosity of n-butane and isobutane. *Ind. Eng. Chem.*, **31**, 223–226.

Venart, J. E. S. (1965). Liquid thermal conductivity measurements. *J. Chem. Eng. Data*, **10**, 239–241.

Wakefield, D. L. & Marsh, K. N. (1987). Viscosities of nonelectrolyte liquid mixtures. I. *n*-hexadecane and *n*-octane. *Int. J. Thermophysics*, **8**, 649–662.

Wei, I-Chein & Rowley, R. L. (1984). Binary liquid mixture viscosities and densities. *J. Chem. Eng. Data*, **29**, 332–335.

Younglove, B. A. & Ely, J. F. (1987). Thermophysical properties of fluids II. Methane, ethane, propane, isobutane and normal butane. *J. Phys. Chem. Ref. Data*, **16**, 577–798.

13

Empirical Estimation

B. E. POLING

University of Toledo, Toledo, Ohio, USA

13.1 Introduction

Numerous techniques to estimate transport properties have been published. In this book, there are separate sections that address empirical estimation and corresponding–states methods. Sometimes a method clearly falls into one of these categories, but often the distinction between these categories is not clear–cut, and, in fact, many methods possess both corresponding–states and empirical features. Empirical correlations that lead to tabulations of constants for individual compounds are not presented in this chapter. Rather, those methods that would be considered empirical estimation methods are summarized. Methods to estimate viscosity and thermal conductivities of pure materials are presented first, followed by a discussion of diffusion coefficients. Finally, methods to estimate the viscosity and thermal conductivity of mixtures are discussed. Within each of these five categories, empirical estimation techniques find their greatest application for low–pressure gases and gaseous mixtures. The dense fluid region is generally described by corresponding–states methods. A complete listing of all methods is not given; for more detail, the reader is referred to Reid *et al.* (1987). There, examples are presented that frequently make the methods easier to use than do the original papers. Methods that have appeared since 1987 are also summarized in this chapter.

13.2 Viscosity of pure fluids

13.2.1 Viscosity of pure gases

Three methods are presented by Reid *et al.* (1987) to estimate the viscosity of a pure gas at low pressure; all require the critical temperature and dipole moment. Two of these, the method of Lucas (1980, 1984) and that of Chung *et al.* (1984, 1988), are corresponding–states methods, while for the third, Reichenberg's method (Reichenberg 1971, 1973,

Table 13.1. *Values of the group contributions C_i for the estimation of a^* in equation (13.1).*

Group		Contribution C_i	Group		Contribution C_i
$-CH_3$		9.04	$-Cl$		10.06
$>CH_2$	(nonring)	6.47	$-Br$		12.83
$>CH-$	(nonring)	2.67	$-OH$	(alkanols)	7.96
$>C<$	(nonring)	-1.53	$>O$	(nonring)	3.59
$=CH_2$		7.68	$>C=O$	(nonring)	12.02
$=CH-$	(nonring)	5.53	HCOO	(formates)	13.41
$>C=$	(nonring)	1.78	$-CHO$	(aldehydes)	14.02
$\equiv CH$		7.41	$-COOH$	(acids)	18.65
$\equiv C-$	(nonring)	5.24	$-COO-$	(esters)	13.41
$>CH_2$	(ring)	6.91	$-NH_2$		9.71
$>CH-$	(ring)	1.16	$>NH$	(nonring)	3.68
$>C<$	(ring)	0.23	$=N-$	(ring)	4.97
$=CH-$	(ring)	5.90	$-CN$		18.13
$>C=$	(ring)	3.59	$>S$	(ring)	8.86
$-F$		4.46			

1975, 1979), the following equation is used in order to calculate the viscosity, $\eta^{(0)}$, for organic compounds

$$\eta^{(0)} = \frac{M^{1/2}T}{a^*\left[1 + (4/T_c)\right]\left[1 + 0.36 T_r(T_r - 1)\right]^{1/6}} \frac{T_r\left(1 + 270\mu_r^4\right)}{T_r + 270\mu_r^4} \qquad (13.1)$$

where $\eta^{(0)}$ is in μP,[*] T is the temperature in K, T_r is the reduced temperature, T_c is the critical temperaure, M is the molar mass and μ_r is the reduced dipole moment defined by

$$\mu_r = 52.46\mu^2 P_c/T_c^2 \qquad (13.2)$$

Here, P_c is the critical pressure in bar, μ is the dipole moment in D, and in equation (13.1), a^* is determined by

$$a^* = \sum_i N_i C_i \qquad (13.3)$$

where N_i represents the number of groups of the ith type and C_i is the group contibution shown in Table 13.1.

It should be noted that the estimation of low–pressure gas viscosities is particularly important to the topic of transport property estimation because the low–pressure gas viscosity is often used in correlations for the other transport properties. Reichenberg's

---

[*] The equations analyzed in this chapter were derived in the context of a certain system of units. In order to use the tabulations and coefficients in the original papers, it is necessary to keep the original units, which sometimes are not SI units.

method, as well as the other methods presented by Reid *et al.* (1987), may be used with the expectation of errors of 0.5 to 1.5% for nonpolar compounds and 2 to 4% for polar compounds.

The same methods have been extended to the dense fluid region by corresponding–states approaches. Errors of 5 to 10% are typical for estimations in the dense fluid region.

13.2.2 *Viscosity of pure liquids*

Values for coefficients in the equation

$$\ln \eta = A + B/T + CT + D/T^2 \tag{13.4}$$

for calculation of liquid viscosity are listed by Reid *et al.* (1987) for approximately 375 different compounds. This equation should be used in preference to an empirical estimation technique if the coefficients for compounds of interest are available. If an estimation method must be used, empirical methods are generally preferred for reduced temperatures below about 0.7, while for higher reduced temperatures, corresponding–states methods similar to those used for dense fluids are often more reliable. Empirical methods that have been described by Reid *et al.* (1987) and evaluated by Joback & Reid (1987) include the method of Orrick and Erbar and the method of van Velzen (1972a,b). In the Orrick–Erbar method

$$\ln \frac{\eta}{\rho M} = A + \frac{B}{T} \tag{13.5}$$

where η is the viscosity in cP, ρ is the density at 20°C in g cm^{-3}, M is the molar mass and T is the temperature in K. The group contributions for determining A and B are given in Table 13.2. For liquids that have a normal boiling point below 20°C, the density at the normal boiling point is to be used, whereas for liquids that have a melting point above 20°C, the density at the melting point temperature must be applied. The method cannot be used for compounds that contain nitrogen or sulfur. Testing for 188 organic liquids gave an average error of 15%.

Several estimation methods have been published since 1987. A method to estimate liquid viscosities over a temperature range from the freezing point up to reduced temperatures of 0.95 has been presented by Sastri and Rao (1992). It is based on the supposition that the liquid viscosity is related to the vapor pressure, P_s, by

$$\eta = \eta_b P_s^{-N} \tag{13.6}$$

where η_b is the viscosity at the normal boiling point. Equations that can be used to estimate the vapor pressure and a group contribution method that can be applied to estimate η_b and N were given by Sastri & Rao (1992). The method appears to give accuracies comparable to methods summarized by Reid *et al.* (1987), and is applicable

Table 13.2. *Orrick & Erbar group contributions for A and B in equation (13.4).*

Group	A	B
Carbon atoms[a]	-(6.95+0.21n)	275+99n
Carbon atoms in $-CR_3$	-0.15	35
Carbon atoms in CR_4	-1.20	400
Double bond	0.24	-90
Five–membered ring	0.10	32
Six–membered ring	-0.45	250
Aromatic ring	0.0	20
Ortho substitution	-0.12	100
Meta substitution	0.05	-34
Para substitution	-0.01	-5
Chlorine	-0.61	220
Bromine	-1.25	365
Iodine	-1.75	400
$-OH$	-3.00	1600
$-COO-$	-1.00	420
$-O-$	-0.38	140
$>C=O$	-0.50	350
$-COOH$	-0.90	770

[a] n = number of carbon atoms, not including those in groups shown below.

over an increased temperature range, but the complexity and ambiguity of the group contribution method is a disadvantage.

Other methods that recently have been introduced for specific classes of compounds include those for the estimation of the liquid viscosity of hydrocarbons (Allan & Teja 1991; Mehrotra 1991a,b) and refrigerants (Dutt & Venugopal 1990; Huber & Ely 1992).

13.3 Thermal conductivity of pure fluids

13.3.1 Thermal conductivity of pure gases

Six methods are recommended by Reid *et al.* (1987) for the estimation of thermal conductivity of nonpolar compounds. These include the corresponding–states methods of Chung *et al.* (1984, 1988), Ely & Hanley (1983) and Hanley (1976). The third method (Roy & Thodos 1968, 1970) is recommended for polar as well as nonpolar compounds. Typical errors for nonpolar compounds are 5 to 7%, with higher errors expected for polar compounds. Both the Chung and Ely–Hanley methods correlate the Eucken factor, $f_E = \lambda M/\eta C_v$, with other variables such as C_v (heat capacity), T_r and ω (acentric factor). Thus the viscosity is required to use these correlations. The Roy–Thodos correlation requires only the critical temperature and the pressure, and employs a group contribution method to account for the effect of internal degrees of freedom.

The group contribution method is fairly involved, and the reader is referred to Reid *et al.* (1987) for further details.

For the dense fluid region, corresponding–states methods are normally used and, in fact, two of three methods described by Reid *et al.* (1987) are merely extensions of corresponding–states approaches according to Chung *et al.* (1984, 1988) and Ely & Hanley (1983). For all these methods the density, rather than the pressure, is used as the system variable, and even for nonpolar molecules errors can be large.

13.3.2 Thermal conductivity of pure liquids

Several methods for the estimation of liquid thermal conductivities are presented by Reid *et al.* (1987), namely, the method of Latini and co–workers (Latini & Pacetti 1978; Barroncini *et al.* 1980, 1981, 1983a,b, 1984), the Sato–Riedel method and the Missenard–Riedel method (Riedel 1949, 1951a,b,c; Missenard 1973). The Latini method employs different parameters for compounds from different families, such as acids, esters, etc., while the other two methods are simply equations that give the thermal conductivity as a function of the molar mass and the reduced temperature. For example, the Sato–Riedel equation reads

$$\lambda = \frac{\left(1.11/M^{1/2}\right)\left[3 + 20\left(1 - T_r\right)^{2/3}\right]}{3 + 20\left(1 - T_{r,b}\right)^{2/3}} \tag{13.7}$$

where $T_{r,b}$ is the reduced normal boiling point temperature, and λ is the thermal conductivity in W m^{-1}K^{-1}.

More recently, the following equation for λ was proposed by Lakshmi & Prasad (1992):

$$\lambda = 0.0655 - 0.0005T + \frac{1.3855 - 0.00197T}{M^{1/2}} \tag{13.8}$$

where T is in K, and λ is in W m^{-1}K^{-1}. Errors for all four of the above mentioned methods are often less than 5% but sometimes as high as 20%. Reid *et al.* (1987) and Lakshmi & Prasad (1992) suggest that the Latini and Sato–Riedel methods are more accurate than the Missenard–Riedel method or equation (13.8), although the latter requires less information.

In addition to the four methods mentioned in the previous paragraph, values of coefficients A, B, C for approximately 75 different compounds are listed in Reid *et al.* (1987) for the calculation of liquid thermal conductivity from the equation

$$\lambda = A + BT + CT^2 \tag{13.9}$$

This equation should be used in preference to an empirical estimation technique if coefficients are available for the compound of interest.

Finally, Myers & Danner (1993) have developed equations to estimate liquid thermal conductivities of silicon compounds. Another approach to estimating liquid thermal conductivities is to use one of the dense fluid corresponding–states correlations. Ely & Hanley (1983), for example, claim a maximum error of 15% over the entire liquid range. However, the accuracy for nonhydrocarbons is in doubt. Dymond, Assael and co–workers (see Chapters 5 and 10) have developed predictive schemes based on kinetic theory and an 'effective core volume' approach and tested their equation for hydrocarbons, alcohols and other compounds.

None of the methods mentioned in this section allows for anomalies in the critical region.

13.4 Diffusion coefficients

13.4.1 Diffusion coefficients for binary gas mixtures at low pressures: Empirical correlations

In this discussion, D_{AB} represents the diffusion coefficient for solute A in solvent B. For ideal gases, however, D_{AB} is independent of concentration, so that $D_{AB} = D_{BA}$. Two empirical methods to estimate D_{AB} for binary gas systems at low pressures are described by Reid *et al.* (1987). These are the Wilke–Lee method (Wilke & Lee 1955) and the Fuller method (Fuller *et al.* 1965, 1966, 1969). The Wilke–Lee method requires only the liquid volume of components A and B at their normal boiling points, the molar mass of A and B, temperature and pressure.

In the Fuller method

$$D_{AB} = \frac{0.00143T^{7/4}}{P\,(M_{AB})^{1/2}\left[(\Sigma_V)_A^{1/3} + (\Sigma_V)_B^{1/3}\right]^2} \tag{13.10}$$

where D_{AB} is the diffusion coefficient in cm^2s^{-1}, P is the pressure in bar, T is the temperature in K, Σ_V is determined by summing the atomic contributions shown in Table 13.3 and M_{AB} is given by

$$M_{AB} = 2\left[1/M_A + 1/M_B\right]^{-1} \tag{13.11}$$

Both the Wilke–Lee and Fuller methods are reliable, often giving errors of only a few percent; occasionally, however, they can give errors as high as 10 to 20%.

13.4.2 Estimation of binary liquid diffusion coefficients at infinite dilution

In liquids, D_{AB} depends on concentration, and D_{AB} is not equal to D_{BA}. Thus, D_{AB}^{∞} is introduced, to represent the value of D_{AB} in an A–B binary when A is infinitely dilute. In practice, the value of D_{AB}^{∞} is typically used for concentrations of *A* up to 5 to 10%; at higher concentrations, however, the effect of concentration must be taken into account, as discussed briefly below.

Four empirical methods are given by Reid *et al.* (1987) to estimate D_{AB}^{∞} in liquid mixtures. These are the methods of Wilke & Chang (1955) and Tyn & Calus (1975),

Table 13.3. *Atomic diffusion contributions for $\sum_V$ in equation (13.10).*

Atomic and structural diffusion volume increments			
C	15.9	F	14.7
H	2.31	Cl	21.0
O	6.11	Br	21.9
N	4.54	I	29.8
Aromatic ring	-18.3	S	22.9
Heterocyclic ring	-18.3		

Diffusion volumes of simple molecules			
He	2.67	CO	18.0
Ne	5.98	CO_2	26.9
Ar	16.2	N_2O	35.9
Kr	24.5	NH_3	20.7
Xe	32.7	H_2O	13.1
H_2	6.12	SF_6	71.3
D_2	6.84	Cl_2	38.4
N_2	18.5	Br_2	69.0
O_2	16.3	SO_2	41.8
Air	19.7		

and the correlations of Hayduk & Minhas (1982) and Nakanishi (1978). Comparisons of the predictions of all these methods with experimental data show wide fluctuations in the relative errors. These 'failures' may be due to inadequacies in the correlation or to poor data, but in general, the Hayduk–Minhas and Tyn–Calus correlations yield the lowest errors. The latter is given by

$$D_{AB}^\infty = 8.93 \cdot 10^{-8} \left(\frac{V_A}{(V_B)^2} \right)^{1/6} \left(\frac{\hat{P}_B}{\hat{P}_A} \right)^{3/5} \frac{T}{\eta_B} \tag{13.12}$$

where D_{AB}^∞ is in $cm^2 s^{-1}$, η_B is the viscosity in cP, V_B is the liquid molar volume of the solvent at the normal boiling temperature in $cm^3 mol^{-1}$ and $\hat{P}$ is the parachor and is related to the surface tension, σ, by

$$\hat{P} = V \sigma^{1/4} \tag{13.13}$$

where V, the liquid volume in $cm^3 mol^{-1}$, and σ, in dyn cm^{-1}, are measured at the same temperature. The unit of $\hat{P}$ is $cm^3 g^{1/4} s^{-1/2}$. Values of $\hat{P}$ for a number of chemicals were tabulated by Quale (1953), or $\hat{P}$ may be estimated from the group contribution method detailed in Reid *et al.* (1987). The Hayduk–Minhas correlation gives three different equations depending on whether the system is a normal paraffin solution, an aqueous solution, or a nonaqueous solution. All three equations require the viscosity of the solution, and the latter equation requires the parachor of A and B, as does equation (13.12).

It is recommended by Reid *et al.* (1987) that the effect of concentration of D_{AB} in binary mixtures should be accounted for by the Vignes equation (Vignes 1966)

$$D_{AB} = \alpha \left[(D_{AB}^{\infty})^{x_B} (D_{BA}^{\infty})^{x_A} \right] \tag{13.14}$$

where $\alpha = (\partial \ln a / \partial \ln x)_{T,P}$ and a is the activity, x_i is the mole fraction; because of the Gibbs–Duhem equation, the derivative is the same whether written for A or B. The application of equation (13.14) to associating systems is discussed by McKeigue & Gulari (1989), whereas a method to extend (13.14) to multicomponent mixtures that does not require any adjustable parameters was proposed by Kooijman & Taylor (1991).

13.5 Viscosity of fluid mixtures

13.5.1 Low–pressure gases and dense fluids

Three empirical techniques are presented by Reid *et al.* (1987) to estimate the viscosity of a mixture of gases at low pressure. These are Wilke's method (Wilke 1950), the method of Herning & Zipperer (1936) and that of Reichenberg (1974, 1975, 1977). All three methods require values of the pure component viscosities and are fairly accurate for nonpolar molecules, with expected errors of less than 2 to 3%. Reichenberg's method includes the dipole moment and would be expected to be more successful for mixtures involving polar molecules. This claim is supported by comparisons presented by Reid *et al.* (1987).

Recently, Davidson (1993) presented a method for the estimation of low–pressure viscosities of mixtures that is based on the use of momentum fraction rather than mole fraction. The momentum fraction, m_i, of component i is the fraction of the total momentum within a mixture that is associated with a particular component and is calculated by

$$m_i = \frac{x_i M_i^{1/2}}{\sum x_j M_j^{1/2}} \tag{13.15}$$

where x_i is the mole fraction and M_i is the molar mass of component i. The mixture viscosity is then calculated by means of the equation

$$\frac{1}{\eta^{(0)}} = \sum_j \sum_i \frac{m_i m_j}{\left(\eta_i^{(0)} \eta_j^{(0)} \right)^{1/2}} E_{ij}^{1/3} \tag{13.16}$$

The quantity, E_{ij}, represents an efficiency with which momentum is transferred, and is defined as

$$E_{ij} = \frac{2 (M_i M_j)^{1/2}}{(M_i + M_j)} \tag{13.17}$$

For a binary mixture equation (13.16) leads to

$$\frac{1}{\eta^{(0)}} = \frac{m_1^2}{\eta_1^{(0)}} + 2\frac{m_1 m_2}{\left(\eta_1^{(0)}\eta_2^{(0)}\right)^{1/2}}E_{12}^{1/3} + \frac{m_2^2}{\eta_2^{(0)}} \qquad (13.18)$$

This was tested for low molecular weight, nonpolar mixtures; for this application, the method was quite accurate, with errors frequently less than 1%. With the exception of systems that contain ammonia or carbon dioxide, the error was always less than 4%.

For extension of corresponding–states methods to mixtures that can be used for both low and high pressures, the reader is referred to Chapter 12 or Reid *et al.* (1987).

13.5.2 Viscosities of liquid mixtures

Methods for liquid mixture viscosities summarized by Reid *et al.* (1987) include that of Grunberg & Nissan (1949) and a two–reference fluid corresponding–states method attributed to Teja & Rice (1981a,b). In the Grunberg–Nissan method the low–temperature liquid viscosity for mixtures is given as

$$\ln \eta = \sum_i x_i \ln \eta_i + \frac{1}{2}\sum_i \sum_j x_i x_j G_{ij} \qquad (13.19)$$

where G_{ij} is an interaction parameter that is a function of the components i and j as well as of temperature. Isdale (1979) has presented a group contribution to estimate G_{ij}, and in extensive testing of over 2000 data points found the root mean square deviation for the mixtures tested to be 1.6%.

Several methods have been published since 1987 (Chhabra & Sridhar 1989; Liu *et al.* 1991), and recent methods for specific situations include those for hydrocarbon mixtures (Chhabra 1992), for solutions with suspended particles (Sudduth 1993) and for phosphorus–containing compounds (Dutt *et al.* 1993).

13.6 Thermal conductivities of mixtures

13.6.1 Low–pressure gases and dense fluids

For the thermal conductivities of gas mixtures at low pressure, two empirical methods and one corresponding–states scheme are presented by Reid *et al.* (1987). Both empirical methods employ the Wassiljewa equation (Wassiljewa 1904)

$$\lambda^{(0)} = \sum_i \frac{x_i \lambda_i^{(0)}}{\sum_j x_j A_{ij}} \qquad (13.20)$$

where $A_{ii} = 1$ and following a suggestion by Mason & Saxena (1958),

$$A_{ij} = \frac{\left[1 + \left(\lambda_{\mathrm{tr},i}^{(0)}/\lambda_{\mathrm{tr},j}^{(0)}\right)^{1/2} (M_i/M_j)^{1/4}\right]^2}{[8(1 + M_i/M_j)]^{1/2}} \tag{13.21}$$

The recommended equation for $\lambda_{\mathrm{tr},i}^{(0)}/\lambda_{\mathrm{tr},j}^{(0)}$ is

$$\frac{\lambda_{\mathrm{tr},i}^{(0)}}{\lambda_{\mathrm{tr},j}^{(0)}} = \frac{\Gamma_j \left[\exp\left(0.0464 T_{\mathrm{r},i}\right) - \exp\left(-0.2412 T_{\mathrm{r},i}\right)\right]}{\Gamma_i \left[\exp\left(0.0464 T_{\mathrm{r},j}\right) - \exp\left(-0.2412 T_{\mathrm{r},j}\right)\right]} \tag{13.22}$$

where

$$\Gamma = 210 \left(\frac{T_c M^3}{P_c^4}\right)^{1/6} \tag{13.23}$$

Errors using the above equations will generally be less than 3 to 4% for nonpolar gas mixtures. If one of the components is polar, errors greater than 5 to 8% may be expected.

Several corresponding–states methods are presented by Reid *et al.* (1987) for dense fluids (see also Chapter 12).

13.6.2 *Thermal conductivity of liquid mixtures*

Methods to estimate the thermal conductivity of liquid mixtures have been reviewed by Reid *et al.* (1977, 1987) and Rowley *et al.* (1988). Five methods are summarized by Reid *et al.* (1987), but three of these can be used only for binary mixtures. The two that can be extended to multicomponent mixtures are the Li method (Li 1976), and Rowley's method (Rowley *et al.* 1988). According to the latter the Li method does not accurately describe ternary behavior. Furthermore, it was indicated that the power law method (Reid *et al.* 1977; Rowley *et al.* 1988) successfully characterizes ternary mixture behavior when none of the pure component thermal conductivities differ by more than a factor of 2. But, the power law method should not be used when water is present in the mixture. Rowley's method is based on a local composition concept, and it uses NRTL parameters from vapor–liquid equilibrium data as part of the model. These parameters are available for a number of binary mixtures (Gmehling & Onken 1977). When tested for 18 ternary systems, Rowley's method gave an average absolute deviation of 1.86%.

References

Allan, J. M. & Teja, A. S. (1991). Correlation and prediction of the viscosity of defined and undefined hydrocarbon liquids. *Can J. Chem. Eng.*, **69**, 986–991.

Barroncini, C. P., Di Filippo P., Latini, G. & Pacetti, M. (1980). An improved correlation for the calculation of liquid thermal conductivity. *Int. J. Thermophys.*, **1**, 159–175.

Barroncini, C. P., Di Filippo P., Latini, G. & Pacetti, M. (1981). Organic liquid thermal conductivity: A prediction method in the reduced temperature range 0.3 to 0.8. *Int. J. Thermophys.*, **2**, 21–38.

Barroncini, C. P., Di Filippo P. & Latini, G. (1983a). Comparison between predicted and experimental thermal conductivity values for the liquid substances and the liquid mixtures at different temperatures and pressures. Paper presented at the Workshop on thermal conductivity measurement, IMECO, Budapest, March 14–16, 1983.

Barroncini, C. P., Di Filippo, P. & Latini, G. (1983b). Thermal conductivity estimation of the organic and inorganic refrigerants in the saturated liquid state. *Int. J. Refrig.*, **6**, 60–67.

Barroncini, C. P., Latini, G. & Pierpaoli, P. (1984). Thermal conductivity of organic liquid binary mixtures: Measurements and prediction method. *Int. J. Thermophys.*, **5**, 387–402.

Chhabra, R. P. (1992). Prediction of viscosity of liquid hydrocarbon mixtures. *AIChE J.*, **38**, 1657–1661.

Chhabra, R. P. & Sridhar, T. (1989). Prediction of viscosity of liquid mixtures using Hildebrand's fluidity model. *Chem. Eng. J. (Lausanne)*, **40**, 39–43.

Chung, T.–H., Ajlan, M., Lee, L. L. & Starling K. E. (1988). Generalized multiparameter correlation for nonpolar and polar fluid transport properties. *Ind. Eng. Chem. Res.*, **27**, 671–679.

Chung, T.–H., Lee, L. L. & Starling K. E. (1984). Application of kinetic gas theories and multiparameter correlation for prediction of dilute gas viscosity and thermal conductivity. *Ind. Eng. Chem. Fundam.*, **23**, 8–13.

Davidson, T. A. (1993). A simple and accurate method for calculating viscosity of gaseous mixtures. *US Bureau of Mines, RI9456.*

Dutt, N. V. K., Ravikumar, Y. V. L. & Rajiah, A. (1993). Prediction of liquid viscosities of phosphorus-containing compounds by Souder's method. *Chem. Eng. J. (Lausanne)*, **51**, 41–44.

Dutt, N. V. K., & Venugopal, D. (1990). Prediction of liquid viscosities of fluorinated hydrocarbons by the extended Thomas' method. *Chem. Eng. J. (Lausanne)*, **44**, 173–175.

Ely, J. F. & Hanley H. J. M. (1983). Prediction of transport properties. 2. Thermal conductivity of pure fluids and mixtures. *Ind. Eng. Chem. Fundam.*, **22**, 90–97.

Fuller, E. N., Ensley, K. & Giddings, J. C. (1969). Diffusion of halogenated hydrocarbons in helium. *J. Phys. Chem.*, **73**, 3679–3685.

Fuller, E. N. & Giddings, J. C. (1965). A comparison of methods for predicting gaseous diffusion coefficients. *Gas Chromatogr.*, **3**, 222–227.

Fuller, E. N., Schettler, P. D. & Giddings, J. C. (1966). A new method of prediction of binary gas-phase diffusion coefficients. *Ind. Eng. Chem.*, **75**, 18–27.

Gmehling, J. & Onken, U. (1977). *Vapour-liquid equilibrium data collection, DECHEMA Chem. Data Series*, Vol. 1. Frankfurt: Verlag & Druckerei Friedrich Bishoff.

Grunberg, L. & Nissan, A. H. (1949). Mixture law for viscosity. *Nature*, **164**, 799–800.

Hanley, H. J. M. (1976). Prediction of the viscosity and thermal conductivity coefficients of mixtures. *Cryogenics*, **16**, 643–651.

Hayduk, W. & Minhas, B. S. (1982). Correlations for prediction of molecular diffusivities in liquids. *Can. J. Chem. Eng.*, **60**, 295–299.

Herning, F. & Zipperer (1936). Calculation of the viscosity of technical gas mixtures from the viscosity of the individual gases. *Gas Wasserfach*, **79**, 49–54.

Huber, M. L. & Ely, J. F. (1992). Prediction of viscosity of refrigerants and refrigerant mixtures. *Fluid Phase Equil.*, **80**, 239–248.

Isdale, J. D. (1979). *Symp. Transp. Props. Fluids and Fluid Mixtures, Natl. Eng. Lab.*, East Kilbride, Glasgow, Scotland.

Joback, K. G. & Reid, R. C. (1987). Estimation of pure-component properties from group-contributions. *Chem. Eng. Comm.*, **57**, 233–243.

Kooijman, H. A. & Taylor, R. (1991). Estimation of diffusion coefficients in multicomponent liquid systems. *Ind. Eng. Chem. Res.*, **30**, 1217–1222.

Lackshmi, D. S. & Prasad, D. H. L. (1992). A rapid estimation method for thermal conductivity of pure liquids. *Chem. Eng. J. (Lausanne)*, **48**, 211–214.

Latini, G. & Pacetti, M. (1978). The thermal conductivity of liquids – A critical survey. *Therm. Cond.*, **15**, 245–253 (1977, pub. 1978).

Li, C. C. (1976). Thermal conductivity of liquid mixtures. *AIChE J.*, **22**, 927–930.

Liu, H., Wenchuan, W. & Chang, C.-H. (1991). Model with temperature–independent parameters for the viscosities of liquid mixtures. *Ind. Eng. Chem. Res.*, **30**, 1617–1624.

Lucas, K. (1980). *Phase Equilibria and Fluid Properties in the Chemical Industry*, Frankfurt: DECHEMA.

Lucas, K. (1984). Berechnungsmethoden für Stoffeigenschaften, in *VDI– Wärmeatlas*. Abschnitt DA, Düsseldorf: Verein Deutscher Ingenieure.

Mason, E. A. & Saxena, S. C. (1958). Approximate formula for the thermal conductivity of gas mixtures. *Phys. Fluids*, **1**, 361–369.

McKeigue, K. & Gulari, E. (1989). Effect of molecular association on diffusion in binary liquid mixtures. *AIChE J.*, **35**, 300–310.

Mehrotra, A.K. (1991a). A generalized viscosity equation for pure heavy hydrocarbons. *Ind. Eng. Chem. Res.*, **30**, 420–427.

Mehrotra, A.K. (1991b). Generalized one-parameter viscosity equation for light and medium liquid hydrocarbons. *Ind. Eng. Chem. Res.*, **30**, 1367–1372.

Missenard, F. A. (1973). Thermal conductivity of organic liquid series. *Rev. Gen. Thermodyn.*, **141**, 751–759.

Myers, K. H. & Danner R. P. (1993). Prediction of properties of silicon, boron, and aluminium compounds. *J. Chem. Eng. Data*, **38**, 175–200.

Nakanishi, K. (1978). Prediction of diffusion coefficient of nonelectrolytes in dilute solution based on generalized Hammond-Stokes plot. *Ind. Eng. Chem. Fundam.*, **17**, 253–256.

Quayle, O. R. (1953). The parachors of organic compounds. An interpretation and catalogue. *Chem. Rev.*, **53**, 439–589.

Reichenberg, D. (1971). *DCS report 11*. Teddington, England: National Physics Laboratory.

Reichenberg, D. (1973). The indeterminacy of the values of potential parameters as derived from transport and virial coefficients. *AIChE J.*, **19**, 854–856.

Reichenberg, D. (1974). The viscosity of gas mixtures at moderate pressures. *NPL Rept. Chem. 29*. Teddington, England: National Physics Laboratory.

Reichenberg, D. (1975). New methods for the estimation of the viscosity coefficients of pure gases at moderate pressures (with particular reference to organic vapours). *AIChE J.*, **21**, 181–183.

Reichenberg, D. (1977). New simplified methods for the estimation of the viscosities of gas mixtures at moderate pressures. *Natl. Eng. Lab. Rept. Chem. 53*. East Kilbride, Scotland: National Engineering Laboratory.

Reichenberg, D. (1979). *Symp. Transp. Props. Fluids and Fluid Mixtures, Natl. Eng. Lab.*, East Kilbride, Scotland.

Reid, R. C., Prausnitz, J. M. & Poling, B. E. (1987). *The Properties of Gases and Liquids*, 4th Ed. New York: McGraw–Hill.

Reid, R. C., Prausnitz, J. M. & Sherwood, T. K. (1977). *The Properties of Gases and Liquids*, 3rd Ed. New York: McGraw–Hill.

Riedel, L. (1949). Wärmeleitfähigkeitsmessungen an Zuckerlösungen, Fruchtsäften und Milch. *Chem. Ing. Tech.*, **21**, 340–341.

Riedel, L. (1951a). Die Wärmeleitfähigkeit von wäßrigen Lösungen starker Elektrolyte. *Chem. Ing. Tech.*, **23**, 59–64.

Riedel, L. (1951b). Neue Wärmeleitfähigkeitsmessungen an organischen Flüssigkeiten. *Chem. Ing. Tech.*, **23**, 321–324.

Riedel, L. (1951c). Wärmeleitfähigkeitsmessungen an Mischungen verschiedener
    organischer Verbindungen mit Wasser. *Chem. Ing. Tech.*, **23**, 465–469.

Rowley, R. L., White, G. L. & Chiu, M. (1988). Ternary liquid mixture thermal
    conductivities. *Chem. Eng. Sci.*, **43**, 361–371.

Roy, D. & Thodos, G. (1968). Thermal conductivity of gases. Hydrocarbons at normal
    pressures. *Ind. Eng. Chem. Fundam.*, **7**, 529–534.

Roy, D. & Thodos, G. (1970). Thermal conductivity of gases. Organic compounds at
    atmospheric pressure. *Ind. Eng. Chem. Fundam.*, **9**, 71–79.

Sastri, S. R. S. & Rao K. K. (1992). A new group contribution method for predicting
    viscosity of organic liquids. *Chem. Eng. J. (Lausanne)*, **50**, 9–25.

Sudduth, R. D. (1993). A generalized model to predict the viscosity of solutions with
    suspended particles. *J. Appl. Polym. Sci.*, **48**, 25–36.

Teja, A. S. & Rice, P. (1981a). The measurement and prediction of the viscosities of some
    binary liquid mixtures containing n-hexane. *Chem. Eng. Sci.*, **36**, 7–10.

Teja, A. S. & Rice, P. (1981b). Generalized corresponding states method for the viscosities of
    liquid mixtures. *Ind. Eng. Chem. Fundam.*, **20**, 77–81.

Tyn, M. T. & Calus, W. F. (1975). Diffusion coefficients in dilute binary liquid mixtures.
    J. Chem. Eng. Data, **20**, 106–109.

van Velzen, D., Cardozo, R. L. & Langenkamp, H. (1972a). A liquid viscosity-temperature-
    chemical constitution relation for organic compounds. *Ind. Eng. Chem. Fundam.*, **11**,
    20–25.

van Velzen, D., Cardozo, R. L. & Langenkamp, H. (1972b). Liquid viscosity and chemical
    constitution of organic compounds: A new correlation and a compilation of literature
    data. *Euratom*, 4735e, Ispra Establishment, Italy: Joint Nuclear Research Centre.

Vignes, A. (1966). Diffusion in binary solutions. *Ind. Eng. Chem. Fundam.*, **5**, 189–199.

Wassiljewa, A. (1904). Wärmeleitung in Gasgemischen. *Phys. Z.*, **5**, 737–742.

Wilke, C. R. (1950). A viscosity equation for gas mixtures. *J. Chem. Phys.*, **18**, 517–519.

Wilke, C. R. & Chang, P. (1955). Correlation of diffusion coefficients in dilute solutions.
    AIChE J., **1**, 264–270.

Wilke, C. R. & Lee, C. Y. (1955). Estimation of diffusion coefficients for gases and vapours.
    Ind. Eng. Chem., **47**, 1253–1257.

Part five

APPLICATION TO SELECTED SUBSTANCES

14

Pure Fluids

14.1 Monatomic fluids–Argon

E. P. SAKONIDOU and H. R. van den BERG

Van der Waals–Zeeman Laboratory,
University of Amsterdam, The Netherlands

and

J. V. SENGERS

Institute for Physical Science and Technology,
University of Maryland, College Park, MD, USA

14.1.1 Introduction

Among the family of monatomic fluids, argon is undoubtedly the fluid studied most extensively. This gas is often used experimentally as a reference fluid for calibration purposes and also as a model system for theoretical calculations and computer simulations. In general the thermal conductivity contains contributions from the translational energy of the molecules, from the internal energy and from contributions due to energy exchange of translational and internal modes of motion as discussed in Chapter 4. The calculation of the latter two contributions is often very complicated. Since these contributions are lacking in noble gases, the thermal conductivity data for these substances are more amenable to full theoretical analysis.

As elucidated in Section 6.4, both the thermal conductivity λ and the viscosity η are decomposed into background contributions $\bar{\lambda}$ and $\bar{\eta}$ and critical enhancements $\Delta\lambda_c$ and $\Delta\eta_c$, according to Sengers & Keyes (1971) and Sengers (1971, 1972)

$$\lambda = \bar{\lambda} + \Delta\lambda_c, \qquad \eta = \bar{\eta} + \Delta\eta_c \tag{14.1}$$

The background contributions in turn are decomposed as

$$\bar{\lambda} = \lambda^{(0)}(T) + \Delta\lambda(\rho, T), \qquad \bar{\eta} = \eta^{(0)}(T) + \Delta\eta(\rho, T) \tag{14.2}$$

where $\lambda^{(0)}(T)$ and $\eta^{(0)}(T)$ are the thermal conductivity and the viscosity in the dilute–gas limit at temperature T. The contributions $\Delta\lambda(\rho, T)$ and $\Delta\eta(\rho, T)$ are referred to as excess thermal conductivity and excess viscosity at density ρ and temperature T. The decomposition for the thermal conductivity is illustrated in Figure 3.1 in Chapter 3 of this volume.

In this chapter correlations are presented for each of the three parts of which the thermal conductivity and viscosity of argon are composed. These correlations are based on an extensive set of experimental data and involve the theoretical expressions for the dilute–gas parts $\lambda^{(0)}$ and $\eta^{(0)}$ and for the critical enhancements $\Delta\lambda_c$ and $\Delta\eta_c$, as presented in Chapters 4 and 6. Recently for several fluids (Vesovic *et al.* 1990, 1994; Krauss *et al.* 1993) the excess contribution to the thermal conductivity has been derived from experimental data simultaneously with the critical enhancement part by using an iterative method. The necessity for the application of this procedure was mainly due to a lack of noncritical data. However, in the case of argon, where the range of available data is much wider, a correlation for the excess thermal conductivity $\Delta\lambda(\rho, T)$ can be determined directly.

The derivation of the correlation for the excess viscosity $\Delta\eta(\rho, T)$ from experimental data in general is much easier, since for the viscosity the critical enhancement is just restricted to a very narrow region around the critical point. After the correlations for the excess thermal conductivity and the excess viscosity have been determined, the coincidence of the theoretical results calculated for the thermal conductivity in the critical region with those obtained experimentally could be optimized by adjusting the value of the parameter q_D^{-1}, a microscopic cutoff distance in the crossover function $\Omega - \Omega_0$ in equation (6.41) of Chapter 6. Moreover, from the correlations for $\Delta\lambda$ and $\Delta\eta$ temperature–dependent initial–density coefficients for the thermal conductivity and the viscosity are deduced and then compared with the theoretical predictions presented in Chapter 5.

14.1.2 Dilute–gas contribution

Equations for the viscosity and thermal conductivity of fluids in the dilute–gas limit have been derived from exact kinetic theory in Chapter 4. For the viscosity, it can be written

$$\eta^{(0)}(T) = \frac{1}{4} (\pi m k_B T)^{1/2} \frac{1}{\pi \sigma^2 \mathfrak{S}_\eta^*} \tag{14.3}$$

and for the thermal conductivity of *monatomic* gases

$$\lambda^{(0)}(T) = \frac{5}{8} \left(\frac{\pi k_B^3 T}{m} \right)^{1/2} \frac{1}{\pi \sigma^2 \mathfrak{S}_\lambda^*} \tag{14.4}$$

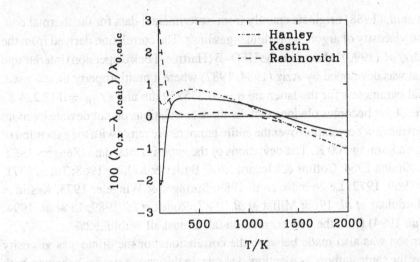

Fig. 14.1. Percentage deviations of various correlations for the thermal conductivity in the dilute–gas limit from the new correlation given by equations (14.4) and (14.7); complete references in text.

where the functionals $\mathfrak{S}_\eta^*$ and $\mathfrak{S}_\lambda^*$ are reduced effective cross sections which for *monatomic gases* can be defined as (see Chapter 4)

$$\mathfrak{S}_\eta^* = \frac{\mathfrak{S}(2000)}{\pi\sigma^2 f_\eta}, \qquad \mathfrak{S}_\lambda^* = \frac{\mathfrak{S}(1010)}{\pi\sigma^2 f_\lambda} = \frac{2\mathfrak{S}(2000)}{3\pi\sigma^2 f_\lambda} \qquad (14.5)$$

Here, f_η and f_λ are correction factors for higher–order approximations, while σ is a length scaling (or potential) parameter, as also introduced in Chapter 4.

In Figure 14.1, a comparison is presented between three older correlations for $\lambda^{(0)}$ (Hanley 1974; Kestin *et al.* 1984; Rabinovich *et al.* 1988) and a correlation recently derived from the work of Bich *et al.* (1990). The latter correlation serves as the baseline in the temperature range 100–2000 K. In each of the four cases the correlations, which originated from equation (14.4) for $\lambda^{(0)}$, were not expressed in terms of the reduced effective collision cross section $\mathfrak{S}_\lambda^*$ but instead in terms of a reduced collision integral $\Omega^{(2,2)*}$. Furthermore, for the higher–order correction the second–order Kihara approximation (Maitland *et al.* 1987) was applied. As detailed in the Appendix to Chapter 4, the conversion from collision integrals to effective cross sections is simple for isotropic interatomic potentials, namely,

$$\mathfrak{S}^*(2000) = \frac{\mathfrak{S}(2000)}{\pi\sigma^2} = \frac{4}{5}\Omega^{(2,2)*}, \qquad \mathfrak{S}^*(1010) = \frac{\mathfrak{S}(1010)}{\pi\sigma^2} = \frac{8}{15}\Omega^{(2,2)*} \qquad (14.6)$$

The correlation earlier developed by Hanley (1974) for the dilute–gas thermal conductivity was calculated for the 11–6–8 interatomic potential with parameters σ and ϵ/k_B from fitting both transport and equilibrium properties. The correlation of Kestin *et al.* (1984) is a 'universal' correlation based on the extended principle of corresponding states (see Chapter 11) for transport and equilibrium properties. The correlation of

Rabinovich *et al.* (1988) originates purely from experimental data for the thermal conductivity and viscosity of argon in the dilute–gas limit. The correlation derived from the work of Bich *et al.* (1990) is based on the HFD–B (Hartree–Fock dispersion) interatomic potential that was developed by Aziz (1984, 1987) where a multiproperty fit was used. The potential parameters for the latter are $\sigma = 0.335279$ nm and $\epsilon/k_B = 143.224$ K. From Figure 14.1 it becomes obvious that the three correlations do not deviate by more than 1% from the new correlation over the entire temperature range with the exception of a small range adjoining 100 K. The deviations of the experimental data (Sengers 1962; Vargaftik & Zimina 1964; Collins & Menard 1966; Bailey & Kellner 1968; Tufeu 1971; Le Neindre 1969, 1972; Le Neindre *et al.* 1989; Springer & Wingeier 1973; Kestin *et al.* 1980; Mardolcar *et al.* 1986; Millat *et al.* 1987; Roder *et al.* 1989; Li *et al.* 1994; Tiesinga *et al.* 1994) from the new correlation fall almost all within $\pm 2\%$.

A comparison was also made between the correlations for the dilute–gas viscosity developed by the same authors as mentioned above. In this case a similar behavior, both qualitatively and quantitatively, was found.

The correlations for $\eta^{(0)}$ and $\lambda^{(0)}$ deduced from the HFD–B potential have been accepted as the basis for the description of the dilute–gas transport data. The functionals $\mathfrak{S}_\eta^*$ and $\mathfrak{S}_\lambda^*$, defined by equations (14.5), are monotonically decreasing functions of the reduced temperature $T^* = k_B T/\epsilon$ in the range $0.5 < T^* < 15$ and are to within $\pm 0.1\%$ represented by

$$\ln \mathfrak{S}_\eta^* = \sum_{i=0}^{5} a_{\eta,i} \left(\ln T^*\right)^i \,, \qquad \ln \mathfrak{S}_\lambda^* = \sum_{i=0}^{5} a_{\lambda,i} \left(\ln T^*\right)^i \qquad (14.7)$$

The coefficients $a_{\eta,i}$ and $a_{\lambda,i}$ are given in Table 14.1.

14.1.3 Excess contribution

The excess thermal conductivity $\Delta\lambda(\rho, T)$ in equations (14.1) and (14.2) has been determined previously from experimental data, for instance, for fluids such as carbon dioxide (Vesovic *et al.* 1990), the refrigerant R134a (Krauss *et al.* 1993) and ethane (Vesovic *et al.* 1994). For each of these fluids the excess contribution was calculated simultaneously with the critical enhancement $\Delta\lambda_c$, which is described by the theoretical expressions for the crossover behavior of the transport properties in the critical region, as discussed in Section 6.5. In the iterative calculation of $\Delta\lambda$ and $\Delta\lambda_c$ both the background $\bar{\lambda}$ for the thermal conductivity and the enhancement parameter q_D^{-1} for a microscopic cutoff distance are adjustable. Moreover, $\Delta\lambda_c$ vanishes for temperatures above a reference temperature T_r, introduced in Chapter 6. For the amplitude R_D in equation (6.41) the values 1.03 and 1.01 were applied, which are within the range given in equation (6.24). This procedure provided, then, a background consistent with the enhancement parameters. However, since possible uncertainties in the theoretical description of the critical enhancement by applying this procedure are partly

Table 14.1. *Coefficients of the fits for the functionals $\mathfrak{S}_\lambda^*$ and $\mathfrak{S}_\eta^*$ in equations (14.7) to be used for the dilute–gas thermal conductivity and viscosity; coefficients of the temperature–dependent excess correlations, equations (14.8) and (14.10), for the thermal conductivity and the viscosity.*

Coefficient	Value	Coefficient	Value
$a_{\lambda,0}$	-1.8804 (-1)	$a_{\eta,0}$	2.1740 (-1)
$a_{\lambda,1}$	-4.7273 (-1)	$a_{\eta,1}$	-4.7253 (-1)
$a_{\lambda,2}$	4.7334 (-2)	$a_{\eta,2}$	4.8798 (-2)
$a_{\lambda,3}$	6.4518 (-2)	$a_{\eta,3}$	6.5096 (-2)
$a_{\lambda,4}$	-3.1750 (-2)	$a_{\eta,4}$	-3.2503 (-2)
$a_{\lambda,5}$	4.3013 (-3)	$a_{\eta,5}$	4.4486 (-3)
$\lambda_{1,0}$	1.02688	$\eta_{1,0}$	9.84574 (-2)
$\lambda_{1,1}$	-1.32791 (-5)	$\eta_{1,1}$	1.19990 (-3)
$\lambda_{2,0}$	-4.47380 (-2)	$\eta_{2,0}$	1.06976 (-1)
$\lambda_{2,1}$	3.19347 (-5)	$\eta_{2,1}$	-6.79621 (-5)
λ_3	1.79971 (-2)	η_3	-8.28656 (-3)
λ_4	-2.31464 (-3)	η_4	7.87279 (-4)
λ_5	1.63071 (-4)	η_5	-3.44562 (-5)
λ_6	-6.21532 (-6)	η_6	7.52582 (-7)
λ_7	1.23122 (-7)	η_7	-5.50680 (-9)
λ_8	-9.87233 (-10)		

λ in mW m^{-1} K^{-1}, η in μPa s, T in K, ρ in mol L^{-1}, (-n) means 10^{-n}.

transmitted to the background, this background is not necessarily consistent with the background as it is defined in equation (14.1), namely, as the thermal conductivity without the critical enhancement due to the long–range fluctuations. The region in $\rho - T$ coordinates where in the case of argon the ratio $\Delta\lambda_c/\lambda$ of the critical enhancement for the thermal conductivity relative to the total thermal conductivity is less than 1% is indicated in Figure 14.2. The shadowed range is considered as the 'noncritical region.' For the limits of this region the value of the reference temperature T_r has been chosen as $2T_c$, where the critical temperature is taken as $T_c = 150.663$ K, so that the enhancement covers a range of 150 K above T_c. Moreover, it is seen from this figure that the enhancement is present at densities up to twice the critical density. Since in the indicated noncritical region an extensive set of experimental data for the thermal conductivity of argon is available, the conclusion was drawn that an independent determination of the excess thermal conductivity and the critical enhancement is possible (Tiesinga *et al.* 1994).

It turned out that the excess transport coefficients $\Delta\lambda$ and $\Delta\eta$ of simple fluids depend only weakly on temperature and they are often represented by polynomials in density ρ with coefficients λ_i and η_i taken to be independent of temperature (Le Neindre 1972;

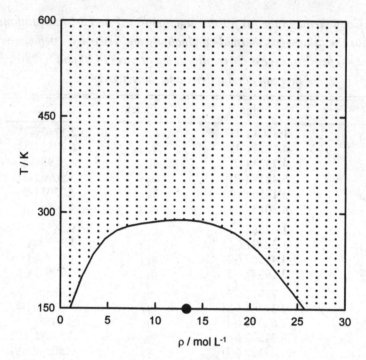

Fig. 14.2. Region in temperature and density where the ratio of the critical thermal conductivity enhancement relative to the total thermal conductivity is below 1%, ● - critical point.

Mostert *et al.* 1990; Vesovic *et al.* 1990; Perkins *et al.* 1991; Krauss *et al.* 1993)

$$\Delta\lambda = \sum_{i=1}^{n} \lambda_i \rho^i, \qquad \Delta\eta = \sum_{i=1}^{m} \eta_i \rho^i \qquad (14.8)$$

The assumption that the coefficients λ_i and η_i in equations (14.8) can be treated as independent of the temperature is valid only approximately (see Chapter 5). There exists experimental (Hanley *et al.* 1969) and theoretical evidence (Nieto de Castro *et al.* 1990) that the leading coefficients λ_i and η_i do depend on temperature. Hence, in the present analysis the coefficients of the first three terms in the representations (14.8) were allowed to vary with temperature.

To determine the coefficients λ_i in equation (14.8) for the excess thermal conductivity the following data sets were considered: parallel–plate data from Sengers (1962) and Tiesinga *et al.* (1994); concentric–cylinder data from Ikenberry & Rice (1963), Bailey & Kellner (1968), Tufeu (1971), Le Neindre (1969, 1972) and Le Neindre *et al.* (1989); transient hot–wire data from Kestin *et al.* (1980) and Roder *et al.* (1988, 1989) supplemented with data of Perkins *et al.* (1991). As discussed above, only those data points that are situated in the shadowed region in Figure 14.2 were taken into account, that is, those points for which the relative critical enhancement contribution was estimated to be well below 1%. Also a few points with an exceptional experimental error were omitted, leaving a total of 477 data points. To reduce the effect of any systematic errors (Vesovic *et al.* 1990), the values for the excess thermal conductivity were deduced by subtracting dedicated zero–density values $\lambda^{(0)}$, that is, the particular values implied by the individual data sets mentioned above.

The densities of all data points were recalculated from the measured temperatures and pressures with a recently developed equation of state (Tiesinga *et al.* 1994). This equation consists, for temperatures and densities in the critical region, of a crossover equation of state based on a six–term Landau expansion in parametric form as proposed by Luettmer–Strathmann *et al.* (1992) and, for outside that region, of a global equation of state proposed by Stewart & Jacobsen (1989). The values that were adopted for the critical temperature, pressure and density are

$$T_c = 150.663 \,\text{K}, \qquad P_c = 4.860 \,\text{MPa}, \qquad \rho_c = 14.395 \,\text{mol}\,\text{L}^{-1} \qquad (14.9)$$

The values for T_c and P_c are those given by Stewart & Jacobsen (1989), whereas the value for ρ_c is derived from isochoric specific–heat–capacity data of Anisimov *et al.* (1978). Since the global equation of state is based on temperatures in terms of IPTS–68, all temperatures are consistently expressed in this temperature scale.

A fit with equal weights of the experimental excess thermal conductivities with an eighth–degree polynomial in density, equation (14.8), but with temperature–independent coefficients λ_i yielded a standard deviation of $\sigma_\lambda = 1.2\%$ for $\Delta\lambda/\bar{\lambda}$. Introducing a linear temperature dependence of λ_1 yielded a substantial decrease of σ_λ by about 30%, and a simultaneous linear temperature dependence of λ_1 and λ_2 yielded an additional decrease of σ_λ by about 20% to $\sigma_\lambda = 0.6\%$. However, a quadratic temperature dependence of λ_1 or an additional linear temperature dependence of λ_3 did not yield any significant decrease of σ_λ. Hence, only a linear temperature dependence of λ_1 and λ_2 was retained (as it was for viscosity; see below)

$$X_i = X_{i,0} + X_{i,1}T \qquad\qquad (X = \eta, \lambda; i = 1, 2) \qquad (14.10)$$

The values of the coefficients in equations (14.8) and (14.10) are included in Table 14.1. From the above mentioned values for the standard deviation σ_λ of different fits for the excess thermal conductivity and from the magnitude of the coefficients $\lambda_{1,1}$ and $\lambda_{2,1}$ the following conclusions can be drawn

- The introduction of temperature–dependent coefficients λ_1 and λ_2 in equation (14.8) gives a significant improvement of the representation for the excess thermal conductivity data.
- The initial–density coefficient λ_1 shows a very weak temperature dependence, so that the temperature dependence of the excess thermal conductivity is practically completely accounted for by the coefficient λ_2. For instance, due to the temperature dependence of the coefficients λ_1 and λ_2, a temperature change of 200 K causes, at a density $\rho = 20 \,\text{mol}\,\text{L}^{-1}$, a change of the separate terms $\lambda_1\rho$ and $\lambda_2\rho^2$ of about 0.1% and 6%, respectively, relative to the total excess thermal conductivity value.

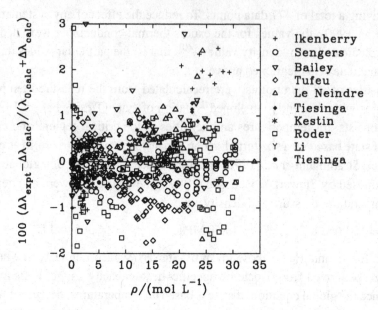

Fig. 14.3. Percentage deviations of various experimental excess thermal–conductivity data from the excess correlation, equations (14.8) and (14.10), relative to the calculated total thermal conductivity.

A comparison between the experimental excess thermal conductivities and the values calculated from equations (14.8) and (14.10), relative to the total thermal conductivity λ, is shown in Figure 14.3. The total thermal conductivity is calculated from the correlating equations (14.4) and (14.7) for $\lambda^{(0)}$ and equations (14.8) and (14.10) for $\Delta\lambda$. It is seen that most of the 477 experimental data, independent of the method by which the data were obtained, agree with the correlation within $\pm 1.5\%$. In Figure 14.3 the results (filled circles) of the most recent measurements obtained at 302 K (Tiesinga *et al.* 1994) are also displayed, as well as a representative selection of recent measurements obtained by Li *et al.* (1994). It turns out that both data sets agree with the correlating equation within $\pm 0.5\%$, although these data were not included in developing the correlations. It is estimated that equations (14.8) and (14.10) represent the excess thermal conductivity of argon in a temperature and density range bounded by $150 \le T \le 650$ K and $0 \le \rho \le 35$ mol L^{-1}.

In Figure 14.4 the excess thermal conductivity values of argon at 300 K according to three previous correlations are compared with the values calculated from the new correlation defined by equations (14.8) and (14.10). The correlations of Le Neindre (1972) and of Perkins *et al.* (1991) are both temperature–independent and based on concentric–cylinder and transient hot–wire data at high and low temperatures respectively. The temperature–dependent correlation of Younglove & Hanley (1986) is based on a large variety of experimental data and covers a wide range of temperatures and densities. It is seen from this figure that at 300 K the deviations for the three correlations

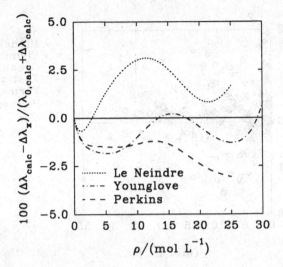

Fig. 14.4. Percentage deviations of various correlations for the excess thermal conductivity at 300 K from the new excess correlation, equations (14.8) and (14.10), relative to the calculated total thermal conductivity.

from the new correlation lie within ±3%. However, at lower temperatures the deviations for the correlation of Younglove & Hanley exceed −15% at the low–density side. It is seen that the excess thermal conductivity values according to the correlations of Le Neindre and of Perkins *et al.* show a systematic difference of about 4%. In view of the conclusion made from the deviation plot in Figure 14.3, it must be stressed that this discrepancy between the two correlations has to be attributed to deficiencies in these correlations and not to the difference in the underlying measuring methods.

For the determination of the coefficients η_i in equation (14.8) for the excess viscosity $\Delta\eta$ the following data sets were taken into account: capillary–flow data of Michels *et al.* (1954), Flynn *et al.* (1962), Gracki *et al.* (1969) and Vermesse & Vidal (1973); oscillating–disk data of Kestin *et al.* (1971); torsional–crystal data of Haynes (1973), omitting a few points left a total of 346 data. None of these points contained a critical contribution, since the viscosity enhancement $\Delta\eta_c$ is restricted to a much smaller range of temperatures and densities than the enhancement for the thermal conductivity. The values for the excess viscosity were obtained by applying dedicated zero–density values $\eta^{(0)}$.

In order to represent the experimental excess viscosities, a seventh–degree polynomial in the density was adopted, and again equal weights were applied to all data points. If all of the coefficients η_i in equation (14.8) are taken as constants independent of temperature, the equation yields a standard deviation σ_η for $\Delta\eta/\bar{\eta}$ of 0.9%. Introduction of a linear temperature dependence of η_1 and η_2 in equation (14.10) reduces σ_η to 0.5%. Therefore, this polynomial fit has been chosen to represent the viscosity excess

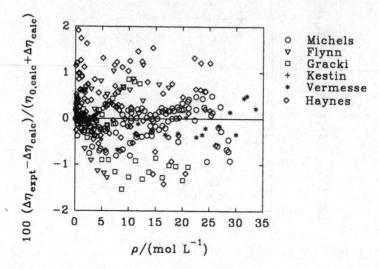

Fig. 14.5. Percentage deviations of various experimental excess viscosity data from the excess correlation, equations (14.8) and (14.10), relative to the calculated total viscosity.

contribution. The values of the coefficients in equations (14.8) and (14.10) are also included in Table 14.1.

A comparison between the experimental excess viscosities and the values calculated from equations (14.8) and (14.10), relative to the total viscosity η, is shown in Figure 14.5. The total viscosity is calculated from the correlation (14.3) and (14.7) for $\eta^{(0)}$ and equations (14.8) and (14.10) for $\Delta\eta$. It is seen that most of the experimental data agree with this correlation within $\pm 1.25\%$. Therefore, it is estimated that equations (14.8) and (14.10) represent the excess viscosity of argon in a temperature and density range bounded by $150 \leq T \leq 400$ K and $0 \leq \rho \leq 40$ mol L^{-1}.

14.1.4 Initial–density contribution

Theoretical expressions for the reduced viscosity and thermal conductivity second virial coefficients are presented in Section 5.3. An important contribution arises from the effect of the formation of dimers, even though the concentration of dimers is still small. For the evaluation of these transport property virial coefficients it was assumed that both the interaction potential between two monomers and between a monomer and a dimer is of the Lennard–Jones (12–6) form but with different potential parameters. The ratios of these parameters, which are characterized by the constants δ and θ defined in Section 5.3, are determined from a large selection of experimental data. According to Stogryn & Hirschfelder (1959), the mole fraction of dimers in argon is 0.2% at 0.1 MPa and 200 K, while this fraction reduces to 0.04% when the temperature rises to 600 K.

The initial–density coefficient $\lambda^{(1)}$ for the thermal conductivity of argon here is identified with λ_1 introduced in equation (14.8). On the other hand, it was calculated as

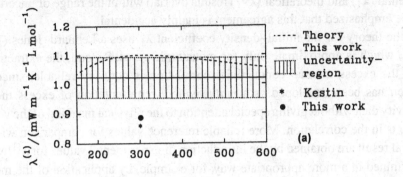

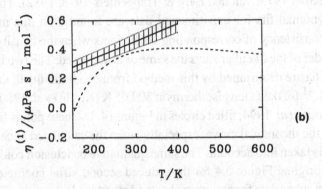

Fig. 14.6. Initial–density coefficient for thermal conductivity (a) and viscosity (b). Solid curves, are deduced from excess correlations, equation (14.10), and dashed curves are calculated from theory, equation (5.19). Data points for λ_1 are directly derived from measurements along isotherms.

a function of temperature by using the practical formula (5.19) for the reduced second thermal conductivity virial coefficient B_λ^* and the values for the potential parameters σ and ϵ/k_B from Table 5.2. The result is represented by a dashed line in Figure 14.6a. This figure covers a range between about 1 and 4 in reduced temperatures. In this range the theoretically calculated coefficient $\lambda^{(1)}$ shows a maximum (see Section 5.2) which does not appear in the corresponding Figure 5.4 for B_λ^*. This is due to the definitions of $\lambda^{(1)}$ and of the second thermal conductivity virial coefficient B_λ in equations (5.6) and (5.7), which imply that $\lambda^{(1)}$ contains both the temperature–dependence of B_λ^* and of the dilute–gas contribution $\lambda^{(0)}$ given by equations (14.4) and (14.7).

The result for the coefficient λ_1 derived from the excess thermal conductivity, which is represented by the linear relation according to equation (14.10), is indicated in Figure 14.6a by a solid straight line together with the uncertainty region of about $\pm 7\%$, which corresponds to one standard deviation σ_λ. It is obvious from this figure that the experimental (λ_1) and theoretical ($\lambda^{(1)}$) results overlap within the range of uncertainty. It must be emphasized that this agreement is mainly accidental.

First, the theory for the initial–density coefficient λ_1 uses a Lennard–Jones (12–6) potential, which is approximate only for monatomic gases. Second, the extraction of λ_1 from the excess thermal conductivity correlation is not very reliable, since this correlation has been developed as a *mathematical* representation of excess thermal conductivity data without giving special attention to the physical meaning of the various coefficients in the correlation. More reliable reference values for comparison with the theoretical result are obtained by the introduction of experimental values for $\lambda^{(1)}$ which are determined in a more appropriate way, for example, by application of the method of consistent coefficients (Hanley *et al.* 1969; Hanley & Haynes 1975; Kestin *et al.* 1971; Kestin 1974; Snel 1973; van den Berg & Trappeniers 1978, 1982). The various coefficients of polynomial fits for experimental data are found with this method by verification of the consistency of corresponding coefficients when the density range is extended and the order of the fit either stays the same or is increased. The two individual points in Figure 14.6a are determined by this method from quadratic fits in the density range up to $12 \, \text{mol} \, \text{L}^{-1}$ for data along isotherms at 300.65 K (Kestin *et al.* 1980) and 302 K (taken from Tiesinga *et al.* 1994; filled circles in Figure 14.3). These points fall clearly (about 20%) below the theoretical curve, especially when the calculated uncertainty of no more than ±3% is taken into account. The same qualitative conclusion could already be drawn from the original Figure 5.4 for the reduced second virial coefficient of the thermal conductivity, when data for argon are observed at reduced temperatures between 1 and 4.

The initial–density dependence η_1 for the viscosity, defined in equation (14.8) – and represented by $\eta^{(1)}$ in equation (5.6) – is shown versus temperature in Figure 14.6b. The theoretical curve for $\eta^{(1)}$ is calculated as in the case for the thermal conductivity, now with the use of equations (14.3) and (14.7) for the dilute–gas viscosity. The experimental curve is again a straight line, which is extracted from the excess correlation according to equation (14.10) and is valid within the temperature range given above. Also in this case the calculated uncertainty region is shown. The fact that the experimental results are substantially higher than the theoretical results is confirmed in Figure 5.3, which presents the reduced second virial coefficient for the viscosity, when argon data are examined in the reduced temperature range between 1 and 3.

14.1.5 Critical contribution

Having determined the background transport coefficients $\bar{\lambda}$ and $\bar{\eta}$, it is possible to compare experimental data for argon in the critical region with theoretical expressions for $\Delta\lambda_c$ and $\Delta\eta_c$ presented in Chapter 6

$$\Delta\lambda_c = \frac{R_D k_B T}{6\pi\eta\xi} \rho c_P \, (\Omega - \Omega_0) \tag{14.11}$$

$$\Delta\eta_c = \bar{\eta}\left[\exp(zcH) - 1\right] \tag{14.12}$$

where R_D, z and $c \approx 1$ are constants, ξ is a correlation length defined in Chapter 6, c_P is the isobaric specific heat capacity and $(\Omega - \Omega_0)$ and H are crossover functions also specified in the Appendix to Chapter 6. These equations contain just one system–dependent parameter q_D^{-1}, a microscopic cutoff distance, since for the amplitude R_D the value 1.03 has been taken which is consistent with the range specified in Chapter 6. Primarily, the evaluation was carried out by using 142 thermal conductivity data from Tiesinga *et al.* (1994), which were obtained with a guarded parallel–plate cell (Mostert *et al.* 1989) along ten isotherms at temperatures up to 0.14 K above T_c. The estimated accuracies for these measurements vary between 2% outside the critical region up to 15% for temperatures within 0.4 K of T_c. A fit to the data points with weights based on these accuracies gave the result

$$q_D^{-1} = 0.191\,\text{nm} \tag{14.13}$$

The various constants used in this analysis are given in Table 14.2. The result for q_D^{-1} is close to the value $q_D^{-1} = 0.200$ nm deduced by Perkins *et al.* (1991) from a set of data points further above T_c.

Table 14.2. *Constants in equations (14.11) and (14.12) for $\Delta\lambda_c$ and $\Delta\eta_c$.*

Critical exponents: $z = 0.063$, $\nu = 0.630$, $\gamma = 1.239$	Constant in equation (14.12): $c = 1$
Critical amplitudes: $R_D = 1.03$, $\xi_0 = 0.164$ nm, $\Gamma = 0.0616$	Cutoff wavenumber: $q_D^{-1} = 0.191$ nm.

The values calculated for the critical enhancement of the thermal conductivity are represented by the solid curves in Figure 14.7. Although the deviations of the experimental points increase significantly for temperatures within 0.4 K of T_c, in view of the experimental accuracies, the calculated behavior of the critical enhancement is consistent with the experimentally observed behavior. In the above mentioned primary analysis only thermal conductivity data of Tiesinga *et al.* (1994) in the critical region were taken into account. Subsequently, the results of this analysis are compared with data of Bailey & Kellner (1967, 1968) and of Roder *et al.* (1989), who investigated the critical enhancement of the thermal conductivity with a concentric–cylinder (CC) and a transient hot–wire (THW) method, respectively. Figure 14.8 presents the thermal conductivity data obtained by these authors in the critical region along isotherms which in reduced temperature $T/T_c - 1$ are more than one decade away from the critical temperature of most of the isotherms of Tiesinga *et al.* This limitation in the approach of the critical temperature is due to the occurrence of heat transfer by convection under near–critical circumstances in both the CC and THW method (Wakeham *et al.* 1991). The solid

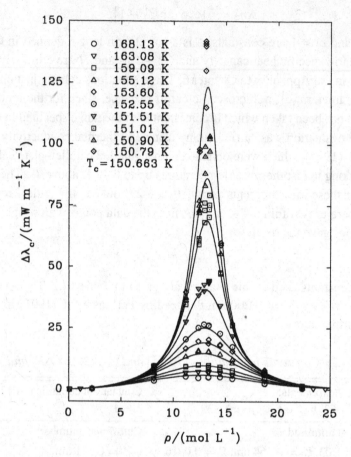

Fig. 14.7. Critical thermal–conductivity enhancement deduced from data obtained by Tiesinga *et al.* (1994). The curves represent the values for the enhancement calculated from the crossover theory, equation (14.11), with constants from Table 14.2.

curves in this figure refer again to the results of the primary analysis. Therefore, it may be concluded that these results are also consistent with the two experimental data sets within the combined accuracy of the measurements. To appreciate the magnitude of the critical thermal conductivity enhancement, it is interesting to compare the thermal conductivity scale in Figure 14.8 with that shown in Figure 14.7.

Roder (1984) and Roder *et al.* (1989) have presented an empirical 'engineering function' to calculate the critical enhancement $\Delta\lambda_c(\rho, T)$ of the thermal conductivity for practical purposes. Since this function is derived from their experimental THW data, the range of validity is restricted to temperatures more than 7 K above the critical temperature and, moreover, the dependence on temperature and density is too approximate. Therefore, a more accurate engineering function has been adopted, which can

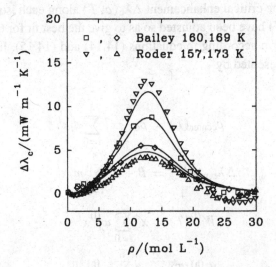

Fig. 14.8. Critical thermal–conductivity enhancement deduced from data obtained by Bailey & Kellner (1968) and Roder *et al.* (1989). The curves represent the values for the enhancement calculated from the crossover theory, equation (14.11), with constants from Table 14.2.

be written

$$\Delta\lambda_c(\rho, T) = \Delta\lambda_{c,max}(T) \, 2^{-\left[(\rho_{center}(T) - \rho)/W^{(l)}(T)\right]^{E^{(l)}(T)}}$$

$$(\rho \leq \rho_{center}) \qquad (14.14)$$

$$\Delta\lambda_c(\rho, T) = \Delta\lambda_{c,max}(T) \, 2^{-\left[(\rho - \rho_{center}(T))/W^{(h)}(T)\right]^{E^{(h)}(T)}}$$

$$(\rho \geq \rho_{center}) \qquad (14.15)$$

This function contains six parameters which all depend on temperature; the density $\rho_{center}(T)$ where the critical enhancement reaches its maximum value $\Delta\lambda_{c,max}(T)$; the low–density half–width $W^{(l)}(T)$, which means that at the density $\rho = \rho_{center}(T) - W^{(l)}(T)$ the enhancement equals $(1/2)\Delta\lambda_{c,max}(T)$; the high–density half–width $W^{(h)}(T)$; the powers $E^{(l)}(T)$ and $E^{(h)}(T)$, which govern the shape of the more or less bell–shaped low– and high–density branches of the curve for $\Delta\lambda_c$ *versus* ρ.

By use of equation (14.11), with the corresponding constants in Table 14.2, a total of 4340 values for $\Delta\lambda_c(\rho, T)$ were calculated, spread over the density range 0–30 mol L^{-1} along 31 isotherms in the temperature range 150.8–300 K. With these 'experimental' values, fits were made for the given temperature–dependent parameters in terms of x, where

$$x = \ln\left(\frac{T}{T_c}\right) \qquad (14.16)$$

Subsequently, for the critical enhancement $\Delta\lambda_c(\rho, T)$ along each isotherm the powers $E^{(l)}(T)$ and $E^{(h)}(T)$ have been adjusted so as to give the best fit for the density dependence of the isotherm according to equations (14.14) and (14.15). In this way the six parameters are represented by

$$\rho_{center}(T) = \rho_c + x^a \sum_{i=0}^{4} d_i x^i \tag{14.17}$$

$$\Delta\lambda_{c,max}(T) = B + x^b \sum_{i=0}^{5} m_i x^i \tag{14.18}$$

$$W^{(l)}(T) = x^c \sum_{i=0}^{5} w_i^{(l)} x^i \tag{14.19}$$

$$W^{(h)}(T) = x^d \sum_{i=0}^{6} w_i^{(h)} x^i \tag{14.20}$$

$$E^{(l)}(T) = F + x^f \sum_{i=0}^{4} e_i^{(l)} x^i \tag{14.21}$$

$$E^{(h)}(T) = G + x^g \sum_{i=0}^{5} e_i^{(h)} x^i \tag{14.22}$$

where the various coefficients are given in Table 14.3.

The function that describes the temperature dependence of the density where the critical enhancement reaches its maximum shows a clear minimum. From the half–width functions it can be seen that the enhancement curve is asymmetric, where, except for the lowest temperatures, the high–density branch of the curve is wider than the low–density branch. The power of the high–density branch is for nearly all temperatures smaller than the power of the low–density branch, and both branches of the enhancement are not Gaussian, in contradiction with the assumption of Roder *et al.* (1989), since the values of the powers vary between 1.3 and 2.6. In the above mentioned ranges of temperature and density where the engineering function for the critical enhancement $\Delta\lambda_c(\rho, T)$ has been developed, this function represents well the theoretical crossover function for $\Delta\lambda_c$. The standard deviation of this representation at 150.8 K amounts to 2.2% relative to the total thermal conductivity values $\lambda(\rho_c, T)$, whereas the maximum deviation is 5.7%. These percentages reduce to 0.5% and 0.9% at $T = 155$ K and are both less than 0.2% for $T > 200$ K. Since the deviations of the available experimental data from the theoretical crossover function exceed these percentages appreciably, it may be concluded that the engineering function offers a useful tool to calculate the critical enhancement of the thermal conductivity.

Table 14.3. *Coefficients in equations (14.17)–(14.22) of the engineering function for the empirical representation of the critical thermal conductivity enhancement* $\Delta\lambda_c$.

Coeff.	Value	Coeff.	Value	Coeff.	Value
a	0.3735	b	-0.5606	c	0.3060
		B	-3.3588		
d_0	-2.3145	m_0	2.3876	$w_0^{(l)}$	13.2700
d_1	16.3518	m_1	3.4421	$w_1^{(l)}$	-17.8417
d_2	-127.8145	m_2	-10.1111	$w_2^{(l)}$	8.9532
d_3	313.4899	m_3	4.4522	$w_3^{(l)}$	-35.1502
d_4	-220.2045	m_4	26.6275	$w_4^{(l)}$	190.0812
		m_5	-29.4924	$w_5^{(l)}$	-184.9309
d	0.3588	f	0.3500	g	-0.4695
		F	1.0937	G	1.0489
$w_0^{(h)}$	17.1761	$e_0^{(l)}$	2.0388	$e_0^{(h)}$	0.0077
$w_1^{(h)}$	-49.4336	$e_1^{(l)}$	4.0214	$e_1^{(h)}$	3.7725
$w_2^{(h)}$	43.4646	$e_2^{(l)}$	-20.0618	$e_2^{(h)}$	-25.8897
$w_3^{(h)}$	700.7929	$e_3^{(l)}$	35.3881	$e_3^{(h)}$	78.9252
$w_4^{(h)}$	-2247.7372	$e_4^{(l)}$	-23.9794	$e_4^{(h)}$	-91.6265
$w_5^{(h)}$	2435.6885			$e_5^{(h)}$	34.5935
$w_6^{(h)}$	-885.2105				

14.1.6 Conclusions

Representative equations have been developed for the thermal conductivity and viscosity of argon as a function of temperature and density. An extensive set of experimental data for the thermal conductivity in the critical region is consistent with the behavior deduced from the theory for the effects of critical fluctuations on the thermal conductivity. For easy calculation of the critical enhancement of the thermal conductivity an empirical engineering form is presented, which represents approximately the theoretical equations for the enhancement with experimentally adjusted parameters. In Figure 14.9 the thermal conductivity surface of argon in the near–supercritical region as calculated from the correlations for the background and the engineering form for the critical enhancement is shown.

The full tables of values of the thermal conductivity of argon are presented in the paper of Tiesinga *et al.* (1994).

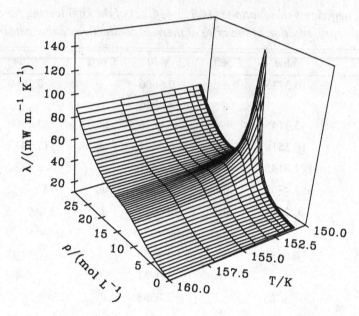

Fig. 14.9. Thermal conductivity surface of argon in the near–supercritical region.

References

Anisimov, M. A., Koval'chuk, B. V., Rabinovich, V. A. & Smirnov, V. A. (1978). *Thermophysical Properties of Substances and Materials* (Russ.), **12**, 86–106. Moscow: GOSSTANDARD.

Aziz, R. A. (1984). Interatomic potentials for rare–gases: Pure and mixed interactions, in *Inert Gases. Potentials, Dynamics and Energy Transfer in Doped Crystals*, ed. Klein, M. L., pp. 5–86. Berlin: Springer–Verlag.

Aziz, R. A. (1987). Accurate thermal conductivity coefficients for argon based on a state–of–the–art interatomic potential. *Int. J. Thermophys.*, **8**, 193–204.

Bailey, B. J. & Kellner, K. (1967). The thermal conductivity of argon near the critical point. *Brit. J. Appl. Phys.*, **18**, 1645–1647.

Bailey, B. J. & Kellner, K. (1968). The thermal conductivity of liquid and gaseous argon. *Physica*, **39**, 444–462.

Bich, E., Millat, J. & Vogel, E. (1990). The viscosity and thermal conductivity of pure monatomic gases from their normal boiling point up to 5000 K in the limit of zero density and at 0.101325 MPa. *J. Phys. Chem. Ref. Data*, **19**, 1289–1305.

Collins, D. J. & Menard, W. A. (1966). Measurement of the thermal conductivity of noble gases in the temperature range 1500 to 5000 deg Kelvin. *Trans. ASME*, 52–56.

Flynn, G. P., Hanks, R. V., Lemaire, N. A. & Ross, J. (1962). Viscosity of nitrogen, helium, neon, and argon from -78.5°C to 100°C below 200 atmospheres. *J. Chem. Phys.*, **38**, 154–162.

Gracki, J. A., Flynn, G. P. & Ross, J. (1969). Viscosity of nitrogen, helium, hydrogen, and argon from -100 to 25°C up to 150–250 atm. *J. Chem. Phys.*, **51**, 3856–3863.

Hanley, H. J. M. (1974). The viscosity and thermal conductivity coefficients of dilute argon, krypton, and xenon. *J. Phys. Chem. Ref. Data*, **2**, 619–642.

Hanley, H. J. M. & Haynes, W. M. (1975). The density expansion of the viscosity coefficient. *J. Chem. Phys.*, **63**, 358–361.

Hanley, H. J. M., McCarty, R. D. & Sengers, J. V. (1969). Density dependence of experimental transport coefficients of gases. *J. Chem. Phys.*, **50**, 857–870.

Haynes, W. M. (1973). Viscosity of gaseous and liquid argon. *Physica*, **67**, 440–470.

Ikenberry, L. D. & Rice, S. A. (1963). On the kinetic theory of dense fluids XIV. Experimental and theoretical studies of thermal conductivity in liquid Ar, Kr, Xe, and CH_4. *J. Chem. Phys.*, **39**, 1561–1571.

Kestin, J., Knierim, K., Mason, E. A., Najafi, B., Ro, S. T. & Waldman, M. (1984). Equilibrium and transport properties of the noble gases and their mixtures at low density. *J. Phys. Chem. Ref. Data*, **13**, 229–303.

Kestin, J. (1974). Experimental study of the existence of the logarithmic term in viscosity and thermal conductivity of gases at high pressure, in *Proceedings Fourth Int. Conf. on High Pressure*, pp. 518–522. Kyoto.

Kestin, J., Paul, R., Clifford, A. A. & Wakeham, W. A. (1980). Absolute determination of the thermal conductivity of the noble gases at room temperature up to 35 MPa, *Physica*, **100A**, 349–369.

Kestin, J., Paykoç, E. & Sengers, J. V. (1971). On the density expansion for viscosity in gases. *Physica*, **54**, 1–19.

Krauss, R., Luettmer–Strathmann, J., Sengers, J. V. & Stephan, K. (1993). Transport properties of 1,1,1,2–tetrafluoroethane (R134a). *Int. J. Thermophys.*, **14**, 951–988.

Le Neindre, B. (1969). Contribution à l' étude expérimentale de la conductivité thermique de quelques fluides a haute température et à haute pression. Ph.D. Thesis, Université de Paris.

Le Neindre, B. (1972). Contribution à l' étude expérimentale de la conductivité thermique de quelques fluides à haute température et à haute pression. *Int. J. Heat Mass Transfer*, **15**, 1–24.

Le Neindre, B., Garrabos, Y. & Tufeu, R. (1989). Thermal conductivity of dense noble gases, *Physica*, **156A**, 512–521.

Li, S. F. Y., Papadaki, M. & Wakeham, W. A. (1994). Thermal conductivity of low–density polyatomic gases, in *Proc. 22nd International Thermal Conductivity Conference*, ed. T. W. Tong, pp. 531–542. Lancaster: Technomic Publ.

Luettmer–Strathmann, J., Tang, S. & Sengers, J. V. (1992). A parametric model for the global thermodynamic behavior of fluids in the critical region. *J. Chem. Phys.*, **97**, 2705–2717.

Maitland, G. C., Rigby, M., Smith, E. B. & Wakeham, W. A. (1987). *Intermolecular Forces: Their Origin and Determination*. Oxford: Clarendon Press.

Mardolcar, U. V., Nieto de Castro, C. A. & Wakeham, W. A. (1986). Thermal conductivity of argon in the temperature range 107 to 423 K. *Int. J. Thermophys.*, **7**, 259–272.

Michels, A., Botzen, A. & Schuurman, W. (1954). The viscosity of argon at pressures up to 2000 atmospheres. *Physica*, **20**, 1141–1148.

Millat, J., Mustafa, M., Ross, M., Wakeham, W. A. & Zalaf, M. (1987). The thermal conductivity of argon, carbon dioxide and nitrous oxide. *Physica*, **145A**, 461–497.

Mostert, R., van den Berg, H. R. & van der Gulik, P. S. (1989). A guarded parallel–plate instrument for measuring the thermal conductivity of fluids in the critical region. *Rev. Sci. Instrum.*, **60**, 3466–3474.

Mostert, R., van den Berg, H. R., van der Gulik, P. S. & Sengers, J. V. (1990). The thermal conductivity of ethane in the critical region. *J. Chem. Phys.*, **92**, 5454–5462.

Nieto de Castro, C. A., Friend, D. G., Perkins R. A. & Rainwater, J. C. (1990). Thermal conductivity of a moderately dense gas. *Chem. Phys.*, **145**, 19–21.

Perkins, R. A., Friend, D. G., Roder, H. M. & Nieto de Castro, C. A. (1991). Thermal conductivity surface of argon: A fresh analysis. *Int. J. Thermophys.*, **12**, 965–983.

Rabinovich, V. A., Vasserman, A. A., Nedostup, V. I. & Veksler, L. S. (1988). A method of correlation of experimental data on viscosity and thermal conductivity, in *Thermophysical Properties of Neon, Argon, Krypton, and Xenon*, ed. T. B. Selover, Jr., pp. 202–214. Washington, D.C.: Hemisphere.

Roder, H. M. (1984). Thermal conductivity of methane for temperatures between 110 and 310 K with pressures to 70 MPa. *Int. J. Thermophys.*, **6**, 119–142.

Roder, H. M., Perkins, R. A. & Nieto de Castro, C. A. (1988). Experimental thermal conductivity, thermal diffusivity and specific–heat values of argon and nitrogen, NIST IR 88–3902 (National Institute of Standards and Technology, Boulder, CO).

Roder, H. M., Perkins, R. A. & Nieto de Castro, C. A. (1989). The thermal conductivity and heat capacity of gaseous argon. *Int. J. Thermophys.*, **10**, 1141–1164.

Sengers, J. V. (1962). Thermal conductivity measurements at elevated gas densities including the critical region. Ph.D. Thesis, University of Amsterdam.

Sengers, J. V. (1971). Transport properties of fluids near critical points, in *Critical Phenomena, Varenna Lectures, Course LI.*, ed. M.S. Green, pp. 445–507. New York: Academic Press.

Sengers, J. V. & Keyes, P.H. (1971). Scaling of the thermal conductivity near the gas–liquid critical point. *Phys. Rev. Lett.*, **26**, 70–73.

Sengers, J.V. (1972). Transport processes near the critical point of gases and binary liquids in the hydrodynamic regime. *Ber. Bunsenges. Phys. Chem.*, **76**, 234–249.

Snel, J. A. A. (1973). De dichtheidsontwikkeling van de warmtegeleidingscoefficient van kooldioxyde en krypton. Ph.D. Thesis, University of Amsterdam.

Springer, G. S. & Wingeier, E. W. (1973). Thermal conductivity of neon, argon, and xenon at high temperatures. *J. Chem. Phys.*, **59**, 2747–2750.

Stewart, R. B. & Jacobsen, R. T. (1989). Thermodynamic properties of argon from the triple point to 1200 K with pressures to 1000 MPa. *J. Phys. Chem. Ref. Data*, **18**, 639–678.

Stogryn, D. E. & Hirschfelder J. O. (1959). Contribution of bound, metastable, and free molecules to the second virial coefficient and some properties of double molecules. *J. Chem. Phys.*, **31**, 1531–1545.

Tiesinga, B. W., Sakonidou, E. P., van den Berg, H. R., Luettmer–Strathmann, J. & Sengers, J. V. (1994). The thermal conductivity of argon in the critical region. *J. Chem. Phys.*, **101**, 6944–6963.

Tufeu, R., (1971). Etude expérimentale en fonction de la température et de la pression de la conductivité thermique de l'ensemble des gaz rares et des mélanges hélium–argon. Ph.D. Thesis, Université de Paris.

van den Berg, H. R. & Trappeniers, N. J. (1978). Experimental determination of the logarithmic term in the density expansion of the viscosity coefficient of krypton at 25°C. *Chem. Phys. Lett.*, **58**, 12–17.

van den Berg, H. R. & Trappeniers, N. J. (1982). The density dependence of the viscosity of krypton including the logarithmic term, in *Proceedings 8th Symposium on Thermophysical Properties*, ed. J. V. Sengers, Vol. I, pp. 172–177. New York: ASME.

Vargaftik, N. B. & Zimina, N. Kh. (1964). Thermal conductivity of argon at high temperatures (Russ.). *Teplofiz. Vys. Temp.*, **2**, 716–723.

Vermesse, J. & Vidal, D. (1973). Mesure du coefficient de viscosité de l'argon à haute pression. *C. R. Acad. Sc. Paris*, **277**, 191–193.

Vesovic, V., Wakeham, W. A., Luettmer–Strathmann, J., Sengers, J. V., Millat, J., Vogel, E. & Assael, M. J. (1994). The transport properties of ethane: II. Thermal conductivity. *Int. J. Thermophys.*, **15**, 33–66.

Vesovic V., Wakeham, W. A., Olchowy, G. A., Sengers, J. V., Watson, J. T. R. & Millat, J. (1990). The transport properties of carbon dioxide. *J. Phys. Chem. Ref. Data*, **19**, 763–808.

Wakeham, W.A., Nagashima, A. & Sengers, J. V. (1991). Measurement of the transport properties of fluids. *Experimental Thermodynamics, Volume III*. Oxford: Blackwell Scientific.

Younglove, B. A. & Hanley, H. J. M. (1986). The viscosity and thermal conductivity coefficients of gaseous and liquid argon. *J. Phys. Chem. Ref. Data*, **15**, 1323–1337.

14.2 Diatomic fluids – Nitrogen

J. MILLAT
NORDUM Institut für Umwelt und Analytik, Kessin/Rostock, Germany

V. VESOVIC
Imperial College, London, UK

14.2.1 Introduction

The nitrogen diatom is one of the simplest stable molecules that exhibits all the features
which are typical of polyatomic molecules compared to the structureless, monatomic
species. Furthermore, this diatomic molecule has been found simple enough for the
results of the rigorous kinetic theory detailed in Chapter 4 to be applied to the analysis
of its dilute–gas transport properties. Hence for the first time this kind of analysis
is applied to real molecules rather than having to use simplifying molecular models
such as rough spheres, loaded spheres, rigid spherocylinders *etc*. This also provides
the opportunity to compare the applicability of different approximate schemes for the
evaluation of dilute–gas transport properties. Accordingly, this section is focused on
the dilute–gas transport properties $X^{(0)}(T)$, which are given as the first term of the
customary residual concept

$$X(\rho, T) = X^{(0)}(T) + \Delta X(\rho, T) + \Delta X_c(\rho, T) \tag{14.23}$$

where $\Delta X(\rho, T)$ is the excess contribution and $\Delta X_c(\rho, T)$ represents the critical enhancement ($X = \eta,\ \lambda$). The latter two contributions will be discussed briefly in 14.2.5
and 14.2.6 in order to complete the relevant transport property surface. Methods to
describe the representations for the excess and critical enhancement contributions are
discussed in greater detail in Chapters 6 and 7 as well as in Sections 14.1 and 14.3.

Finally, the complete surface has been compared (Millat & Wakeham 1995) with
other representations given in the literature, the most recent being that of Stephan *et al.*
(1987).

14.2.2 Semiempirical correlation of the dilute–gas transport properties

The dilute–gas contribution to the transport properties is for most practical purposes
an experimentally accessible quantity. Some techniques were developed in order to
measure it directly, but in view of the possible effect of the initial density dependence
(see Chapter 5) it is advisable, especially at subcritical temperatures, to analyze transport
coefficients as a function of density along isotherms. Appropriate extrapolation schemes
(see 14.1) can be used to deduce the zero–density value of the transport property, which
can be identified with the dilute–gas value (see Chapter 4). It is thus always possible to
analyze $X^{(0)}(T)$ independently of the remaining terms in equation (14.23).

Based on experimental dilute–gas data derived using this scheme, correlations for the dilute–gas viscosity and thermal conductivity of nitrogen were published by Vogel *et al.* (1989) and Millat & Wakeham (1989a,b) respectively. Since more recent experimental findings (Millat *et al.* 1989b) confirm the results presented there, these correlations are recommended here as well.

14.2.2.1 Viscosity

As detailed in Chapter 4, the viscosity is related to just one reduced effective collision cross section. In practical form the expression reads

$$\eta^{(0)}(T) = \frac{0.06706227}{\pi} \frac{(TM)^{1/2}}{\sigma^2 \mathfrak{S}_\eta^*} \qquad (14.24)$$

where $\mathfrak{S}_\eta^* = \mathfrak{S}(2000)/\pi\sigma^2 S_\eta f_\eta$ is a reduced effective cross section, T the temperature in K, M the molar mass in $\mathrm{g\,mol^{-1}}$, σ a length scaling parameter in nm, S_η a correction factor for spin polarization (angular momentum polarization), f_η a dimensionless higher-order correction factor and $\eta^{(0)}$ is in units of μPa s.

The effective cross-section $\mathfrak{S}_\eta^*$ for nitrogen is expressed in the functional form

$$\ln \mathfrak{S}_\eta^* = \sum_{i=0}^{4} a_i (\ln T^*)^i \qquad (14.25)$$

where $T^* = k_B T/\epsilon$ is the reduced temperature, and ϵ/k_B an energy scaling parameter in K. In developing the correlation, experimental values of $\mathfrak{S}_\eta^*(T^*)$ were derived from the given primary viscosity data (see Vogel *et al.* 1989) and then fitted to equation (14.25). The details concerning the selection of primary data which cover the temperature range from 110 K to 2000 K and the correlation procedure are given in the original paper and will not be repeated here. However, for completeness, optimized coefficients a_i, together with values for the scaling factors, are presented in Table 14.4.

Table 14.4. *Coefficients for the representation of the dilute–gas viscosity and thermal conductivity of nitrogen.*

i	a_i	b_i
0	0.2077338	4. 5384086
1	-0.472230	-0. 71858394
2	0.0878508	0. 74042225
3	0.0107001	-0. 91728276
4	-0.00518589	0. 68036729
5		-0. 27205905
6		0. 055323448
7		-0. 0045342078

$\epsilon/k_B = 104.2$ K, $\sigma = 0.3632$ nm, $M = 28.0134 \mathrm{\,g\,mol^{-1}}$

14.2.2.2 Thermal conductivity

The uncertainty of experimental data for thermal conductivity is usually much higher than that for viscosity (Millat & Wakeham 1989a). This makes it difficult to assess and assign data to one of the two categories of primary and secondary data based on quoted accuracies alone. Hence, a development of a correlation was based on a scheme (Millat & Wakeham 1989a,b; Millat *et al.* 1989a) which takes advantage of kinetic theory results within the two–flux approach detailed in Chapter 4.

In order to perform such an analysis, equation (14.25) for $\mathfrak{S}_\eta^*$ (which was identified with $\mathfrak{S}^*(2000)$) is used as a subcorrelation. Similarly, a representation for the rotational collision number was derived based on experimental information and the Brau–Jonkman formula (Brau & Jonkman 1970). Furthermore, it has been assumed that vibrational collision numbers for nitrogen are very large (order of magnitude of 10^8) and will not influence significantly the subcorrelation for $\mathfrak{S}^*(0001)$.

The remaining quantity, $\mathfrak{S}^*(1001)$, is not experimentally accessible but could be deduced using kinetic theory expressions also given in Chapter 4, and the subcorrelations for $\mathfrak{S}^*(0001)$ and $\mathfrak{S}^*(2000)$ as well as data for the thermal conductivity. A consistency test using D_{int}/D as a criterion then could be applied to exclude experimental data from the primary data set which are burdened with systematic errors (Millat & Wakeham 1989a,b; Millat *et al.* 1989b).

Naturally, this approach (see 14.2.3) is rather cumbersome for practical purposes. Therefore, in a final form the correlated data have been represented using the practical Thijsse expression (see Chapter 4)

$$\lambda^{(0)}(T) = 139.4627\,(1+r^2)\left(\frac{T}{M}\right)^{1/2}\frac{1}{\pi\sigma^2\mathfrak{S}_\lambda^*} \tag{14.26}$$

where $\mathfrak{S}_\lambda^* = \mathfrak{S}(10E)/(\pi\sigma^2 S_\lambda\,f_\lambda)$ is the reduced effective collision cross section for nitrogen that is expressed as

$$\ln\mathfrak{S}_\lambda^* = \sum_{i=0}^{7} b_i\,(\ln T^*)^i \tag{14.27}$$

Here, $r^2 = 2C_{int}/5R$, with C_{int} being the contribution of internal degrees of freedom to the heat capacity at constant volume, S_λ and f_λ are dimensionless correction factors for angular momentum polarization and higher-order corrections respectively and $\lambda^{(0)}$ is in units of mW m^{-1} K^{-1}. The primary thermal conductivity data for nitrogen cover the temperature range from 282 K to 942 K (Millat & Wakeham 1989a). In developing the thermal conductivity correlation, values of $\mathfrak{S}_\lambda^*(T^*)$ have been deduced from each of the experimental data points by means of scaling parameters derived from the analysis of viscosity data. They were then fitted applying appropriate statistical weights (Millat & Wakeham 1989a) to the functional form given in equation (14.27). Values for the coefficients b_i are included in Table 14.4. It has been demonstrated (Millat *et al.* 1989b) that equations (14.26) and (14.27) are applicable in the temperature range from 110 to 2000 K.

14.2.3 Calculation of dilute–gas transport properties using the classical trajectory method

Only very recently it became feasible to carry out classical trajectory calculations of transport properties of nitrogen (Heck & Dickinson 1994; Heck *et al.* 1994). In these calculations the nitrogen molecule has been treated as a rigid rotor, and the intermolecular pair potential was based on the anisotropic *ab initio* potential energy surface of van der Avoird *et al.* (1986). The details of these computations as well as the uncertainties of their results are described in the given papers and will not be repeated here.

More recently, Heck *et al.* (1994) extended the calculations to include second–order corrections. Thus, it became possible to compare correlated data, as given in 14.2.2, with calculated second–order kinetic theory results.

The results of a comparison between correlated and calculated data are demonstrated for viscosity and thermal conductivity in Figure 14.10. The agreement for viscosity is excellent (within ±0.3%), whereas for thermal conductivity a systematic difference is found (about ±1%). Nevertheless, differences found for the latter property are within the combined uncertainties of the two methods.

The good agreement between experimental (correlated) and calculated (predicted) data enables a comparison between different approximations to the kinetic theory to be carried out.

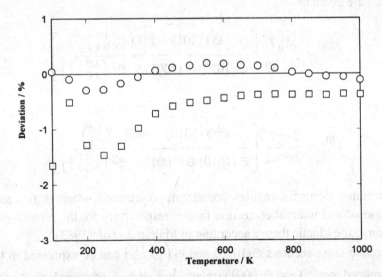

Fig. 14.10. Comparison of correlated data with those predicted using classical trajectory calculations. O - viscosity, □ - thermal conductivity.

14.2.4 Comparison of different approximations to the kinetic theory based on classical trajectory calculations

As detailed in Chapter 4, the viscosity of a polyatomic gas can be expressed as

$$\eta^{(0)} = \frac{k_B T}{\langle c \rangle \, \mathfrak{S}(2000)} S_\eta f_\eta \tag{14.28}$$

where

$$\langle c \rangle = 4 \left(\frac{k_B T}{\pi m} \right)^{1/2} \tag{14.29}$$

$\mathfrak{S}(2000)$ is the effective cross section that can be calculated from the intermolecular pair potential. S_η and f_η are correction factors for angular momentum polarization and for the *second* approximation to the kinetic theory (Maitland *et al.* 1983) respectively. Both correction factors can be related to different effective cross sections (see Heck *et al.* 1994 and Section 4.2) and are therefore obtainable by classical trajectory calculations.

Here, two approximate formulas are also considered that have been used to analyze the viscosity data for nitrogen in order to compare the achieved accuracy. Table 14.5a summarizes the relevant approximations.

The thermal conductivity within the two-flux approach is given by

$$\lambda^{(0)} = \left(\lambda_{tr}^{(0)} f_{\lambda,tr} + \lambda^{(0)}_{int} f_{\lambda,int} \right) S_\lambda \tag{14.30}$$

where the isotropic translational and internal contributions within the first–order approximation are given by

$$\lambda_{tr}^{(0)} = \frac{5k_B^2 T}{2m \langle c \rangle} \left[\frac{\mathfrak{S}(1001) - r\mathfrak{S}\left(\begin{smallmatrix} 1 0 1 0 \\ 1 0 0 1 \end{smallmatrix} \right)}{\mathfrak{S}(1010)\,\mathfrak{S}(1001) - \mathfrak{S}^2\left(\begin{smallmatrix} 1 0 1 0 \\ 1 0 0 1 \end{smallmatrix} \right)} \right] \tag{14.31}$$

and

$$\lambda^{(0)}_{int} = \frac{5k_B^2 T}{2m \langle c \rangle} \left[\frac{r^2\mathfrak{S}(1010) - r\mathfrak{S}\left(\begin{smallmatrix} 1 0 1 0 \\ 1 0 0 1 \end{smallmatrix} \right)}{\mathfrak{S}(1010)\,\mathfrak{S}(1001) - \mathfrak{S}^2\left(\begin{smallmatrix} 1 0 1 0 \\ 1 0 0 1 \end{smallmatrix} \right)} \right] \tag{14.32}$$

S_λ is a correction factor for angular momentum polarization, whereas $f_{\lambda,tr}$ and $f_{\lambda,int}$ are translational and internal correction factors respectively for the *second–order* approximation to the kinetic theory according to Maitland *et al.* (1983).

The effective cross sections $\mathfrak{S}(1010)$ and $\mathfrak{S}\left(\begin{smallmatrix} 1 0 1 0 \\ 1 0 0 1 \end{smallmatrix} \right)$ can be expressed in terms of $\mathfrak{S}(2000)$ (related to $\eta^{(0)}$) and $\mathfrak{S}(0001)$ (related to κ or ζ_{int}), whereas both $\mathfrak{S}(0001)$ and $\mathfrak{S}(1001)$ are used to define D_{int} (equation (4.31)). By rearranging equations (14.30)–(14.32), it is possible to express the thermal conductivity in terms of experimentally

Table 14.5. *Different approximations to the kinetic theory analyzed.*

(a) Viscosity

Approach	f_η	S_η
1st order (MPM)	1	1
KA/VMS	1	$\neq 1$
MMW	$\neq 1$	1
2nd order	$\neq 1$	$\neq 1$

(b) Thermal conductivity

Approach	f_η	S_η	$f_{\lambda,\mathrm{tr}}$; $f_{\lambda,\mathrm{int}}$	S_λ	Δ_{int}	$\frac{\rho D_{\mathrm{int}}}{\eta}$
Eucken	1	1	1	1	0	1
Modified Eucken	1	1	1	1	0	$\frac{6A^*}{5}$
Chapman/Cowling	1	1	1	1	0	$\frac{D_{\mathrm{int}}}{D}\frac{6A^*}{5}$
1st order (MPM)	1	1	1	1	$\neq 0$	$\frac{D_{\mathrm{int}}}{D}\frac{6A^*}{5}$
KA/VMS	1	$\neq 1$	1	$\neq 1$	$\neq 0$	$\frac{D_{\mathrm{int}}}{D}\frac{6A^*}{5}$
MMW	$\neq 1$	1	$\neq 1$	1	$\neq 0$	$\frac{D_{\mathrm{int}}}{D}\frac{6A^*}{5}$
2nd order	$\neq 1$	$\neq 1$	$\neq 1$	$\neq 1$	$\neq 0$	$\frac{D_{\mathrm{int}}}{D}\frac{6A^* f_D}{5}$
van den Oord/Korving 1st order	1	1	1	1	$\neq 0^a$	implicit[a]

[a] This scheme is based on the total-energy flux approach and assumes $\mathfrak{S}\left(\begin{smallmatrix}1\,0\,\mathrm{E}\\1\,0\,\mathrm{D}\end{smallmatrix}\right)=0$.

accessible quantities

$$\lambda^{(0)} = \frac{R\eta^{(0)}}{M}\frac{S_\lambda}{f_\eta S_\eta}\left[\frac{5}{2}\left(\frac{C_{V,\mathrm{tr}}}{R} - \Delta_{\mathrm{int}}\right) f_{\lambda,\mathrm{tr}}\right.$$

$$\left. + \left[\frac{\rho D_{\mathrm{int}}}{\eta}\right]^{(0)}\left(\frac{C_{V,\mathrm{int}}}{R} + \Delta_{\mathrm{int}}\right) f_{\lambda,\mathrm{int}}\right] \tag{14.33}$$

where the correction term for internal energy relaxation Δ_{int} is a function of $\mathfrak{S}(2000)$, $\mathfrak{S}(0001)$ and $\mathfrak{S}(1001)$ or, in other words, it absorbs $\eta^{(0)}$, ζ_{int} and D_{int}.

Equation (14.33) can be used to analyze the different levels of approximation within the two–flux approach. The Monchick–Pereira–Mason (MPM) approximation (Monchick *et al.* 1965) considers only the isotropic, first–order result; the Kagan–Afanas'ev (Kagan & Afanas'ev 1961) or Viehland–Mason–Sandler (Viehland *et al.* 1978)

(KA/VMS) approximation takes into account the correction for angular momentum polarization only, while the MMW approximation following Maitland et al. (1983) considers the second–order correction only.

Historically, a number of different, low–order approximations (see Chapter 4) has been used to predict the thermal conductivity. In principle, they can be deduced from the general formula (14.33) by making appropriate substitutions. Table 14.5b summarizes the plethora of these formulas and approaches.

Finally, the approximation proposed by van den Oord & Korving (1988) has been included. It is based on the first–order theory and the assumption that the so–called Mason–Monchick approximation holds exactly (see 4.2.2.4). It is also possible within this approach to derive an expression for $\lambda^{(0)}$ in terms of macroscopic quantities

$$\lambda^{(0)} = \frac{C_P}{M} \left\{ \frac{2}{3\eta^{(0)}} + \frac{2}{9} \left(\frac{C_{int}}{C_V} \right)^2 \frac{1}{\kappa^{(0)}} \right\}^{-1} \tag{14.34}$$

where $\kappa^{(0)}$ is the dilute–gas bulk viscosity.

The physical assumptions made in deriving the approximate formulas given in Table 14.5 are discussed in detail in Chapter 4.

All the comparisons between different approaches have been made by assuming that the 'full' second–order results of Heck et al. (1994) represent 'true' values of the transport coefficients. Then, the consistent set of effective cross sections published by Heck & Dickinson (1994) was used to calculate values for the different approximations. The results of the comparison are summarized in Figures 14.11, 14.12 and 14.13. From these graphs the following conclusions can be drawn:

For viscosity:

- Above 500 K the MPM approximation (first–order result) to viscosity leads to values which are about 0.8% smaller than second–order results, decreasing with decreasing temperature.
- The spin polarization correction amounts to about 0.2%.
- The MMW results deviate from the full second–order results by about 0.3%.

For thermal conductivity:

- The first–order results for thermal conductivity deviate from the full second–order results by an almost constant 1.2%. This value is within the uncertainty of the present correlation (Millat & Wakeham 1989a).
- The KA/VMS result, which includes the correction for spin polarization, is within ±0.8% of the second–order result. The MMW results show very similar behavior.
- The van den Oord–Korving approach (equation (4.64)) shows reasonable agreement. The largest differences (−3%) appear at low temperatures. The agreement can be improved by using the expression (14.34) in terms of experimentally accessible quantities. These results are also included in Figure 14.13 for both η and κ included from

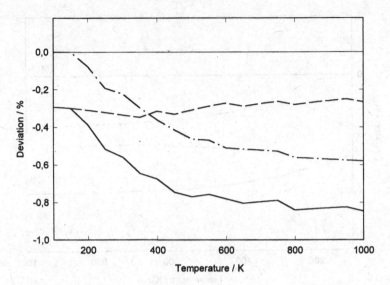

Fig. 14.11. Comparison of results for different approximations – Viscosity: —— MPM, - - - - MMW, - · - · - KA/VMS.

the first– and second–order approximation respectively. The relatively large differences between the calculated values for first– and second–order input data are mainly due to the fact that Heck *et al.* (1994) found that the second–order correction to bulk viscosity can be up to 20% at high temperatures.

- The simple Eucken formula leads to relatively good agreement only for low temperatures. The differences above 300 K are about 5%, increasing to about 8% at 1000 K.
- On the contrary, the modified Eucken expression is worse at the lowest temperatures but then shows differences of only 3%, decreasing to 2% in the temperature range 400 to 1000 K.
- The situation at low temperatures is slightly improved if the Chapman–Cowling approximation is used.

14.2.5 The excess contribution

14.2.5.1 Initial–density dependence

The excess contribution $\Delta X(\rho, T)$ can – at least in principle – be separated into the initial–density dependence $X^{(1)}\rho$ and higher–density contributions $\Delta_h X(\rho, T)$ (see Chapter 5).

For viscosity this means

$$\Delta\eta = \eta^{(1)}\rho + \Delta_h\eta \tag{14.35}$$

It is customary (Friend & Rainwater 1984; Rainwater & Friend 1987; Bich & Vogel

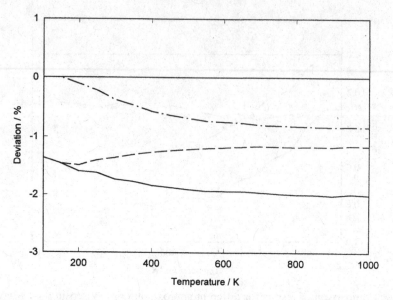

Fig. 14.12. Comparison of results for different approximations – Thermal conductivity (a):
——— MPM, - - - - - KA/VMS, - - - - MMW.

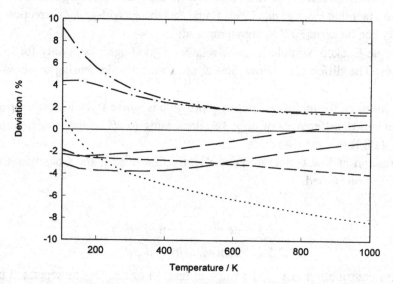

Fig. 14.13. Comparison of results for different approximations – Thermal conductivity (b):
· · · · · · Eucken formula, - ·· - modified Eucken expression, - - - - - Chapman-Cowling modified Eucken expression, van den Oord/Korving expression: - - - via $\mathfrak{S}(10E)$, ——— — using η and κ (1st approximation), – – – using η and κ (2nd approximation).

1991) to express the coefficient $\eta^{(1)}$ in terms of the second viscosity virial coefficient B_η by means of the definition

$$B_\eta(T) = \frac{\eta^{(1)}(T)}{\eta^{(0)}(T)} \qquad (14.36)$$

The latter can be predicted from the 'universal' expression (5.19) for the reduced second viscosity virial coefficient by using

$$B_\eta = 0.6022137\sigma^3 B_\eta^*(T^*) = \sum_{i=0}^{12} c_i \left(\sqrt{T^*}\right)^i \qquad (14.37)$$

B_η with coefficients c_i listed in Table 5.3 is in units of L mol^{-1}. The nitrogen scaling parameters for the Lennard–Jones (12–6) potential needed in equation (14.37) were taken from Table 5.2 ($\epsilon_{12-6}/k_B = 90.9$ K, $\sigma_{12-6} = 0.368$ nm).

Because of the existence of internal degrees of freedom and the anisotropy of the intermolecular potential the prediction of the initial–density dependence of thermal conductivity is still very uncertain (see Section 5.3). For this reason, an equation analogous to equation (14.35) was not applied for thermal conductivity.

14.2.5.2 The viscosity higher–density contribution

Although some successful models exist for the prediction of transport properties in the liquid region (see Chapters 5 and 10), it has not proved possible, so far, to incorporate these semiempirical formulations into a scheme for correlating the excess transport property covering the complete dense fluid range. Hence, the development of the higher–density contribution to the excess viscosity for nitrogen, like that for other polyatomic fluids, is based entirely on correlating experimental data.

Here, it is assumed that the critical enhancement is 'known' (see, for instance, Section 14.2.6.1). A critical review of the available data based on notations of primary instruments and consistency among data of different authors has established that 14 sets (see Millat & Wakeham 1995) can be classified as primary for the purposes of developing the excess viscosity correlation. These data cover a temperature range 66K $\leq$ T $\leq$ 570K and pressures up to about 100 MPa. Densities have been calculated using an equation of state proposed by Jacobsen *et al.* (1986) (ALLPROPS Package). The upper density limit for viscosity is about 30 mol L^{-1}. Since the temperature range for practical purposes may not be sufficient, the excess contribution at supercritical conditions was analyzed first. As for other fluids, it turned out to be temperature independent. This result was used to create quasi–experimental data and to extend the temperature range of the correlation up to 1100 K (Millat & Wakeham 1995).

The experimental excess viscosity has been determined for each datum by the use of equation (14.23) and subtracting the dilute-gas value, $\eta^{(0)}$, and the critical enhancement, $\Delta\eta_c$, from the measured value, $\eta(\rho, T)$. Once the excess viscosity data were generated,

Table 14.6. *Coefficients for the representation of the high-density contribution* $\Delta_h\eta$.

i	d_{i0}	d_{i1}	d_{i2}
2	0.0	1.182998	0.0
3	0.10402351	-1.2559502	2.5143303
4	-0.19622091(-1)	0.22553635	-0.52280720
5	0.14429076(-2)	-0.16314979(-1)	0.40221026(-1)
6	-0.46336936(-4)	0.52287461(-3)	-0.13353156(-2)
7	0.54426353(-6)	-0.61710605(-5)	0.16134957(-4)

i	d_{i3}	d_{i4}
2	-8.2197401	7.0445039
3	0.0	-1.3454738
4	0.23691788	0.76673384(-1)
5	-0.25021186(-1)	0.0
6	0.95204759(-3)	-0.10677951(-3)
7	-0.12415627(4)	0.21032420(-5)

the initial–density dependence $\eta^{(1)}\rho$ was subtracted to obtain the higher–density contribution, $\Delta_h\eta$. There is no theoretical guidance to the functional form of $\Delta_h\eta$, but it is customary to express it in terms of power series in the density and in the reciprocal reduced temperature. Thus,

$$\Delta_h\eta = \sum_{i=2}^{n}\sum_{j=0}^{m} d_{ij}\frac{\rho^i}{T^{*j}} \tag{14.38}$$

The fitting has been performed by use of the SEEQ algorithm based on the stepwise least-squares technique (de Reuck & Armstrong 1979). The appropriate statistical weights have been generated from the experimental uncertainties and were taken to be inversely proportional to density. The optimum values of coefficients d_{ij} are given in Table 14.6. No systematic trends have been observed when analyzing the deviations of the primary experimental data from the correlated values, and all the data have been fitted within $\pm 3\%$ or better.

14.2.5.3 The thermal conductivity excess contribution

Most of the available measurements of the thermal conductivity of nitrogen at elevated pressures have been carried out either in concentric cylinder or in transient hot-wire instruments. Both of these instruments can yield data of high accuracy and as such have to be considered as primary. Seven sets of data (see Millat & Wakeham 1995) were selected for the purpose of developing the excess thermal conductivity correlation. The primary data set consists of 864 data points, covering a temperature range 81 K $\leq$ T $\leq$ 973 K and densities up to 32 mol L^{-1}. The ascribed uncertainty of the data varies from $\pm 0.8\%$ to $\pm 3.0\%$.

Table 14.7. *Coefficients for the representation of the excess thermal conductivity.*

i	e_{i0}	e_{i1}	e_{i2}
1	0.92676029	0.44047999	0.0
2	0.84730662(-1)[a]	-0.83393801(-1)	0.0
3	0.0	0.0	0.41695457(-2)
4	-0.11670526(-3)	0.0	0.0
5	0.19844501(-4)	-0.14709950(-4)	0.0
6	-0.49114709(-6)	0.63510569(-6)	-0.18494283(-6)

[a] (-n) means 10^{-n}.

The excess function has been generated using equation (14.23) by subtracting the dilute-gas value $\lambda^{(0)}$ and the critical enhancement $\Delta\lambda_c$ from the reported experimental value $\lambda(\rho, T)$. For this purpose the dilute-gas values reported by the experimentalist have been used consistently, while the thermal conductivity critical enhancement, $\Delta\lambda_c$, has been calculated by means of equation (14.41). All the primary excess thermal conductivity data have been fitted by the use of the SEEQ algorithm to

$$\Delta\lambda = \sum_{i=1}^{n} \sum_{j=0}^{m} e_{ij} \frac{\rho^i}{T^{*j}} \tag{14.39}$$

The appropriate statistical weights have been generated from the experimental uncertainties and were taken to be inversely proportional to density. The optimum values of coefficients e_{ij} are given in Table 14.7. The final correlation represents the data to within $\pm 4\%$ over the whole phase space investigated.

14.2.6 The critical enhancements

Up to now there has been no rigorous analysis of the critical enhancement as carried out, for instance, for argon (see Section 14.1), ethane (Section 14.3) and R134a (Section 14.5). Furthermore, the equation of state applied here cannot represent the data accurately within the temperature range $0.99T_c < T < 1.01T_c$. Therefore, the range close to the critical temperature has been excluded from this correlation.

14.2.6.1 Viscosity

The viscosity of nitrogen near the critical point has been analyzed approximately by Basu & Sengers (1979). But, since the extent of the region where this contribution is significant has been found very small, it should be sufficient for most practical purposes to apply

$$\Delta\eta_c(\rho, T) = 0 \tag{14.40}$$

Table 14.8. *Coefficients and critical constants for the representation of the thermal conductivity critical enhancement (Perkins et al. 1991).*

C_1	C_2	C_3	C_4	C_5
8.35308(-2)	-1.2500(2)	2.23744(-3)	-8.23063(-6)	2.0798(-1)

$T_c = 126.193$ K $\rho_c = 11.177$ mol L^{-1}

14.2.6.2 Thermal conductivity

The thermal conductivity critical enhancement of nitrogen has been analyzed by Perkins *et al.* (1991). An empirical expression with five adjustable coefficients was used in order to represent the experimental data in the vicinity but not very close to the critical point. Accordingly, the following expressions should not be applied in the temperature range 123 to 129 K and in the density range 10.9 to 11.4 mol L^{-1}.

$$\Delta\lambda_c(\rho, T) = A(T) \exp(-x^2) \tag{14.41}$$

where

$$A(T) = C_1(T' + C_2)^{-1} + C_3 + C_4 T' \tag{14.42}$$

$$x = C_5(\rho - \rho_c) \tag{14.43}$$

and

$$T' = T \qquad (T > T_c) \tag{14.44}$$

$$T' = 2T_c - T \qquad (T < T_c) \tag{14.45}$$

The coefficients C_i and the critical parameters used for nitrogen are summarized in Table 14.8. A comparison between experimental and calculated results is illustrated in Figure 14.14.

14.2.7 The overall representation

The final representations of the transport properties of nitrogen are given by equation (14.23) together with the subcorrelations for the individual contributions as detailed above. In general, the range of validity is from 70 to 1100 K and up to 100 MPa (30 mol L^{-1}), with a reduced pressure (density) range at higher temperatures. The detailed ranges of applicability of the present representations for the viscosity and thermal conductivity of nitrogen as well as the estimated uncertainties in various thermodynamic regions will be detailed elsewhere (Millat & Wakeham 1995).

Table 14.9 contains the values of the viscosity and the thermal conductivity of nitrogen along the saturation line. This table also serves the purpose of providing a check for

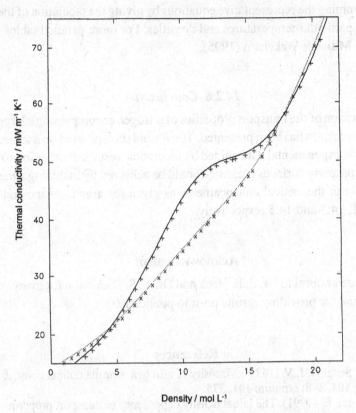

Fig. 14.14. A comparison between experimental and correlated data for the thermal conductivity. · · · · · 152 K, —— 131 K.

Table 14.9. *Transport properties of nitrogen along the saturation line.*

$\frac{T}{K}$	$\frac{P}{MPa}$	$\frac{\rho_g}{mol\ L^{-1}}$	$\frac{\eta_g}{\mu Pa\ s}$	$\frac{\lambda_g}{mW\ m^{-1}\ K^{-1}}$	$\frac{\rho_l}{mol\ L^{-1}}$	$\frac{\eta_l}{\mu Pa\ s}$	$\frac{\lambda_l}{mW\ m^{-1}\ K^{-1}}$
80	0.1370	0.2180	5.490	6.889	28.3483	142.3	139.79
85	0.2290	0.3518	5.879	7.816	27.4906	119.9	130.45
90	0.3607	0.5399	6.248	8.769	26.5853	102.4	120.58
95	0.5408	0.7972	6.561	9.774	25.6200	88.60	110.53
100	0.7788	1.1436	6.779	10.868	24.5831	77.47	100.51
105	1.0842	1.6085	6.890	12.111	23.4510	67.74	90.60
110	1.4672	2.2387	6.953	13.619	22.1710	58.31	80.65
115	1.9387	3.1212	7.169	15.682	20.6397	48.16	70.43
120	2.5125	4.4637	8.092	19.504	18.6446	36.82	60.09

those programming the representative equations by giving the tabulation of the transport properties at particular temperatures and densities. For more detailed tables the reader is referred to Millat & Wakeham (1995).

14.2.8 Conclusion

The representation of the transport properties of nitrogen encompassing a large region of thermodynamic states has been presented. The formulation is based on a critical analysis of the available experimental data guided by theoretical results. Further improvement of the transport property surfaces presented could be achieved by including a theoretically based analysis of the critical enhancements as given for argon, ethane and R134a in Sections 14.1, 14.3 and 14.5 respectively.

Acknowledgment

The authors are grateful to Dr. E. L. Heck and Dr. A. S. Dickinson for many stimulating discussions and for providing results prior to publication.

References

Basu, R. S. & Sengers, J. V. (1979). Viscosity of nitrogen near the critical point. *J. Heat Transfer*, **101**, 3–8; erratum: **101**, 575.

Bich, E. & Vogel, E. (1991). The initial density dependence of transport properties: Noble gases. *Int. J. Thermophys.*, **12**, 27–42.

Brau, C. A. & Jonkman, R. M. (1970). Classical theory of rotational relaxation in diatomic gases. *J. Chem. Phys.*, **52**, 477–484.

de Reuck, K.M.R. & Armstrong, B.A. (1979). A method of correlation using a search procedure based on a step–wise least–squares technique, and its application to an equation of state for propylene. *Cryogenics*, **19**, 505–512.

Friend, D.G. & Rainwater, J.C. (1984). Transport properties of a moderately dense gas. *Chem. Phys. Lett.*, **107**, 590–604.

Heck, E. L. & Dickinson A.S. (1994). Transport and relaxation properties of N_2. *Mol. Phys.*, **81**, 1325–1352.

Heck, E. L., Dickinson, A.S. & Vesovic, V. (1994). Second-order corrections for transport properties of pure diatomic gases. *Mol. Phys.*, **83**, 907–932.

Jacobson R. T., Stewart, R. B. & Jahangiri, M. (1986). Thermodynamic properties of nitrogen from the freezing line to 2000 K at pressures to 1000 MPa. *J. Phys. Chem. Ref. Data*, **15**, 735–909.

Kagan, Yu. & Afanas'ev, A. M. (1961). On the kinetic theory of gases with rotational degrees of freedom. *Zh. Eksp. i Theor. Fiz.*, **41**, 1536–1545. (Russ). Engl. transl. in: *Soviet Phys.–JETP*, **14** (1962), 1096–1101.

Maitland, G.C., Mustafa, M. & Wakeham, W.A. (1983). Second order approximations for the transport properties of dilute polyatomic gases. *J. Chem. Soc., Faraday Trans. 2*, **79**, 1425–1441.

Millat, J., Ross M. J. & Wakeham, W. A., (1989b). Thermal conductivity of nitrogen in the temperature range 177 to 270 K. *Physica*, **159A**, 28–43.

Millat, J., Vesovic, V. & Wakeham, W.A. (1989a). Theoretically–based data assessment for the correlation of the thermal conductivity of dilute gases. *Int. J. Thermophys.*, **10**, 805–818.

Millat, J. & Wakeham, W.A. (1989a). The thermal conductivity of nitrogen and carbon monoxide in the limit of zero density. *J. Phys. Chem. Ref. Data*, **18**, 565–581.

Millat, J. & Wakeham, W.A. (1989b). The correlation and prediction of thermal conductivity and other properties at zero density. *Int. J. Thermophys.*, **10**, 983–993.

Millat, J. & Wakeham, W.A. (1995). The transport property surfaces of fluid nitrogen. *To be published*.

Monchick, L., Pereira, A.N.G. & Mason, E.A. (1965). Heat conductivity of polyatomic and polar gases and gas mixtures. *J. Chem. Phys.*, **42**, 3241–3256.

Perkins, R.A., Roder, H.M., Friend, D.G. & Nieto de Castro, C.A. (1991). Thermal conductivity and heat capacity of nitrogen. *Physica*, **173A**, 332–362.

Rainwater, J.C. & Friend, D.G. (1987). Second viscosity and thermal conductivity virial coefficients of gases: Extension to low reduced temperatures. *Phys. Rev.*, **36A**, 4062–4066.

Stephan, K., Krauss, R. & Laesecke, A. (1987). Viscosity and thermal conductivity of nitrogen for a wide range of fluid states. *J. Phys. Chem. Ref. Data*, **16**, 993–1023.

van den Oord, R.J. & Korving J. (1988). The thermal conductivity of polyatomic gases. *J. Chem. Phys.*, **89**, 4333–4338.

van der Avoird, A., Wormer P. E. S. & Jansen A. P. J. (1986). An improved intermolecular potential for nitrogen. *J. Chem. Phys.*, **84**, 1629–1635.

Viehland, L.A., Mason, E.A. & Sandler, S.I. (1978). Effect of spin polarization on the thermal conductivity of polyatomic gases. *J. Chem. Phys.*, **68**, 5277–5282.

Vogel, E., Strehlow, T., Millat, J. & Wakeham, W.A. (1989). On the temperature function of the viscosity of nitrogen in the limit of zero density. *Z. Phys. Chem., Leipzig*, **270**, 1145–1152.

14.3 Polyatomic fluids – Ethane

V. VESOVIC
Imperial College, London, UK

and

J. MILLAT
NORDUM Institut für Umwelt und Analytik, Kessin/Rostock, Germany

14.3.1 Introduction

The basic philosophy of the approach presented here for ethane is to make use of the best available experimental data and theory to produce accurate, consistent and theoretically sound representations of transport properties over the widest range of thermodynamic states possible. So far, the transport properties of three polyatomic fluids, nitrogen (see Section 14.2), carbon dioxide (Vesovic *et al.* 1990) and ethane (Hendl *et al.* 1994; Vesovic *et al.* 1994), have been correlated in such a fashion.

In order to illustrate the correlation techniques as applied to real polyatomic fluids, the development of the representations of viscosity and thermal conductivity of ethane here serves as an example. It is not possible or desirable for that matter to present more than a review here, with special emphasis on novel techniques that could also be adopted for other polyatomic fluids.

It is customary (Vesovic *et al.* 1990; Vesovic & Wakeham 1991) for both fundamental and practical reasons, to represent the transport properties of a fluid as,

$$X(\rho, T) = X^{(0)}(T) + \Delta X(\rho, T) + \Delta X_c(\rho, T) = \bar{X}(\rho, T) + \Delta X_c(\rho, T) \quad (14.46)$$

Here $X^{(0)}(T)$ is the dilute–gas transport property, $\Delta X(\rho, T)$ is the excess transport property and $\Delta X_c(\rho, T)$ a critical enhancement and $X = \eta, \lambda, (\rho D)$. Furthermore, it is useful to define the background contribution $\bar{X}(\rho, T)$ as the sum of the first two terms as given in equation (14.46).

The initial step in developing any correlation is the analysis of the available experimental data. This analysis is carried out in order to establish a primary data set (see Chapter 3) over as wide a range of thermodynamic states as possible for each contribution in equation (14.46). For the specific case of ethane there is a plethora of experimental data on the viscosity and thermal conductivity measured in a number of different instruments over the last 50 years (for a summary see Hendl *et al.* 1994; Vesovic *et al.* 1994). Not surprisingly, there have also been a number of correlations developed for its transport properties (Golubev 1970; Makita *et al.* 1974; Vargaftik 1975; Hanley *et al.* 1977; Tarzimanov *et al.* 1987; Younglove & Ely 1987; Friend *et al.* 1991). The most recent one (Hendl *et al.* 1994; Vesovic *et al.* 1994) is considered here in detail, since it incorporates the most up–to–date developments, including new experimental information *and* recent theoretical developments that allow for a more secure analysis of the available experimental data than has been possible hitherto. For the purposes of the analysis the experimental transport property data must be available at a specified temperature and density. It is necessary to evaluate the density of the fluid from the temperature and pressure reported by experimentalists. In order to accomplish this task for ethane, the most recent classical equation of state (EOS) outside the critical region (Friend *et al.* 1991) and a new parametric crossover EOS in the critical region (Luettmer–Strathmann *et al.* 1992) have been employed. The switching between the two equations of states has been performed along the rectangular boundary in the temperature–density plane given by $302.5 \text{ K} \leq T \leq 316 \text{ K}$, $3.82 \text{ mol L}^{-1} \leq \rho \leq 8.65 \text{ mol L}^{-1}$.

In subsequent sections each contribution to equation (14.46) is treated separately, discussing the appropriate correlation techniques and their possible use for other fluids.

14.3.2 Dilute–gas contribution

The dilute–gas contribution is an experimentally accessible quantity and can always be analyzed independently of other terms in equation (14.46). It is, therefore, the natural starting point for the development of any analysis of this kind. Furthermore, some guidance to the form of the representation can be discerned by use of the results of the kinetic theory (see Chapter 4).

When applying the results of the kinetic theory of polyatomic gases, one has to take into account the influence of vibrational and sometimes hindered rotational modes on the transport properties. The kinetic theory of molecules possessing vibrational modes of motion is still in its formative stages and is nowhere near as well–developed as that of rigid rotor molecules (see Section 14.2). Thus, the type of analysis that has been carried out for nitrogen is not possible in its entirety for polyatomic fluids. Nevertheless, a modified analysis, based on an empirical extension of the kinetic theory of diatomic gases, turned out to be possible (Hendl *et al.* 1991, 1994; Vesovic *et al.* 1994).

14.3.2.1 Viscosity

Formally, the relationship between the coefficient of viscosity of a dilute polyatomic gas and the related effective cross section is identical to that for monatomic gases (Chapter 4). In a practical, engineering form it is given by

$$\eta^{(0)}(T) = \frac{0.021357(TM)^{1/2}}{\sigma^2 \mathfrak{S}_\eta^*} \tag{14.47}$$

where $\mathfrak{S}_\eta^* = \mathfrak{S}(2000)/\pi\sigma^2 f_\eta$ is a reduced effective cross section, T the temperature in K, M the molar mass in g mol^{-1}, σ a length scaling–parameter in nm, f_η a dimensionless higher–order correction factor and $\eta^{(0)}$ is in units of μPa s.

The effective cross section $\mathfrak{S}_\eta^*$ is usually expressed in the functional form

$$\ln \mathfrak{S}_\eta^* = \sum_{i=0}^{4} a_i (\ln T^*)^i \tag{14.48}$$

where $T^* = k_B T / \epsilon$ is the reduced temperature and ϵ / k_B an energy–scaling parameter in K.

The selected primary data on the dilute–gas viscosity of ethane cover a temperature range from 200 K to 633 K (see Hendl *et al.* 1994). In developing the viscosity correlation, values of $\mathfrak{S}_\eta^*(T^*)$ have been derived from the given experimental viscosity data and were fitted, applying appropriate statistical weights, to equation (14.48). The fitting procedure is carried out in two steps. In the first step, values for the scaling parameters ϵ / k_B and σ are deduced by fitting all the primary data to a universal correlation based on the corresponding–states principle (see Chapter 11). In the second step, the scaling parameters are used to generate 'experimental' $\mathfrak{S}_\eta^*$, T^* pairs from the primary data, which are then fitted applying appropriate statistical weights (Hendl *et al.* 1994), to the functional form given by equation (14.48) in order to derive a correlation which is specific for ethane ('individual' correlation). Values for the coefficients a_i are given in Table 14.10.

The resulting correlation is valid in the temperature range from 200 K to 1000 K, and its uncertainty is estimated to be $\pm 0.5\%$ in the range $300\,\text{K} \le T \le 600\,\text{K}$, increasing to $\pm 1.5\%$ and $\pm 2.5\%$ at 200 K and 1000 K respectively.

14.3.2.2 Thermal conductivity

As outlined in Chapter 4, the thermal conductivity in the dilute–gas limit, $\lambda^{(0)}$, is related to a number of effective cross sections, which are associated with the transport of translational and internal energy, and with their interaction. In principle, a similar analysis as given for nitrogen in Section 14.2 can be performed for polyatomic molecules. In practice, such an analysis is often hampered by a lack of experimental information and insufficient knowledge of the behavior of cross sections describing the diffusion of internal energy at high temperatures.

For the specific case of ethane, the lack of experimental data on rotational collision numbers as a function of temperature makes the full analysis impossible. Thus, the thermal conductivity has been analyzed in terms of just one effective cross section related to the total–energy flux (see Chapter 4; Hendl *et al.* 1991; Vesovic *et al.* 1994). In usable form the thermal conductivity in the dilute–gas limit is then given by

$$\lambda^{(0)}(T) = 0.177568 \frac{C_P^{id}}{R} \left(\frac{T}{M}\right)^{1/2} \frac{1}{\sigma^2 \mathfrak{S}_\lambda^*} \tag{14.49}$$

where $\mathfrak{S}_\lambda^* = \mathfrak{S}(10E)/(\pi \sigma^2 f_\lambda S_\lambda)$ is the reduced effective collision cross section, C_P^{id}/R is the reduced isobaric ideal–gas heat capacity, f_λ and S_λ are dimensionless correction factors for higher–order corrections and angular momentum polarization, respectively, and the thermal conductivity $\lambda^{(0)}$ is in units of mW m^{-1} K^{-1}.

The primary thermal conductivity data for ethane cover the temperature range from 225 K to 725 K (Vesovic *et al.* 1994). In developing the thermal conductivity correlation, values of $\mathfrak{S}_\lambda^*(T^*)$ have been deduced from each of the experimental data points by means of scaling parameters derived from the analysis of viscosity data. They were then fitted applying appropriate statistical weights (Vesovic *et al.* 1994) to the functional form

$$\mathfrak{S}_\lambda^* = \sum_{i=0}^{2} b_i \left(\frac{1}{T^*}\right)^i \tag{14.50}$$

In general this form is to be preferred to the expansion in $\ln T^*$, since the results for a number of fluids (Millat *et al.* 1989) indicate that the effective cross sections $\mathfrak{S}_\lambda^*$ are nearly inversely proportional to $1/T^*$. In order to evaluate $\lambda^{(0)}$ using equation (14.49), one needs a subsidiary representation for the heat capacity. This can be represented by (Vesovic *et al.* 1994)

$$\frac{C_P^{id}}{R} = c_8 + u \sum_{i=1}^{7} c_i Y^{4-i} \tag{14.51}$$

where

$$u = \exp(-c_9 Y) \quad \text{and} \quad Y = T/100\text{K} \tag{14.52}$$

The coefficients b_i and c_i are given in Table 14.10. The resulting correlation is valid in the temperature range 225 K to 725 K, and its uncertainty is estimated to be $\pm 2.0\%$ in the range 300 K $\leq T \leq$ 500 K, increasing to a maximum of $\pm 3.0\%$ at either end.

There are a number of ways of extrapolating the thermal conductivity to higher temperatures. They are based either on the near proportionality of $\mathfrak{S}_\lambda^*$ to $1/T^*$ (Millat *et al.*

Table 14.10. *Coefficients for the representation of the viscosity and thermal conductivity of ethane (dilute gas and initial–density dependence).*

i	a_i	b_i	c_i	d_i
0	0.221882	0. 444358		-0.3597244
1	-0.5079322	0. 327867	0.0036890096	4.525644
2	0.1285776	0. 1936835	-0.17196907	-5.474280
3	-0.00832817		3.159226	3.396994
4	-0.00271317		-8.0459942	-4.986360
5			7.4237673	2.760371
6			0.0	-0.6827789
7			-2.0724572	
8			4.0	
9			0.02	

$\epsilon/k_B = 264.7$ K, $\quad \sigma = 0.43075$ nm, $\quad M = 30.069$ g mol^{-1}

1989) or on a known high–temperature asymptotic behavior of particular dimension-less numbers (Hendl *et al.* 1991). For ethane, the extrapolation has been based on the behavior of the Prandtl number, which for a dilute gas gives

$$Pr = \frac{C_P^{id} \eta^{(0)}}{M \lambda^{(0)}} = \frac{\mathfrak{S}_\lambda^*}{\mathfrak{S}_\eta^*} \tag{14.53}$$

At high temperatures the available experimental evidence (van den Oord & Korving 1988; Hendl *et al.* 1991), supported by a few numerical calculations and theoretical analyses (Wong *et al.* 1989), leads to the conclusion that the Prandtl number monoton-ically decreases with increasing temperature to an almost constant value. For ethane, the experimental data indicate that the Prandtl number above 600 K can be taken as a constant, $Pr \rightarrow 0.7$. This observation allows the extension of the thermal conductivity representation with a maximum uncertainty of $\pm 5\%$ to its upper practical temperature limit of 1000 K by

$$\lambda^{(0)} = \frac{11.877 \, \eta^{(0)} \, C_P^{id}}{M} \frac{C_P^{id}}{R} \quad (T > 725\text{K}) \tag{14.54}$$

14.3.3 *Initial–density dependence*

Recent theoretical advances in understanding the behavior of transport properties of gases at moderate densities as described in Section 5.3 have allowed a development of a separate treatment of the transport properties of monatomic gases and of the viscosity of polyatomic gases in this region.

At present for polyatomic gases, this is possible only for viscosity, since the results for the thermal conductivity are not yet at the stage where they can be used for correlation or prediction purposes. In principle, the best approach to produce the correlation of viscosity at low densities is to analyze the available experimental data in conjunction with theory. Unfortunately, for ethane the available experimental data on the viscosity in the vapor phase at low density are very scarce (Hendl *et al.* 1994), and it has not been possible to take advantage of these data in the development of the initial–density contribution. Thus the theory has been used in a predictive mode to generate the initial–density dependence of the viscosity. This was deemed necessary for ethane, since the vapor phase covers an industrially important and easily accessible region where the need for accurate transport properties is significant.

In order to separate the initial–density dependence of the viscosity from higher–density terms, the excess viscosity is expressed as

$$\Delta \eta = \eta^{(1)} \rho + \Delta_h \eta \tag{14.55}$$

where $\eta^{(1)} \rho$ is the initial–density contribution, while $\Delta_h \eta$ accounts for higher–density contributions. It is customary (Friend & Rainwater 1984; Rainwater & Friend 1987;

Bich & Vogel 1991) to express the coefficient $\eta^{(1)}$ in terms of the second viscosity virial coefficient B_η by means of the definition

$$B_\eta(T) = \frac{\eta^{(1)}(T)}{\eta^{(0)}(T)} \tag{14.56}$$

As a result of both theoretical and experimental studies of the second viscosity virial coefficient, it has been possible to develop a generalized representation of its temperature dependence based on the Lennard–Jones (12–6) potential (Rainwater & Friend 1987; Bich & Vogel 1991). A particular advantage of this approach is that it is possible to estimate the coefficient $B_\eta(T)$ for a gas for which no experimental viscosity data as a function of density exist, given a knowledge of the Lennard–Jones (12–6) potential parameters as derived by an analysis of dilute–gas viscosity data. Such an estimation of B_η has been performed for ethane by use of the recommended parameters $\epsilon/k_B = 251.1$ K and $\sigma = 0.4325$ nm (Hendl *et al.* 1994).

This procedure based on the Rainwater–Friend theory is valid only for reduced temperatures $T^* \geq 0.7$ ($T = 175$ K for ethane). This lower limit will not be exceeded by the viscosity representation of ethane in the vapor phase, since the range of validity of its zero–density contribution has a lower limit of $T = 200$ K. Nevertheless, experimental data in the liquid phase are available at much lower temperatures extending to $T = 100$ K. In order to use a single overall viscosity correlation it must be ensured that the initial–density contribution extrapolates satisfactorily to low temperatures. For this purpose, for temperatures below $T^* = 0.7$, the second viscosity virial coefficient has been estimated by use of the modified Enskog theory (see Chapter 5), which relates B_η to the second and third pressure virial coefficients. Although this method enables B_η to be evaluated, it is cumbersome for practical applications. Therefore, the calculated B_η values using both methods have been fitted to the functional form

$$B_\eta = 0.6022137\sigma^3 B_\eta^*(T^*) = \sum_{i=0}^{6} d_i \left(\frac{1}{T^*} \right)^i \tag{14.57}$$

The coefficients d_i are listed in Table 14.10, and B_η is in units of L· mol^{-1}. For the sake of consistency, the scaling parameters ϵ and σ used in equation (14.57) have been chosen to coincide with the ones used for the dilute–gas viscosity representation.

Figure 14.15 illustrates the behavior of the viscosity of ethane in the vapor phase. Since B_η is negative for temperatures below 313 K, the viscosity in the vapor phase along an isotherm first decreases followed by an increase owing to the influence of higher–order terms in density.

14.3.4 The critical enhancements

It has been observed for a number of fluids (for instance, Mostert *et al.* 1990 and references therein) that the transport properties show an enhancement in a region around

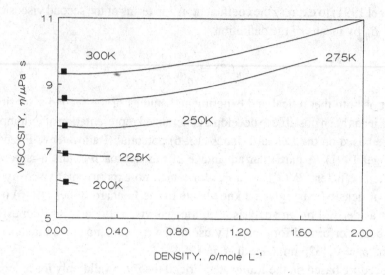

Fig. 14.15. Behavior of the viscosity of ethane in the vapor phase along selected isotherms; ■ viscosity at the pressure of 1 bar.

the critical point and tend to become infinite at the critical point itself. The theory of the critical enhancements is described in detail in Chapter 6, so this section concentrates on its application for the purpose of correlating the viscosity and the thermal conductivity. The critical enhancement of the thermal conductivity is analyzed first because it is much stronger than that for viscosity and covers a substantial region around the critical point.

14.3.4.1 Thermal conductivity

The mode–coupling theory of dynamic critical phenomena can be used to generalize the asymptotic equation for the thermal conductivity to give (Chapter 6; Sengers 1985; Mostert et al. 1990)

$$\Delta\lambda_c = \frac{7.32466 \cdot 10^{-4} R_D T \rho c_P}{\eta \xi}(\Omega - \Omega_0) \qquad (14.58)$$

In addition to the symbols introduced before, ξ is the correlation length in nm, R_D is the universal amplitude and $\Delta\lambda_c$ is in mW m^{-1} K^{-1}. The quantities Ω and Ω_0 are the crossover functions, which depend on thermodynamic properties and on the background contributions to the thermal conductivity, $\bar{\lambda}$, and the viscosity, $\bar{\eta}$.

Hence, in order to evaluate the critical enhancement of the thermal conductivity by means of equation (14.58) apart from the background transport properties, a knowledge of three thermodynamic properties is needed, namely, that of the isochoric and isobaric heat capacities and of the correlation length, all of which can be evaluated from the relevant equation of state. Furthermore, one needs to determine four adjustable parameters (see Chapter 6) which enter the expressions for ξ, Ω and Ω_0. Two of these, namely, the system–dependent amplitudes ξ_0 and Γ, have been obtained by the application of the parametric crossover model to the equation of state for ethane (Luettmer–Strathmann *et al.* 1992). The cutoff wavenumber q_D and the reference temperature T_r have to be determined by fitting experimental data on the critical enhancement of the thermal conductivity to equation (14.58). T_r is chosen so that for temperatures greater than T_r the critical enhancement is negligible. For ethane a value of $T_r = 1.5T_c$ has been adopted as the most appropriate (Vesovic *et al.* 1994).

In order to determine the cutoff wavenumber q_D, first a primary data set has to be established. In principle viscosity and thermal–diffusivity data can be used to supplement the thermal conductivity data. For ethane this was not necessary, since there exists a large amount of accurate thermal conductivity data in the critical region. Six data sets (see Vesovic *et al.* 1994) comprising more than 800 data points covering pressures and temperatures where the critical enhancement is observed have been selected. In order to determine the cutoff wavenumber q_D, the critical enhancement has been separated from the background term according to equation (14.46). It is impossible to perform this separation unequivocally, so that an iterative approach must be applied. An initial estimate of the background thermal conductivity is made by use of thermal conductivity data for which the critical enhancement is negligible. This estimate is then used to evaluate the first iterate of $\Delta\lambda_c$ by subtracting the estimated background from data in the critical region. The background transport properties are then used in the generation of Ω and Ω_0 (see Chapter 6), while q_D is treated as an adjustable parameter in fitting $\Delta\lambda_c$. Once a first iterate of q_D has been obtained, a refinement of the background contributions can be undertaken. In subsequent iterations the whole set of primary thermal conductivity data is used to redetermine the background contribution by subtracting from each experimental thermal conductivity value an appropriate critical enhancement. The iterative process is then continued until there is no significant change in either the background or the critical contribution. In practice, the process converges very rapidly; for ethane only two iterations were necessary. Using the procedure described above, the optimum value of the parameter q_D has been found. Its inverse is listed in Table 14.11 together with other parameters which are necessary in order to evaluate the critical enhancements. The agreement between the proposed correlation and the available experimental results is satisfactory (Vesovic *et al.* 1994). The correlation has additionally been used to predict the thermal diffusivity ($a = \lambda/\rho C_P$) of ethane in the critical region. The agreement with the experimental data of Jany & Straub (1987) is excellent. It should be emphasized that the critical enhancement of the thermal conductivity generally is present over a large range of densities and temperatures. For ethane it is only outside the region approximately bounded by 225 K $\leq$ T $\leq$ 457 K and 0.3 mol L^{-1} $\leq \rho \leq$ 15.6 mol L^{-1} that the relative critical thermal conductivity enhancement, $\Delta\lambda_c/\lambda$, is smaller than 1%. Figure 14.16 illustrates the behavior of the thermal conductivity of ethane in the vicinity of the critical point.

Table 14.11. *Constants in the equations for the critical enhancement of the transport properties of ethane.*

Critical parameters		Critical exponents	
T_c	305.33 K	z	0.063
P_c	4.8718 MPa	ν	0.63
ρ_c	6.87 mol L^{-1}	γ	1.239
Critical amplitudes		**Cutoff wave number**	
R_c	1.03	q_D^{-1}	0.187 nm
ξ_0	0.19 nm		
Γ	0.0541		

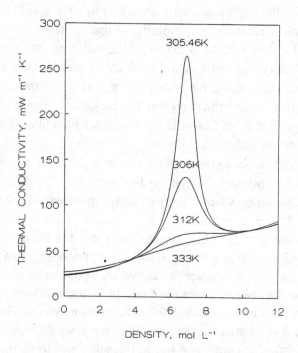

Fig. 14.16. Behavior of the thermal conductivity of ethane in the critical region along selected isotherms.

14.3.4.2 Viscosity

The viscosity critical enhancement, $\Delta\eta_c$, is represented by an equation of the form

$$\Delta\eta_c = \bar{\eta}[\exp(zH) - 1] \tag{14.59}$$

where H is a crossover function (see Chapter 6) and z a critical exponent. H depends upon the thermodynamic properties of the fluid and upon the background contributions

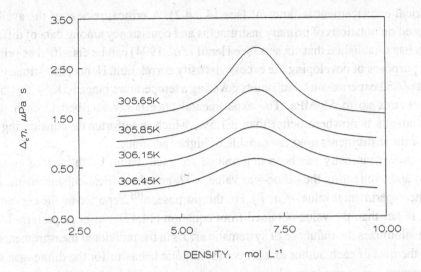

Fig. 14.17. Behavior of the viscosity of ethane in the critical region along selected isotherms. To separate the isotherms, the viscosity values have been displaced by 0.5μPa s each.

$\bar{\eta}$ and $\bar{\lambda}$. Rather than rely upon uncertain experimental data, the critical enhancement of the viscosity, $\Delta \eta_c$ of ethane is predicted by means of equation (14.59), together with parameters given in Table 14.11 (Hendl *et al.* 1994). Equation (14.59) should be used inside the region approximately bounded by 302.5 K and 311 K in temperature and 4.3 mol L^{-1} and 8.6 mol L^{-1} in density. Outside of this region the relative critical viscosity enhancement, $\Delta \eta_c / \eta$, is smaller than 1% and for engineering purposes can be safely neglected.

Figure 14.17 illustrates the size and the extent of the critical enhancement of the viscosity of ethane.

14.3.5 *The excess transport properties*

Although some successful models exist for the prediction of transport properties in the liquid region (see Chapters 5 and 10), it has not proved possible, so far, to incorporate these semiempirical formulations into a scheme for correlating the excess transport property covering the complete dense fluid range. Hence, the development of the excess transport properties for ethane, like that for other polyatomic fluids, is based entirely on fitting experimental data.

14.3.5.1 *Viscosity*

The form of equation (14.46) indicates that it is relatively straightforward to calculate the excess transport property from the experimental data by subtracting the dilute–gas contribution and the critical enhancement. As mentioned before, in practice an iterative procedure is often necessary. Here, for the purpose of discussion it will be assumed that the critical enhancement is 'known' (see 14.3.4.2). A critical review of the available data based on notations of primary instruments and consistency among data of different authors has established that six sets (see Hendl *et al.* 1994) can be classified as primary for the purposes of developing the excess viscosity correlation. Hence, the primary data consist of 552 experimental data points covering a temperature range $95K \leq T \leq 500K$ and pressures up to 55 MPa. The experimental uncertainty ascribed to each set of measurements is nowhere better than $\pm 1.5\%$, which is unfortunate considering that some of the instruments used are capable of higher accuracy.

The excess viscosity has been determined for each datum by the use of equation (14.46) and subtracting the dilute–gas value, $\eta^{(0)}$, and the critical enhancement, $\Delta \eta_c$, from the experimental value, $\eta(\rho, T)$. For this purpose, $\eta^{(0)}$, reported by the experimentalists, rather than the value obtained from equation (14.47), has been preferred. This choice minimizes the influence of systematic errors in the individual measurements and forces the data of each author to a proper asymptotic behavior for the dilute–gas state. Furthermore, the 'experimental' excess viscosity obtained in this fashion is independent of the choice for a dilute–gas viscosity correlation. Unfortunately, the majority of measurements on viscosity of ethane have been performed at pressures above 0.7 MPa and hence only a few authors reported a $\eta^{(0)}$ value. Therefore, in order to estimate the 'experimental' zero–density viscosity of each isotherm, again an iterative procedure had to be used (Hendl *et al.* 1994). The correction introduced by the extrapolation to zero density is small and in general does not exceed $\pm 0.5\%$.

Once the excess viscosity data have been generated, the initial density dependence can be evaluated as $\eta^{(1)} \rho$ and subtracted from the excess to obtain the higher–density contribution, $\Delta_h \eta$, by use of equation (14.55). There is no theoretical guidance to the functional form of $\Delta_h \eta$, but it is customary to express it in terms of power series in the density and in the reciprocal reduced temperature. Thus,

$$\Delta_h \eta = \sum_{i=2}^{6} \sum_{j=0}^{4} e_{ij} \frac{\rho^i}{T^{*j}} \qquad (14.60)$$

The fitting has been performed by use of the SEEQ algorithm based on the step–wise least–squares technique (de Reuck & Armstrong 1979). The appropriate statistical weights have been generated from the experimental uncertainties and were taken to be inversely proportional to density. The optimum values of coefficients e_{ij} are given in Table 14.12. No systematic trends have been observed when analyzing the deviations of the primary experimental data from the correlated values, and all the data have been fitted to better than 2.5%.

14.3.5.2 Thermal conductivity

All the available measurements (see Vesovic *et al.* 1994) of the thermal conductivity of ethane at elevated pressures have been carried out either in concentric cylinder or in transient hot–wire instruments. Both of these instruments can yield data of high

Table 14.12. *Coefficients for the representation of the high–density contribution $\Delta_h \eta$.*

j	e_{2j}	$10 \cdot e_{3j}$	$10^2 \cdot e_{4j}$	$10^3 \cdot e_{5j}$	$10^5 \cdot e_{6j}$
0	0.71783704	0.13303049	0.0	0.0	0.0
1	-0.97504949	-3.9334195	1.9415794	0.0	0.0
2	0.0	8.1481112	-1.6768122	-2.6635987	7.7495402
3	1.1573719	-6.1320940	0.0	3.3513388	-9.7839334
4	0.0	0.0	1.5270339	-1.4891405	3.5935183

accuracy and as such have to be considered as primary. Five sets of data (see Vesovic *et al.* 1994) were selected for the purpose of developing the excess thermal conductivity correlation. The primary data set consists of 1143 experimental data points covering a temperature range 112 K $\leq$ T $\leq$ 800 K and pressures up to 70 MPa. The ascribed uncertainty of the data varies from $\pm 0.8\%$ to $\pm 3.0\%$.

The excess function has been generated using equation (14.46) by subtracting the dilute–gas value $\lambda^{(0)}$ and the critical enhancement $\Delta \lambda_c$ from the reported experimental value $\lambda(\rho, T)$. For this purpose again the dilute–gas values $\lambda^{(0)}$ reported by the experimentalist have been used consistently, while the thermal conductivity critical enhancement, $\Delta \lambda_c$, has been calculated by means of equation (14.58). All the primary excess thermal conductivity data have been fitted by the use of the SEEQ algorithm to the equation

$$\Delta \lambda = \sum_{i=1}^{6} \sum_{j=0}^{2} f_{ij} \frac{\rho^i}{T^{*j}} \tag{14.61}$$

The appropriate statistical weights have been generated from the experimental uncertainties and were taken to be inversely proportional to density. The optimum values of coefficients f_{ij} are given in Table 14.13. The final correlation represents the data to within $\pm 4\%$ over the whole phase space investigated.

Table 14.13. *Coefficients for the representation of the excess thermal conductivity.*

i	f_{i0}	f_{i1}	f_{i2}
1	1.1795365	-1.5320900	2.0159682
2	3.1188977	-4.7166037	0.0
3	-8.3572937(-1)[a]	1.4575942	0.0
4	8.5729762(-2)	-1.6354312(-1)	0.0
5	-3.5751570(-3)	7.9301012(-3)	-1.6496369(-4)
6	4.9626960(-5)	-1.3652796(-4)	6.6052581(-6)

[a] $(-n)$ means 10^{-n}.

14.3.6 The overall representation

The final representation of the transport properties of ethane is given by equation (14.46) together with the subcorrelations for the individual contributions as detailed above. Figure 14.18 illustrates the range of applicability of the present representation for the viscosity of ethane as well as the estimated uncertainty in various thermodynamic regions. In the vapor phase and at low densities in the supercritical region the correlation can be used between 200 K and 1000 K. At higher densities in the supercritical region the upper temperature limit is lowered to 500 K owing to a lack of experimental data. Although the lower temperature limit in the gaseous phase is 200 K, the overall correlation can be used to estimate the liquid phase viscosities down to 100 K. The upper limit in pressure has been set at 60 MPa.

Figure 14.19 illustrates the range of applicability of the present representation of the thermal conductivity as well as the estimated uncertainty in various thermodynamic regions. The correlation can be used up to 625 K and 70 MPa. At pressures below 10 bars, the representation can be used to estimate the thermal conductivity up to 1000 K. The lower temperature limit in the gaseous phase is 225 K, while the overall correlation can be used to estimate the liquid phase thermal conductivity down to 100 K.

Any extrapolation to higher temperatures or pressures of either correlation can lead to a rapid reduction in the accuracy of the predicted values and is not recommended.

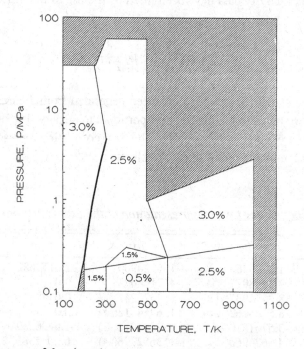

Fig. 14.18. The extent of the viscosity representation and its estimated uncertainty. No representation is available in the hatched region.

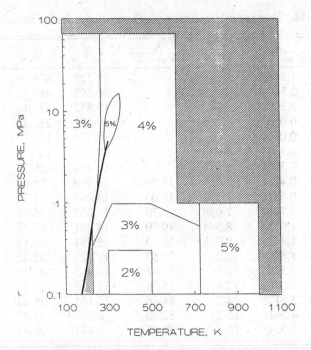

Fig. 14.19. The extent of the thermal conductivity representation and its estimated uncertainty. No representation is available in the hatched region.

Table 14.14 contains the values of the viscosity and the thermal conductivity of ethane along the saturation line. This table also serves the purpose of providing a check for those programming the representative equations by giving the tabulation of the transport properties at particular temperatures and densities. For more detailed tables the reader is referred to Hendl *et al.* (1994) and Vesovic *et al.* (1994).

The uncertainty ascribed to the viscosity correlation for ethane is nowhere greater than ±3%, while that ascribed to thermal conductivity is nowhere greater than ±5%, which for many engineering purposes will prove adequate. Nevertheless, it would be advantageous to have more measurements of the viscosity of ethane in the vapor phase, especially along the saturation line and in the immediate vicinity of the critical point, in order to advance further the understanding of the behavior of the viscosity in these regions and also to decrease the ascribed uncertainties of the correlations. Furthermore, measurements of thermal conductivity of ethane in the vapor phase should also be carried out, since it is in this region that the uncertainties in the thermal conductivity are larger than for other gases studied.

14.3.7 Conclusion

The representation of the transport properties of ethane encompassing a large region of thermodynamic states has been presented. The formulation is based on a critical analysis

Table 14.14. *Transport properties of ethane along the saturation line.*

$\frac{T}{K}$	$\frac{P}{MPa}$	$\frac{\rho_g}{mol\ L^{-1}}$	$\frac{\eta_g}{\mu Pa\ s}$	$\frac{\lambda_g}{mW\ m^{-1}\ K^{-1}}$	$\frac{\rho_l}{mol\ L^{-1}}$	$\frac{\eta_l}{\mu Pa\ s}$	$\frac{\lambda_l}{mW\ m^{-1}\ K^{-1}}$
100	1.1 (-5)[a]	1.3(-5)			21.323	876.02	239.76
120	3.55(-4)	3.56(-4)			20.597	486.51	229.69
140	3.83(-3)	3.30(-3)			19.852	322.16	212.15
160	0.02145	0.01631			19.081	232.45	191.55
180	0.07872	0.05412			18.276	175.74	170.75
200	0.2174	0.1387	6.020		17.423	136.73	150.71
220	0.4923	0.3001	6.684		16.498	108.24	132.00
230	0.7005	0.4221	7.055	14.49	15.999	96.62	123.21
240	0.9671	0.5810	7.472	16.00	15.467	86.27	114.78
250	1.3012	0.7867	7.958	17.70	14.892	76.91	106.73
260	1.7120	1.0534	8.544	19.70	14.261	68.29	99.07
270	2.2097	1.4038	9.278	22.19	13.551	60.17	91.82
280	2.8058	1.8787	10.25	25.59	12.723	52.28	85.14
290	3.5144	2.5703	11.64	31.07	11.683	44.19	79.39
295	3.9169	3.0723	12.67	35.78	11.010	39.77	76.97
300	4.3560	3.8129	14.23	44.68	10.101	34.63	75.12
302	4.5432	4.2616	15.23	52.00	9.5867	32.09	75.30
304	4.7377	4.9547	16.91	71.97	8.8412	28.85	79.81

[a] (-n) means 10^{-n}.

of the available experimental data guided by theoretical results. The formal procedure for correlating the data, described here, can be easily applied to other polyatomic fluids, providing sufficient experimental data are available.

References

Bich, E. & Vogel, E. (1991). The initial density dependence of transport properties: Noble gases. *Int. J. Thermophys.*, **12**, 27–42.

de Reuck, K.M. & Armstrong, B.A. (1979). A method of correlation using a search procedure based on a step–wise least–squares technique, and its application to an equation of state for propylene. *Cryogenics*, **19**, 505–512.

Friend, D.G. & Rainwater J.C. (1984). Transport properties of a moderately dense gas. *Chem. Phys. Lett.*, **107**, 590–594.

Friend, D.G., Ingham, H. & Ely, J.F. (1991). Thermophysical properties of ethane. *J. Phys. Chem. Ref. Data*, **20**, 275–347.

Golubev, I.F. (1970). *The Viscosity of Gases and Gas Mixtures.* Jerusalem: Israel Program for Scientific Translations.

Hanley, H.J.M., Gubbins, K.E. & Murad, S. (1977). A correlation of the existing viscosity and thermal conductivity data of gaseous and liquid ethane. *J. Phys. Chem. Ref. Data*, **6**, 1167–1180.

Hendl, S., Millat, J., Vesovic, V., Vogel, E. & Wakeham, W.A., Luettmer–Strathmann, J., Sengers, J.V. & Assael, M.J. (1994). The transport properties of ethane I. Viscosity. *Int. J. Thermophys.*, **15**, 1–31.

Hendl, S., Millat, J., Vogel, E., Vesovic, V. & Wakeham, W.A. (1991). The viscosity and thermal conductivity of ethane in the limit of zero density. *Int. J. Thermophys.*, **12**, 999–1012.

Jany, P. & Straub, J. (1987). Thermal diffusivity of fluids in a broad region around the critical point. *Int. J. Thermophys.*, **8**, 165–180.

Kawasaki, K. (1976). Mode coupling and critical dynamics, in *Phase Transitions and Critical Phenomena*, eds. C. Domb & M.S. Green, **5A**, pp. 165–403. New York: Academic Press.

Luettmer–Strathmann, J., Tang, S. & Sengers, J.V. (1992). A parametric model for the global thermodynamic behaviour of fluids in the critical region. *J. Chem. Phys.*, **97**, 2705–2717.

Maitland, G.C., Rigby, M., Smith, E.B., & Wakeham, W.A. (1987). *Intermolecular Forces: Their Origin and Determination*. Oxford: Clarendon Press.

Makita, T., Tanaka, Y. & Nagashima, A. (1974). Evaluation and correlation of viscosity data. The most probable values of the viscosity of gaseous ethane. *Rev. Phys. Chem. (Japan)*, **44**, 98–111.

Millat, J., Vesovic, V. & Wakeham, W.A. (1989). Theoretically–based data assessment for the correlation of the thermal conductivity of dilute gases. *Int. J. Thermophys.*, **10**, 805–818.

Mostert, R., van den Berg, H.R., van der Gulik, P.S. & Sengers, J.V. (1990). The thermal conductivity of ethane in the critical region. *J. Chem. Phys.*, **92**, 5454–5462.

Rainwater, J.C. & Friend, D.G. (1987). Second viscosity and thermal conductivity virial coefficient of gases: Extension to low temperatures. *Phys. Rev.*, **36A**, 4062–4066.

Sengers, J.V. (1985). Transport properties of fluids near critical points. *Int. J. Thermophys.*, **6**, 203–232.

Tarzimanov, A.A., Lusternik, V.E. & Arslanov, V.A. (1987). *Viscosity of Gaseous Hydrocarbons, a Survey of Thermophysical Properties of Compounds*, No. 1, 63. Moscow: Institute of High Temperatures.

van den Oord, R.J. & Korving J. (1988). The thermal conductivity of polyatomic gases. *J. Chem. Phys.*, **89**, 4333–4338.

Vargaftik, N.B. (1975). *Tables of the Thermophysical Properties of Liquids and Gases*. New York: Halsted Press.

Vesovic, V. & Wakeham, W.A. (1991). Transport properties of supercritical fluids and fluid mixtures, in *Critical Fluid Technology*, eds. T.J. Bruno, & J.F. Ely. Boca Raton: CRC Press, pp. 245–289.

Vesovic, V., Wakeham, W.A., Luettmer–Strathmann, J., Sengers, J.V., Millat, J., Vogel, E. & Assael M. J. (1994). The transport properties of ethane II. Thermal conductivity. *Int. J. Thermophys.* **15**, 33–66.

Vesovic, V., Wakeham, W.A., Olchowy, G.A., Sengers, J.V., Watson, J.T.R. & Millat, J. (1990). The transport properties of carbon dioxide. *J. Phys. Chem. Ref. Data*, **19**, 763–808.

Wong, C.C.K., McCourt, F.R.W. & Dickinson, A.S. (1989). A comparison between classical trajectory and infinite–order sudden calculation of transport and relaxation cross–sections for N_2–Ne mixtures. *Mol. Phys.*, **66**, 1235–1260.

Younglove, B.A. & Ely, J.F. (1987). Thermophysical properties of fluids II. Methane, ethane, propane, iso–butane, and normal butane. *J. Phys. Chem. Ref. Data*, **16**, 577–798.

14.4 Polar fluids – Water and steam

A. NAGASHIMA
Keio University, Yokohama, Japan

and

J. H. DYMOND
The University, Glasgow, UK

14.4.1 Introduction

The thermophysical properties of water and steam exhibit behavior which is peculiar to the water molecule and different from that of ordinary fluids. This is due to the molecular structure and to the resultant hydrogen bonding. Water has critical parameters which are much higher than those of ordinary organic fluids; as a result, experimental study at supercritical conditions is not easy. It is therefore important to develop theoretical methods especially for properties at high temperature and high pressure. A theoretical treatment is required also to predict the isotopic effect among the various isotopic forms of water, to provide consistency among the different transport properties and to explain the negative pressure effect on the viscosity of steam at subcritical temperatures.

In view of the problems in describing accurately the interactions between water molecules, theoretical predictions of thermophysical properties have not been quantitatively very successful for water and steam. An investigation by Thoen–Hellemans & Mason (1973) of the mutual consistency of the thermal conductivity and viscosity data at low density for steam, which does not require detailed knowledge of the intermolecular pair potential energy function, showed that the thermal conductivity data in the skeleton tables at that time were consistent with the viscosity data within the given tolerances, but it was considered that the thermal conductivities were a few percent too low. However, more recent recommendations for the thermal conductivity data of steam (Sengers *et al.* 1984) are in fact lower than those earlier values.

It is therefore necessary to have extensive accurate experimental data in order to develop equations for the transport properties. Since water substance is the most important working fluid for a power plant, its properties have been studied most precisely and most extensively since the nineteenth century. In the last 60 years, the best–quality data were accumulated by the International Association for the Properties of Water and Steam, IAPWS (denoted initially, in 1929, as ICPS and later as IAPS). IAPWS issued equations or tables of the best data sets. The historical background and the current recommendations are given in references by Nagashima (1977) and by Sengers & Watson (1986). The first internationally agreed recommendations of the transport properties viscosity and thermal conductivity for water and steam by IAPS (IAPWS) appeared in 1964. The temperature range covered was from 273 K to 973 K and up to 80 MPa in pressure for viscosity, and from 273 K to 873 K at pressures up to 50 MPa for thermal conductivity. These recommendations were revised in 1975 for viscosity and in 1977 for thermal conductivity, with an increase in the upper temperature to 1073 K and pressures up to 100 MPa. These recommendations contained equations for the properties as functions of temperature and pressure. The density was then based on the IFC 68 Formulation for Scientific and General Use. This was revised in 1982, and a much enlarged density range was covered (Haar *et al.* 1984). Current IAPWS recommendations for the viscosity and thermal conductivity of water and steam were issued in 1985, covering a much wider range, as described below. The IAPWS recommendations give the most accurate representation of these transport properties; they are described in Sections 14.4.2 and 14.4.3. In the case of heavy water substance, D_2O, the IAPWS adopted recommendations for the viscosity and thermal conductivity in 1982. The background and details of the formulation are given in the reference by Matsunaga & Nagashima (1983). A most important report which appeared later, and was therefore not included in the recommendations mentioned above, is a paper by Øye and coworkers which describes a new absolute determination of the viscosity of water at 293 K and at 0.1 MPa (Berstad *et al.* 1988). All current recommendations of the viscosity of water (and of heavy water substance too) by the IAPWS and also by other international bodies are relative to a standard value for the viscosity of water at 293 K and 0.1 MPa which is based on the measurement by Swindells *et al.* (1952). This means that adoption of a new standard value under these conditions would require adjustment of all values in the IAPWS recommendations, although the change is expected to be of the order of 0.1%.

14.4.2 *Viscosity of water and steam*

In accordance with the approach outlined in earlier chapters, the available accurate experimental viscosity data for water substance have been represented (Watson *et al.* 1980) by the equation

$$\eta(\rho, T) = \eta^{(0)}(T) + \Delta\eta(\rho, T) + \Delta\eta_c(\rho, T) \tag{14.62}$$

$\eta^{(0)}(T)$ represents the *dilute–gas viscosity*, and $\Delta\eta(\rho, T)$ gives the excess viscosity, that is, the increase in viscosity at density ρ over the dilute gas value at the same temperature. $\Delta\eta_c(\rho, T)$ is the extra enhancement observed in the immediate region around the critical point. It should be noted that although the excess viscosity can sometimes be represented to a very good approximation by a function of density alone (as, for example, for supercritical methane, in Chapter 5), for an accurate representation over a wide range of temperature and pressure the excess viscosity of water must be treated as a function of both density and temperature. The reason is that the excess function displays some abnormal behavior at subcritical temperatures in both the liquid

Table 14.15. *Coefficients H_i and H_{ij} for equations (14.63) and (14.66).*

i	j	H_{ij}	i	j	H_{ij}
0	0	0.513 204 7	0	3	0.177 806 4[a]
1	0	0.320 565 6	1	3	0.460 504 0
4	0	-0.778 256 7	2	3	0.234 037 9
5	0	0.188 544 7	3	3	-0.492 417 9
0	1	0.215 177 8	0	4	-0.041 766 10
1	1	0.731 788 3	3	4	0.160 043 5
2	1	1.241 044	1	5	-0.015 783 86
3	1	1.476 783	3	6	-0.003 629 481
0	2	-0.281 810 7			
1	2	-1.070 789			
2	2	-1.263 184			

$H_0 =$	1.000 000	$H_1 =$	0.978 197
$H_2 =$	0.579 829	$H_3 =$	-0.202 354

[a] Coefficients H_{ij} omitted from the table are all equal to zero identically.

and vapor phases. The equation for $\eta^{(0)}(T)$ must represent the viscosity data when extrapolated to zero density. A careful study of the viscosity data for steam at low pressure was made by Aleksandrov *et al.* (1975). They started from the expression for viscosity given by the kinetic theory of gases but replaced the collision integral – which cannot be calculated exactly because of imprecise knowledge of the intermolecular pair potential – by an empirical power series expansion in inverse temperature to give

$$\frac{\eta^{(0)}(\bar{T})}{\eta^*} = \frac{\sqrt{\bar{T}}}{\sum\limits_{i=0}^{3}\left(H_i/\bar{T}^i\right)} \tag{14.63}$$

Quantities are expressed in dimensionless form with $\bar{T} = T/T^*$ and reference values $T^* = 647.27$ K and $\eta^* = 55.071 \cdot 10^{-6}$ Pa s. The reference temperature is close, but not identical, to the critical temperature. Values for the coefficients H_i in equation (14.63) are given in Table 14.15.

The *initial–density dependence* can be described as follows (Section 5.2):

$$\eta(\rho, T) = \eta^{(0)}(T) + \eta^{(1)}(T)\rho + \ldots = \eta^{(0)}(T)\left[1 + B_\eta(T)\rho + \ldots\right] \tag{14.64}$$

where $B_\eta(T)$ is the second viscosity virial coefficient. Analysis of experimental viscosity data for water vapor gave values for $\eta^{(1)}(T)$ (Watson *et al.* 1980), which have been converted to $B_\eta(T)$ and shown in Figure 14.20. Although water molecules exhibit a much stronger attractive interaction than do comparable non–hydrogen–bonded

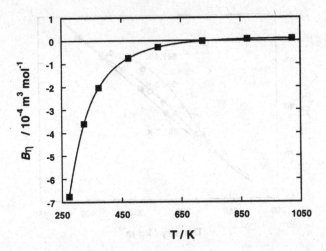

Fig. 14.20. Second viscosity virial coefficient as a function of temperature.

molecules, there is a remarkable general similarity between Figure 14.20 and Figure 5.3.

To account for the viscosity up to higher pressures, it is necessary to consider the *excess viscosity*. This can be determined by analysis of the primary experimental data sets listed by Watson *et al.* (1980). However, it was found that instead of expressing this excess viscosity by an equation in temperature and density, a more concise relationship was obtained by relating the background viscosity $\bar{\eta}(\rho, T)$, equal to the sum of the first two terms in equation (14.62), to the dilute–gas value by an expression of the form

$$\frac{\bar{\eta}(\bar{\rho}, \bar{T})}{\eta^{(0)}(\bar{T})} = \exp\left\{F\left(\bar{\rho}, \bar{T}\right)\right\} \tag{14.65}$$

where

$$F\left(\bar{\rho}, \bar{T}\right) = \bar{\rho} \sum_{i=0}^{5} \sum_{j=0}^{6} H_{ij} \left(\bar{T}^{-1} - 1\right)^{i} (\bar{\rho} - 1)^{j} \tag{14.66}$$

The equation is given in dimensionless form with $\bar{\rho} = \rho/\rho^*$ and $\rho^* = 317.763 \text{ kg m}^{-3}$. Values for the coefficients H_{ij} are listed in Table 14.15. The *critical enhancement* of the viscosity of steam occurs in a region which covers approximately 1% in absolute temperature and 30% in density relative to the critical temperature and critical density, as shown in Figure 14.21. Inside this critical region, the equations given above are used to represent the background viscosity. In the case of water substance, it has been the practice to consider the relative critical enhancement, or viscosity ratio, rather than the absolute enhancement. The asymptotic behavior of the viscosity is written (Chapter 6) in terms of the correlation length ξ, which characterizes the spatial

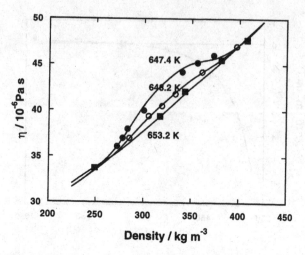

Fig. 14.21. Viscosity data along near–critical isotherms.

extent of the large–scale density fluctuations

$$\frac{\eta(\bar{\rho}, \bar{T})}{\bar{\eta}(\bar{\rho}, \bar{T})} = (Q\xi)^z \tag{14.67}$$

for $Q\xi > 1$, and $\eta(\bar{\rho}, \bar{T}) = \bar{\eta}(\bar{\rho}, \bar{T})$ for $Q\xi < 1$.

It was found by Watson *et al.* (1980) that for the range of conditions

$$0.9970 \leq \bar{T} \leq 1.0082, \qquad 0.755 \leq \bar{\rho} \leq 1.290 \tag{14.68}$$

the experimental viscosity data for steam were found to be consistent with $z = 0.05$, in general agreement with theoretical estimates. Parameter Q was taken to be adjustable and found to have the value $Q^{-1} = 26.6 \cdot 10^{-10}$ m. The correlation length was deter- mined by relating it to the symmetrical compressibility $\bar{\chi}_T$

$$\bar{\chi}_T = \bar{\rho} \left(\frac{\partial \bar{\rho}}{\partial \bar{P}} \right)_{\bar{T}} \tag{14.69}$$

where dimensionless quantities are used and the reduced pressure $\bar{P} = P/P^*$ and $P^* = 22.115 \cdot 10^6$ Pa. As a result, equation (14.67) is written as

$$\frac{\eta(\bar{\rho}, \bar{T})}{\bar{\eta}(\bar{\rho}, \bar{T})} = 0.922 (\bar{\chi}_T)^{0.0263} \qquad \text{for} \qquad \bar{\chi}_T \geq 21.93 \tag{14.70}$$

Outside this range, there is no critical enhancement. It should be noted that, in fact, in most industrial calculations the background contribution is all that has to be considered.

In practice, therefore, equation (14.62), which is based upon theoretical considera- tions, is written as

$$\eta(\bar{\rho}, \bar{T}) = \eta^{(0)}(\bar{T}) \cdot \frac{\bar{\eta}(\bar{\rho}, \bar{T})}{\eta^{(0)}(\bar{T})} \cdot \frac{\eta(\bar{\rho}, \bar{T})}{\bar{\eta}(\bar{\rho}, \bar{T})} \tag{14.71}$$

Table 14.16. *Viscosity ($/10^{-6}$ Pa s) of water substance obtained from equation (14.71).*

$\dfrac{P}{MPa}$	Temperature/°C						
	0	50	100	200	300	350	375
0.1	1793	547.0	12.27	16.18	20.29	22.37	23.41
1.0	1791	547.2	282.1	15.89	20.18	22.31	23.37
5.0	1781	547.9	283.1	135.2	19.80	22.12	23.25
10.0	1769	548.8	284.5	136.4	86.51	22.15	23.33
20.0	1748	550.8	287.2	138.8	90.11	69.33	25.92
30.0	1729	552.8	289.8	141.2	93.21	75.51	64.57
40.0	1712	555.0	292.5	143.4	95.98	79.79	71.31
50.0	1698	557.4	295.1	145.6	98.52	83.26	75.85
60.0	1685	559.8	297.8	147.8	100.9	86.25	79.46
70.0	1674	562.4	300.5	149.9	103.1	88.92	82.54
80.0	1666	565.0	303.1	151.9	105.2	91.36	85.27
90.0	1658	567.8	305.7	154.0	107.2	93.61	87.74
100.0	1653	570.7	308.4	155.9	109.1	95.72	90.01

$\dfrac{P}{MPa}$	Temperature/°C						
	400	425	450	500	600	700	800
0.1	24.45	25.49	26.52	28.57	32.61	36.55	40.37
1.0	24.42	25.47	26.51	28.58	32.64	36.59	40.41
5.0	24.37	25.46	26.55	27.62	32.81	36.78	40.60
10.0	24.49	25.62	26.73	28.91	33.09	37.07	40.88
20.0	26.03	26.85	27.81	29.85	33.92	37.84	41.58
30.0	43.99	31.86	30.85	31.73	35.17	38.86	42.47
40.0	61.31	48.61	39.02	35.11	36.90	40.13	43.52
50.0	68.01	59.39	50.50	40.48	39.14	41.65	44.72
60.0	72.64	65.60	58.45	47.03	41.85	43.38	46.04
70.0	76.32	70.13	63.97	53.12	44.93	45.30	47.47
80.0	79.45	73.79	68.25	58.22	48.20	47.34	48.98
90.0	82.21	76.91	71.80	62.45	51.49	49.46	50.54
100.0	84.70	79.67	74.86	66.06	54.66	51.61	52.12

This is the form of equation which is recommended in the *Release on the IAPS (IAPWS) Formulation 1985 for the Viscosity of Ordinary Water Substance, Appendix I* (Sengers & Watson 1986). Examples of viscosity values calculated in this way are given in Table 14.16 for a wide range of temperature and pressure, and in Table 14.17 for water substance along the saturation line. The density values used in these calculations were taken from the *IAPS Formulation 1984 for the Thermodynamic Properties of Ordinary Water Substance for Scientific and General Use* (Haar *et al.* 1984; Kestin & Sengers

Table 14.17. *Viscosity (/10^{-6} Pa s) and thermal conductivity (/mW m^{-1} K^{-1}) of water substance along the saturation line calculated from equations (14.71) and (14.73).*

$\frac{t}{°C}$	$\frac{P}{MPa}$	η_l	η_v	λ_l	λ_v
0.01	0.0006117	1792.	9.22	561.0	17.07
10.	0.001228	1307.	9.46	580.0	17.62
20.	0.002339	1002.	9.73	598.4	18.23
30.	0.004246	797.7	10.01	615.4	18.89
40.	0.007381	653.2	10.31	630.5	19.60
50.	0.01234	547.0	10.62	643.5	20.36
60.	0.01993	466.5	10.93	654.3	21.18
70.	0.03118	404.0	11.26	663.1	22.07
80.	0.04737	354.4	11.59	670.0	23.01
90.	0.07012	314.5	11.93	675.3	24.02
100.	0.1013	281.8	12.27	679.1	25.09
110.	0.1432	254.8	12.61	681.7	26.24
120.	0.1985	232.1	12.96	683.2	27.46
130.	0.2700	213.0	13.30	683.7	28.76
140.	0.3612	196.6	13.65	683.3	30.14
150.	0.4757	182.5	13.99	682.1	31.59
160.	0.6177	170.3	14.34	680.0	33.12
170.	0.7915	159.6	14.68	677.1	34.74
180.	1.002	150.2	15.02	673.4	36.44
190.	1.254	141.8	15.37	668.8	38.23
200.	1.554	134.4	15.71	663.4	40.10
220.	2.318	121.6	16.41	649.8	44.15
240.	3.345	110.9	17.12	632.0	48.70
260.	4.689	101.7	17.88	609.4	53.98
280.	6.413	93.56	18.70	581.4	60.52
300.	8.584	85.95	19.65	547.7	69.49
320.	11.279	78.45	20.84	509.4	83.59
340.	14.594	70.45	22.55	468.6	110.2
360.	18.655	60.39	25.71	427.2	178.0
370.	21.030	52.25	29.57	428.0	299.4

1986). However, for calculation of the saturation density at 370°C and above, some changes are necessary to the original computer program of Haar *et al.* (1984) (Sengers & Watson 1986). Examples of the density calculated from the formulation are given in Table 14.18. For most industrial calculations, the IFC Formulation for Industrial Use, the equation of state used for preparing many currently available steam tables (Schmidt

Table 14.18. *Density ($/\mathrm{kg\,m^{-3}}$) calculated from the IAPS Formulation 1984 for the Thermodynamic Properties of Ordinary Water Substance for Scientific and General Use.*

$\dfrac{P}{\text{MPa}}$	Temperature/°C						
	0	50	100	200	300	350	375
0.1	999.83	988.03	0.5896	0.4604	0.3790	0.3483	0.3348
1.0	1000.3	988.42	958.81	4.8566	3.8771	3.5402	3.3984
5.0	1002.3	990.16	960.68	867.35	22.073	19.255	18.203
10.0	1004.8	992.31	962.98	871.03	715.58	44.611	40.763
20.0	1009.7	996.53	967.48	878.06	735.02	600.72	130.42
30.0	1014.5	1000.7	971.86	884.70	750.93	644.27	558.25
40.0	1019.2	1004.7	976.12	891.00	764.58	672.10	609.56
50.0	1023.8	1008.7	890.27	897.02	776.64	693.39	641.32
60.0	1028.3	1012.6	984.33	902.77	787.51	710.93	665.16
70.0	1032.7	1016.4	988.30	908.30	797.44	726.00	684.57
80.0	1037.0	1020.1	992.18	913.61	806.62	739.31	701.09
90.0	1041.2	1023.8	995.98	918.75	815.18	751.29	715.58
100.0	1045.3	1027.4	999.70	923.71	823.21	762.21	728.54

$\dfrac{P}{\text{MPa}}$	Temperature/°C						
	400	425	450	500	600	700	800
0.1	0.3223	0.3107	0.2999	0.2805	0.2483	0.2227	0.2019
1.0	3.2617	3.1394	3.0263	2.8241	2.4932	2.2331	2.0228
5.0	17.299	16.505	15.798	14.586	12.709	11.299	10.189
10.0	37.867	35.547	33.611	30.503	26.068	22.941	20.564
20.0	100.54	87.227	78.732	67.711	55.039	47.319	41.871
30.0	358.05	188.66	148.45	115.26	87.481	73.234	63.919
40.0	523.67	394.56	270.91	177.97	123.81	100.71	86.682
50.0	577.99	498.16	402.28	256.95	163.99	129.64	110.09
60.0	612.45	550.90	479.87	338.44	207.20	159.77	134.02
70.0	638.30	586.40	528.62	405.76	251.76	190.65	158.30
80.0	659.27	613.48	563.69	456.99	295.45	221.74	182.72
90.0	677.05	635.55	591.14	496.53	336.53	252.48	207.03
100.0	692.58	654.30	613.80	528.21	373.93	282.36	231.03

1979; JSME 1980), for example, may be used with the final term in equation (14.71) set equal to unity. The effective range of temperature and pressure for equation (14.71) and the accuracy of the calculated viscosities are shown in Figure 14.22.

Since equation (14.71) is rather complicated, simpler equations might be more convenient for limited purposes. Such equations have been reported by Nagashima (1977) for the viscosity of water at atmospheric pressure.

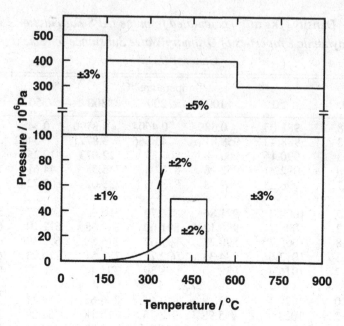

Fig. 14.22. Estimate of accuracy and range of applicability of equation (14.71).

14.4.3 Thermal conductivity of water and steam

The thermal conductivity of water and steam can be represented by the equation developed in earlier chapters

$$\lambda(\rho, T) = \lambda^{(0)}(T) + \Delta\lambda(\rho, T) + \Delta\lambda_c(\rho, T) = \bar{\lambda}(\rho, T) + \Delta\lambda_c(\rho, T) \quad (14.72)$$

Here, $\lambda^{(0)}(T)$ gives the low–density thermal conductivity and $\Delta\lambda(\rho, T)$ represents the excess thermal conductivity, that is, the difference between the actual thermal conductivity at density ρ and the low–density value at the same temperature. The sum of these terms is the background thermal conductivity $\bar{\lambda}(\rho, T)$, and $\Delta\lambda_c(\rho, T)$ gives the critical enhancement. This form of equation has been used, in dimensionless form, as the recommended interpolating equation for industrial use (Sengers & Watson 1986). Individual terms were fitted empirically as functions of temperature and density or (first term) temperature alone. This representation has the disadvantage of a limited range of applicability (maximum pressure of 100 MPa at temperatures from 0°C to 500°C). It gives a finite thermal conductivity at the critical point. As in the case of viscosity described above, so here also it proved advantageous from practical considerations to write the equation in a slightly different form

$$\lambda(\rho, T) = \lambda^{(0)}(T) \cdot \frac{\bar{\lambda}(\rho, T)}{\lambda^{(0)}(T)} + \Delta\lambda_c(\rho, T) \quad (14.73)$$

This form of equation was recommended in the *Release on the IAPS Formulation 1985 for the Thermal Conductivity of Ordinary Water Substance, Appendix II* (Sengers &

Table 14.19. *Coefficients L_i and L_{ij} for equations (14.74) and (14.77).*

			i		
	0	1	2	3	4
j					
0	1.329304	1.701836	5.224615	8.712767	-1.852599
1	-0.4045243	-2.215684	-10.12411	-9.500061	0.9340469
2	0.2440949	1.651105	4.987468	4.378660	0.0
3	0.01866075	-0.7673600	-0.2729769	-0.9178378	0.0
4	-0.1296106	0.3728334	-0.4308339	0.0	0.0
5	0.04480995	-0.1120316	0.1333384	0.0	0.0

$L_0 =$	1.000 000	$L_1 =$	6.978 267
$L_2 =$	2.599 096	$L_3 =$	-0.998 254

Watson 1986) as the recommended interpolating equation for scientific use. Equations for the different contributing terms were based on an analysis of critically evaluated thermal conductivity data. Since precise calculation of the effective cross section is not possible, the *thermal conductivity in the dilute–gas limit* is represented by the equation

$$\frac{\lambda^{(0)}(\bar{T})}{\lambda^*} = \frac{\sqrt{T}}{\sum_{i=0}^{3} L_i/\bar{T}^i} \tag{14.74}$$

λ^* is introduced to make the terms in the equation dimensionless. It has the value 0.4945 W m^{-1} K^{-1}. The reduced temperature is equal to T/T^*, where T^* equals 647.27 K. The coefficients in the equation, which were originally determined by Aleksandrov & Matveev (1978), are given in Table 14.19. The *initial–density dependence* of the thermal conductivity in the gaseous phase at low pressures can be represented by

$$\lambda(\rho, T) = \lambda^{(0)}(T)\,[1 + B_\lambda(T)\rho + \ldots] \tag{14.75}$$

Values for B_λ, the second virial coefficient of thermal conductivity, have been derived from experimental thermal conductivity data for steam. The temperature dependence is shown in Figure 14.23. It should be noted that the values below 100°C are less reliable because there are insufficient experimental data at these temperatures to determine the values of B_λ with confidence.

The temperature dependence of B_λ should be contrasted with that of B_η, which becomes negative at low temperatures (see Section 5.2).

The background term was obtained by fitting the thermal conductivity of water substance over a wide range of pressure and temperature, outside the critical region (see below), to a double power series

$$\frac{\bar{\lambda}(\bar{\rho}, \bar{T})}{\lambda^{(0)}(\bar{T})} = \exp\{G(\bar{\rho}, \bar{T})\} \tag{14.76}$$

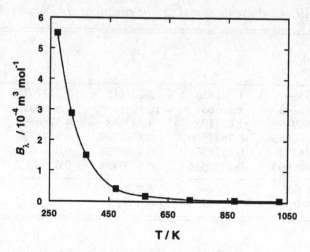

Fig. 14.23. Second thermal conductivity virial coefficient for water vapor as a function of temperature.

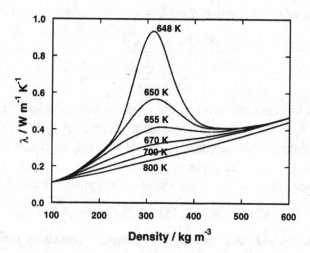

Fig. 14.24. Thermal conductivity as a function of density in the critical region (calculated from equation (14.73)).

where

$$G(\bar{\rho}, \bar{T}) = \bar{\rho} \sum_{i=0}^{4} \sum_{j=0}^{5} L_{ij}(\bar{T}^{-1} - 1)^i (\bar{\rho} - 1)^j \qquad (14.77)$$

The equation is given in dimensionless form with $\bar{\rho} = \rho/\rho^*$ and $\rho^* = 317.763$ kg m^{-3}. Values for the coefficients L_{ij} are included in Table 14.19.

For thermal conductivity, the magnitude and range of the critical anomaly, shown in Figure 14.24, are much more significant than in the case of viscosity (Figure 14.21).

The critical enhancement term $\Delta\lambda_c(\rho, T)$ was based on the theory of dynamic critical phenomena, which predicts that the critical part of the thermal diffusivity $\lambda/\rho c_P$, where c_P is the heat capacity at constant pressure, close to the critical point satisfies a Stokes–Einstein diffusion law of the form (Chapter 6)

$$\frac{\Delta\lambda_c(\rho, T)}{\rho c_P} = \frac{R_D k_B T}{6\pi\eta\xi} \tag{14.78}$$

Here, k_B is the Boltzmann constant, R_D is a coefficient of order unity and ξ is the correlation length associated with the critical fluctuations. ρc_P can be expressed in terms of the symmetrized compressibility (equation (14.69)). By consideration of the divergence of this quantity, and also that of the correlation length, at the critical density, the following expression was obtained for the critical enhancement in the thermal conductivity (Sengers *et al.* 1984)

$$\frac{\Delta\lambda_c(\bar{\rho}, \bar{T})}{\lambda^*} = 0.0013848 \frac{\eta^*}{\bar{\eta}(\bar{\rho}, \bar{T})} \left(\frac{\bar{T}}{\bar{\rho}}\right)^2 \left(\frac{\partial\bar{P}}{\partial\bar{T}}\right)^2_{\bar{\rho}} (\bar{\chi}_T)^{0.4678} Q(\Delta\bar{T}, \Delta\bar{\rho}) \tag{14.79}$$

where $\Delta\bar{T} = (T - T_c)/T_c$ and $\Delta\bar{\rho} = (\rho - \rho_c)/\rho_c$.

$Q(\Delta\bar{T}, \Delta\bar{\rho})$ is a crossover function which has the value of unity at the critical temperature and density and zero under conditions far removed from the critical conditions. Since the critical enhancement is present in a large range of temperatures and densities around the critical point, it was necessary initially to predict the background thermal conductivity in this range and to subtract this from the experimental values. The amplitude R_D of the critical enhancement was then determined, and parameters in the equation for the background were adjusted. The equation selected by Basu & Sengers (1977) for the crossover function was written in the form

$$Q(\Delta\bar{T}, \Delta\bar{\rho}) = \sqrt{\bar{\rho}}\left\{\exp\left[-18.66(\bar{T} - 1)^2 - (\bar{\rho} - 1)^4\right]\right\} \tag{14.80}$$

Values for the thermal conductivity of ordinary water substance calculated from the above equations are given in Table 14.20 for a wide range of temperature and pressure, and in Table 14.17 values are given along the saturation line. More extensive tabulations are given by Sengers & Watson (1986). The recommended ranges of applicability of this equation are as follows:

$$P \le 400 \text{ MPa for } 0°C \le t \le 125°C$$
$$P \le 200 \text{ MPa for } 125°C \le t \le 250°C$$
$$P \le 150 \text{ MPa for } 250°C \le t \le 400°C$$
$$P \le 100 \text{ MPa for } 400°C \le t \le 800°C$$

Detailed comparisons with experimental data are given by Sengers *et al.* (1984).

14.4.4 Viscosity and thermal conductivity of heavy water substance

The viscosity and thermal conductivity of heavy water and steam, the heavy water substance D_2O, show trends which are similar to those of ordinary water substance.

Table 14.20. *Thermal conductivity* ($/mW\ m^{-1}\ K^{-1}$) *of water substance obtained with the aid of equation (14.73).*

$\frac{P}{MPa}$	Temperature/°C						
	0	50	100	200	300	350	375
0.1	561.0	643.6	25.08	33.28	43.42	48.96	51.83
1.0	561.5	644.0	679.6	37.21	44.95	50.06	52.79
5.0	563.5	645.8	681.8	666.4	53.86	55.99	57.87
10.0	566.5	648.2	684.5	670.7	550.9	68.11	67.35
20.0	572.0	652.8	690.0	679.1	571.6	463.3	142.1
30.0	577.5	657.4	695.3	687.3	589.1	496.3	437.9
40.0	582.9	662.1	700.7	695.3	604.6	521.0	473.1
50.0	588.1	666.7	706.1	703.2	618.5	541.7	498.4
60.0	593.3	671.2	711.2	710.9	631.3	559.7	519.4
70.0	598.3	675.7	716.4	718.5	643.2	575.9	537.6
80.0	603.1	680.2	721.5	726.0	654.5	590.6	554.1
90.0	607.8	684.6	726.6	733.4	665.1	604.2	569.1
100.0	612.2	688.9	731.6	740.7	675.4	616.8	583.0

$\frac{P}{MPa}$	Temperature/°C						
	400	425	450	500	600	700	800
0.1	54.76	57.74	60.77	66.97	79.89	93.37	107.3
1.0	55.61	58.51	61.48	67.60	80.44	93.87	107.8
5.0	60.06	62.49	65.10	70.74	83.13	96.34	109.9
10.0	67.89	69.19	71.00	75.61	87.14	99.97	113.2
20.0	105.4	95.11	91.03	89.89	97.57	109.1	121.2
30.0	329.7	176.1	136.0	113.7	111.7	120.7	131.3
40.0	414.0	323.5	227.7	151.6	130.134	8.143	1.
50.0	451.5	391.5	315.8	202.8	125.1	150.9	156.4
60.0	477.6	429.9	371.2	255.9	177.0	168.5	170.6
70.0	498.6	456.3	408.3	301.7	202.7	186.7	185.2
80.0	516.7	478.3	435.7	339.1	227.5	204.4	199.6
90.0	533.0	496.4	457.6	369.6	250.6	221.1	213.1
100.0	547.9	512.6	476.0	394.8	271.8	236.2	225.5

Equations for representing these properties are given also by IAPWS and are described by Matsunaga & Nagashima (1983). This reference also gives a detailed explanation of the background to these equations.

References

Aleksandrov, A.A., Ivanov, A.I. & Matveev, A.B. (1975). Absolute viscosity of water and steam in a wide range of temperatures and pressures. *Teploenergetika*, **22**, 59–64. [English translation: *Therm. Eng.*, **22**, 77–82 (1975)].

Aleksandrov, A.A. & Matveev, A.B. (1978). Thermal conductivity of water and water vapour in a wide range of temperatures and pressures. *Teploenergetika*, **25**, 80–85. [English translation: *Therm. Eng.*, **25**, 58–63 (1978)].

Basu, R. S. & Sengers, J.V. (1977). Thermal conductivity of steam in the critical region. *7th Symp. Thermophys. Props.*, ed. A. Cezairliyan. New York: Am. Soc. Mech. Engrs., 822–830.

Berstad, D. A., Knapstad, B., Lamvik, M., Skjolvik, P. A., Tørklep, K. & Øye, H. A. (1988). Accurate measurements of the viscosity of water in the temperature range 19.5–25.5°C. *Physica*, **151A**, 246–280.

Haar, L., Gallagher, J. S. & Kell, G. S. (1984). *NBS/NRC Steam Tables*. New York: Hemisphere.

Kestin, J. & Sengers, J. V. (1986). New international formulations for the thermodynamic properties of light and heavy water. *J. Phys. Chem. Ref. Data*, **15**, 305–320.

Matsunaga, N. & Nagashima, A. (1983). Transport properties of liquid and gaseous D_2O over a wide range of temperature and pressure. *J. Phys. Chem. Ref. Data*, **12**, 933–966.

Nagashima, A. (1977). Viscosity of water substance – New international formulation and its background. *J. Phys. Chem. Ref. Data*, **6**, 1133–1166.

Schmidt, E. (1979). *Properties of Water and Steam in SI–Units*. New York: Springer.

Sengers, J. V. & Watson, J. T. R. (1986). Improved international formulations for the viscosity and thermal conductivity of water and steam. *J. Phys. Chem. Ref. Data*, **15**, 1291–1314.

Sengers, J. V., Watson, J. T. R., Basu, R. S., Kamgar–Parsi, B. & Hendricks, R. C. (1984). Representative equations for the thermal conductivity of water substance. *J. Phys. Chem. Ref. Data*, **13**, 893–933.

Swindells, J. F., Coe, J. R. & Godfrey, T. B. (1952). Absolute viscosity of water at 20 °C. *J. Res. NBS*, **48**, 1–31.

Thoen–Hellemans, J. & Mason, E.A. (1973). Theoretical consistency test of steam transport properties. *Int. J. Engng. Sci.*, **11**, 1247–1253.

Watson, J. T. R., Basu, R. S. & Sengers, J. V. (1980). An improved representative equation for the dynamic viscosity of water substance. *J. Phys. Chem. Ref. Data*, **9**, 1255–1290.

1980 *SI JSME Steam Tables*, (1980). Edited by the Japan Soc. Mech Eng., JSME Steam Tables Publication Committee chaired by I. Tanishita, Tokyo: The Japan Society of Mechanical Engineers. (Published in 1981.)

14.5 Polyatomic fluids – R134a

A. NAGASHIMA

Keio University, Yokohama, Japan

and

R. KRAUSS and K. STEPHAN

Universität Stuttgart, Germany

14.5.1 Introduction

The polar hydrofluorocarbon R134a (1,1,1,2-tetrafluoroethane) is an environmentally acceptable substitute for the ozone–depleting refrigerant R12 (dichlorodifluoromethane) already used in domestic refrigeration and automotive air conditioning. A large number of studies were published after 1988, and since then R134a has become the most studied substance among the so-called alternative refrigerants. The status of experimental studies and some correlations are described, for instance, in the papers by Ueno *et al.* (1991) and Okubo *et al.* (1992). Recently, a comprehensive review article on transport properties of R134a was published by Krauss, Luettmer-Strathmann, Sengers & Stephan (1993), quoted as KLSS in the following section. This article contains not only detailed explanations of available data but also equations for the viscosity and the thermal conductivity of R134a for both the liquid and gaseous phases including the critical region. As it may be considered to be the most comprehensive and reliable evaluation and correlation of the transport properties of R134a at present, the following section is mainly based on this work.

As explained earlier, the transport property $X(\rho, T)$ of a dipolar fluid can be represented as for a nonpolar fluid as

$$X(\rho, T) = X^{(0)}(T) + \Delta X(\rho, T) + \Delta X_c(\rho, T) \qquad (14.81)$$

where X denotes either the viscosity or the thermal conductivity, ρ the density and T the temperature.

The first term in equation (14.81), the dilute-gas term $X^{(0)}$, is a function of temperature only. In principle it can be determined theoretically for simple molecules. For complicated molecules it must be fitted to experimental data in order to obtain quantitatively accurate results. Since often no experimental dilute-gas data are given, data obtained at or near atmospheric pressure are used.

The excess term $\Delta X(\rho, T)$ accounts for the pressure dependence and is usually expressed as a function of density and temperature. It can be explained qualitatively by theoretical treatise. However, quantitatively, ΔX is determined by fitting experimental data, particularly in the case of a polar fluid. The temperature dependence of the residual term may be neglected if the resulting error is smaller than the scattering of the experimental data. This is often true especially for supercritical conditions.

The third term $\Delta X_c(\rho, T)$ of equation (14.81) represents the critical enhancement. Using the concept of the crossover theory of Olchowy & Sengers (1988), the behavior of the fluid in the critical region is expressed quantitatively. While the critical enhancement of the viscosity is small, that of the thermal conductivity is large and a quantitative expression is complicated.

14.5.2 Viscosity

Based on a critical evaluation of the available experimental data and following the basic concept of equation (14.81), the viscosity of R134a is expressed as

$$\eta(\rho, T) = \eta^{(0)}(T) + \Delta\eta(\rho) + \Delta\eta_c(\rho, T) \tag{14.82}$$

The viscosity of R134a in the *dilute-gas region* is given by the expression

$$\eta^{(0)} = \frac{0.2696566\sqrt{T}}{\sigma^2 \Omega_\eta(T^*)} \tag{14.83}$$

where the functional Ω_η is expressed as a function of the dimensionless temperature $T^* = k_B T/\epsilon$. Hence

$$\ln \Omega_\eta(T^*) = \sum_{i=0}^{4} a_i \, (\ln T^*)^i \tag{14.84}$$

The energy scaling factor ϵ/k_B (ϵ represents the potential well depth and k_B the Boltzmann constant), the length scaling factor or collision diameter σ, and the constants a_i in equation (14.84) are given in Table 14.21.

The *excess term* $\Delta\eta(\rho)$ in the KLSS viscosity equation is expressed as a function of density only, neglecting a temperature dependence, because the resulting error is

Table 14.21. *Coefficients of equations (14.84) and (14.85).*

i	a_i	e_i
0	0.4425728	
1	−0.5138403	−0.189758×10^1
2	0.1547566	0.256449
3	−0.2821844×10^{-1}	−0.301641
4	0.1578286×10^{-2}	−0.231648×10^2
5		0.344752×10^1

$\epsilon/k = 279.86$ K; $\sigma = 0.50768$ nm;
$H_c = 25.21$ μPa s; $\rho_c = 515.25$ kg m^{-3}.

smaller than the experimental error in the temperature and density range of interest. Hence

$$\frac{\Delta\eta(\rho)}{H_c} = \sum_{i=1}^{3} e_i \left(\frac{\rho}{\rho_c}\right)^i + \frac{e_4}{\rho/\rho_c - e_5} + \frac{e_4}{e_5} \tag{14.85}$$

where ρ_c is the critical density and H_c is a viscosity reduction factor calculated from the critical constants. They are also given in Table 14.21 together with the constants e_i of equation (14.85).

The third term of equation (14.82), the critical enhancement, is derived from the crossover theory of Olchowy & Sengers (1988) and expressed as

$$\Delta\eta_c(\rho, T) = (\eta^{(0)} + \Delta\eta)[\exp(zH) - 1] \tag{14.86}$$

In equation (14.86) z is a universal critical exponent and H represents a crossover function for the viscosity, converging asymptotically to zero in a region away from the critical point. The calculation of H is described in detail by Mostert *et al.* (1990). There are no viscosity data reported for R134a in the critical region. Furthermore, as the contribution of the critical term is not significant outside the immediate vicinity of the critical point, it may be omitted for practical purposes.

Due to the inconsistencies between the experimental data sets the uncertainty of the correlation for the viscosity cannot be better than about ±4% in the dilute–gas region and about ±5% in the dense fluid region. The correlation is valid in the temperature range 290–430 K for densities up to 1400 kg m^{-3} in both the liquid and gaseous phases. Since the basic form of the viscosity equation is based on theory to some extent, it can be slightly extrapolated to lower temperatures and higher densities – however, with a larger uncertainty. A detailed comparison of equation (14.82) with available experimental data is given in the review by KLSS.

The viscosity of R134a on the saturation line calculated by equation (14.82) is listed in Table 14.23, where subscripts l and v indicate values of the saturated liquid and saturated vapor, respectively. The viscosity as a function of temperature and pressure is given in Table 14.24. The KLSS equation uses densities calculated by the modified BWR equation of state by Huber & McLinden (1992) combined with a crossover for the critical region developed by Tang *et al.* (1991). However, the difference in the calculated viscosity is expected to be small when a different equation of state is used, except in the critical region.

14.5.3 Thermal conductivity

According to the residual concept of equation (14.81) the thermal conductivity is expressed as

$$\lambda(\rho, T) = \lambda^{(0)}(T) + \Delta\lambda(\rho) + \Delta\lambda_c(\rho, T) \tag{14.87}$$

Table 14.22. *Coefficient of equations (14.88) and (14.89).*

i	d_i	l_i
0	-0.165744×10^2	
1	0.124286	0.579388×10^{-1}
2	-0.761769×10^{-4}	-0.254517
3		0.352171×10^1
4		-0.371906

$\Lambda_c = 2.055$ mW m^{-1} K^{-1}; $\rho_c = 515.25$ kg m^{-3}.

As a theoretical, quantitatively correct calculation of the thermal conductivity in the dilute-gas region is difficult for polar fluids, the first term of equation (14.87) was determined empirically by fitting the coefficients of a quadratic polynomial to experimental data. Hence

$$\lambda^{(0)} = d_0 + d_1 T + d_2 T^2 \tag{14.88}$$

with the constants d_i given in Table 14.22.

The residual term, the second term in equation (14.87), can also be represented by a polynomial of the form

$$\frac{\Delta \lambda(\rho)}{\Lambda_c} = \sum_{i=1}^{4} l_i \left(\frac{\rho}{\rho_c} \right)^i \tag{14.89}$$

where Λ_c is a reduction factor for the thermal conductivity, calculated from the critical constants. It is given in Table 14.22 together with the constants l_i of equation (14.89). As for the viscosity, so also for the thermal conductivity a temperature dependence of the residual term could be neglected.

There is a significant difference in the magnitude and the range of the critical enhancement for the thermal conductivity of R134a, compared with that of the viscosity. Therefore, for the thermal conductivity the contribution of the third term in equation (14.87) is larger than that of the viscosity. The expression is derived from the crossover theory of Olchowy & Sengers (1988) and has the form

$$\Delta \lambda_c = \rho c_P \frac{R_D k_B T}{6 \pi \eta \xi} (\Omega - \Omega_0) \tag{14.90}$$

where c_P is the specific heat and R_D a dimensionless universal amplitude in the crossover theory. The correlation length ξ can be calculated from the density and the isothermal compressibility, as explained in the review by KLSS. The crossover function for the thermal conductivity $(\Omega - \Omega_0)$ asymptotically converges to zero far away from the critical point, as does the crossover function H for the viscosity. Detailed information for the calculation of the crossover functions is reported by Mostert *et al.* (1990). The correlation for the thermal conductivity is valid in the temperature range 240–410 K at densities up to 1500 kg m^{-3}. Due to the inconsistencies in the experimental data, the uncertainty of equation (14.87) cannot be better than about 5%, with highest values in the critical region.

The thermal conductivity of R134a on the saturation line calculated with equation (14.87) is listed in Table 14.23. The thermal conductivity as a function of temperature and pressure is given in Table 14.25.

References

Huber, M. L. & McLinden, M. O. (1992). Thermodynamic properties of R134a (1,1,1,2-tetrafluoroethane). *Proc. Int. Refrig. Conf., West Lafayette, U.S.A., July 14–17,* **Vol. 2**, 453–462.

Krauss, R., Luettmer-Strathmann, J. Sengers, J. V. & Stephan, K. (1993). Transport properties of 1,1,1,2-tetrafluoroethane (R134a). *Int. J. Thermophys.*, **14**, 951–988.

Mostert, R., van den Berg, H. R., van der Gulik, P. S. & Sengers, J. V. (1990). The thermal conductivity of ethane in the critical region. *J. Chem. Phys.*, **92**, 5454–5462.

Okubo, T., Hasuo, T. & Nagashima, A. (1992). Measurement of the viscosity of HFC 134a in the temperature range 213-423 K and at pressures up to 30 MPa. *Int. J. Thermophys.*, **13**, 931–942.

Olchowy, G.A. & Sengers, J.V. (1988). Crossover from singular to regular behavior of the transport properties of fluids in the critical region. *Phys. Rev. Lett.*, **61**, 15–18.

Tang, S., Jin, G.J. & Sengers, J. V. (1991). Thermodynamic properties of 1,1,1,2–tetrafluoroethane (R134a) in the critical region. *Int. J. Thermophys.*, **12**, 515–540.

Ueno, Y., Kobayashi, Y., Nagasaka, Y. & Nagashima, A. (1991). Thermal conductivity of CFC alternatives. Measurements of HCFC-123 and HFC-134a in the liquid phase by the transient hot-wire method. *Trans. Japan Soc. Mech. Eng.* (in Japanese), **57B**, 3169–3175.

Appendix – Tables

Table 14.23. *Saturation Properties of R134a.*

$\dfrac{T}{K}$	$\dfrac{P}{MPa}$	ϱ_l	ϱ_v	η_l	η_v	λ_l	λ_v
		kg m^{-3}		μ Pa s		mW m^{-1} K^{-1}	
240	0.0724752	1395.	3.837			108.1	8.866
244	0.0878133	1383.	4.596			106.4	9.216
248	0.105641	1372.	5.470			104.7	9.642
252	0.126216	1360.	6.470			102.9	10.01
256	0.149823	1348.	7.609			101.2	10.37
260	0.176759	1335.	8.902			99.43	10.74
264	0.207332	1323.	10.36			97.69	11.10
268	0.241864	1310.	12.01			95.93	11.48
272	0.280687	1297.	13.85			94.18	11.85
276	0.324144	1284.	15.91			92.41	12.23
280	0.372588	1271.	18.22			90.65	12.61
284	0.426384	1257.	20.78			88.87	13.00
288	0.485879	1243.	23.63			87.09	13.40
292	0.551524	1229.	26.79	211.6	11.86	85.30	13.81
296	0.623662	1214.	30.29	201.6	12.04	83.50	14.23
300	0.702702	1199.	34.16	191.9	12.23	81.70	14.66
304	0.789066	1184.	38.44	182.7	12.43	79.88	15.10
308	0.883166	1168.	43.17	173.7	12.63	78.05	15.57
312	0.985456	1151.	48.41	165.1	12.83	76.22	16.05
316	1.09637	1134.	54.20	156.8	13.05	74.37	16.56
320	1.21643	1117.	60.62	148.8	13.28	72.51	17.09
324	1.34608	1098.	67.74	141.0	13.52	70.65	17.67
328	1.48581	1079.	75.66	133.5	13.78	68.78	18.28
332	1.63622	1058.	84.50	126.2	14.07	66.90	18.94
336	1.79782	1037.	94.40	119.0	14.38	65.03	19.67
340	1.97124	1014.	105.6	112.0	14.73	63.16	20.47
344	2.15707	990.3	118.2	105.2	15.12	61.30	21.37
348	2.35615	964.3	132.7	98.38	15.58	59.46	22.40
352	2.56916	936.0	149.6	91.64	16.13	57.66	23.61
356	2.79699	904.7	169.4	84.85	16.80	55.91	25.08
360	3.04066	869.2	193.5	77.92	17.65	54.21	26.94
364	3.30138	827.3	224.0	70.62	18.81	52.60	29.51
366	3.43842	802.8	243.4	66.75	19.60	51.86	32.53
368	3.58016	775.0	266.8	62.66	20.61	51.24	36.22
370	3.72746	742.0	295.7	58.20	21.96	50.98	40.28
372	3.88128	697.9	336.2	52.82	24.07	51.98	47.45
374	4.04227	604.9	425.9	43.85	30.13	75.02	95.31

A. Nagashima, R. Krauss and K. Stephan

Table 14.24. *Viscosity of R134a (/μ Pa s).*

P/MPa	290	295	300	305	310	315	320
				T/K			
0.1	11.65	11.86	12.06	12.25	12.45	12.65	12.84
0.5	11.76	11.96	12.15	12.35	12.54	12.74	12.93
1.0	218.6	205.6	193.1	181.1	169.7	12.95	13.13
2.0	222.4	209.4	197.0	185.1	173.7	162.8	152.2
3.0	226.1	213.1	200.7	188.9	177.7	166.8	156.4
4.0	229.8	216.8	204.4	192.7	181.5	170.7	160.4
5.0	233.4	220.4	208.1	196.3	185.2	174.5	164.3
6.0	236.9	223.9	211.6	199.9	188.8	178.2	168.0
7.0	240.4	227.5	215.1	203.4	192.3	181.7	171.6
8.0	243.9	230.9	218.6	206.9	195.8	185.2	175.2
9.0	247.4	234.3	222.0	210.3	199.2	188.6	178.6
10.0	250.8	237.7	225.3	213.6	202.5	192.0	182.0
12.0	257.5	244.4	232.0	220.2	209.1	198.5	188.5
14.0	264.1	250.9	238.4	226.6	215.5	204.9	194.8
16.0	270.6	257.3	244.8	232.9	221.7	211.1	201.0
18.0	277.1	263.7	251.1	239.1	227.8	217.1	207.0
20.0	283.5	270.0	257.2	245.2	233.8	223.1	212.9
25.0	299.3	285.4	272.4	260.1	248.5	237.5	227.2
30.0	314.8	300.6	287.3	274.7	262.8	251.6	241.0

P/MPa	325	330	335	340	345	350	360
				T/K			
0.1	13.04	13.23	13.43	13.62	13.81	14.00	14.38
0.5	13.12	13.32	13.51	13.70	13.89	14.07	14.45
1.0	13.31	13.49	13.68	13.86	14.04	14.22	14.59
2.0	142.0	132.0	122.1	112.2	14.86	14.97	15.22
3.0	146.4	136.6	127.1	117.7	108.3	98.79	17.43
4.0	150.5	140.9	131.7	122.6	113.7	104.8	86.67
5.0	154.5	145.1	135.9	127.1	118.5	110.1	93.39
6.0	158.3	149.0	140.0	131.3	123.0	114.8	98.99
7.0	162.0	152.7	143.9	135.3	127.1	119.2	103.9
8.0	165.6	156.4	147.6	139.1	131.0	123.2	108.4
9.0	169.0	159.9	151.2	142.8	134.8	127.1	112.5
10.0	172.4	163.3	154.6	146.3	138.4	130.7	116.4
12.0	179.0	169.9	161.3	153.0	145.2	137.6	123.6
14.0	185.3	176.2	167.6	159.4	151.6	144.1	130.2
16.0	191.4	182.3	173.7	165.5	157.7	150.2	136.4
18.0	197.4	188.3	179.6	171.4	163.6	156.1	142.2
20.0	203.2	194.1	185.4	177.1	169.2	161.8	147.9
25.0	217.3	208.0	199.2	190.8	182.8	175.2	161.1
30.0	230.9	221.4	212.4	203.8	195.7	187.9	173.6

Table 14.24. *(Continued)*

P/MPa	T/K						
	370	380	390	400	410	420	430
0.1	14.75	15.12	15.49	15.85	16.21	16.57	16.93
0.5	14.82	15.19	15.55	15.91	16.27	16.63	16.98
1.0	14.95	15.31	15.67	16.02	16.37	16.72	17.07
2.0	15.50	15.80	16.12	16.43	16.75	17.08	17.40
3.0	16.97	16.95	17.07	17.25	17.48	17.73	17.99
4.0	64.57	20.44	19.21	18.86	18.78	18.83	18.95
5.0	76.03	54.50	26.63	22.48	21.23	20.69	20.47
6.0	83.36	67.16	48.83	32.08	26.14	23.94	22.88
7.0	89.28	74.85	60.46	46.10	34.96	29.30	26.58
8.0	94.36	80.94	67.96	55.74	44.73	36.60	31.71
9.0	98.92	86.11	73.99	62.69	52.54	44.00	37.69
10.0	103.1	90.70	79.12	68.37	58.75	50.40	43.59
12.0	110.6	98.73	87.76	77.70	68.58	60.52	53.69
14.0	117.4	105.8	95.14	85.40	76.58	68.69	61.78
16.0	123.7	112.2	101.7	92.16	83.48	75.69	68.77
18.0	129.7	118.2	107.8	98.28	89.68	81.91	74.97
20.0	135.3	123.8	113.4	104.0	95.37	87.60	80.62
25.0	148.4	136.8	126.3	116.8	108.1	100.3	93.15
30.0	160.6	148.8	138.2	128.4	119.6	111.5	104.2

A. Nagashima, R. Krauss and K. Stephan

Table 14.25. *Thermal Conductivity of R134a* (/mW m^{-1} K^{-1}).

P/MPa	\multicolumn{7}{c}{T/K}
	240	250	260	270	280	290	295
0.1	108.1	9.736	10.59	11.43	12.25	13.06	13.46
0.5	108.3	104.0	99.61	95.20	90.73	13.57	13.94
1.0	108.5	104.2	99.87	95.49	91.06	86.56	84.27
2.0	108.9	104.7	100.4	96.07	91.71	87.29	85.05
3.0	109.4	105.2	100.9	96.64	92.34	87.99	85.80
4.0	109.7	105.6	101.4	97.19	92.95	88.68	86.52
5.0	110.1	106.1	101.9	97.74	93.55	89.34	87.23
6.0	110.5	106.5	102.4	98.27	94.14	89.99	87.91
7.0	111.0	106.9	102.9	98.79	94.71	90.62	88.57
8.0	111.4	107.4	103.3	99.30	95.27	91.24	89.22
9.0	111.8	107.8	103.8	99.81	95.82	91.84	89.85
10.0	112.2	108.2	104.3	100.3	96.36	92.43	90.46
12.0	113.0	109.0	105.1	101.3	97.40	93.56	91.65
14.0	113.8	109.7	106.0	102.2	98.41	94.66	92.79
16.0	114.6	110.6	106.8	103.1	99.39	95.71	93.88
18.0	115.3	111.4	107.6	104.0	100.3	96.72	94.93
20.0	116.0	112.1	108.3	104.8	101.2	97.70	95.94
25.0	117.8	114.0	110.3	106.8	103.4	100.0	98.23
30.0	119.5	115.8	112.2	108.7	105.4	102.2	100.5

P/MPa	\multicolumn{7}{c}{T/K}
	300	305	310	315	320	330	340
0.1	13.86	14.25	14.63	15.02	15.40	16.14	16.88
0.5	14.30	14.66	15.02	15.38	15.73	16.43	17.13
1.0	81.95	79.60	77.20	16.25	16.54	17.13	17.73
2.0	82.78	80.49	78.17	75.81	73.40	68.44	63.20
3.0	83.58	81.34	79.08	76.79	74.47	69.71	64.76
4.0	84.35	82.17	79.96	77.73	75.48	70.89	66.17
5.0	85.10	82.96	80.80	78.63	76.44	72.00	67.46
6.0	85.82	83.72	81.61	79.49	77.35	73.04	68.67
7.0	86.52	84.46	82.39	80.31	78.23	74.03	69.79
8.0	87.20	85.17	83.14	81.11	79.07	74.97	70.85
9.0	87.86	85.86	83.87	81.87	79.88	75.87	71.86
10.0	88.50	86.54	84.58	82.62	80.66	76.73	72.82
12.0	89.74	87.83	85.93	84.03	82.14	78.36	74.61
14.0	90.92	89.07	87.22	85.37	83.54	79.88	76.27
16.0	92.06	90.24	88.44	86.65	84.86	81.32	77.81
18.0	93.15	91.37	89.61	87.86	86.12	82.67	79.27
20.0	94.19	92.46	90.73	89.02	87.32	83.95	80.64
25.0	96.66	95.01	93.36	91.73	90.12	86.93	83.80
30.0	98.94	97.35	95.78	94.22	92.67	89.62	86.64

Table 14.25. *(Continued)*

P/MPa	350	360	370	380	390	400	410
			T/K				
0.1	17.59	18.30	18.98	19.65	20.31	20.95	21.58
0.5	17.81	18.48	19.14	19.79	20.42	21.05	21.66
1.0	18.34	18.95	19.55	20.14	20.73	21.31	21.89
2.0	20.54	20.75	21.07	21.46	21.88	22.31	22.75
3.0	59.57	26.31	24.40	23.94	23.88	23.99	24.19
4.0	61.29	56.26	51.00	30.24	27.59	26.71	26.35
5.0	62.82	58.09	53.29	48.34	37.14	31.51	29.78
6.0	64.22	59.73	55.24	50.75	46.08	39.44	34.53
7.0	65.51	61.21	56.94	52.76	48.68	44.51	39.93
8.0	66.72	62.58	58.48	54.48	50.61	47.05	43.52
9.0	67.84	63.84	59.89	56.04	52.37	48.90	45.79
10.0	68.91	65.03	61.20	57.47	53.91	50.58	47.52
12.0	70.88	67.20	63.58	60.06	56.67	53.47	50.53
14.0	72.69	69.17	65.71	62.35	59.11	56.03	53.15
16.0	74.36	70.97	67.65	64.42	61.30	58.32	55.51
18.0	75.92	72.64	69.44	66.32	63.30	60.42	57.67
20.0	77.39	74.21	71.10	68.08	65.15	62.34	59.67
25.0	80.74	77.75	74.83	72.00	69.26	66.61	64.08
30.0	83.73	80.88	78.11	75.42	72.82	70.30	67.88

Binary Mixtures: Carbon Dioxide–Ethane

W. A. WAKEHAM and V. VESOVIC

Imperial College, London, UK

E. VOGEL and S. HENDL

Universität Rostock, Germany

15.1 Introduction

From a technological and industrial point of view, fluid mixtures are of rather greater importance than pure fluids simply because they are encountered more frequently. It is therefore important that it should be possible to represent, or predict, the transport properties of multicomponent mixtures of an arbitrary composition from a limited set of information. The IUPAC Subcommittee on Transport Properties of Commission I.2 on Thermodynamics has determined that an archetypal system to test the methodology available for the prediction and representation of the viscosity and thermal conductivity of mixtures is the binary mixture of carbon dioxide and ethane. This decision rests upon the facts that accurate representations of the properties of the two pure components are available (Vesovic *et al.* 1990; Hendl *et al.* 1994; Vesovic *et al.* 1994); some experimental data of high quality are already available for the mixtures, while others are in progress. In particular, a significant body of information is available for the thermal conductivity in the vicinity of the critical line (Mostert *et al.* 1992). Here a summary of the study of this system is presented as an illustration of the methodology for the treatment of gas mixtures set out in Chapters 4, 5 and 6. The presentation proceeds through the steps of zero density, $X^{(0)}(T)$, excess property, $\Delta X(\rho, T)$, and the critical enhancement, $\Delta X_c(\rho, T)$, to yield the total value of the property according to the equation

$$X(\rho, T) = X^{(0)}(T) + \Delta X(\rho, T) + \Delta X_c(\rho, T) \tag{15.1}$$

where X may be the viscosity or the thermal conductivity.

15.2 The dilute gas

The kinetic theory results of Chapters 4 and 5 have revealed that the relationship between the transport properties of a binary gas mixture and those of its pure components is formally well defined. Indeed, for the viscosity, the addition of just two quantities that are characteristic of the pair potential between the unlike species is sufficient to be able to calculate the viscosity of a mixture of an arbitrary composition with a high degree of accuracy. For that reason, as well as for reasons of practicability, the representation of the transport properties of the dilute gas state of mixtures is best founded upon the theory set out in Section 4.3 rather than upon empirical equations.

15.2.1 Viscosity

The representations of the viscosity of the pure components carbon dioxide and ethane include explicit contributions for the viscosity in the zero–density limit (Vesovic *et al.* 1990; Hendl *et al.* 1994). The representation of the viscosity of binary mixtures in the same limit requires values of just the interaction viscosity, η_{12}, of equations (4.115) and (4.116) and the quantity A_{12}^* of equation (4.117) in order to evaluate, and thus represent, the viscosity of an arbitrary binary mixture. Since the intermolecular pair potential of the carbon dioxide–ethane interaction is not known, it is not possible to evaluate either of these quantities theoretically.

Hendl & Vogel (1993, 1994) have performed measurements of high accuracy on the viscosity of three different binary mixtures of carbon dioxide and ethane in the temperature range 297 to 625 K at low and moderate densities. From these measurements the zero–density property was derived by extrapolation of a density series as discussed in Chapter 5. These measurements were then used to deduce values of the interaction viscosity by means of equation (4.112) for $N = 2$ with the aid of their own measurements of the pure gas viscosity and an estimate of A_{12}^*. In fact, the assignment $A_{12}^* = 1.100$ was employed. The values of η_{12} have been correlated by the equation

$$\frac{\eta_{12}^{(0)}}{\mu Pa\,s} = 10.0 \exp\left\{A \ln T_r + \frac{B}{T_r} + \frac{C}{T_r^2} + D\right\} \tag{15.2}$$

where

$$T_r = T/298.15\,K \tag{15.3}$$

and

$$A = 0.354821, \quad B = -1.076338, \quad C = 0.242845, \quad D = 1.029471 \tag{15.4}$$

The combination of equations (15.2) and (4.112)–(4.114) with the viscosity of the two pure components given elsewhere (Vesovic *et al.* 1990; Hendl *et al.* 1994) and the assigned value $A_{12}^* = 1.100$ permits the calculation of the viscosity of a mixture of carbon dioxide and ethane of an arbitrary composition within the temperature range of the original measurements of Hendl & Vogel (1993). Figure 15.1 contains a comparison of the calculated values of the mixture viscosity in the limit of zero–density with those obtained from experiment. It can be seen that the deviations do not amount to more than 0.07%, which is even better than the experimental uncertainty.

It is also possible to evaluate the viscosity of mixtures in the dilute–gas state over a wider range of temperature than that for which measurements are available. For this

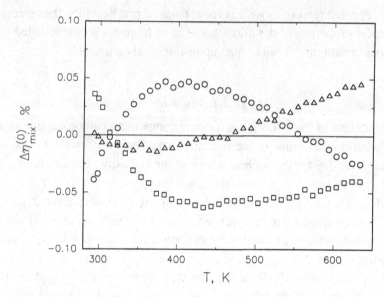

Fig. 15.1. Deviations of the experimental values of the viscosity of carbon dioxide–ethane mixtures at zero density (Hendl & Vogel 1993) from the representation of equations (4.113) and (15.2) for three compositions: $\circ$ - $x_{CO_2} = 0.2500$, $\square$ - $x_{CO_2} = 0.500$, $\triangle$ - $x_{CO_2} = 0.7398$.

purpose one can make use of the results of the extended law of corresponding states (Chapter 11; Maitland *et al.* 1987), which provides representations that enable the interaction viscosity η_{12} and the cross section ratio A_{12}^* to be evaluated for ethane and carbon dioxide over a very wide temperature range. In this case, the scaling parameters ϵ_{12} and σ_{12} have been derived from earlier viscosity data over a limited temperature range (Maitland *et al.* 1987). Naturally, the values of the interaction quantities determined in this way are less reliable than those determined using the most precise measurements, but the representation has the advantage of extending over a much wider range of temperature. It is to be expected that the viscosity predicted in this way will have an error of no more than a few percent within the temperature range of chemical stability of the carbon dioxide–ethane system.

It should be added that the determination of the interaction viscosity and the cross–section ratio in the way described allows the binary diffusion coefficient in the limit of low density to be evaluated through equation (4.116) with an accuracy comparable with that of the best direct measurements.

15.2.2 Thermal conductivity

The thermal conductivity of the binary mixture of carbon dioxide and ethane in the limit of zero density could, in principle, be most accurately evaluated from equation (4.127), including the full inelastic contributions contained within the term $\Delta\lambda$. However, such an evaluation requires that theoretical or experimental values of the collision numbers for the relaxation of internal energy in the mixture are known or that there are accurate thermal conductivity data in the limit of zero density from which they may be derived. At the time of writing, neither collision numbers nor the requisite thermal conductivity data for the carbon dioxide–ethane system are available, although measurements of the latter are in progress. As a consequence, the system provides an example of the manner in which it is possible to make predictions of the thermal conductivity.

The most appropriate form of the theoretical equations for predictive purposes is that based upon the Hirschfelder–Eucken result of equation (4.127) in which all of the various diffusion coefficients for internal energy are replaced by binary diffusion coefficients. That is, equation (4.127) is rewritten as

$$\lambda_{mix}^{(0)} \simeq \lambda_{HE}^{(0)} = \lambda_{mix,tr}^{(0)} + \frac{\lambda_1^{(0)} - \lambda_{1,tr}^{(0)}}{\left[1 + \frac{x_2 \mathcal{D}_{11}^{(0)}}{x_1 \mathcal{D}_{12}^{(0)}}\right]} + \frac{\lambda_2^{(0)} - \lambda_{2,tr}^{(0)}}{\left[1 + \frac{x_1 \mathcal{D}_{22}^{(0)}}{x_2 \mathcal{D}_{12}^{(0)}}\right]} \tag{15.5}$$

Here, the approximate contribution of translational energy to the thermal conductivity is

$$\lambda_{mix,tr}^{(0)} = - \frac{\begin{vmatrix} L_{11} & L_{12} & x_1 \\ L_{12} & L_{22} & x_2 \\ x_1 & x_2 & 0 \end{vmatrix}}{\begin{vmatrix} L_{11} & L_{12} \\ L_{12} & L_{22} \end{vmatrix}} \tag{15.6}$$

In this equation,

$$L_{11} = \frac{x_1^2}{\lambda_{1,tr}^{(0)}} + \frac{x_1 x_2}{2\lambda_{12}^{(0)} A_{12}^*(m_1 + m_2)^2}$$

$$\times \left[\frac{15}{2}m_1^2 + \frac{25}{4}m_2^2 - 3m_2^2 B_{12}^* + 4m_1 m_2 A_{12}^*\right] \tag{15.7}$$

with a similar expression for L_{22} obtained by an exchange of subscripts, and

$$L_{12} = - \frac{x_1 x_2 m_1 m_2}{2\lambda_{12}^{(0)} A_{12}^*(m_1 + m_2)^2} \left[\frac{55}{4} - 3B_{12}^* - 4A_{12}^*\right] \tag{15.8}$$

In addition, the interaction thermal conductivity $\lambda_{12}^{(0)}$ is obtained from the interaction viscosity derived in the previous section by means of the equation

$$\lambda_{12}^{(0)} = \frac{15}{8}k_B \frac{(m_1 + m_2)}{m_1 m_2}\eta_{12}^{(0)} = \frac{25}{8}\frac{k_B}{A_{12}^*}n\mathcal{D}_{12}^{(0)} \tag{15.9}$$

and the translational contribution to the thermal conductivity of the pure gases estimated from the equation

$$\lambda_{i,tr}^{(0)} = \frac{15}{4}\frac{k_B}{m_i}\eta_i^{(0)} = \frac{25}{8}\frac{k_B}{A_{ii}^*}n\mathcal{D}_{ii}^{(0)} \tag{15.10}$$

From these results it can be seen that a number of the quantities, including the diffusion coefficients, required for the evaluation of the mixture thermal conductivity can be obtained from the dilute–gas viscosity of the pure components (Vesovic *et al.* 1990; Hendl *et al.* 1994) and of the binary mixture (Hendl & Vogel 1993, 1994). In fact, the use of equation (15.5) requires, in addition, only values for the total thermal conductivity of the pure components at zero density and of the ratio A^* for the three interactions, and the ratio B^* for the unlike interaction. The thermal conductivity for both pure gases has been represented by Vesovic *et al.* (1990, 1994) over a wide range of temperature. The cross–section ratios are not amenable to calculation, owing to the absence of an intermolecular potential for the system nor indeed to direct measurement. It is therefore necessary to estimate these quantities by means of the correlations of the extended law of corresponding states (Chapter 11), since scaling parameters for all of the interactions of interest here are available (Maitland *et al.* 1987).

The absence of experimental data for the thermal conductivity of the mixtures of carbon dioxide and ethane precludes a comparison with the predictions at this intermediate stage in the discussion.

15.3 The excess properties

The discussion in Chapter 5 has made it clear that the rigorous theory of the transport properties of fluids or fluid mixtures at elevated densities far from the critical point can provide little assistance with the development of a correlation or with the prediction of the properties of mixtures. The one exception to this is connected with the initial density dependence of the viscosity, where the results of Rainwater & Friend provide some insight. As a consequence, the development of the representation of the transport properties of mixtures over a wide range of conditions can be performed either by means of an empirical surface fit of the type described in Chapter 7, where the composition of the mixture adds just one more dimension to the space for the fitting, or by means of an approximate theory. The advantage of the former procedure for binary mixtures is the ability to reproduce a body of experimental data rather precisely (Friend & Roder 1987). The advantages of the latter procedure are that both interpolation and extrapolation are more secure in temperature, density and composition, that rather less input information is required and that the extension to multicomponent mixtures is straightforward, which is not the case for empirical methods. Here we consider only the use of schemes based upon approximate theory.

15.3.1 The viscosity

The Rainwater–Friend theory that has proved so successful in the representation of the initial–density dependence of the viscosity of pure gases has not been extended to mixtures. It is therefore necessary to make use of the Thorne–Enskog equations for hard–sphere systems set out in Section 5.5.1. In the interest of brevity, equations (5.48)–(5.52) will not be repeated here for the special case of a binary mixture. Instead it is sufficient to note that in the limit of zero density these results reduce exactly to those for dilute–gas mixtures given in Chapter 4. This means that the process to be described leads immediately to the sum of the zero density and excess contributions to the viscosity.

The essence of the application of these equations to real fluids consists of three parts: first the replacement of the hard–sphere results for the pure gas viscosity and the interaction viscosity by the values for the real fluid system; second, the evaluation of a pseudo–radial distribution function for each binary interaction to replace the hard–sphere equivalent at contact; and, finally, the selection of a molecular size parameter for each binary interaction to account for the mean free–path shortening in the dense gas.

Hendl & Vogel (1994) have examined a number of schemes for implementing these equations in the limit for which a linear expansion of the mixture viscosity in density is valid. In their most successful approach they employed a hard–sphere expansion for the density dependence of the radial–distribution function. The first density coefficients of those expansions for the two pure components and the unlike interaction were determined by a fit to their experimental mixture viscosity data for one composition. Effective hard–sphere diameters appropriate to each interaction for the mean free–path shortening were derived by the application of the modified Enskog procedure, described in Section 5.2. These diameters were different from those determined from considerations of the distribution function. Although this process is not entirely self–consistent, it was found useful (Hendl & Vogel 1994) to reproduce the experimental viscosity data for all of the mixtures over the temperature range 297 to 625 K to within $\pm 0.2\%$, as is shown in Figure 15.2; a deviation that exceeds the experimental uncertainty by only a small margin. Of course, it is necessary to point out that for this particular system the initial dependence of the viscosity upon density is small (less than 1% at densities up to 0.6 mol l^{-1}), so that the agreement, to a large extent, reflects the performance of the representation of the zero–density viscosity.

An alternative application of the Thorne–Enskog equations is that employed by Vesovic & Wakeham (1989, 1991). The procedure has been described in detail in Section 5.5, so that it is necessary here to describe only the sources of information employed. First the correlation of the viscosity of pure carbon dioxide has been taken from the work of Vesovic *et al.* (1990) and that of ethane from the work of Hendl *et al.* (1994). For each temperature at which it was desired to evaluate the mixture viscosity these correlations were used in equation (5.53) to evaluate the pseudo–radial distribution functions $\tilde{g}_i$ for each pure component, while equation (5.56) was used to determine a consistent value of the mean free–path shortening parameter α_{ii} for i–i interactions. Figure 15.3 contains a plot of the pseudo–radial distribution function determined in this way for the two pure gases at one temperature.

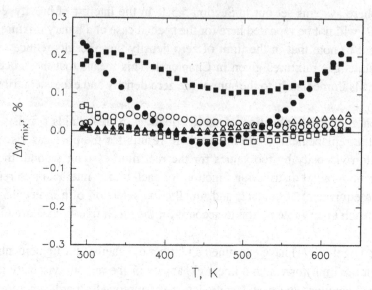

Fig. 15.2. Deviations of the experimental values of the viscosity of carbon dioxide–ethane mixtures from those calculated, using the procedure of Hendl & Vogel (1994).
For an amount–of–substance density of 0.01 mol l^{-1}: o - x_{CO_2} = 0.2500,
□ - x_{CO_2} = 0.5000, △ - x_{CO_2} = 0.7398.
For an amount–of–substance density of 0.05 mol l^{-1}: ● - x_{CO_2} = 0.2500,
■ - x_{CO_2} = 0.5000, ▲ - x_{CO_2} = 0.7398.

In order to evaluate the viscosity of carbon dioxide–ethane mixtures of a particular composition at a particular amount–of–substance density and temperature, the pure component pseudo–radial distribution functions for both gases at the same temperature and density have been combined according to the combination rule of equation (5.57) to yield the pseudo–radial distribution functions for the mixture. At the same time, the mean free–path parameters are combined according to equation (5.58). Together with the information used in the evaluation of the viscosity of the system in the zero–density limit, these results have been used in the Thorne–Enskog equations (5.48)–(5.52) to evaluate the viscosity over a wide range of temperature and density. The calculations make no use of any information on the viscosity of any of the mixtures at elevated densities, so that in this sense the calculations are a prediction. Figure 15.4 contains a comparison with the experimental data of Hendl & Vogel (1993) at modest densities and with the results of Diller et al. (1988) and Diller & Ely (1989) at much higher densities.

The agreement with the results of Hendl & Vogel (1993) is excellent, with discrepancies no greater than 0.6%. At higher densities the agreement with the results of Diller et al. (1988) slowly degrades as the density increases, reaching values of as much as ±17% at the highest densities. The more recent results of Diller and Ely (1989), made on mixtures with almost the same composition as those of Diller et al. (1988), are in

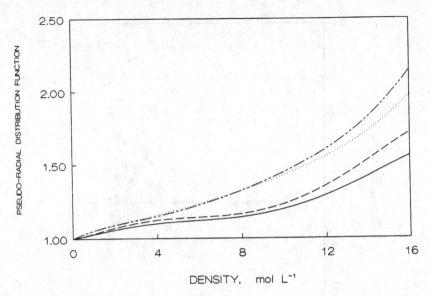

Fig. 15.3. Pseudo–radial distribution functions for ethane and carbon dioxide at a temperature of 350 K.
Ethane:
— – – – — from viscosity; — — from thermal conductivity.
Carbon dioxide:
· · · · · from viscosity; ——— from thermal conductivity.

much closer agreement with the predictions and with the results of Hendl and Vogel; the deviations are no larger than about ±5%. This confirms the observation of Diller & Ely with respect to undetected errors in the results of Diller *et al.* (1988). In addition, the comparison confirms the findings of earlier work that the procedure employed here is capable of predicting the viscosity at elevated densities within a few percent (Vesovic & Wakeham 1989, 1991).

15.3.2 Thermal conductivity

The evaluation of the excess contribution to the thermal conductivity of carbon dioxide–ethane mixtures has been performed by a route parallel to that employed for the viscosity. Again, that route has been described in detail in Section 5.5.1.2 so that here it is merely recorded that for pure carbon dioxide and pure ethane the thermal conductivity has been taken from the work of Vesovic *et al.* (1990, 1994). All other quantities required for the calculation are the same as those employed for the zero–density thermal conductivity. In this context it should be noted that the Thorne–Enskog equations employed for these calculations have, as their zero–density limit, the Hirschfelder–Eucken result in the form of equation (15.5). Figure 15.3 contains the pseudo–radial distribution functions for carbon dioxide and ethane determined

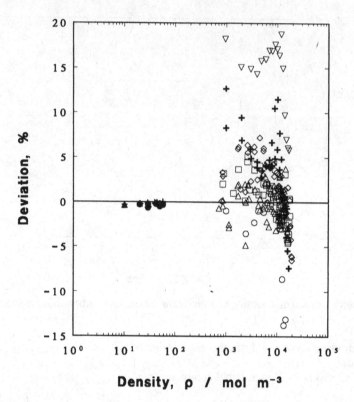

Fig. 15.4. Deviations of the experimental viscosity of carbon dioxide–ethane mixtures from the calculated results, using the method of Vesovic & Wakeham (1989).
Diller *et al.* (1988), temperature range 300-320 K:
○ - $x_{CO_2} = 0.2517$; + - $x_{CO_2} = 0.4925$; ▽ - $x_{CO_2} = 0.7398$.
Diller & Ely (1989), temperature range 320-500 K:
△ - $x_{CO_2} = 0.2517$; □ - $x_{CO_2} = 0.4925$; ◇ - $x_{CO_2} = 0.7398$.
Hendl & Vogel (1993), temperature range 297–620 K:
* - $x_{CO_2} = 0.2500$; ▲ - $x_{CO_2} = 0.5000$; ● - $x_{CO_2} = 0.7398$.

from the pure component thermal conductivity data. It can be seen that the results are quite different from those obtained from the viscosity. This observation illustrates the fundamental limitations of the Thorne–Enskog equations and, indeed, of the Enskog theory, since if the theory were correct the functions derived from the two properties should be the same. In practical terms, the result inhibits the generation of one property from another, as is possible in the zero–density limit.

As was pointed out earlier, there are no published thermal conductivity data for the carbon dioxide–ethane system in which either the excess or the background contribution have been unequivocally identified. Consequently, no comparison with experiment is yet possible.

15.4 The critical region

The behavior of the transport properties of a binary mixture of compressible fluids near to the vapor–liquid critical line has been discussed in Section 6.6. It has been made clear in that discussion, and elsewhere (Luettmer–Strathmann 1994; Luettmer–Strathmann & Sengers 1994), that the behavior of the transport properties for a mixture reflects the fact that the relevant order parameter is now not simply the density but also the species concentration. It is also apparent from the discussion in Section 6.6 that the truly asymptotic behavior of the properties occurs so close to the critical point that it has not yet been possible to observe it. As a consequence, it is necessary to employ, for the description of the behavior of the transport properties in the region of states of practical significance, a theory which deals with an extended region around the critical point known as the crossover region.

The most recent theory of the transport properties of a mixture in the critical region is the mode–coupling theory of Luettmer–Strathmann & Sengers (1994), outlined in Section 6.6. The evaluation of the critical enhancement of the thermal conductivity and viscosity of a binary mixture through this theory depends on the availability of thermodynamic data, of the background values for these transport properties and of the mass diffusion and thermal diffusion coefficients (Luettmer–Strathmann 1994). Luettmer–Strathmann & Sengers (1994) have employed the background viscosity and thermal conductivity for ethane and carbon dioxide deduced as described above to represent the experimental thermal conductivity data of Mostert *et al.* (1992) obtained close to the critical line of the mixture and combined this information with the available data for the thermodynamic properties of the same mixture (Jin 1993). In order to achieve an optimum representation with fixed background contributions for the viscosity and thermal conductivity, they allowed the background contributions from two other kinetic coefficients, used for convenience in place of the mass and thermal diffusion coefficients to be determined by a fit to the available thermal conductivity data. A cutoff wave vector for fluctuations in one of the two combinations of heat and diffusive modes important to the phenomenon was also determined in the fitting. Subsequently, comparisons were made with the entire body of thermal conductivity data and predictions made of the behavior of other transport properties in the critical region using the same values of the parameters. At the time of writing, data exist only for the thermal conductivity. Figure 6.7 (not repeated here) contains a comparison between the experimental thermal conductivity of a carbon dioxide–ethane mixture and the calculated values for one composition. The figure includes the background thermal conductivity determined from the combination of the results of Sections 15.2 and 15.3. The figure illustrates the magnitude and extent of the critical contribution to the thermal conductivity as well as the degree of agreement with the experimental data.

Figure 15.5 contains a plot of the predicted critical enhancement of the viscosity of the same carbon dioxide–ethane mixture, which also includes the background contribution

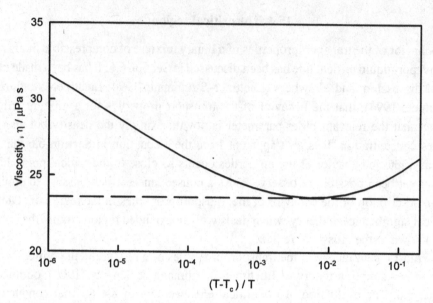

Fig. 15.5. A comparison of the predicted behavior of the viscosity of an ethane–carbon dioxide mixture ($x_{CO_2} = 0.2500$), with the calculated background values (Luettmer-Strathmann 1994; Vesovic & Wakeham 1989, 1991).
—— total viscosity; — — — calculated viscosity background.

calculated from the procedures described above. In this case there are no experimental data against which to assess the accuracy of the results.

15.5 Summary

The chapter has combined several themes of earlier parts of the volume and brought them to bear on the description of the properties of a binary gas mixture over a wide range of temperatures and densities including the critical region. The use of appropriate kinetic theories of varying degrees of rigor has made it possible to carry out the evaluation of the thermal conductivity and viscosity of the mixture from measurements of the same properties of the pure components over a wide range of densities and temperatures and of binary mixtures in the dilute–gas state and near the critical point. In addition it has only been necessary to make estimates of two cross–section ratios of the kinetic theory. The comparisons with experimental data that are possible indicate that the procedure is reasonably accurate and can provide either a compact means of representing the data or a means of predicting properties in regions where no experimental data exist.

References

Diller, D. E., & Ely J. F. (1989). Measurements of the viscosity of compressed gaseous carbon dioxide, ethane, and their mixtures, at temperatures up to 500 K. *High Temp. – High Press.*, **21**, 613–620.

Diller, D.E., Van Poolen, L.J. & Dos Santos, F.V. (1988). Measurements of the viscosities of compressed fluid and liquid carbon dioxide + ethane mixtures. *J. Chem. Eng. Data*, **33**, 460–464.

Friend, D.G. & Roder, H.M. (1987). The thermal conductivity surface for mixtures of methane and ethane *Int. J. Thermophys.*, **8**, 13–26.

Hendl, S., Millat, J., Vogel, E., Vesovic, V., Wakeham, W.A., Luettmer–Strathmann, J., Sengers, J.V. & Assael, M.J. (1994). The transport properties of ethane I. Viscosity. *Int. J. Thermophys.*, **15**, 1–31.

Hendl, S. & Vogel, E. (1993). Temperature and initial density dependence of viscosity of binary mixtures: Carbon dioxide–ethane. *High Temp.–High Press.*, **25**, 279–289.

Hendl, S. & Vogel, E. (1994). Correlation and extrapolation scheme for the composition and temperature dependence of viscosity of binary gaseous mixtures: Carbon dioxide – ethane. *Proc. 12th Symposium on Thermophysical Properties*, Boulder, CO.

Jin, G.X. (1993). *The Effect of Critical Fluctuations on the Thermodynamic Properties of Fluids and Fluid Mixtures*. Ph.D. thesis, University of Maryland.

Luettmer–Strathmann, J. (1994). *Transport Properties of Fluids and Fluid Mixtures in the Critical Region*. Ph.D. Thesis, University of Maryland.

Luettmer–Strathmann, J. & Sengers, J.V. (1994). Transport properties of fluid mixtures in the critical region. *Int. J. Thermophys.*, **15**, 1241–1249.

Maitland, G. C., Rigby, M., Smith, E. B. & Wakeham, W. A. (1987). *Intermolecular Forces: Their Origin and Determination*. Oxford: Clarendon Press.

Mostert, R., van den Berg, H.R., van der Gulik, P.S. & Sengers, J.V. (1992). The thermal conductivity of carbon dioxide–ethane mixtures in the critical region. *High Temp.– High Press.*, **24**, 469–474.

Vesovic, V. & Wakeham, W.A. (1989). The prediction of the viscosity of dense gas mixtures. *Int. J. Thermophys.*, **10**, 125–132.

Vesovic, V. & Wakeham, W.A. (1991). Prediction of the thermal conductivity of fluid mixtures over wide ranges of temperature and pressure. *High Temp.–High Press*, **23**, 179–190.

Vesovic, V., Wakeham, W.A., Luettmer–Strathmann, J., Sengers, J.V., Millat, J., Vogel, E. & Assael, M.J. (1994). The transport properties of ethane II. Thermal conductivity. *Int. J. Thermophys.*, **15**, 33–66.

Vesovic, V., Wakeham, W.A., Olchowy, G.A., Sengers, J.V., Watson, J.T.R. & Millat, J. (1990). The transport properties of carbon dioxide. *J. Phys. Chem. Ref. Data*, **19**, 763–808.

16

Reacting Mixtures at Low Density – Alkali Metal Vapors

P. S. FIALHO

University of Azores, Portugal

M. L. V. RAMIRES and C. A. NIETO DE CASTRO

University of Lisbon, Portugal

J. M. N. A. FARELEIRA

Technical University of Lisbon, Portugal

16.1 Introduction[a]

Previous chapters of this volume dealt with the transport properties of nonreacting mixtures. However, there are several systems of scientific and industrial interest that involve chemical reactions between some of the atoms or molecules present. This fact modifies the values of the transport properties of these systems because there are additional processes of heat and mass transfer caused by the existence of the chemical reaction.

The simplest kind of system where chemical reaction is present is an alkali metal vapor, where dimerization of atoms takes place, with a temperature– and pressure–dependent composition. This fact, allied to the high–temperature domain of the application of these vapors, makes the determination of the transport properties of these systems very challenging from a theoretical point of view.

The thermophysical properties of alkali–metal vapors are very important for several technological processes, namely as working fluids for Rankine cycles, solar power plants and magnetic hydrodynamic generation (Ohse 1985). Their measurement is, however, very difficult to achieve with accuracies similar to those obtainable for other fluids, because of the high temperatures and low pressures, and therefore low densities involved (Wakeham *et al.* 1991). The prediction of these properties is the recommended alternative to the deficient quality and scarcity of the experimental data. In addition to the difficulties associated with the formation of dimers, the alkali metal atoms interact via either one of two intermolecular pair potential energy functions, namely, the ground singlet [$^1U(R)$] and the triplet [$^3U(R)$] potential energy curves, belonging respectively to the $^1\Sigma_g^+$ and $^3\Sigma_u^+$ diatomic molecular states.

The calculation of the transport properties of alkali metal vapors involves the use of the kinetic theory of binary reacting mixtures and its application to a mixture of monomers and dimers. The use of the theory requires the evaluation of the monomer properties, the establishment of some approximations to calculate the viscosity and the

---

[a] This chapter is dedicated to the memory of Professor V. B. Vargaftik (1904–1994) for his outstanding contributions in the field of thermophysical properties of alkali metal vapors.

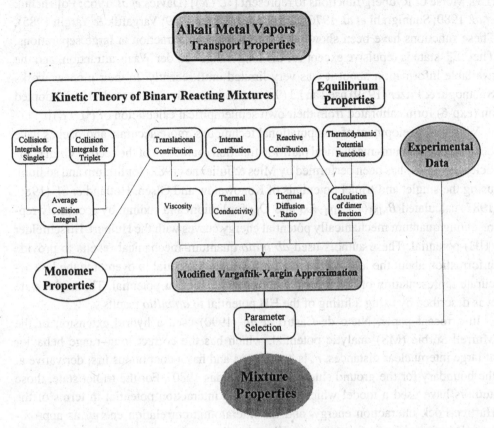

Fig. 16.1. Diagram for the calculation of the transport properties of alkali metal vapors.

contributions for the thermal conductivity and thermal diffusion ratio of dimers and of the mixture, and the evaluation through thermodynamic calculations of the dimer fraction at each temperature and pressure. This procedure is exemplified in Figure 16.1 and will be described in the following sections, with reference to the possible types of approaches.

16.2 Monatomic Systems

A fundamental requirement for the theoretical calculation of the transport properties of a system composed of alkali atoms is the knowledge of sound reliable pair potential energy functions along which any two atoms interact. According to previous works (Davies *et al.* 1965), if the atoms are assumed as neutral, in their ground electronic state, the problem reduces to that of obtaining the ground singlet $[^1U(R)]$ and triplet $[^3U(R)]$ potential energy curves. Most of the previous calculations of the equilibrium (second virial coefficients) and transport properties of these systems used the Lennard-Jones

(LJ), Morse or Rydberg functions to represent $[^1U(R)]$ (Davies *et al.* 1965; Polishchuk *et al.* 1980; Sannigrahi *et al.* 1976; Sinanoglu & Pitzer 1959; Vargaftik & Yargin 1985). These functions have been shown to overestimate the attraction at large separations. The $^3\Sigma_u^+$ state is repulsive except for the long–range van der Waals attraction, and the available information about it was very limited until recently. In their pioneer works Sinanoglu & Pitzer (1959) used an LJ (9-6) potential, while Davies *et al.* (1965) adopted an (exp-6) form calibrated from their own semiempirical calculation of $[^3U(R)]$.

With the development of computational techniques, more accurate approaches have been used. Quantum–mechanical partition function evaluation of the second virial coefficient $B_{MM}(T)$ has been performed by Mies & Julienne (1982) for lithium and sodium, using the singlet and triplet potentials of Konowalow and Olsen. Holland *et al.* (1986, 1987) calculated $B_{MM}(T)$, η_M, λ_M and D_M for lithium and sodium by accurately representing quantum mechanically potential energy curves with the Hulburt-Hirschfelder (HH) potential. These authors used *ab initio* quantum–mechanical results to provide information about the long–range part of the singlet potential to overcome the less accurate representation of the attractive tail provided by that potential. The triplet state was described by using a fitting of the HH potential to *ab initio* results.

In a recent paper, Nieto de Castro *et al.* (1990) used a hybrid extension of the Murrell-Sorbie (MS) analytic potential, which has the correct long–range behavior at large internuclear distances, r, is continuous and has a continuous first derivative at the boundary for the ground singlet state (Varandas 1980). For the triplet state, those authors have used a model which expresses the interaction potential in terms of the Hartree-Fock interaction energy, and the interatomic correlation energy as approximated semiempirically from the second–order dispersion energy calculated, including the effect of charge overlap between the electron clouds of the two interacting species (Varandas & Brandão 1982). The viscosity, thermal conductivity, self–diffusion coefficients and second virial coefficients of lithium, sodium, potassium, rubidium and cesium monatomic vapors, have been calculated between 600 and 5000 K, and the results were compared with previous work (Holland *et al.* 1986, 1987; Mies & Julienne 1982; Polishchuk *et al.* 1980; Stefanov 1980; Vargaftik & Yargin 1985). The results have shown that, in general, there was a significant improvement in the quality of the predictions.

Recently, Varandas & Dias da Silva (1992) have developed a new model potential for the ground molecular state of diatomic molecules, which is a development of the so–called Extended Hartree-Fock Approximate Correlation Energy (EHFACE) potential (Varandas & Dias da Silva 1986). These authors have pointed out the need to include in the latter model '. . . *a more realistic description of the potential energy curve at the highly repulsive regions near the united-atom limit which results from the collapsed diatomic for a vanishing internuclear separation (r $\rightarrow$ 0).*' To accomplish this, Varandas & Dias da Silva (1992) have developed a new model potential (whose acronym is EHFACE2U) for the ground molecular state of diatomic molecules, including the alkali metals, which gives the correct description of the potential energy as $r \rightarrow 0$ (Varandas & Dias da Silva 1992). This is based on the addition of an extended Hartree-Fock-type energy to the dynamical correlation.

When analyzing the experimental data with the former models some inconsistencies were found, namely for cesium, where the amount of dimers present at high temperatures is very small; therefore, a higher accuracy is required for the monatomic systems.

As the transport properties are particularly sensitive to the repulsive part of the intermolecular potential, especially at high temperatures, Fialho *et al.* (1993) have repeated their previous calculations of the transport properties of the atomic alkali metal vapors, using for comparison the EHFACE2U model for the ground singlet intermolecular potential interaction. The calculations presented earlier (Nieto de Castro *et al.* 1990) were repeated using the new potential functions for the ground singlet states of the alkali metals, lithium, sodium, potassium, rubidium and cesium. The excited state triplet potential functions were the same as before. The precision of the calculations has been improved, by the use of more sophisticated numerical procedures. Furthermore, because of the utilization of the monomeric systems results for the purposes of comparison with experimental data at the temperatures at which they have been obtained, a new interpolation scheme for the collision integrals has been developed. Details of the potentials can be found in the work of Varandas & Brandão (1982) and Varandas & Dias da Silva (1992).

The first–order Chapman-Enskog solution of the Boltzmann equation for the viscosity and thermal conductivity of monatomic species are given by the expressions in terms of effective collision cross sections outlined in Chapter 4. However, in order to be consistent with the original papers, here the equivalent expressions in terms of collision integrals are adopted.

$$[\eta]_1 = \frac{5}{16} (\pi m k_B T)^{1/2} \frac{1}{\bar{\Omega}^{(2,2)}(T)} \tag{16.1}$$

$$[\lambda]_1 = \frac{25}{32} \left(\frac{\pi k_B T}{m}\right)^{1/2} \frac{c_V}{\bar{\Omega}^{(2,2)}(T)} \tag{16.2}$$

$$[D]_1 = \frac{3}{8P} \left(\frac{\pi k_B^3 T^3}{m}\right)^{1/2} \frac{1}{\bar{\Omega}^{(1,1)}(T)} \tag{16.3}$$

where m is the molecular mass, $\bar{\Omega}^{(l,s)}(T)$ are weighted average collision integrals defined as (Sinanoglu & Pitzer 1959)

$$\bar{\Omega}^{(l,s)}(T) = \frac{1}{4}\pi\sigma_1^2 \bar{\Omega}_1^{(l,s)*}(T_1^*) + \frac{3}{4}\pi\sigma_3^2 \bar{\Omega}_3^{(l,s)*}(T_3^*) \tag{16.4}$$

$\bar{\Omega}_1^{(l,s)*}(T_1^*)$ and $\bar{\Omega}_3^{(l,s)*}(T_3^*)$ are reduced collision integrals for the singlet (subscript 1) and for the triplet (subscript 3) interactions, and T_i^* are the corresponding reduced temperatures, such that $T_i^* = k_B T / \epsilon_i$. The second–order Kihara correction factors to these solutions are given by Maitland *et al.* (1987) and Hirschfelder *et al.* (1954). The collision integrals $\bar{\Omega}_1^{(l,s)*}(T_1^*)$ and $\bar{\Omega}_3^{(l,s)*}(T_3^*)$ are related to the interatomic potential functions through the usual expressions (Maitland *et al.* 1987).

Fialho *et al.* (1993) calculated the viscosity and thermal conductivity of the monatomic alkali systems using the above mentioned pair potentials. A description of the calculation of the collision integrals can be found therein. The calculation of the transport properties of the monatomic alkali metal vapors at temperature T requires the determination of two reduced collision integrals at the corresponding reduced temperatures T_1^* and T_3^*. This is conveniently carried out using adequate interpolation procedures between tabulated values of the collision integrals for both the singlet and the triplet potentials. For this purpose, polynomials of the form

$$\ln\left[\bar{\Omega}_i^{(l,s)*}(T_i^*)\right] = \sum_{j=0}^{4} C_j T_i^{*(j-2)} \tag{16.5}$$

were fitted to the results obtained for $\bar{\Omega}_i^{(l,s)*}(T_i^*)$, where $i = 1$ or $i = 3$. The results obtained for $\bar{\Omega}_i^{(l,s)*}(T_i^*)$, $l = 1, 2$ and $s = 1, 2, 3$, for the singlet ($i = 1$) and for the triplet ($i = 3$) states in the range $500 < T < 2000$ K are summarized in Tables 16.1 and 16.2 where the fitting parameters of equation (16.5), obtained for each of the alkali metals, are presented together with the rms deviation, σ, of the fittings. The precision with which the values of the collision integrals can be estimated within that temperature interval is better than $\pm 0.6\%$ and $\pm 0.03\%$ for the singlet and triplet states, respectively, at a 95% confidence level. The overall precision of the estimates of the average collision integrals of the monatomic alkali metals using expression (16.5) with the values of Table 16.1 is estimated to be of the order of $\pm 0.6\%$.

The average collision integrals $\bar{\Omega}^{(l,s)}(T)$ given by equation (16.4) and the fitting polynomials (16.5) for each of the metals were compared with the results obtained by Nieto de Castro *et al.* (1990). The agreement is very good for lithium, sodium and cesium, where the maximum deviation is less than $\pm 1\%$ in the temperature range $700 \leq T \leq 2000$ K, and also for rubidium where it amounts to $\pm 1.5\%$. Therefore, the comparisons made by Nieto de Castro *et al.* (1990), in their previous paper for lithium and sodium, with those published by other authors remain essentially unchanged. For potassium, a systematic negative deviation of the previous calculations from this work was found, which amounts to -3.5% at 700 K decreasing with increasing temperature to less than -0.5% at 2000 K, probably due to the fact that the new singlet potential function is slightly less steeply repulsive than the previous one.

The calculated viscosity, thermal conductivity and self-diffusion coefficients (the latter at 0.10 MPa) of nonionized monatomic lithium, sodium, potassium, rubidium and cesium vapors can be consulted for temperatures between 700 and 2000 K in Tables VIII to XII of the work of Fialho *et al.* (1993).

Table 16.1. *Singlet potential parameters for equation (16.5) (Fialho et al. 1993).*

	$10^4 \cdot C_0$	$10^3 \cdot C_1$	C_2	C_3	C_4	$\frac{\sigma}{\%}$
			a) $\Omega_1^{(1,1)*}(T_1^*)$			
Li	-0.5516051	7.729600	2.558459	-2.008881	-1.770730	0.11
Na	0.3513133	3.852063	2.331874	-2.374357	0.1364867	0.16
K	5.985651	-7.206078	2.089676	-2.450936	0.8512206	0.11
Ru	3.931187	-5.047879	2.149760	-2.590676	1.011096	0.12
Cs	5.293546	-7.041881	1.930501	-2.196644	0.6830733	0.12
			b) $\Omega_1^{(1,2)*}(T_1^*)$			
Li	0.2042988	2.577080	2.611257	-3.092686	-2.339541	0.18
Na	2.518076	-7.700390	2.473208	-3.938448	1.272143	0.23
K	12.23200	-30.29785	2.309786	-4.086829	2.311473	0.13
Ru	7.494228	-21.40654	2.327975	-4.158040	2.371899	0.15
Cs	10.09811	-25.26335	2.109774	-3.601405	1.846914	0.17
			c) $\Omega_1^{(1,3)*}(T_1^*)$			
Li	1.828337	-9.054357	2.840996	-5.701037	0.8521955	0.22
Na	5.224332	-23.14067	2.725702	-6.367950	4.284772	0.27
K	9.513925	-31.21800	2.391396	-5.477656	3.871515	0.14
Ru	3.654676	-19.00137	2.389106	-5.550351	3.921720	0.18
Cs	7.474690	-25.70764	2.187178	-4.889873	3.234587	0.22
			d) $\Omega_1^{(2,2)*}(T_1^*)$			
Li	-2.650538	23.80728	2.100460	0.4879992	-3.929917	0.20
Na	-3.454700	26.63395	1.850960	0.2934780	-3.124283	0.13
K	0.6798244	15.25509	1.744732	-0.7737833	-0.7625710	0.15
Ru	13.55447	-17.31024	2.048434	-1.495705	-0.1425657	0.28
Cs	2.427715	8.738779	1.681272	-0.8506149	-0.5733677	0.18
			e) $\Omega_1^{(2,3)*}(T_1^*)$			
Li	-3.088718	27.79483	1.910291	2.093833	-10.01412	0.19
Na	-0.9001423	15.53608	1.919113	-0.0193895	-4.046381	0.21
K	20.16255	-40.61396	2.202697	-2.561157	0.0673081	0.20
Ru	33.41984	-74.01689	2.520579	-3.347017	1.317932	0.30
Cs	20.20223	-41.97838	2.094257	-2.443705	0.6469965	0.25

16.3 The vapor mixture

To a first approximation, the vapor phase of an alkali metal may be considered as a binary mixture of monomer and dimer species, whose molar fractions are dependent on both temperature and pressure. The viscosity of such a binary mixture with mole fractions y_M and y_D (subscripts M and D refer to monomer and dimer respectively) is well described by the first-order Chapman–Enskog theory (Hirschfelder *et al.* 1954;

Table 16.2. *Triplet potential parameters for equation (16.5) (Fialho et al. 1993).*

	$10 \cdot C_0$	C_1	$10 \cdot C_2$	$10^2 \cdot C_3$	$10^3 \cdot C_4$	$\frac{\sigma}{\%}$
			a) $\Omega_3^{(1,1)*}(T_3^*)$			
Li	-2.590547	1.212185	-4.138669	-6.654710	2.191451	0.05
Na	-2.615505	1.153324	-4.081251	-4.666220	1.135139	0.12
K	-2.304266	1.080582	-3.733486	-4.534369	1.313487	0.03
Ru	-2.350945	1.084547	-3.758147	-4.233029	1.151050	0.02
Cs	-2.551676	1.120338	-3.980115	-3.541296	0.7348892	0.04
			b) $\Omega_3^{(1,2)*}(T_3^*)$			
Li	-1.080536	0.8241279	-4.579437	-7.685941	2.444255	0.02
Na	-1.081520	0.7723458	-4.296703	-5.777638	1.521760	0.10
K	-0.8940780	0.7206854	-3.970009	-5.405562	1.574559	0.03
Ru	-0.8876363	0.7148743	-3.914575	-5.234619	1.515485	0.03
Cs	-0.8959670	0.7137547	-3.913260	-4.954174	1.404486	0.02
			c) $\Omega_3^{(1,3)*}(T_3^*)$			
Li	-0.5811200	0.6083763	-4.603172	-9.596871	3.576178	0.003
Na	-0.7662706	0.6038852	-4.492843	-6.839778	1.977117	0.012
K	-0.5344234	0.5437907	-4.052455	-6.574065	2.169167	0.006
Ru	-0.5201589	0.5370695	-3.978914	-6.404623	2.119163	0.005
Cs	-0.5047070	0.5310475	-3.937206	-6.159051	2.045112	0.005
			d) $\Omega_3^{(2,2)*}(T_3^*)$			
Li	-2.796898	1.269097	-3.320011	-3.717834	0.3520759	0.004
Na	-2.426747	1.146071	-2.949714	-3.053893	0.3819975	0.031
K	-2.385169	1.123730	-2.939140	-2.428154	0.1376324	0.015
Ru	-2.341062	1.111948	-2.884526	-2.360277	0.1359558	0.014
Cs	-2.281250	1.100439	-2.855358	-2.288036	0.1495502	0.011
			e) $\Omega_3^{(2,3)*}(T_3^*)$			
Li	-1.199342	0.8934929	-3.101604	-5.165962	0.9138658	0.018
Na	-0.9349136	0.7934329	-2.684982	-4.358730	0.9132173	0.024
K	-0.8964713	0.7722816	-2.624224	-3.735935	0.7181930	0.026
Ru	-0.8768487	0.7648212	-2.579292	-3.621804	0.6904539	0.026
Cs	-0.8500697	0.7582051	-2.567246	-3.477965	0.6626365	0.025

Maitland *et al.* 1987; Chapter 4 of this volume)

$$[\eta_{\text{mix}}]_1 = \frac{1 + Z_\eta}{X_\eta + Y_\eta} \tag{16.6}$$

with

$$X_\eta = \frac{y_M^2}{[\eta_M]_1} + \frac{2 y_M y_D}{[\eta_{MD}]_1} + \frac{y_D^2}{[\eta_D]_1} \tag{16.7}$$

$$Y_\eta = \frac{3}{5} A^*_{MD} \left\{ \frac{y^2_M}{2\,[\eta_M]_1} + \frac{9\, y_M y_D\, [\eta_{MD}]_1}{4\, [\eta_M]_1\, [\eta_D]_1} + \frac{2 y^2_D}{[\eta_D]_1} \right\} \tag{16.8}$$

$$Z_\eta = \frac{3}{5} A^*_{MD} \left\{ \frac{y^2_M}{2} + 2 y_M y_D \left[\frac{9}{8} \left(\frac{[\eta_{MD}]_1}{[\eta_M]_1} + \frac{[\eta_{MD}]_1}{[\eta_D]_1} \right) - 1 \right] + 2 y^2_D \right\} \tag{16.9}$$

where

$$A^*_{MD} = \frac{\left\langle \Omega^{(2,2)}_{MD}(T) \right\rangle}{\left\langle \Omega^{(1,1)}_{MD}(T) \right\rangle} \tag{16.10}$$

and $[\eta_M]_1$, $[\eta_D]_1$ and $[\eta_{MD}]_1$ are the first–order solutions for the viscosity coefficients of the monomer, the dimer and monomer-dimer interaction, given by the expressions

$$[\eta_M]_1 = \frac{5}{16} (\pi m_M k_B T)^{1/2} \frac{1}{\bar\Omega^{(2,2)}_M(T)} \tag{16.11}$$

$$[\eta_D]_1 = \frac{5}{16} (\pi m_D k_B T)^{1/2} \frac{1}{\left\langle \Omega^{(2,2)}_D(T) \right\rangle} \tag{16.12}$$

$$[\eta_{MD}]_1 = \frac{5}{16} \left(\frac{2\pi m_M m_D k_B T}{m_M + m_D} \right)^{1/2} \frac{1}{\left\langle \Omega^{(2,2)}_{MD}(T) \right\rangle} \tag{16.13}$$

The quantity $\bar\Omega^{(2,2)}_M(T)$ is the average collision integral for the monomer systems (equation (16.4)). The quantities $\left\langle \Omega^{(2,2)}_D(T) \right\rangle$ and $\left\langle \Omega^{(2,2)}_{MD}(T) \right\rangle$ appearing in equations (16.12) and (16.13) are angle–averaged collision integrals for the polyatomic interactions between monomer–dimer and dimer–dimer, respectively (Maitland *et al.* 1987). According to an approximate scheme originally proposed by Stogryn & Hirschfelder (1959), the thermal conductivity of a mixture of this kind can be assumed to be the sum of two contributions: a 'frozen' (nonreacting) coefficient, λ_f, and a chemical reaction coefficient, λ_r, associated with the transport of the energy involved in the formation of the dimers. The 'frozen' thermal conductivity is itself composed of a translational part, λ_{tr}, and an internal part, λ_{int}, due to the contribution of the internal degrees of freedom of the dimers. Therefore,

$$\lambda_{mix} = \lambda_{tr} + \lambda_{int} + \lambda_r \tag{16.14}$$

The translational contribution is assumed to be the same as that for a monatomic gas and thus can be obtained from the Chapman-Enskog theory of binary mixtures of monatomic species (Chapter 4)

$$[\lambda_{tr}]_1 = \frac{1 + Z_\lambda}{X_\lambda + Y_\lambda} \tag{16.15}$$

with

$$X_\lambda = \frac{y_M^2}{[\lambda_M]_{1,\text{tr}}} + \frac{2y_M y_D}{[\lambda_{MD}]_{1,\text{tr}}} + \frac{y_D^2}{[\lambda_D]_{1,\text{tr}}} \tag{16.16}$$

$$Y_\lambda = \frac{y_M^2}{[\lambda_M]_{1,\text{tr}}} U^{(1)} + \frac{2y_M y_D}{[\lambda_{MD}]_{1,\text{tr}}} U^{(y)} + \frac{y_D^2}{[\lambda_D]_{1,\text{tr}}} U^{(2)} \tag{16.17}$$

$$Z_\lambda = y_M^2 U^{(1)} + 2y_M y_D U^{(z)} + y_D^2 U^{(2)} \tag{16.18}$$

$$U^{(1)} = \frac{4}{15} A_{MD}^* - \frac{1}{24}\left(\frac{12}{5} B_{MD}^* + 1\right) + \frac{1}{4} \tag{16.19}$$

$$U^{(2)} = \frac{4}{15} A_{MD}^* - \frac{1}{6}\left(\frac{12}{5} B_{MD}^* + 1\right) + \frac{1}{4} \tag{16.20}$$

$$U^{(y)} = \frac{3}{10} A_{MD}^* \frac{[\lambda_{MD}]_{1,\text{tr}}^2}{[\lambda_M]_{1,\text{tr}} [\lambda_D]_{1,\text{tr}}} - \frac{1}{12}\left(\frac{12}{5} B_{MD}^* + 1\right)$$
$$- \frac{5}{64 A_{MD}^*}\left(\frac{12}{5} B_{MD}^* - 5\right) \tag{16.21}$$

$$U^{(z)} = \frac{4}{15} A_{MD}^* \left[\frac{9}{8}\left(\frac{[\lambda_{MD}]_{1,\text{tr}}}{[\lambda_M]_{1,\text{tr}}} + \frac{[\lambda_{MD}]_{1,\text{tr}}}{[\lambda_D]_{1,\text{tr}}}\right) - 1\right] - \frac{1}{12}\left(\frac{12}{5} B_{MD}^* + 1\right) \tag{16.22}$$

where

$$B_{MD}^* = \frac{5\left\langle \Omega_{MD}^{(1,2)}(T)\right\rangle - 4\left\langle \Omega_{MD}^{(1,3)}(T)\right\rangle}{\left\langle \Omega_{MD}^{(1,1)}(T)\right\rangle} \tag{16.23}$$

and $[\lambda_M]_1$, $[\lambda_D]_1$ and $[\lambda_{MD}]_1$ are the first order solutions of the Chapman-Enskog theory for the monomer, dimer and monomer-dimer interaction thermal conductivity coefficients respectively

$$[\lambda_M]_{1,\text{tr}} = \frac{25}{32}\left(\frac{\pi k_B T}{m_M}\right)^{1/2} \frac{c_{V,\text{tr}}}{\bar{\Omega}_M^{(2,2)}(T)} \tag{16.24}$$

$$[\lambda_D]_{1,\text{tr}} = \frac{25}{32}\left(\frac{\pi k_B T}{m_D}\right)^{1/2} \frac{c_{V,\text{tr}}}{\left\langle \Omega_D^{(2,2)}(T)\right\rangle} \tag{16.25}$$

$$[\lambda_{MD}]_{1,\text{tr}} = \frac{25}{32}\left(\frac{\pi k_B T (m_M + m_D)}{2 m_M m_D}\right)^{1/2} \frac{c_{V,\text{tr}}}{\left\langle \Omega_{MD}^{(2,2)}(T)\right\rangle} \tag{16.26}$$

where $c_{V,\text{tr}} = 3k_B/2$ is the translational contribution to the molecular heat capacity.

The dimer is a diatomic molecule and therefore has internal degrees of freedom that contribute to energy transfer, as explained in Chapter 4. In addition, it creates anisotropy in the intermolecular interaction with the monomer. Lacking information on the details of the internal molecular energy transfer in the dimers in order to employ more accurate theoretical expressions, one may use the approximate Eucken expression for λ_{int} (Stogryn & Hirschfelder 1959), which, for simple molecules, does not overestimate the value of this contribution by more than 3%

$$\lambda_{\text{int}} = \rho D_2 y_D C_{V,D,\text{int}}^{\text{id}} \tag{16.27}$$

with

$$D_2 = \left(\frac{y_D}{[D_D]_1} + \frac{y_M}{[D_{MD}]_1} \right)^{-1} \tag{16.28}$$

D_D and D_{MD} are the dimer self-diffusion coefficient and the binary diffusion coefficient of the gaseous mixture, respectively, given by

$$[D_D]_1 = \frac{3}{8P} \left(\frac{\pi k_B^3 T^3}{m_D} \right)^{1/2} \frac{1}{\left\langle \Omega_D^{(1,1)}(T) \right\rangle} \tag{16.29}$$

$$[D_{MD}]_1 = \frac{3}{16P} \left(\frac{\pi k_B^3 T^3 (m_M + m_D)}{2m_M m_D} \right)^{1/2} \frac{1}{\left\langle \Omega_{MD}^{(1,1)}(T) \right\rangle} \tag{16.30}$$

Here, ρ is the molar density of the mixture, $C_{V D,\text{int}}^{\text{id}}$ is the internal heat capacity at constant volume of the dimer, which for dilute gases is given by

$$C_{V,D,\text{int}}^{\text{id}} = C_{V,D,\text{rot}}^{\text{id}} + C_{V,D,\text{vib}}^{\text{id}} \tag{16.31}$$

$C_{V,D,\text{rot}}^{\text{id}}$ and $C_{V,D,\text{vib}}^{\text{id}}$ are the contributions of rotational and vibrational degrees of freedom, respectively, to the molar heat capacity at constant volume. The rotational contribution to the heat capacity can be considered classical, because $T \gg \theta_{\text{rot}}$, where θ_{rot} is the characteristic rotational temperature, and the dimer is assumed to be a rigid rotor – a hypothesis commensurate with the use of the Eucken approximation. Hence,

$$C_{V,D,\text{rot}}^{\text{id}} = R \tag{16.32}$$

The vibrational contribution can be calculated assuming that the vibration is that of a linear harmonic oscillator

$$C_{V,D,\text{vib}}^{\text{id}} = R \left(\frac{\theta_{\text{vib}}}{T} \right)^2 \frac{\exp(\theta_{\text{vib}}/T)}{\left[\exp(\theta_{\text{vib}}/T) - 1 \right]^2} \tag{16.33}$$

where θ_{vib} is a characteristic vibrational temperature, given in Table 16.3 for each of the alkali metal dimers.

The vapor of the alkali metal is considered to be a chemical reacting mixture of monomers and dimers, in equilibrium

$$M + M \rightleftharpoons D$$

Table 16.3. *Spectroscopic values for the wavenumber and the characteristic vibrational temperature for the dimers of the alkali metals.*

Metal	$\dfrac{\bar{\nu}_e}{cm^{-1}}$	$\dfrac{\theta_{vib}}{K}$
Lithium	351.39	505.58
Sodium	159.11	228.93
Potassium	92.021	132.40
Rubidium	57.747	83.086
Cesium	42.0194	60.4571

It is well known that these vapors cannot be considered as ideal gases, as the second virial coefficients are different from zero, except in the case of cesium at high temperatures (Nieto de Castro *et al.* 1990). However, the quasi–chemical equilibrium hypothesis can be used that states that the imperfection has its origin in the atom association, and therefore the mixture of monomers and dimers can be considered a perfect gas mixture (Ewing *et al.* 1967). Assuming also local chemical equilibrium, Stogryn & Hirschfelder (1959) considered that the heat of reaction could affect the 'reactive' contribution and found that the chemical reaction component is given by

$$\lambda_r = \rho \, [D_{MD}]_1 \, \frac{\left(\Delta H^0(T)\right)^2 \, y_M y_D}{RT^2 \, (1 + y_D)^2} \tag{16.34}$$

where R is the universal gas constant and $\Delta H^0(T)$ is the standard enthalpy of reaction at temperature T. The viscosity coefficient can be calculated from equations (16.6)–(16.10). It is convenient to write it in a dimensionless form by dividing by $[\eta_M]_1$, in which case equations (16.7)–(16.9) are rewritten in terms of the ratios

$$\beta_1^2 = \frac{\left\langle \Omega_{MD}^{(2,2)}(T) \right\rangle}{\left\langle \Omega_D^{(2,2)}(T) \right\rangle}, \quad \beta_{12}^2 = \frac{\left\langle \Omega_{MD}^{(2,2)}(T) \right\rangle}{\bar{\Omega}_M^{(2,2)}(T)}, \quad \beta_2^2 = \frac{\beta_{12}^2}{\beta_1^2} \tag{16.35}$$

Correspondingly, the thermal conductivity of the mixture can be rewritten in the dimensionless form

$$\lambda^*(y_D, T) = \frac{[\lambda_{mix}]_1}{[\lambda_M]_1} = \lambda_{tr}^*(y_D) + \lambda_{int}^*(y_D, T) + \lambda_r^*(y_D, T) \tag{16.36}$$

The various contributions can be rearranged into the forms:

a) Translational:

The translational contribution is given by equation (16.15), reduced by $[\lambda_M]_1$. Expressions (16.16)–(16.22) apply, and where ratios of thermal conductivities for different species occur, these are rewritten in terms of the ratios given by equation (16.35).

b) *Internal:*

$$\lambda_{int}^* = \frac{2\sqrt{2}A_D^* \left(C_{V,D,rot}^{id} + C_{V,D,vib}^{id} \right)}{25R \left(\beta_2^2 + \frac{\sqrt{2}A_D^* y_M}{\sqrt{3}A_{MD}^* y_D} \beta_{12}^2 \right)} \tag{16.37}$$

c) *Chemical Reaction:*

$$\lambda_r^* = \frac{2\sqrt{3}}{25} A_{MD}^* \left(\frac{\Delta H^0(T)}{RT} \right)^2 \frac{y_M y_D}{(1+y_D)^2} \beta_{12}^{-2} \tag{16.38}$$

The prediction of the viscosity and thermal conductivity coefficients of the mixtures on purely theoretical grounds is not possible at present, as the interaction viscosity, η_{MD}, thermal conductivity, λ_{MD}, and the ratios of collision integrals, A_{MD}^* and B_{MD}^*, cannot be theoretically evaluated in the absence of reliable pair potentials for the monomer-dimer interactions. Similarly, the 'pure' dimer viscosity, η_D, and thermal conductivity, λ_D, cannot be determined either experimentally or theoretically, without resort to an accurate pair potential for the dimer-dimer interaction.

However, a data assessment procedure based on the above model is devisable, which is endowed with a fairly sound theoretical basis. Testing the compliance of the data and the mutual consistency of viscosity and thermal conductivity to the outlined model needs the establishment of a practical procedure. Namely, the collision integrals, or rather their ratios, involving the dimer species will have to be calculated by fitting the model expressions to the experimental data. For this purpose, some further simplifications of the model were necessary:

(i) the ratios A_{MD}^*, B_{MD}^* are taken as independent of the type of interaction, and calculated by replacing the ratio of collision integrals in (16.10) and (16.23) by the corresponding monomer-monomer collision integrals, such that $A_{MD}^* = A_M^* = A_D^*$ and $B_{MD}^* = B_M^* = B_D^*$. This approximation is supported by the well–known result that these ratios are weakly dependent on the intermolecular pair potential for spherical models.

(ii) the parameters β_i^2 (where $i = 1, 2, 12$) were assumed to be temperature–independent, as proposed by Vargaftik & Yargin (1985). This is probably the most questionable approximation at the moment, but the unavailability of information about the interaction M-D makes it necessary.

In spite of these assumptions, one cannot proceed without the use of some experimental information. Most of the published data on the viscosity and thermal conductivity of the alkali metal vapors are collected in the works of Vargaftik & Yargin (1985) and Fialho *et al.* (1993). It has been shown that the viscosity and the thermal conductivity data must be treated on different grounds, because of the much higher theoretical complexity and measurement difficulty of the latter. In particular, a poorer precision and, especially, a much higher level of corrections to the raw data, affecting the accuracy of the results, are found for the thermal conductivity, when compared to the viscosity measurements. In order to make the best use of the minimum necessary experimental information on the collision integral ratios, Fialho *et al.* (1993) applied the following procedure to each viscosity data set:

(i) for each datum at a certain temperature and pressure, the monatomic viscosity, $[\eta_M]_1$, has been calculated, using the equations described in Section 16.2,

(ii) the respective dimer mole fractions have been calculated using the values published by Vargaftik & Yargin (1985). Interpolation between those values was performed using the fitting equation

$$\ln\left(K'_d\right) = \ln\left(\frac{y_M^2 P}{y_D P^0}\right) = E - \frac{\Delta H_0^0}{RT} + \frac{A}{R}\ln\left(T\right)$$
$$+ \frac{B}{R}T + \frac{C}{2R}T^2 + \cdots \tag{16.39}$$

where P is the pressure, P^0 is the standard state reference pressure ($P^0 = 0.1$ MPa), $y_M \approx 1 - y_D$ is the monomer mole fraction, and the coefficients A, B, C and E are fitting coefficients (Table 16.4); ΔH_0^0 is the standard enthalpy of formation of the dimer at 0 K (Vargaftik & Voljak 1985), whose values are tabulated in Table 16.5. The calculated values for y_D and y_M are dependent on

Table 16.4. *Parameters necessary for the use of equation (16.39). Limits of validity of the parameters in the temperature range. Pressures should be less than 1.0 MPa.*

Metal	$\frac{T}{K}$	$\frac{A}{\text{J mol}^{-1}\text{K}^{-1}}$	$\frac{B}{\text{J mol}^{-1}\text{K}^{-2}}$	$\frac{10^6 \cdot C}{\text{J mol}^{-1}\text{K}^{-3}}$	$10^{-1} \cdot E$
Lithium	1200 – 2000	9.945835	-6.634476	1.864750	1.355096
Sodium	800 – 1300	7.711819	-5.631470	1.652149	1.492893
Potassium	800 – 1900	3.939518	-1.732501	1.804059	1.686225
Rubidium	700 – 1200	6.502770	-6.845959	4.091932	1.504084
Cesium	800 – 1200	5.557522	-5.990512	4.135396	1.545305

Table 16.5. *Ionization energy and dissociation enthalpy for the alkali metals (Vargaftik & Voljak 1985).*

Metal	$\frac{10^{-4} \cdot I/k_B}{K}$	$\frac{\Delta H_0^0}{\text{kJ mol}^{-1}}$
Lithium	6.256	107.8
Sodium	5.964	71.38
Potassium	5.036	53.8
Rubidium	4.847	48.57
Cesium	4.514	44.38

the validity of the assumptions made by Vargaftik & Voljak (1985), namely, that the systems behave as a perfect gas mixture, and that the ionization of the atomic species is comparatively of no practical consequence in the temperature ranges covered by the data. This can be shown by using the result of Vargaftik & Voljak (1985) for the equilibrium constant K_e for the ionization process

$$K_e = 3.33 \cdot 10^{-7} T^{5/2} \exp\left(-I/k_B T\right) \tag{16.40}$$

where I is the ionization energy (see Table 16.5). For the present calculations, the maximum error due to ionization is of the order of 0.15% of y_D.

(iii) The 'best values' of β_1^2 and β_{12}^2 were calculated by a nonlinear multiparametric fitting procedure to the whole set of data, using the Marquard method (Press *et al.* 1986). For the purpose of calculating β_1^2 and β_{12}^2, the experimental data are expressed as the relative deviation from the corresponding monomer property

$$\frac{\Delta \eta}{[\eta_M]_1} = \frac{\eta_{exp}(y_D, T) - [\eta_M]_1}{[\eta_M]_1} \tag{16.41}$$

which is a function of β_1^2 and β_{12}^2. These parameters have been determined for each data set.

This model, which is a modification of the procedure adopted by Vargaftik & Yargin (1985), in order to apply the kinetic theory for binary reacting mixtures of gases to the study of alkali metal vapors, will herewith be referred to as Modified Vargaftik and Yargin Approximation (MVYA). The viscosity data were analyzed and used to calculate those parameters, which were impossible to evaluate theoretically. The procedure is illustrated here with sodium vapor.

The available data for the viscosity coefficients of sodium vapor are plotted in the form of deviations from the monomer viscosities versus the corresponding dimer mole fractions in Figure 16.2. Data giving rise to positive deviations greater than the nominal precision of the experimental method have been discarded from the analysis, as the presence of dimers decreases the viscosity of the mixture in comparison with the monomer. The quality of the data was primarily assessed by testing their compliance to the model. Subsequently the highest–quality data sets for each were selected according to that criterion. The best parameters selected are shown in Table 16.6. In case of doubt, the data obtained with the more precise experimental method have been selected. Details of the discussion can be found in the work of Fialho *et al.* (1994a,b). It was not possible to optimize the fits for cesium, because both sets of data available are in mutual disagreement, deviating from the MVYA approximation; therefore this metal is excluded from the present analysis.

The deviations between the experimental data for the viscosity of sodium vapor and values calculated using the MVYA model as a function of the dimer mole fraction are presented in Figure 16.3. In this figure, Stefanov *et al.* (1966) data are in good agreement at low dimer fractions and deviate by almost −15% at a dimer mole fraction of 0.1.

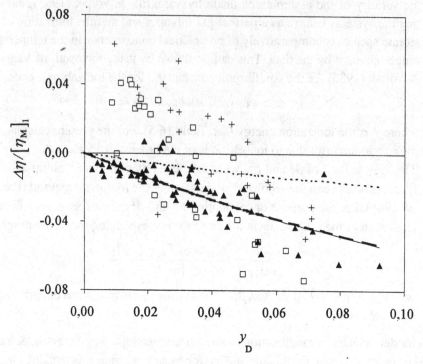

Fig. 16.2. Deviations from the monomeric behavior for the viscosity of the vapor of sodium. The symbols represent the experimental data and the lines correspond to the fitting of the MVYA model. □, - - - annular gap, Vargaftik *et al.* (1992). +, · · · falling body, Vargaftik & Yargin (1985). ▲, —— oscillating disc, Timrot & Varava (1977).

Table 16.6. *Parameters obtained by the best fit to the viscosity data, of the alkali metal vapors available in the literature, to the model MVYA (Fialho et al. 1994a).*

Metal	Ref.	β_1^2	β_{12}^2
Lithium	Stepanenko *et al.* (1986)	1.19±0.04	1.42±0.02
Sodium	Timrot & Varava (1977)	0.87±0.02	1.092±0.008
Potassium	Stefanov *et al.* (1966),		
	Vargaftik *et al.* (1975)	1.18±0.05	1.64±0.02
Rubidium	Sidorov *et al.* (1975)	1.3±0.2	1.48±0.06

The data obtained with higher–precision techniques, namely, oscillating disc (Timrot & Varava 1977) and concentric cylinders (Vargaftik *et al.* 1992) used to determine β_1^2 and β_{12}^2, agree well with the prediction, within the claimed uncertainties.

The assessment of the thermal conductivity data was performed, comparing the experimental values with the results calculated through the model equations using the values of β_1^2 and β_{12}^2, obtained from the selected viscosity data, and the spectroscopic

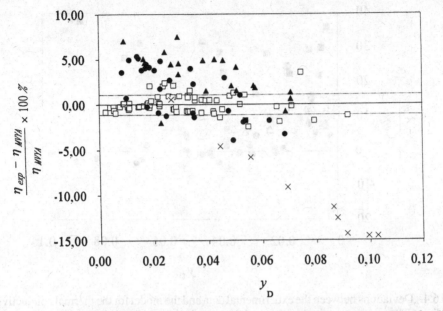

Fig. 16.3. Deviations between the experimental data and the model for the viscosity of sodium. The parameters used in the model were taken from the viscosity data of Timrot & Varava (1977). The dotted lines (+1.1 and -1.1 %) represent the claimed experimental accuracy. □ - oscillating disc, Timrot & Varava (1977). ▲ - falling body, Vargaftik & Yargin (1985). x - capillary tube, Stefanov *et al.* (1966). ● - annular gap, Vargaftik *et al.* (1992).

information collected in Tables 16.3 and 16.5. As for viscosity the thermal conductivity results are expressed as deviations from the monomer properties, in the form

$$\frac{\Delta\lambda}{[\lambda_M]_1} = \frac{\lambda_{\exp}(y_D, T) - [\lambda_M]_1}{[\lambda_M]_1} \tag{16.42}$$

A comparison of the thermal conductivity data with the model predictions, using the selected β_{12}^2 and β_1^2 parameters for sodium, is presented in Figure 16.4.

This figure shows that the deviations from the data obtained by Vargaftik & Voshchinin (1967) agree well with the predictions within the mutual uncertainties involved in the comparison. Higher positive deviations are found for the data published with the steady-state hot-wire technique by Timrot *et al.* (1976), especially for low y_D. The results obtained by Timrot & Totskii (1967) show large systematic discrepancies with the present model. However, it should be noted that, despite the systematic deviations found for the latter data, and its large scatter (±10%), the variation with the dimer concentration does not differ substantially from the one predicted by this model. Applying the model previously described (MVYA) with the parameters defined in Tables 16.3 to 16.6, the transport property surfaces for lithium, sodium and potassium can be generated (Fialho *et al.* 1994a,b).

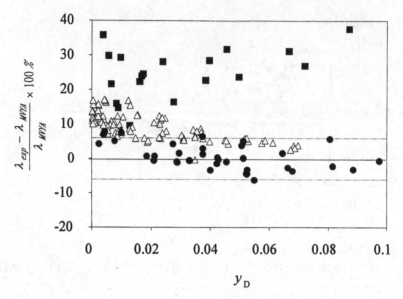

Fig. 16.4. Deviations between the experimental data and the model for the thermal conductivity of sodium. The parameters used in the model were taken from the viscosity data of Timrot & Varava (1977). The dotted lines (+6% and −6%) represent the claimed experimental accuracy. ■- steady–state concentric cylinders, Timrot & Totskii (1967). △ - steady–state hot wire, Timrot et al. (1976). ● - steady–state concentric cylinders, Vargaftik & Voshchinin (1967).

16.4 Conclusions

The preceding sections have shown how it is possible to use the available kinetic theory of low–density reacting binary mixtures to calculate the viscosity and thermal conductivity of the alkali metal vapors. The model can also be applied to the binary diffusion coefficient in the mixture. It can be easily shown that, for this model, the ratio $[D_{MD}]_1 / [D_M]_1 = \sqrt{3}\beta_{12}^{-2}/2$. This result says that the density and temperature dependence of both diffusion coefficients is the same. From the results of the self–diffusion coefficient of the monomer reported by Fialho et al. (1993) and the value of β_{12}^{-2} from Table 16.6, the binary diffusion coefficient can be obtained. It is also possible on the basis of this model to verify the internal consistency of all the experimental information available on the transport properties of these vapors and to select precise data and propose accuracy classes (Fialho et al. 1994). The results obtained suggest that there is still a lot to be done on the experimental measurement of thermal conductivity of these vapors, where radiation effects at high temperatures play a significant role, most of the time difficult to estimate (Wakeham et al. 1991). The data selected can be used to produce a consistent set of transport property tables for the alkali metal vapors, lithium, sodium and potassium, with a confidence level commensurate with the uncertainty of the selected data. This work was performed by the authors and will be the subject of a future publication. For the case of rubidium it is still possible to produce such tables, but with a decrease in the confidence level. For cesium it was not possible to construct such tables (Fialho *et al.* 1994a,b). In order to illustrate the applications presented in this chapter, a small sample of the tables is shown in the Appendix below (Table 16.7).

Appendix

Table 16.7. *Calculated values for the viscosity and thermal conductivity for sodium vapor using the MVYA model (Fialho 1993).*

$\frac{T}{K}$	$\frac{\eta_M}{\mu Pa\ s}$	$P = 1$ kPa	$P = 10$ kPa	$P = 100$ kPa	$P = 1000$ kPa
900	18.18	18.00			
1000	19.80	19.72	19.18		
1200	22.91	22.89	22.73	21.64	
1500	27.39	27.38	27.34	26.91	25.07
$\frac{T}{K}$	$\frac{\lambda_M}{mW\ m^{-1}K^{-1}}$	$P = 1$ kPa	$P = 10$ kPa	$P = 100$ kPa	$P = 1000$ kPa
900	24.66	28.93			
1000	26.86	28.31	38.57		
1200	31.08	31.35	33.60	47.50	
1500	37.15	37.20	37.60	41.19	55.35

References

Davies, R. H., Mason, E. A. & Munn, R. J. (1965). High temperature transport properties of alkali metal vapours. *Phys. Fluids*, **8**, 444–452.

Ewing, C. T., Stone, J. P., Spann, J. R. & Miller, R. R. (1967). Molecular association in sodium, potassium and cesium vapors at high temperature. *J. Phys. Chem.*, **71**, 473–477.

Fialho, P. S. (1993). *Propriedades Termofísicas de Fluidos. Sua Previsão, Estimativa e Correlação*. Ph.D. thesis, Faculty of Sciences, University of Lisbon, Portugal.

Fialho, P. S., Fareleira, J. M. N. A., Ramires, M. L. V. & Nieto de Castro, C. A. (1993). The thermophysical properties of alkali metal vapours. Part IA - Prediction and correlation of transport properties for monatomic systems. *Ber. Bunsenges. Phys. Chem.*, **97**, 1487–1492.

Fialho, P. S., Ramires, M. L. V., Fareleira, J. M. N. A. & Nieto de Castro, C. A. (1994b). Viscosity, thermal conductivity and thermal diffusion ratio of the alkali metal vapours, in *Thermal Conductivity* 22, ed. T. W. Tong, Lancaster, USA: Technomic Pubs. Co., 126–135.

Fialho, P. S., Ramires, M. L. V., Nieto de Castro, C. A., Fareleira, J. M. N. A. & Mardolcar, U. V. (1994a). The thermophysical properties of alkali metal vapours. Part II - Assessment of experimental data on thermal conductivity and viscosity. *Ber. Bunsenges. Phys. Chem.*, **98**, 92–102.

Hirschfelder, J.O., Curtiss, C. F. & Bird, R.B. (1954). *Molecular Theory of Gases and Liquids*. New York: John Wiley & Sons.

Holland, P. N. & Biosi, L. (1987). Calculation of the transport properties of ground state sodium atoms. *J. Chem. Phys.*, **87**, 1261–1266.

Holland, P. N., Biosi, L. & Rainwater, J. C. (1986). Theoretical calculations of the transport properties of monatomic lithium vapor. *J. Chem. Phys.*, **85**, 4011–4018.

Maitland, G.C., Rigby, M., Smith E.B. & Wakeham, W.A. (1987). *Intermolecular Forces: Their Origin and Determination*. Oxford: Clarendon Press.

Mies, F. H. & Julienne, P. S. (1982). The thermodynamic properties of diatomic molecules at elevated temperatures: Role of continuum and metastable states. *J. Chem. Phys.*, **77**, 6162–6176.

Nieto de Castro, C.A., Fareleira, J.M.N.A., Ramires, M.L.V., Canelas Pais, A.A.C. & Varandas, A.J.C. (1990). The thermophysical properties of alkali metal vapours. Part I - Theoretical calculation of the properties of monatomic systems. *Ber. Bunsenges. Phys. Chem.*, **94**, 53–59.

Ohse, R. W. (Ed.). (1985). *IUPAC Handbook of Thermodynamic and Transport Properties of Alkali Metals*. Oxford: Blackwell Scientific Publications.

Polishchuk, A. J., Sphilrain, E. E. & Yacubov, I. T. (1980). Thermal conductivity and viscosity of lithium vapours. *J. Eng. Phys.*, **38**, 247–250.

Press, W. H., Flannery, S. A., Teukolvsky, S. A. & Vetterling, W. T. (1986). *Numerical Recipes. The Art of Scientific Computing*. Cambridge: University Press.

Sannigrahi, A. B., Mohammed, S. N. & Mookhejee, D. C. (1976). Second virial coefficients of alkali vapours. *Mol. Phys.*, **31**, 963–970.

Sidorov, N. I., Tarkalov, Y. V. & Yargin, Y. S. (1975). *Izv. VUZOV. Energetica*, **4**, 96–101, quoted in Vargaftik, V. B. & Yargin, V. S. (1985).

Sinanoglu, O. & Pitzer, K. S. (1959). Equation of state and thermodynamic properties of gases at high temperatures I. Diatomic molecules. *J. Chem. Phys.*, **31**, 960–967.

Stefanov, B. (1980). Generalisation of thermal conductivity and viscosity data for monatomic alkali metal vapours. *High Temp.-High Press.*, **12**, 189–194.

Stefanov, B. I., Timrot, D. L., Totskii, E. E. & When-Hao, C. (1966). Viscosity and thermal conductivity of the vapours of sodium and potassium. *Teplofiz. Vys. Temp.*, **4**, 141–142.

Stepanenko, I. F., Sidorov, N. I., Tarkalov, Y. V. & Yargin, V. S. (1986). Experimental study of the viscosity of lithium vapour at high temperatures. *Int. J. Thermophys.*, **7**, 829–835.

Stogryn, D. E. & Hirschfelder, J. O. (1959). Initial pressure dependence of thermal conductivity and viscosity. *J. Chem. Phys.*, **31**, 1545–1554.

Timrot, D. L., Makhrov, V. V. & Sviridenko, V. I. (1976). Method of the heated filament (incorporating a new section) for use in corrosive media, and determination of the thermal conductivity of sodium vapour. *Teplofiz. Vys. Temp.*, **14**, 67–94.

Timrot, D. L. & Totskii, E. E. (1967). Measurements of the thermal conductivity of sodium and potassium vapours as a function of temperature and pressure. *Teplofiz. Vys. Temp.*, **5**, 793–801.

Timrot, D. L. & Varava, A. N. (1977). Experimental investigation of the viscosity of sodium vapors. *Teplofiz. Vys. Temp.*, **15**, 750–757.

Varandas, A. J. C. (1980). Hybrid potential function for bound diatomic molecules. *J. Chem. Soc. Faraday Trans. II*, **76**, 129–135.

Varandas, A. J. C. & Brandaõ, J. (1982). A simple semiempirical approach to the intermolecular potential of van der Waals systems I. Isotropic interactions: Application to the lowest triplet state of the alkali dimers. *Mol. Phys.*, **45**, 857–875.

Varandas, A. J. C. & Dias da Silva, J. (1986). Hartree-Fock approximate correlation energy (HFACE) potential for diatomic interactions. Molecules and van der Waals molecules. *J.Chem. Soc. Faraday Trans. II*, **82**, 593–608.

Varandas, A. J. C. & Dias da Silva, J. (1992). Potential model for diatomic molecules including the United-Atom Limit, and its use in a multiproperty fit for argon. *J. Chem. Soc. Faraday Trans.*, **88**, 941–954.

Vargaftik, V. B., Vinogradov, Y. K., Dolgov, V. I., Dzis, V. G., Stepanenko, I. F., Yakimovitch, Y. K. & Yargin, V. S. (1992). Viscosity and thermal conductivity of alkali metal vapours at temperatures up to 2000 K. *Int. J. Thermophys.*, **12**, 85–103.

Vargaftik, V. B. & Voljak, L. D. (1985). Thermodynamic properties of alkali metal vapours at low pressures. Chapter 7.6.6.1 of *IUPAC Handbook of Thermodynamic and Transport Properties of Alkali Metals*, ed. R. W. Ohse. Oxford: Blackwell Scientific Pubblications.

Vargaftik, V. B. & Voshchinin, A. A. (1967). Experimental study of the thermal conductivity of sodium and potassium as vapours. *Teplofiz. Vys. Temp.*, **5**, 802–811.

Vargaftik, N. B., Sidorov, N. I., Tarkalov, Y. V. & Yargin, V. S. (1975). Viscosity of potassium vapour. (Russ.). *Teplofiz. Vys. Temp.*, **13**, 974–978.

Vargaftik, V. B. & Yargin, V. S. (1985). Thermal conductivity and viscosity of the gaseous phase. Chapter 7.4 of *IUPAC Handbook of Thermodynamic and Transport Properties of Alkali Metals*, ed. R. W. Ohse. Oxford: Blackwell Scientific Publications.

Wakeham, W. A., Nagashima, A. & Sengers, J. V. (Eds.). (1991). *Measurement of the Transport Properties of Fluids*. Oxford: Blackwell Scientific Publications.

Part six

DATA BANKS AND PREDICTION PACKAGES

17

Data Collection and Dissemination Systems

17.1 MIDAS database

R. KRAUSS and K. STEPHAN

Universität Stuttgart, Germany

17.1.1 Introduction

Thermophysical properties – thermodynamic and transport properties – of pure fluids and mixtures are of essential importance for apparatus design and energy and process technology. A safe, ecologically harmless and economical use of substances can be achieved only if their thermophysical properties are known to a sufficiently high degree of accuracy. Practical experience, however, shows that it is often costly in terms of time and labor to get reliable information. In addition, even for experts it is sometimes difficult to overview and critically assess all available data of interest. Inevitably, unreliable and inconsistent data are often used.

Based on a comprehensive evaluation of the viscosity of 50 fluids (Stephan & Lucas 1979) the MIDAS database for transport properties was founded at the institute of the authors in 1979 in order to overcome the above mentioned problems by providing users from industry, universities and other institutions with reliable, thermodynamically consistent data covering a wide range of fluid state (Laesecke *et al.* 1986).

In view of the environmental problems such as the greenhouse effect and stratospheric ozone destruction, mainly caused by chlorofluorocarbons (CFC), the database was extended to thermophysical properties of environmentally acceptable refrigerants (Stephan & Krauss 1990).

17.1.2 Structure of the database system

The database is managed by a relational database management system developed by the ORACLE Corporation, Belmont, California, USA. Unlike hierarchical database

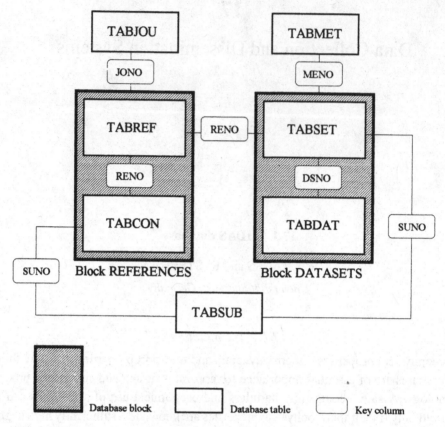

Fig. 17.1. Structure of the MIDAS database system.

systems, relational database systems consist of tables which are built of columns and rows. For each entry a new row is added to the database table. The tables may be linked by particular key columns that exist in either table.

The structure of the MIDAS database system is outlined in Figure 17.1. The kernel of the system is built by the database blocks REFERENCES and DATASETS, each consisting of two main tables and representing the bibliographic and numerical part of the database, respectively. Both database blocks are connected by the key column RENO that represents a serial, unique reference number. The auxiliary database tables TABJOU, TABMET, TABSUB complete the MIDAS database system.

17.1.2.1 The bibliographic part

The database block REFERENCES with the main tables TABREF and TABCON represents the bibliographic part of the database. Both database tables are linked by the key column RENO, the reference number. A typical entry in TABREF for a publication with the reference number 5733 is shown in Figure 17.2. Key columns are in bold type. For the sake of a clearer depiction, columns and rows have been interchanged. In addition,

Database table **TABREF**

Column	Entry
RENO:	5733
JONO:	162
VOLUME:	14
NUMBER:	2
PAGE:	183
YEAR:	1993
AUTHOR1:	ASSAEL M.J.
AUTHOR2:	KARAGIANNIDIS E.
AUTHOR3:	
AUTHOR4:	
AUTHOR5:	
TITLE1:	MEASUREMENTS OF THE THERMAL CONDUCTIVITY
TITLE2:	OF R22, R123, AND R134A IN THE TEMPERATURE
TITLE3:	RANGE 250-340 K AT PRESSURES UP TO 30 MPA.
TOPIC:	TC,VOL
FLAG:	

Database table **TABCON**

Column	Entry 1	Entry 2	Entry 3
RENO:	5733	5733	5733
SUB1:	1010	1029	2233
SUB2:			
SUB3:			
SUB4:			
SYST:	1	1	1
PROP:	TC,VOL	TC,VOL	TC,VOL
FLAG:			

Fig. 17.2. Database block REFERENCES.

TABREF is connected to the auxiliary database table TABJOU for journal names by the journal number JONO with the current value 162. The meaning of the other columns may be deduced from their names.

Details about the substances and properties investigated in the publications are stored in TABCON. The example shows that the topic of the work is the thermal conductivity TC and density or specific volume VOL of three different pure substances with the substance numbers 1010, 1029 and 2233. That means that three different entries and rows respectively are stored in TABCON for reference number 5733. Mixtures with up to four components, represented by the key columns SUB1 to SUB4, are possible. These key columns are linked with the auxiliary database table TABSUB for substance names where the corresponding key column is SUNO. The system parameter SYST indicates the number of components of a mixture. The value 1 indicates a pure fluid. In the example we proceed with the third entry in TABCON, the entry for a substance with the substance number 2233.

Database table **TABSET**

Column	Entry 1	Entry 2
RENO:	5733	5733
DSNO:	3898	3899
SUB1:	2233	2233
SUB2:		
SUB3:		
SUB4:		
SYST:	1	1
PROP:	TC	TC
MENO:	15	15
PHST:	SL	L
PMIN:	1	6
PMAX:	17	224
TMIN:	253	253
TMAX:	333	333
POINTS:	5	36
FLAG:		

Database table **TABDAT**

Column	Entry 1	Entry 2	Entry 3
DSNO:	3899	3899	3899
PRES:	6.4	27.1	50.8
TEMP:	253.15	253.15	253.15
DENS:			
DATA:	100.6	101.8	103.1
MOL1:			
MOL2:			
MOL3:			
MOL4:			
FLAG:			

Fig. 17.3. Database block DATASETS.

17.1.2.2 The numerical part

The numerical data published in the filed literature from the bibliographic part of the database are stored in the numerical part that is built by the database block DATASETS with the main tables TABSET and TABDAT. Both database tables are linked by the key column DSNO, which represents a serial, unique data set number. TABDAT contains the numerical data, and TABSET provides a summary of the data sets. Proceeding with the example considered above, Figure 17.3 shows two entries for reference number 5733 and substance number 2233 in TABSET. Each entry has its own data set number. In the current example there are two entries, because one series of measurement was made in the liquid state L and the other one was made at saturated liquid state SL, indicated by the column PHST for the physical state. The measured property PROP is the thermal conductivity TC. Instead of the measuring method, the method number

MENO is shown. This is a key column that links TABSET with the auxiliary database table TABMET for the names of measuring methods. In the example the current value is 15. Finally the pressure and temperature range and the number of points of the data sets are stored.

The user proceeds with table TABDAT and the second data set with the data set number 3899. The property data, in the current example the thermal conductivity, is listed in the column DATA together with the pressure and temperature coordinates PRES and TEMP. Additionally, the corresponding density DENS and, in case of mixtures, the mole fractions MOL1 to MOL4 can be stored. In the example only 3 out of 36 points are shown. All points belonging to the same data set are identified by the same data set number DSNO. The data are converted to S.I. units and checked for typing or printing errors. Conversion to S.I. units makes it easier for different data sets to be merged.

17.1.2.3 The auxiliary database tables

The auxiliary database tables TABJOU, TABMET and TABSUB have been built up to avoid typing mistakes and to guarantee a unified spelling. They are shown in Figure 17.4 to complete the current example.

TABJOU is linked to TABREF by the key column JONO, which represents a journal number. In the example the number 162 stands for the *International Journal of Thermophysics*. At present more than 800 different journals, proceedings, monographs etc. are stored.

In TABMET, which is linked to TABSET by the method number MENO, more than 50 measuring methods are stored. The current value 15 belongs to HOT WIRE, INSTAT., the transient hot–wire method.

In the most important auxiliary database table TABSUB, more than 2400 different substance names are stored together with two synonyms, Chemical–Abstracts number CANO and other characteristic information. The meaning of the columns can be deduced from the column names and from Table 17.3. TABSUB is connected with TABCON and TABSET by the substance number SUNO. In the example, the current value 2233 stands for Refrigerant 134a (HFC-134a).

17.1.2.4 Statistics

Statistical information about the MIDAS database is given in Table 17.1, split up into information on pure fluids, binary, ternary and quaternary mixtures. The database table TABREF contains 5100 publications. More than 16,500 data sets published in them are registered in TABCON. They all have to be entered in TABSET where up to now nearly 3900 data sets are already registered. These data sets consist of more than 280,000 points.

```
┌─────────────────────────────────────┐
│     Database table TABJOU            │
│                                      │
│     Column    Entry                  │
│     ─────────────────────────        │
│     JONO:     162                    │
│     JOURNAL:  INT.J.THERMOPHYSICS    │
└─────────────────────────────────────┘
```

```
┌─────────────────────────────────────┐
│     Database table TABMET            │
│                                      │
│     Column    Entry                  │
│     ─────────────────────────        │
│     MENO:     15                     │
│     METHOD:   HOT WIRE,INSTAT.       │
└─────────────────────────────────────┘
```

```
┌─────────────────────────────────────────┐
│       Database table TABSUB              │
│                                          │
│     Column    Entry                      │
│     ─────────────────────────────        │
│     SUNO:     2233                       │
│     NAME1:    Refrigerant 134a           │
│     NAME2:    1,1,1,2-Tetrafluoroethane  │
│     NAME3:    HFC-134a                    │
│     CANO:     811-97-2                    │
│     GROUP:    Alkane C2 (halogenated)    │
│     FORM:     C2H2F4                      │
│     MOLE:     102.032                     │
│     NBP:      247.05                      │
│     PCRIT:    40.65                       │
│     TCRIT:    374.274                     │
│     TTRIP:    172                         │
│     ODP:      0                           │
│     GWP:      420                         │
│     ALT:      16                          │
│     LFL:      none                        │
│     TOX:      1000                        │
│     FLAG:     1                           │
└─────────────────────────────────────────┘
```

Fig. 17.4. Auxiliary database tables.

Table 17.1. *Number of entries in the database tables.*

Table	total	pure	binary	ternary	quaternary
TABREF	5,100	2,678	1,522	255	40
TABCON	16,576	10,734	5,246	541	55
TABSET	3,879	1,721	2,021	122	15
TABDAT	283,389	197,935	79,146	5,809	499

17.1.3 Transport properties of pure fluids and mixtures

As already mentioned, the database project was initiated with the aim to collect and evaluate the viscosity and the thermal conductivity of pure fluids and mixtures. Up to now, the viscosity and thermal conductivity of about 50 pure substances have been evaluated based on a comparative study of experimental results, using graphical and

Table 17.2. *Publications on substance properties resulting from the MIDAS database project.*

Author(s)	Year	Properties	Substances
Stephan & Lucas	1979	η,λ	50 pure fluids
Stephan & Laesecke	1985	λ	air
Laesecke & Stephan	1986	η	water
Stephan et al.	1987	η,λ	N_2
Stephan & Heckenberger	1988	η,λ	30 binary mixtures
Krauss & Stephan	1989	λ	CFC-12/113/114,FC-C318
Laesecke et al.	1990	η,λ	O_2
Krauss	1991	phy	air,N_2,CO_2,NH_3
Krauss & Stephan	1992	phy	HFC-134a
Stephan & Krauss	1993	η,λ	HFC-134a
Krauss et al.	1993	η,λ	HFC-134a

η: viscosity; λ: thermal conductivity; phy: thermophysical.

interpolation methods at the beginning and the residual concept in recent years. The substances studied include the noble gases, alkanes up to C_{20}, alkenes, alcohols, aromatic hydrocarbons, some simple molecules like oxygen, nitrogen and carbon dioxide; air, water, ammonia, HFC-134a and other refrigerants are also being studied. Generally, these recommended data sets are comprehensive and describe the whole fluid pressure and temperature range where measurements were carried out.

Based on the data of the pure fluids, the viscosity and thermal conductivity of 30 binary mixtures of noble gases, simple molecules like oxygen, nitrogen and carbon dioxide, as well as alcohols and water were evaluated. Whenever possible, the evaluation was done for a wide range of pressure and temperature, although in many cases only measurements at atmospheric pressure were available. Mixing rules were established which describe the isotherms of the transport properties as a function of the mole fraction at atmospheric pressure and which need the transport properties of the pure components as input data.

Some publications resulting from the database project are compiled in Table 17.2, ordered by the year of publication.

17.1.4 Thermophysical properties of environmentally acceptable refrigerants and their mixtures

Chlorofluorocarbons (CFCs) are major contributors to the depletion of the ozone layer; additionally, they contribute to the greenhouse effect. Consequently the phaseout of CFCs and halons was scheduled and written down in the Montreal Protocol 1987, the amendments of the London Meeting 1990 and the Copenhagen Conference 1992, with the result that CFCs will be phased out by the end of 1995. Of course these restrictions led to a worldwide search for adequate environmentally acceptable replacements in all areas of application and to investigations of their properties (Stephan & Krauss 1993).

In view of these environmental problems the MIDAS database system was extended to thermophysical properties of environmentally acceptable refrigerants and their mixtures.

17.1.4.1 Substances

Several environmentally acceptable hydrofluorocarbons (HFCs) and hydrochlorofluorocarbons (HCFCs) have been identified as possible substitutes for the regulated CFCs. However, HCFCs can be regarded only as transitional replacements. Their phaseout is scheduled for the year 2030. Possible replacement fluids are compiled in Table 17.3, together with their environmental properties, toxicity and flammability. Substances belonging to categories A and B are of main interest for the database project because, besides the above mentioned properties, their thermodynamic properties also are promising.

Nonazeotropic or near–azeotropic mixtures may expand the number of possible replacements. They make it possible to 'tailor' special environmental or thermophysical properties. For instance, the combination of flammable and nonflammable fluids may yield a nonflammable mixture with even better thermophysical properties than the pure fluids. It is also conceivable to reduce the ODP and GWP by mixing pure fluids. Near–azeotropes, having only a small temperature variation during the phase change, are very promising in case a 'drop–in' substitute for existing equipment is required. On the other hand, the handling of mixtures is more complicated in case of leakage and recycling. In Table 17.4 promising mixtures to replace currently used CFCs are listed.

17.1.4.2 Properties

Important thermophysical properties for some of the most promising HFCs, HCFCs and ammonia are listed in Table 17.5, together with the number of publications registered in the MIDAS database. The properties of main interest are vapor pressure (including normal boiling point and critical point), P, V, T–data (including saturated liquid density), heat capacities, viscosity, thermal conductivity and surface tension. The substance studied most often is HCFC-22. More than 400 publications underline the importance of this working fluid in refrigeration. Among the new refrigerants, HFC-134a was investigated the most. As a result, there are many new and accurate thermodynamic data. In the field of transport properties the situation is less favorable. Besides these substances, the thermophysical properties of ammonia, HFC-23, HFC-152a, HCFC-123 and HCFC-142b have been investigated quite often. At present, investigations are carried out on the thermophysical properties of HFC-32.

Table 17.3. *Possible environmentally acceptable refrigerants (CFCs and CO_2 for comparison only).*

Refrigerant	Formula	ALT	ODP	GWP	LFL	TOX	CAT
ammonia	NH_3	<1	0	<<	15.0	25	A
propane	C_3H_8	<1	0	3	2.1	as	A
butane	C_4H_{10}	<1	0	3	1.5	800	A
isobutane	C_4H_{10}	<1	0	3	1.8	1000	A
HFC-23	CHF_3	310	0	12000	none	1000	A
HFC-32	CH_2F_2	6	0	220	14.6		A
HFC-125	C_2HF_5	28	0	860	none	1000	A
HFC-134	$C_2H_2F_4$	12	0		none		A
HFC-134a	$C_2H_2F_4$	16	0	420	none	1000	A
HFC-143a	$C_2H_3F_3$	41	0	1000	7.1		A
HFC-152a	$C_2H_4F_2$	2	0	47	3.7	1000	A
HFC-227	C_3HF_7	30	0	1100	none		A
E-134	$C_2H_2F_4O$	<12	0		none		A
HCFC-22	$CHClF_2$	15	0.05	510	none	1000	B
HCFC-123	$C_2HCl_2F_3$	2	0.02	29	none	10	B
HCFC-124	C_2HClF_4	7	0.02	150	none	500	B
HCFC-141b	$C_2H_3Cl_2F$	8	0.10	150	7.4	500	B
HCFC-142b	$C_2H_3ClF_2$	19	0.06	540	6.9	1000	B
CFC-11	CCl_3F	60	1	1500	none	1000	C
CFC-12	CCl_2F_2	130	0.9	4500	none	1000	C
CFC-13	$CClF_3$	400	0.45		none	1000	C
CFC-113	$C_2Cl_3F_3$	90	0.8	2100	none	1000	C
CFC-114	$C_2Cl_2F_4$	200	0.7	5500	none	1000	C
CFC-115	C_2ClF_5	400	0.35	7400	none	1000	C
carbon dioxide	CO_2	120	0	1	none	5000	

ALT: atmospheric lifetime
ODP: ozone depletion potential (rel. to CFC-11)
GWP: global warming potential (rel. to CO_2; integration time = 500 a)
LFL: lower flammability limit (volume % in dry air)
TOX: toxicity (threshold limit value in ppm; as = asphyxiant)
CAT: category
— A = environmentally acceptable replacements
— B = transitional replacements
— C = CFCs covered by the Montreal Protocol

Table 17.4. *Current refrigerants and alternative mixtures.*

Current refrigerant	Alternative mixtures
CFC-12	HCFC-22/HFC-152a/HCFC-124
	HCFC-22/HCFC-142b
HCFC-22	HFC-32/HFC-125
	HFC-32/HFC-134a
	HFC-32/HFC-152a
	HCFC-22/HFC-23
	HCFC-22/HCFC-142b
	above mixtures + HFC-143a
R-502	HCFC-22/propane/FC-218
	HCFC-22/propane/HFC-125

Table 17.5. *Number of publications on thermophysical properties of alternative refrigerants.*

Refrigerant	P_s	PVT	c_P	c_P^{id}	dyn	η	λ	σ	phy	total
HFC-23	10	48	9	5	15	35	28	1	8	139
HFC-32	8	10	2	2	5	13	14	0	5	48
HFC-125	3	5	1	1	5	3	2	1	6	20
HFC-134	2	6	0	0	1	0	0	2	1	10
HFC-134a	38	59	16	10	18	24	38	8	13	162
HFC-143a	4	5	3	3	4	4	4	1	4	22
HFC-152a	24	47	10	7	15	25	22	5	6	127
HFC-227	1	0	1	0	2	0	0	0	0	3
HCFC-22	50	133	19	9	62	115	100	3	16	433
HCFC-123	22	32	9	4	9	10	22	5	8	93
HCFC-124	4	6	1	1	1	2	7	0	3	24
HCFC-141b	8	9	4	4	2	3	9	2	3	31
HCFC-142b	12	28	3	1	7	10	19	2	7	72
Ammonia	5	27	3	2	19	70	66	1	3	161

P_s: vapor pressure; PVT: P, V, T–data, density, equation of state;
c_P: heat capacity; c_P^{id}: ideal–gas heat capacity;
dyn: miscellaneous thermodynamic properties;
η: viscosity; λ: thermal conductivity;
σ: surface tension; phy: miscellaneous thermophysical properties;
total: total amount of stored literature.

17.1.5 Access and dissemination of data

The MIDAS database system is installed on a DEC VAX Workstation 3200. On–line access is not feasible at present, but the data may be mailed by listing, 3.5– or 5.25–inch diskettes or electronic mail. Requests should be addressed to:

Universität Stuttgart, Institut für Technische Thermodynamik und Thermische Verfahrenstechnik (1305), 70550 Stuttgart
Tel.: +49.711.685.6103, Fax: +49.711.685.6140,
e–mail: krauss@itt.verfahrenstechnik.uni–stuttgart.d400.de

In 1995, the part of the database concerning the thermophysical properties of environmentally acceptable refrigerants became available commercially on diskette for use with WINDOWS on a PC.

Acknowledgments

The database project is financed by the German Bundesminister für Forschung und Technologie BMFT in cooperation with the Fachinformationszentrum FIZ in Karlsruhe (Contract No. 0329255A). We are indebted to the members of the Subcommittee on Transport Properties of Commission I.2 of the International Union of Pure and Applied Chemistry (IUPAC) and to the members of Annex 18 of the International Energy Agency (IEA) for providing the database with their data.

References

Krauss, R. (1991). Stoffwerte von Luft, Stickstoff, Kohlendioxid und Ammoniak. In *VDI–Wärmeatlas, Berechnungsblätter für den Wärmeübergang, 6., erweiterte Auflage*, pp. Db16–Db71. Düsseldorf: VDI–Verlag.

Krauss, R. & Stephan, K. (1989). Thermal conductivity of refrigerants in a wide range of temperature and pressure. *J. Phys. Chem. Ref. Data*, **18**, 43–76.

Krauss, R., Luettmer-Strathmann, J., Sengers, J.V. & Stephan, K. (1993). Transport properties of 1,1,1,2-tetrafluoroethane (R134a). *Int. J. Thermophys.*, **14**, 951–988.

Krauss, R. & Stephan, K. (1992). *Tables and Diagrams for the Refrigeration Industry. Thermodynamic and Physical Properties. R134a.* Paris: International Institute of Refrigeration.

Laesecke, A., Krauss, R., Stephan, K. & Wagner, W. (1990). Transport properties of oxygen. *J. Phys. Chem. Ref. Data*, **19**, 1089–1122.

Laesecke, A. & Stephan, K. (1986). Representation of the viscosity of water in terms of pressure and temperature. In *The Properties of Steam, Meeting of the International Association for the Properties of Steam, Moscow, 1984*, pp. 398–414. Moscow: Mir Publishers.

Laesecke, A., Stephan, K. & Krauss, R. (1986). The MIDAS data-bank system for the transport properties of fluids. *Int. J. Thermophys.*, **7**, 973–986.

Stephan, K. & Heckenberger, T. (1988). Thermal conductivity and viscosity data of fluid mixtures. *DECHEMA Chemistry Data Series, Volume X, Part 1*. Frankfurt am Main: DECHEMA.

Stephan, K. & Krauss, R. (1990). A database system for thermophysical properties of pure
    fluids and fluid mixtures. In *Thermophysical Properties of Pure Substances and Mixtures
    for Refrigeration, Meeting of the International Institute of Refrigeration, Commission
    B1, Herzlia (Israel)*, pp. 49–55. Paris: International Institute of Refrigeration.
Stephan, K. & Krauss, R. (1993). Regulated CFCs and their alternatives. *Heat Recovery
    Systems & CHP*, **13**, 373–381.
Stephan, K., Krauss, R. & Laesecke, A. (1987). Viscosity and thermal conductivity of
    nitrogen for a wide range of fluid states. *J. Phys. Chem. Ref. Data*, **16**, 993–1023.
Stephan, K. & Laesecke, A. (1985). The thermal conductivity of fluid air. *J. Phys. Chem.
    Ref. Data*, **14**, 227–234.
Stephan, K. & Lucas, K. (1979). *Viscosity of Dense Fluids*. New York: Plenum Press.

17.2 The PPDS system

A. I. JOHNS and J. T. R. WATSON

National Engineering Laboratory, East Kilbride, UK

17.2.1 Introduction

The actual values of the thermodynamic and transport properties of materials form a crucial part in a wide variety of process engineering design calculations. In general, the values of the thermodynamic variables will determine the feasibility of a chemical process. For example, the standard Gibbs free energy of reaction will tell the engineer if a given reaction is thermodynamically possible. In contrast to the thermodynamic properties, the transport properties usually give information on what size of equipment is required to perform a process task.

With any calculation, however, the engineer is faced with the question of what actual value of a specified property to use (assuming he or she can find a value). In many cases there is usually a paucity of data; furthermore, any existing data are probably at a set of conditions different from those required. There is also the question of how reliable the data are, and how any uncertainty in the value propagates through the design.

The Physical Properties Data Service (PPDS) system is a combination of calculational software and associated databases which has been produced to satisfy these common engineering calculation requirements outlined above. The databanks and associated software have been developed over many years by the NEL Executive Agency, based at East Kilbride, Scotland. The complete PPDS system covers the full set of thermo-dynamic properties and contains some safety information in addition to the transport properties. The following sections of this chapter describe the main elements of the system and their operation.

Figure 17.5 is a schematic diagram showing the relation between the software and the associated databanks in the PPDS system. In the diagram, the ellipses represent software executable programs and the rectangles databanks, which are accessed by the software. The system has five software packages, three of which are of primary importance: (a) PPDS–2 (*the main calculation package*), (b) LOADER–2 (*the pure component data estimation and fitting package*) and (c) PFIT–2 (*the binary interaction optimization package*). The remaining two software packages, called DDE–2 and MFIT–2, act as maintenance utilities for two of the databanks. All of the three main programs utilize one or more of the databanks, which are grouped together to form what is known as 'bank sets,' with each set composed of (a) PC–BANK, a pure component databank of recom-mended values of physical constants and correlation coefficients, (b) NAMES–BANK, a databank of names and aliases and (c) MP–BANK, a model–parameter databank which contains some pure component model–dependent characteristic parameters for the corresponding component in the pure component bank. The main bank set supplied

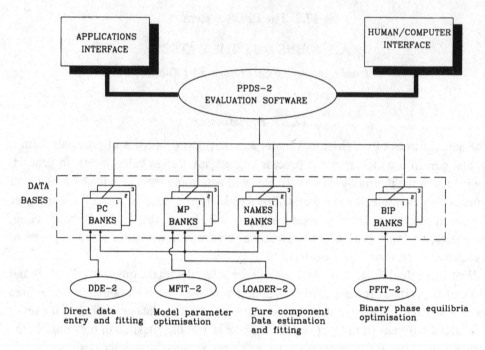

Fig. 17.5. An overview of the PPDS calculation system.

with the calculational software is the SYSTEM bank set, but this need not be the only one, as users can define their own banks to the software. The other databank which appears on Figure 17.5 but which has not yet been mentioned is the binary interaction parameter (BIP) databank. This databank stores optimization parameters which assist in making specific phase equilibria models represent the phase behavior of a given binary system more accurately. The PPDS suite has a standard BIP bank supplied, but up to five different BIP banks can be used by the software at any one time.

Thus the PPDS–2 program retrieves information from all the databanks just described to perform the various phase equilibria and property calculations. LOADER–2 writes information to the pure component, model parameter and names databanks, and PFIT–2 writes information to the BIP databank. The system maintains an internal check on which pure component banks were used to optimize a given BIP (all BIP optimization assumes some pure component data knowledge).

17.2.2 *The pure component databank*

The databanks can be regarded as the fuel for the PPDS–2 calculation engine. The most important bank in any bank set is the PC–BANK. It provides a set of physical constants and regression coefficients for the temperature–dependent properties. The PPDS SYSTEM bank has information for some 1500 pure components, in a form which

Table 17.6. *Chemical types stored in the PPDS system.*

Group	Group
Alkanes	Halo–alkanes
Alkenes	Halo–alkenes
Alkynes	Aromatics
Cycloalkanes	Halo–cycloalkanes
Alcohols	Esters
Glycols	Thiols
Aldehydes	Ketones
Carboxylic acids	Acid anhydrides
Ethers	Glycol ethers
Amines	Nitriles
Sulphides	Nitro–compounds
Oxygen heterocyclics	Hetrocyclic sulphur compounds
Silanes	Boron compounds
Inorganic fluids	Quantum fluids
Metals	Small molecules

enables mixture calculations to be carried out over a wide range of temperatures and pressures. Table 17.6 shows the chemical coverage of the bank which is composed of about 90% organic and 10% inorganic compounds. In addition to storing the regression coefficients, the databank has information on the temperature range over which each of the temperature–dependent property correlations are valid, together with a date stamp and eight–character quality code. These are expanded upon in Section 17.2.5.

17.2.3 Constants

The data constants which are stored for each compound are listed in Table 17.7. The stored constants shown in Table 17.7 have been divided into groups broadly based on the function of the properties they contain. The properties, generally speaking, do not need any amplification, except for SMILES, acentric factor of the homomorph, and solubility parameter.

The SMILES descriptor is a simple way of encoding the molecular structure of a molecule, including any chiral centers. The method uses a line notation, with the ordinary alphabetic character set that is commonly found on every computer keyboard and results in a simple string commonly less than 15 characters.

The acentric factor of the homomorph is defined as the acentric factor of the hydrocarbon molecule, which is obtained by replacing all functional groups of the target molecule with carbon atoms (including any noncarbon atoms in the main structure). This parameter is used in some estimation methods.

Table 17.7. *Characteristic physical constants stored in the pure component databank.*

Identification:	
Chemical name	CAS no. (Chemical Abstracts Registry Number)
Molecular formula	Relative molecular mass
SMILES (Simplified Molecular Input Line Entry System)	
Structure descriptor	
NEL chemical class identifier	NEL polarity code
Thermodynamic characteristic temperatures, pressures and density:	
Triple point temperature	Melting point
Triple point pressure	Normal boiling point
Critical temperature	Critical density and volume
Critical pressure	
Safety information:	
Flash point temperature	Lower flammability limit
Autoignition temperature	Upper flammability limit
Standard thermodynamic functions at 298.15 K and 1 atmosphere:	
Standard entropy (298.15 K, 1 atm)	Standard vapor Gibbs energy of formation
Standard vapor enthalpy of formation	Standard liquid Gibbs energy of formation
Standard liquid enthalpy of formation	
Other useful characterization parameters:	
Pitzer's acentric factor	Parachor
Acentric factor of the homomorph	Radius of gyration
Dipole moment	Solubility parameter

The solubility parameter, δ, is defined by the following equation:

$$\delta = \left(\frac{\Delta U}{V}\right)^{1/2} \tag{17.1}$$

where ΔU is the energy of vaporization, and V is the molar volume of the liquid both normally at 298.15 K.

17.2.4 Temperature–dependent properties

The temperature–dependent properties stored in the databank are listed in Table 17.8. They are segregated into three groups: an ideal–gas (or essentially zero–density) set for the gas phase, a saturated liquid set and, last, a change of phase set containing the enthalpy and entropy of vaporization.

The representative equation forms chosen to fit these properties have been carefully selected to give realistic behavior over the whole fluid range. Further, where possible,

Table 17.8. *Temperature–dependent properties stored in the databank.*

Zero–density or ideal–gas properties	Saturated liquid properties
Isobaric specific heat capacity	Saturated vapor pressure
Enthalpy	Surface tension
Entropy	Isobaric specific heat capacity
Thermal conductivity	Density
Viscosity	Enthalpy
	Entropy
	Thermal conductivity
	Viscosity
Change of state properties	
Enthalpy of vaporization	
Entropy of vaporization	

the equation forms have been constrained to allow sensible extrapolation outside the physical fluid limits. This feature is especially important when calculating the properties of mixtures, for example, where a mixture may be a liquid above the critical point of one or more of its components.

The temperature ranges over which the databank correlations are valid are, in general, from 0 to 3000 K for ideal properties and from the melting temperature to the critical temperature, T_c, for liquid properties, except for the case of liquid specific heat capacity, where an upper limit of $0.98T_c$ applies. The temperature limits over which a correlation is valid are also stored as part of the data record for the property.

17.2.5 Quality coding system

The range of components present in the databank is reasonably wide; also, because of the variability of the quantity of experimental data available, the accuracy of the information stored varies from property to property and also from compound to compound. To be useful to the practicing scientist, it is therefore necessary to give some tolerance information on the data that the equations reproduce. This is done by assigning an eight–character quality code to each constant and to each property coefficient set for every compound. Within the quality code, each position has a range of character codes which signify various quality attributes concerned with the data or coefficients.

In this way, any piece of data retrieved from the bank will have a tolerance indicator which will show the uncertainty band in the data and also how the data were obtained. Figure 17.6 shows how the quality code is assembled. It is shown as a set of eight blocks forming a character string, with each block in the string having a specific meaning

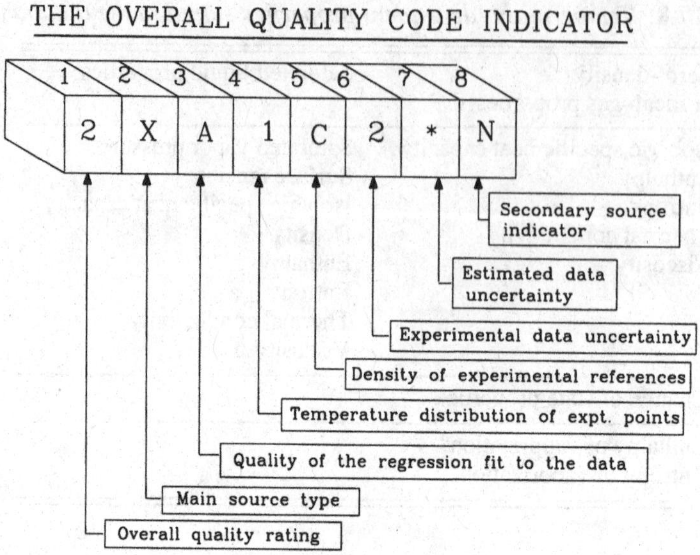

THE OVERALL QUALITY CODE INDICATOR

Fig. 17.6. Databank quality code.

and containing a character code. In the example, '2XA1C2*N' represents a typical entry, which can be easily decoded by the user. Thus, the user would know that this particular string means that: (a) the overall uncertainty of the data to which the string was attached was 0.1–1.0%; (b) the data were derived from measurements which gave (c) an excellent fit to the equation form selected; further, (d) experimental data were available over the full liquid range, (e) there were between five and ten separate primary sources of experimental data and finally (f) no secondary sources or estimated data were used.

17.2.6 *Mixture and pressure corrected property calculations*

The pure component databank can only return data for the saturated liquid, some changes of state properties and ideal–gas properties for a given substance. To obtain a wider range of data for mixtures, including values at pressure, the PPDS system uses a computer program to access the databank and perform additional calculations. This property calculation program, called PPDS–2, is shown as the main element in the PPDS computer suite in Figure 17.5. The mechanism by which the program carries out these calculations is conceptually a two–stage process where the pure component properties are first mixed according to a specified prescription and then a pressure correction, or sometimes a density correction, depending on the property, is applied. This approach means that, in practice, it must be possible to extrapolate the pure component properties in order to provide sensible pseudo–values for mixture calculations. The PPDS system has such extrapolations built into the mixture calculations, and the user is always informed if an extrapolation has occurred.

17.2.6.1 *Pressure corrections*

The pure component databank only stores correlation coefficients for the ideal–gas or zero–density temperature dependency. In the vapor phase, properties are corrected by means of generalized equations. In the case of the thermodynamic properties, the equation of state developed by Lee & Kesler (1975) is employed. For the transport properties the correlation of Stiel & Thodos (1964a,b) is used for thermal conductivity and that of Jossi *et al.* (1962) for viscosity. Both transport property corrections employ mechanisms for differentiating between polar and nonpolar streams.

17.2.6.2 *Mixing rules*

The calculation of the properties of mixtures is usually the end goal for thermophysical calculations in engineering design, since it is unusual for a process stream to contain a pure component. The approach taken by the PPDS software depends on which phase is under consideration and which property. In the vapor phase case, the mixing is performed by applying a mixing rule to the ideal–gas equation form and then applying a pressure or density correction, obtained by considering the mixture as a hypothetical pure fluid, using the Lee and Kesler (1975) equation of state.

For the liquid phase properties, the system uses mole–fraction averages of the pure components to obtain the thermodynamic properties and has separate procedures developed at the NEL Executive Agency for evaluating the thermal conductivity and viscosity.

17.2.7 *Phase equilibria calculations*

A very important set of calculations in process design is the determination of the phase of a mixture. Despite the fact that the thermodynamic conditions for the coexistence of more than one phase can be formally written down exactly, the models which are used to calculate the compositions of any phases present are not so precise. Furthermore, there is no general model which can be applied in all cases. This being so, most calculation packages abdicate responsibility to some extent by offering the engineer a variety of possible methods and letting him or her decide which to use. The PPDS system is no exception to this. The following sections give a summary of the calculations that the PPDS offers and also the phase equilibrium models available.

17.2.7.1 Calculation types

The software package can perform the phase equilibria calculations summarized in Table 17.9 for systems containing up to four phases:

Table 17.9. *Possible phase equilibria calculations.*

Calculation type	Calculation type
Fixed T, P flash	
Bubble point at defined T	Bubble point at defined P
Dew point at defined T	Dew point at defined P
Isenthalpic flash	Isentropic flash
Isochoric flash	Phase envelope
Retrograde bubble point at defined T	Retrograde bubble point at defined P
Constant phase fraction	

17.2.7.2 Phase equilibria models

The PPDS system encompasses a wide range of phase equilibria models which are routinely used for engineering design work, including both equation of state models and the so–called activity models. Table 17.10 shows the range available for the user to choose from.

Table 17.10. *Phase equilibria models implemented.*

Model	
NRTL	Wilson original
Wilson–A	UNIFAC
UNIQUAC	Lee, Kesler, Plocker EOS
Peng–Robinson EOS	Redlich, Kwong, Soave EOS
Henry's Law	

17.2.8 Special packages

The PPDS system also provides for several special packages for treating fluids which either have a more accurate representation or which cannot be treated by the methods described above. There are four of these packages, which are detailed in the following sections.

17.2.8.1 Water substance

The importance of water in process engineering and power generation is such that its properties are probably the most well known of any substance. PPDS has coded the formulation of 1984 recommended by the International Association for the Properties of Steam (IAPS) for the thermodynamic and transport properties of water substance. This package returns values for pure water over a wide range of temperature and pressure.

17.2.8.2 IUPAC equation of state package

There is also another small group of important fluids for which there is a large body of experimental data. IUPAC has set up the Thermodynamic Tables Project Centre at Imperial College, London, UK, which has published a number of monographs setting out recommended values for some 14 fluids. These formulations have been internationally approved and are accompanied by recommended formulations, which have been encoded by the IUPAC Thermodynamics Tables Project Centre and are included in the PPDS system. Table 17.11 gives the names of the fluids covered by this package.

Table 17.11. *Substances included in the IUPAC EOS package.*

Fluid			
Helium-4	Methane	Hydrogen	Methanol
Argon	Ethane	Carbon dioxide	Ammonia
Air	Ethene	Chlorine	R123
Nitrogen	Propane	Fluorine	R134a
Oxygen	Propene		

17.2.8.3 Refrigerant package

The PPDS refrigerant package evaluates thermodynamic and transport properties of selected refrigerants and is based on the recommendations produced by the International Institute of Refrigeration (IIR). Table 17.12 lists the refrigerants covered.

Table 17.12. *Substances included in the PPDS refrigerant package.*

Fluid	
Ammonia	1,1,2-Trichlorotrifluoroethane (R113)
Chlorodifluoromethane (R22)	Bromotrifluoromethane (R13b1)
Dichlorodifluoromethane (R12)	Tetrafluoromethane (R14)
Trichlorofluoromethane (R11)	1,2-Dichlorotetrafluoroethane (R114)

17.2.8.4 Petroleum fractions

There are some classes of mixture for which the standard approach used by the PPDS does not provide good answers. An important class of such mixtures are the so–called petroleum fractions, which are encountered in the processing of crude oil. Here, the various product cuts obtained by distillation and cracking of the crude are, in fact, complex mixtures of many components in unspecified concentrations. An alternative way of treating these mixtures is to use the concept of a pseudo–component, which has certain characterization factors associated with it. This is the mechanism developed by the PPDS system. It allows the properties of these mixtures to be calculated and mixed with the other components available in the main pure component databank. The fraction is usually characterized by specifying some average boiling temperature and a specific gravity or, increasingly more commonly, by specifying temperatures at which a specified amount of a sample has been recovered from a distillation carried out under standard conditions (for example, by carrying out a simulated distillation according to the American Standards for the Testing of Materials method, ASTM–D86).

17.2.9 Typical output for property calculation

The previous sections have given a very brief overview of the PPDS–2 calculation program. The examples given in the Appendix illustrate typical outputs obtained from the system. Example 1 shows the calculated pure component properties of tert–amyl methyl ether, which is commonly used as a lead–free additive to petrol to increase its octane number. Example 2 gives the results of a phase envelope calculation, illustrated in Figure 17.7, for a liquified natural gas mixture, typical of those which are pumped ashore from the North Sea. It contains a large amount of dissolved carbon dioxide. These examples have only scratched the surface of the available calculations which, due to the lack of space, are not possible to review here.

17.2.10 Data prediction package for pure components

No matter what size of pure component databank is available to the PPDS system, there is always the case that information is required for a substance which is not present. Usually there is not very much experimental information available for the component under investigation. The PPDS system has a powerful utility package called LOADER–2 which allows a data record to be created for any component and subsequently used by the main program. The following sections describe the principles behind the program.

17.2.10.1 Principle of operation

The principle of the program is to provide a complete data record in PPDS system format for the compound under investigation. The program has the ability to incorporate any experimental data for each property and/or to estimate data where no experimental information is available.

17.2.10.2 Property hierarchy

Estimation methods that have been developed for the various physical properties and which are reported in the open literature usually do not take account of the exact relations

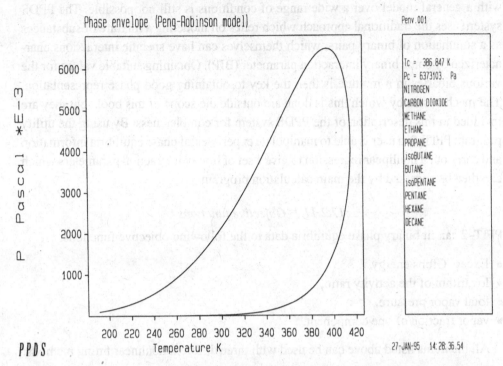

Fig. 17.7. Phase envelope plot for a liquified petroleum gas mixture (Example 2 of Appendix).

that are imposed by thermodynamic theory. The PPDS LOADER package improves upon the estimation of properties by utilizing a hierarchy of estimation methods which are used in conjunction with any experimental data entered by the user. The estimation method most suited to a given compound is determined from two simple characterization parameters – a chemical class code and a polarity code.

17.2.10.3 Thermodynamic consistency

In addition to estimating individual properties, LOADER–2 checks whenever possible that the thermodynamic consistency between properties is maintained. For example, the liquid heat capacity will be compatible with the enthalpy of vaporization and liquid enthalpy. The ideal–gas thermal conductivity will be compatible with the ideal–gas specific heat capacity. By fitting well–behaved representative equations to each property, the program produces a self–contained set of thermophysical property data, which can be used with the main PPDS system.

17.2.11 Binary interaction parameter optimization

The problem of modeling phase equilibria is one which has occupied a large volume of the scientific literature over many years. Even today, representing a specific system with a general model over a wide range of conditions is still not possible. The PPDS system uses the traditional approach which relies on modeling a mixture of substances as a summation of binary pairs, which themselves can have specific interactions characterized by some binary interaction parameter (BIP). Obtaining suitable values for the various binaries in a mixture is then the key to obtaining good phase representation. The mechanisms by which this is done are outside the scope of this book, but they are included in this description of the PPDS system for completeness. By using the utility program PFIT–2, a user is able to manipulate experimental phase equilibria information and carry out a nonlinear regression to give a set of binary interaction parameters which can then be accessed by the main calculation program.

17.2.11.1 *Objective functions*

PFIT–2 can fit binary phase equilibria data to the following objective functions:

- Excess Gibbs energy,
- logarithm of the activity ratio,
- total vapor pressure,
- vapor fraction of one component.

All the items listed above can be used with three different nonlinear fitting methods:

- Simplex,
- Marquardt and
- a Quasi–Newton method.

17.2.11.2 *Thermodynamic models*

The resulting BIPs can be obtained for all the thermodynamic phase equilibia models, which the main calculation program can handle.

17.2.12 *Conclusion*

The PPDS calculational system has the full range of capabilities for performing thermophysical property calculations on both pure components and mixtures of fluids. The system provides reliable estimates to the user of both the data values and the associated tolerances involved in carrying out these calculations and, where applicable, gives the user the opportunity to use specific high–accuracy information.

References

Jossi, J.A., Stiel, L.I. & Thodos, G. (1962). The viscosity of pure substances in the dense gaseous and the liquid phases. *AIChE J.*, **8**, 59–63.

Lee, B.I., & Kesler, M.G. (1975). A generalised thermodynamic correlation based on three–parameter corresponding states. *AIChE J.*, **21**, 510–527.

Stiel, L.I. & Thodos, G. (1964a). The thermal conductivity of non–polar substances in the dense gaseous and liquid regions. *AIChE J.*, **10**, 26–30.

Stiel, L.I. & Thodos, G. (1964b). The viscosity of polar substances in the dense gaseous and liquid regions. *AIChE J.*, **10**, 275–277.

Appendix – Examples

Example 1: *Pure component data for tert-amyl methyl ether*

```
-------------------------------------------------------------------
      P P D S 2    version 3.0M              02-JUN-94   20:48:04.31

run number    4
Title:-
       Example_1


Compound    tertAMYL METHYL ETHER
User input 905
Formula     C6H14O1
CARN        [000994-05-8]
SMILES      CCC(C)(C)OC
Code number      905
Bank number      1
  Molecular Weight          102.177
  Melting point                 N/A K
  Boiling point             359.260 K
  Critical temperature      531.200 K
  Critical pressure       0.325300E+07 Pascal
  Critical volume         0.367647E-02 m**3/kg
  Critical density          272.000 kg/m**3
  Acentric factor           0.346700
  Homomorph ac. factor      0.267000
  Dipole moment                 N/A debye
  Parachor                  290.782 ((d/m)**.25)m**3/mol
  Solubility parameter      16307.1 sqrt(J/m**3)
  Entropy at 298K           3.89520 J/g.K
  Hliq of formation        -3277.65 J/g
  Hvap of formation        -2923.36 J/g
  Flash point                   N/A K
  Autoignition temp.            N/A K
  Lower flash limit %           N/A
  Upper flash limit %           N/A
  Gliq of formation        -1058.95 J/g
  Gvap of formation        -1017.84 J/g


-------------------------------------------------------------------
```

```
Saturation line properties
Temperature  =        40.0000 C
Pressure     =        19538.0 Pascal
Source and Quality code (V-B-3M)
```

Property		VAPOUR		LIQUID	
SVP	Pascal	19538.0	(V-B-3M)	19538.0	(V-B-3M)
Rho	kg/m**3	0.777711	(V-B-4E)	750.380	(V-B-3M)
Cp	J/g.K	1.62513	(V-B-4E)	2.23369	(V-B-3C)
Cv	J/g.K	1.53774	(V-B-4E)	N/A	
Cp/Cv		1.05682	(V-B-4E)	N/A	
H	J/g	22.4507	(V-B-4E)	-339.516	(V-B-4E)
S	J/g.K	0.208545	(V-B-4E)	-0.947345	(V-B-4E)
Del-H	J/g	361.967	(V-B-5M)	361.967	(V-B-5M)
Del-S	J/g.K	1.15589	(V-B-5M)	1.15589	(V-B-5M)
Gibbs	J/g	-42.8553	(V-B-4E)	-42.8553	(V-B-4E)
Hform	J/g	-2900.91	(V-B-4E)	-3262.88	(V-B-4E)
Eta	N s/m**2	0.720818E-05	(V-B-5E)	0.344752E-03	(V-B-6E)
Lambda	W/m.K	0.155736E-01	(V-B-5E)	0.116186	(V-B-5E)
Expans	/K	N/A		0.137094E-02	(V-B-3M)
Surf	N/m	N/A		0.207263E-01	(V-B-3M)
Z		0.985893	(V-B-4E)	0.102180E-02	(V-B-3M)

Example 2: *Phase envelope calculation for a natural liquid petroleum gas mixture*

```
---------------------------------------------------------------------
        P P D S 2    version 3.0M              03-JUN-94   08:45:34.75

run number    4
Title:-       Example_2

Internal   Bank Name                                  User input name
 number

 1 NITROGEN                                           n2
 2 CARBON DIOXIDE                                     co2
 3 METHANE                                            methane
 4 ETHANE                                             ethane
 5 PROPANE                                            propane
 6 isoBUTANE                                          isobutane
 7 BUTANE                                             butane
 8 isoPENTANE                                         isopentane
 9 PENTANE                                            pentane
10 HEXANE                                             hexane
11 DECANE                                             decane
```

Int	Code	Bank Formula numbers	CARN	% mole input	Composition mole-fraction *NORMALISED*
1	51	1 N2	[007727-37-9]	0.000000	0.000000
2	77	1 C1O2	[000124-38-9]	1.42900	0.142900E-01
3	41	1 C1H4	[000074-82-8]	0.644400	0.644401E-02
4	30	1 C2H6	[000074-84-0]	40.8215	0.408215
5	57	1 C3H8	[000074-98-6]	36.0250	0.360250
6	48	1 C4H10	[000075-28-5]	3.63440	0.363440E-01
7	8	1 C4H10	[000106-97-8]	9.28000	0.928001E-01
8	42	1 C5H12	[000078-78-4]	1.69660	0.169660E-01
9	56	1 C5H12	[000109-66-0]	2.34730	0.234730E-01
10	37	1 C6H14	[000110-54-3]	2.03910	0.203910E-01
11	20	1 C10H22	[000124-18-5]	2.08260	0.208260E-01

```
                     Total flow   =    100.000        % mole
                     Total moles  =    1.00000
              Molecular weight =     44.0379
```

```
------------------------------------------------------------------
MODEL           PR (Peng-Robinson) EOS        (Vapour phase)
                PR (Peng-Robinson) EOS        (Liquid phase)

BULK  CALCULATION    PHASE ENVELOPE       DEFINING PRESSURE
KEY COMPOUND         CARBON DIOXIDE

      TEMPERATURE          PRESSURE          STABILITY ANALYSIS
          K                Pascal

          144.3990              10000.0       Stable
          150.3686              14049.7       Stable
          163.2757              27940.1       Stable
          179.0870              59743.2       Stable
          198.7525              136610.       Stable
          223.3623              326238.       Stable
          253.8027              779419.       Stable
          289.9105          0.175416E+07      Stable
          327.8998          0.339132E+07      Stable
          340.9070          0.408441E+07      Stable
          373.9775          0.589032E+07      Stable
          394.6044          0.647622E+07      Stable
          397.0766          0.646348E+07      Stable
          413.5172          0.514235E+07      Stable
          410.6554          0.260396E+07      Stable
          397.7016          0.137099E+07      Stable
          373.2953              488590.       Stable
          366.5846              366604.       Stable
          344.9674              137266.       Stable
          322.2690              42491.1       Stable
          300.9046              11707.6       Marginally unstable
          274.8143              1751.09       Stable
          254.4150              286.574       Marginally unstable
          235.5656              38.7835       Marginally unstable
          219.6357              5.24878       Marginally unstable
          205.9543              0.710345      Marginally unstable
          194.0504          0.961348E-01      Indeterminate

      TEMPERATURE          PRESSURE             SPECIAL POINT
          K                Pascal

          386.8494          0.637279E+07      Critical point
          415.2317          0.418129E+07      Maximum Temperature
          394.6044          0.647622E+07      Maximum Pressure
```

```
P P D S 2    version 3.0M          Program finish 03-JUN-94   09:40:43.68
```

17.3 Imperial College Thermophysical Properties Data Centre

K. M. de REUCK, R. J. B. CRAVEN

and

A. E. ELHASSAN

Imperial College Thermophysical Properties Data Centre,
Imperial College, London, UK

17.3.1 Introduction

The Imperial College Thermophysical Properties Data Centre incorporates the International Union of Pure and Applied Chemistry (IUPAC) Thermodynamic Tables Project Centre and the IUPAC Transport Properties Project Centre. It publishes both books and scientific papers which contain extensive tables for each correlated fluid and give all the necessary equations for calculating relevant properties; accuracy limits are given for each property. It prepares computer packages available worldwide which calculate the fluid thermophysical properties from the correlating equations. The center also writes special programs, on a consultancy basis, for individual industrial problems. The center maintains a computerized on–line bibliographic database which is tailored specifically to the thermophysical properties of fluids. This is a specialized collection of citations maintained for the last 30 years by the IUPAC Thermodynamic Tables Project Centre. It now contains some 20,000 references and is updated regularly. It may be used by any microcomputer capable of running dBaseIII+ and upward software.

17.3.2 Computer packages

EOSPAC and EOSMIX are two computer packages written by the center. They are available as part of the main Physical Property Data Service (PPDS) software package developed by the National Engineering Laboratory Executive Agency, East Kilbride, Glasgow, UK, or can be purchased as stand-alone packages. Currently, EOSPAC contains accurate equations of state for 19 pure fluids, from which all the equilibrium thermodynamic properties can be calculated. Transport properties for most of them are also available. EOSMIX calculates the thermodynamic and transport mixture properties of the pure fluids available in EOSPAC. It employs the accurate two-fluid corresponding–states model with shape factors, which is augmented by optimized mixing parameters.

17.3.2.1 EOSPAC

The development of this package began more than 10 years ago; it is continually updated and improved, and new equations of state are added as they become available. The majority of the equations of state in the package have been internationally validated by

IUPAC and have been published in the series of International Thermodynamic Tables of the Fluid State. These are:

- Carbon Dioxide (Angus *et al*. 1976)
- Ethylene (Jacobsen *et al*. 1988)
- Fluorine (de Reuck 1990)
- Helium-4 (Angus *et al*. 1977)
- Methanol (de Reuck & Craven 1993)
- Nitrogen (Angus *et al*. 1979)
- Oxygen (Wagner & de Reuck 1987)
- Propylene (Angus *et al*. 1980)

Each equation of state has been fitted to an equation in the form of a dimensionless Helmholtz energy function using multiproperty fitting of the most accurate experimental data. The Helmholtz function ensures that all the properties are thermodynamically consistent and enables the calculations to be performed by differentiation alone. The properties which can be calculated are:

   (i) compression factor
  (ii) density
 (iii) isobaric coefficient of cubical expansion
 (iv) enthalpy
  (v) enthalpy of evaporation
 (vi) entropy
 (vii) entropy of evaporation
(viii) Gibbs energy
 (ix) heat capacity at constant pressure
  (x) heat capacity at constant volume
 (xi) pressure
 (xii) temperature
(xiii) thermal conductivity
(xiv) viscosity
 (xv) speed of sound
(xvi) isothermal Joule-Thomson coefficient
(xvii) ratio of heat capacities

From within the main PPDS package, for specified values of P and T, properties (i)–(xii) can be retrieved and values for properties (xiii) and (xiv), the viscosity and thermal conductivity, are available for most of the fluids; these transport properties are thermodynamically consistent with the equilibrium properties because the same density is used for the calculation of both. Individual equations of state can be purchased separately from which all the properties listed above can be calculated.

Published equations of state with accuracy comparable to those which have received international validation have been selected by the IUPAC Thermodynamic Tables Project Centre and are included in the packages. These include:

- Air as a pure fluid (Sytchev *et al.* 1978; Jacobsen *et al.* 1992)
- Ammonia (Haar & Gallagher 1978)
- Argon (Bender 1970)
- Chlorine (Angus *et al.* 1985)
- Ethane (Friend *et al.* 1991)
- Hydrogen (McCarty 1974)
- Methane (Setzmann & Wagner 1991)
- Propane (Bühner *et al.* 1981)
- R123 (Younglove, B. A., personal communication)
- R134a (Tillner-Roth & Baehr 1994)
- Water (Saul & Wagner 1989)

17.3.2.2 EOSMIX

EOSMIX is the most recent extension of the EOSPAC package where the very accurate equations of state for the pure fluids are used to predict the properties of selected mixtures using the 'two-fluid' corresponding–states model first proposed by Rowlinson & Watson (1969). In this approach each component of the mixture is replaced by a hypothetical substance, and the hypothetical components are then mixed ideally. Following the corresponding–states principle, each hypothetical substance can then, in turn, be expressed in terms of a reference substance. It follows that each component in the mixture can have a separate reference substance, and each component can be its own reference. The model is implemented by using a modified form of the van der Waals two-fluid mixing rules of Rowlinson & Watson together with the shape factors by Leach *et al.* (1968).

The same properties which can be calculated for pure fluids can be calculated by EOSMIX for mixtures consisting of up to ten components. Phase equilibrium properties are also calculated, and stability tests are performed. At present, the number of mixtures for which the mixing parameters have been optimized is limited, but additional optimizations are being added annually.

Future extensions will include the implementation of corresponding–states prediction of pure fluids so as to widen the choice of mixture components beyond those which are currently available.

References

Angus, S., Armstrong, B. & de Reuck, K. M. (1976). *International Thermodynamic Tables of the Fluid State–volume 3–Carbon Dioxide*. Oxford: Pergamon Press.

Angus, S., Armstrong, B. & de Reuck, K. M. (1977). *International Thermodynamic Tables of the Fluid State–volume 4–Helium*. Oxford: Pergamon Press.

Angus, S., Armstrong, B. & de Reuck, K. M. (1980). *International Thermodynamic Tables of the Fluid State–volume 7–Propylene*. Oxford: Pergamon Press.

Angus, S., Armstrong, B. & de Reuck, K. M. (1985). *International Thermodynamic Tables of the Fluid State–volume–8–Chlorine–Tentative Tables*. Oxford: Pergamon Press.

Angus, S., de Reuck, K. M. & Armstrong, B. (1979) *International Thermodynamic Tables of the Fluid State–volume 6–Nitrogen*. Oxford: Pergamon Press.

Bender, E. (1970). Equations of state exactly representing the phase behavior of pure substances. *Proceedings of 5th. Symposium on Thermophysical Properties*. New York: ASME.

Bühner, K., Maurer, G. & Bender, E. (1981). Pressure–enthalpy diagrams for methane, ethane, propane, ethylene and propylene. *Cryogenics*, **21**, 157–164.

de Reuck, K. M. (1990). *International Thermodynamic Tables of the Fluid State–volume 11–Fluorine*. Oxford: Blackwell Scientific Publications.

de Reuck, K. M. & Craven, R. J. B. (1993). *International Thermodynamic Tables of the Fluid State–volume 12–Methanol*. Oxford: Blackwell Scientific Publications.

Friend, D. G., Ingham, H. & Ely, J. F. (1991). Thermophysical properties of ethane. *J. Phys. Chem. Ref. Data*, **20**, 275–347.

Haar, L. & Gallagher, J.S. (1978) Thermodynamic properties of ammonia. *J. Phys. Chem. Ref. Data*, **7**, 635–792.

Jacobsen, R. T., Jahangiri, M., Stewart, R. B., McCarty, R. D., Levelt Sengers, J. M. H., White, H. J., Sengers, J. V. & Olchowy, G. A. (1988). *International Thermodynamic Tables of the Fluid State–volume 10–Ethylene*. Oxford: Blackwell Scientific Publications.

Jacobsen, R.T., Penoncello, S.G., Beyerlein, S.W., Clarke, W.P. & Lemmon, E.W. (1992). A thermodynamic property formulation for air. *Fluid Phase Equilib.*, **79**, 113–124.

Leach, J.W., Chappelear, P.S. & Leland, T.W. (1968). Use of molecular shape factors in vapor-liquid equilibrium calculations with the corresponding states principle. *AIChE J.*, **14**, 568–576.

McCarty, R. D. (1974). *A Modified Benedict-Webb-Rubin Equation of State for Parahydrogen*, NBSIR, 74–357.

Rowlinson, J.S. & Watson, I.D. (1969). The prediction of the thermodynamic properties of fluids and fluid mixtures I. The principle of corresponding states and its extensions. *Chem. Eng. Sci.*, **24**, 1565–1574.

Saul, A. & Wagner, W. (1989). A fundamental equation for water covering the range from the melting line to 1273 K at pressures up to 25 000 MPa. *J. Phys. Chem. Ref. Data*, **18**, 1537–1564.

Setzmann, U. & Wagner, W. (1991). A new equation of state and tables of thermodynamic properties for methane covering the range from the melting line to 625 K at pressures up to 1000 MPa. *J. Phys. Chem. Ref. Data*, **20**, 1061–1155.

Sytchev, V. V., Vasserman, A. A., Kozlov, A. D., Spiridonov, G. A. & Tsymarny, V. A. (1978) *Thermodynamic Properties of Air*. Washington, Berlin: Hemisphere-Springer Verlag.

Tillner-Roth, R. & Baehr, H. D. (1994). An international standard formulation for the thermodynamic properties of 1,1,1,2-tetrafluoroethane (HFC-134a) for temperatures from 170 K to 455 K at pressures up to 70 MPa. *J. Phys. Chem. Ref. Data*, **23**, 657–729.

Wagner, W. & de Reuck, K. M. (1987). *International Thermodynamic Tables of the Fluid State–volume 9–Oxygen*. Oxford: Blackwell Scientific Publications.

17.4 Thermodynamics Research Center Databases

K. N. MARSH and R. C. WILHOIT

Thermodynamics Research Center,
The Texas A&M University System, College Station, TX, USA

17.4.1 Introduction

For nearly 50 years the Thermodynamics Research Center and its immediate prede-
cessors, the API Research Project 44 and Manufacturing Chemists Association Data
Project, have selected and published recommended values of the thermodynamic and
physical properties of organic and nonmetallic inorganic compounds. The evaluated
data are published as the *TRC Thermodynamic Tables – Hydrocarbons and Non–*
Hydrocarbons with four updating supplements per year. As well as thermodynamic
data, the tables contain recommended viscosity and (recently) thermal conductivity
data for both gases and liquids over a temperature and pressure range. The selection
of data is a multistep process. It starts with a search of the world's scientific literature
and includes the extraction and storage of pertinent data, the estimation of experimental uncertainties, the comparison of related data from different sources, the selection
of the most accurate values and the publication of the results in some suitable form.
The selection process accommodates all thermodynamic constraints in order to achieve
internal consistency. For much of this period, publication has been in the form of tables
printed in loose–leaf format. In recent years an increasing emphasis has been placed on
computer–readable forms of publication.

Because of the continual appearance of new experimental data, the selection is a
never–ending process. It requires not only the consideration of new compounds, properties and ranges of variables, but also a reanalysis of all existing data and a revision of
previous selections in the light of new information. In order to keep abreast of the accumulation of published data, the TRC has adopted a series of procedures to streamline
and automate portions of the selection process. A major element was the installation of
two large computer–resident databases containing numerical values of thermophysical
properties. The TRC SOURCE Database contains directly measured values of thermophysical properties of pure compounds and mixtures along with associated information.
The goal is to accumulate all data relevant to TRC projects including the TRC Thermodynamic Tables, International Data Series, Series A, Selected Data on Mixtures and
the DIPPR 882 Project. The TRC THERMODYNAMIC TABLES Database (1994) is
a collection of the selected values in the current version of the TRC Thermodynamic
Tables – Hydrocarbons and TRC Thermodynamic Tables – Non–Hydrocarbons in electronic form. At present it does not contain the references available in the hard copy.
The SOURCE Database supplies the input data required in the selection process for
preparing the TRC THERMODYNAMIC TABLES Database and the printed publication. The DIPPR Project 882 is sponsored by the Design Institute for Physical Property

Data of the American Institute of Chemical Engineers. The primary aim of that project is to develop and make available to the sponsors in electronic form evaluated data on the viscosity, thermal conductivity, diffusivity, surface tension, density, excess volume and critical temperature, pressure and volume of binary mixtures, with the emphasis on classes of organic mixtures selected by the sponsors.

17.4.2 TRC SOURCE Database

The TRC SOURCE Database serves as a permanent multipurpose archive for recording both factual and numerical values of equilibrium thermodynamic, thermochemical and transport properties primarily of organic compounds and their binary, ternary and multicomponent mixtures.

References to molecular structure and energies as derived from spectroscopy, electron diffraction and X–ray diffraction and theoretical calculation are included. They are indexed by compound, but the numerical data are not included. These references form the basis for the work on statistical–mechanical calculations of ideal–gas functions. References to publications on theoretical calculations and correlations, molecular dynamics, equations of state, compilations, error analysis and experimental techniques are also included and can be searched by topic area. It is possible to extract from the database recent publications containing experimental data selected on the basis of the properties measured.

The TRC SOURCE Database furnishes a mechanism for permanently storing experimental values of properties which can be used for comparing new results with previous measurements, as the starting point for evaluation and selection of property values, and for the testing of theories and correlations. Sufficient information is included to permit interpretation and evaluation of the numbers and to identify the sources of origin of the values.

While the TRC SOURCE Database stores numerical values of properties, the numbers alone are not meaningful. The numbers must be associated with sufficient descriptive information called "metadata" to give them meaning. The contents of the database are contained in about 45 kinds of records, and each kind of record can contain several items of information. Some of the items serve as index keys for the rapid location and retrieval of the corresponding records.

Some data consist of single numbers only. Examples are critical constants and temperature and pressure for three phase equilibria of pure substances (triple points). Some kinds of data are single numbers by definition, such as normal boiling temperatures and standard enthalpies of formation at 298.15 K. Other data are functions of one or more independent variables. In general, the number of independent variables for a system is determined by the Gibbs phase rule. The database structure must allow for the maintenance of information on the phase or phases present, any constraints and other essential metadata, as discussed below. The distinction between a property and the associated independent variables is arbitrary; almost any choice can be made for storing data in database records. The identity of the property and the variables is specified by the metadata.

Some kinds of metadata are required to give meaning to the numbers, called 'essential metadata.' Examples are the identity of the compound (or compounds for a mixture), the identity of the property and its associated variables, its units and the identity of the phases. The reaction must also be specified for thermochemical data. For values that are relative to some reference state, the reference state must be described.

Other kinds of metadata give important additional information or can be used for search and retrieval. These include the references to the source of the data, descriptions of the purification method used, method for determination of the purity of the sample and estimates of the uncertainty of the measurements. A specification of whether the value is obtained by direct measurement, by derivation, by selection from previous literature or by estimation is included.

Within the TRC SOURCE Database other metadata are also allowed. These include an indication of the form of reporting the data in the original document (direct values, graph, equation), the kind of investigation made, the original units and the number of significant digits originally reported. The data are converted to a consistent set of SI units automatically on entry.

The ability to store textual comments is included. Such comments can describe the method of measurement, any special conditions, equations provided by the authors for fitting the data, or any other useful information that helps in the evaluation or interpretation of the values.

Metadata related to data integrity such as the identification of the person who added the information, the organization that contributed the data and the date of incorporation of the data into the database can be recorded along with the date of any corrections or revision. A field is available to indicate whether the value was selected or rejected in an evaluation. Properties and variables are identified by a set of 50 codes for pure compounds. Fifty more are defined for mixtures and 8 more for reactions. These are described in the TRC SOURCE Database Document published by TRC, which is available upon request. Estimates of uncertainties can be given in either or both of two ways. A field exists for recording a numerical measure of uncertainty of a property value, assuming that all sources of error have been propagated to the property. Uncertainty estimates may also be recorded as an equation for a particular set of data. It allows the assignment of uncertainties to each variable and the property independently. These equations can be given for both an imprecision and an overall uncertainty which includes any known systematic errors. This information is contained in comment records associated with a set of data using computer interpretable coding. Uncertainty estimates are an important item in the TRC SOURCE Database, as they are used for data selection and for the generation of weight factors for fitting to smoothing equations.

Complete citations to references are included. Authors' names are stored in a separate record type. Reciprocal links connect authors and references. Provision for titles of articles, Chemical Abstract Services references, CODEN, ISSN or ISBN numbers are included. A field for storing comments to references is available. Codes that indicate if the reference contains theory, experimental data or correlation or has a review have been added with recent references. For recent references that contain experimental data, codes to indicate the particular properties measured are also included. This allows extraction of current content information. Prior to adding the data, the references containing experimental data can be linked to the compound or compounds and to the property code. Hence the database also serves as a bibliographic database.

Chemical Abstract Services Registry Numbers (CASRN) provide the primary identification of compounds and mixtures. These are linked to the empirical formula, to the SMILES notation for structural formulas and to one or more names. Codes have been added to identify names recommended by IUPAC, Chemical Abstract Services and TRC.

Metadata is classified according to whether or not it is easily interpretable by computer software. Those intended for computer processing are formally defined in the Database Document. All essential metadata and all metadata used as index keys are computer–interpretable. Examples of metadata that are not intended for computer processing are sample descriptions, comments for references and comments on data, measurements method, temperature scales *etc.*

The TRC SOURCE Database schema defines each type of record and each of the fields it contains. The design of the schema for a large database such as this requires many arbitrary decisions and compromises. The intent is to permit unambiguous interpretation of the numbers to allow evaluation and selection of the best values without the necessity of consulting the original documents, and to accommodate all kinds of thermodynamic and related data reported in the published literature. It is also designed for compatibility with various retrieval software. However, the addition of new data items exacts a cost in incorporating it in the database and in developing retrieval software. Excessive complexity also creates a difficulty in interacting with other database systems.

17.4.2.1 Database structure and functions

The 45 kinds of records in the database can be classified into five groups. One group is concerned with compound identification and associated names and formulas. The primary index key for these is the Chemical Abstract Services Registry Number. Another group stores information on authors, references and related information. References are indexed by a key consisting of the year of publication, the first three letters of the first two authors' names, and a sequence number to resolve conflicts.

Another group of records contains the values of properties of pure compounds. Their primary index consists of the CASRN and the reference key. These records include types for storing single value data, fixed value data and data which are functions of one and two variables.

Other records store data for mixtures. Binary mixture data are indexed by the registry numbers of two compounds, ternary data for three compounds *etc.* These are followed by index keys for the references.

The final group is for reaction data. They are indexed by the registry numbers for reactants and products, and then the reference key. The reaction coefficients are also included in the indices.

The current schema is the result of ten years of testing, development and use at TRC. It supports many kinds of searches and retrievals. The most common kind of retrieval is the extraction of a particular property, or related group of properties, for a specified compound or mixture. Usually related information such as references, uncertainty estimates, sample descriptions and comments are retrieved.

TRC has developed software for the retrieval of the contents, bibliographic information and data. Most programs are written in the C language and can readily be ported to different computer systems. Most of the development work has been done on the Hewlett–Packard 9000 Series 835 and on a 486 PC using the Scientific Information Retrieval (SIR) database system. In order to be able to run the TRC SOURCE Database on a variety of platforms, alterative ways to accessing the TRC SOURCE using B–tree index files are to be developed. This method will allow faster access to information in the database.

The following data (or references to data) can be stored in the database:

(i) Equilibria

(a) Vapor–liquid or vapor–solid equilibria for pure compounds and mixtures (including infinite–dilution activity coefficients, eutectic and azeotropic temperatures and compositions)

(b) Liquid–liquid and liquid–solid equilibria

(c) Gas–liquid critical points

(ii) Volumetric properties, including density and volumetric derivatives (for pure compounds and mixtures)

(a) Virial coefficients

(b) On phase boundary (gas + liquid, gas + solid, solid + liquid)

(c) In single phase–liquid, gas or fluid

(d) At critical point

(iii) Excess volumes (mixtures only)

(a) Gas or fluid

(b) Liquid

(iv) Calorimetric properties (pure compounds and mixtures)

(a) Heat capacity and enthalpy along phase boundary

(b) Heat capacity and enthalpy in single phase

(c) Enthalpy of mixing and dilution

(d) Excess heat capacity

(v) Thermochemical properties (reactions)

    (a) Combustion and formation

    (b) Enthalpy of reactions

    (c) Equilibrium constants of reactions

(vi) Transport properties (pure compounds and mixtures)

    (a) Viscosity

    (b) Thermal conductivity

    (c) Diffusion

(vii) Miscellaneous (pure compounds and mixtures)

    (a) Surface tension

    (b) Refractive Index

    (c) Dielectric constants (also electric and magnetic properties)

    (d) Joule–Thompson coefficient

    (e) Speed of sound

    (f) Flash points

(viii) Derived from statistical mechanics

    (a) Ideal–gas properties calculated from spectroscopic assignments

    (b) References to spectroscopic and structural data

    (c) Intermolecular potentials

    (d) Virial coefficients

The main emphasis is on the properties of organic compounds and simple inorganic compounds and their mixtures. The database can be readily extended to systems containing electrolytes and molten salts, and to additional properties that are required for environmental purposes. This has not been done at TRC because of funding constraints.

The SOURCE Database at present contains a large collection of experimental data for pure compounds and mixtures, and it is expanding rapidly by the addition of both new data and data available in alternate electronic format at TRC. Persons from outside TRC can contribute to the SOURCE database by sending ASCII files on disk or other computer–readable media. A description of the preferred organization and format is available from TRC on request. Files that contain the essential information in a defined format can usually be processed without reference to the original article. TRC will extract data from the SOURCE database for contributors and provide the infomation either on diskette or by e–mail. Data in the SOURCE and THERMODYNAMIC TABLES Databases are routinely extracted for authors publishing in the *Journal of Chemical and Engineering Data* during the review process.

17.4.3 TRC THERMODYNAMIC TABLES Database

The TRC THERMODYNAMIC TABLES Database contains the evaluated data published in the TRC Thermodynamic Tables – Hydrocarbons and Non–Hydrocarbons. The contents of this database are updated quarterly in conjunction with the publication of the hard–copy supplements. The database is accessible on–line through TRCTHERMO module on the STN international network. From 1994, a version of this database developed using B–tree index files will be available for IBM PC or compatible computers on a yearly lease basis with a subscription to the hard copy of the TRC Thermodynamic Tables. Multiple copies will be available with a single hard–copy subscription. Its features include search for compound by name, formula or CASRN, with flexible name and formula searching capabilities. Data for a selected series of compounds can be retrieved and saved to external files for later analysis.

17.4.4 DIPPR Project 882

DIPPR Project 882 is organized to develop, maintain and make available to its sponsors a computer databank of selected and evaluated physical, thermodynamic and transport properties for binary mixtures. The properties include: viscosity, thermal conductivity, mutual–diffusion coefficient, excess volume and density, surface tension, critical temperature, pressure and density and the solubility of sparingly soluble materials. The data from the original literature have been compiled in their original units. Computer routines have been developed to provide the data in SI units for final dissemination. Assessments of the imprecisions and inaccuracies for each of the variables (temperature, pressure, composition and property) are made, and the results have been screened and adjusted, where applicable, to be consistent with the pure component data calculated from a variety of reliable sources. The data may be drawn from electronic database as tables and plots of the experimental data in the original or SI units.

The properties chosen by the DIPPR Project 882 Steering Committee are those that are not well covered for mixtures by other data services at the inception of the project. The development of a computer databank for this project was begun in April 1988. TRC established a formal procedure to locate, acquire, assess, enter and check the literature data. After review by the members of the Project 882 Committee and incorporation of their suggestions, the results were released to the sponsors for proprietary use for 1 year before the results were released to the public in the form of books and electronic databases. The first release of data to the sponsors was in December 1989 and consisted of data for 130 property–binary–mixture tables. These give comprehensive coverage for binary pairs from the following combinations: alcohols + water, alcohols + hydrocarbons, organic acids + water, organic acids + hydrocarbons. The second release contained data for 300 property–binary–mixture tables which expanded on the coverage of the mixture combinations mentioned and additionally heavy hydrocarbons + light hydrocarbons, nitrogen and hydrogen; aldehydes and ketones + water; and amines + water.

17.4.4.1 Literature search and maintenance of literature files

After establishing the general class of mixtures to be covered in a particular literature search, specific compounds within these classes were chosen for Chemical Abstract Services searches. These compounds included those class members which had significant amounts of data. In subsequent searches for each compound chosen, all papers with data that had the substance as a component and had properties pertinent to the project were put into the evaluation list to avoid searching more than one time over a selected substance name. This process gleaned all papers for any mixture involving the chosen substance to avoid having to rescan the literature for each different binary mixture containing the chosen compound. To reduce ultimate costs, this philosophy of sweeping diverse materials clean of any information that may ever be pertinent to the project when such information was encountered was maintained through the efforts of literature retrieval, off–line evaluation and data entry.

In addition to Chemical Abstract Services searches, the following bibliographies were searched for relevant data on the selected mixtures; these were used in initial searches to locate data that were ultimately evaluated by inspection of the original journal article:

- Jamieson D. J., Irving J. B. & Tudhope J. S. (1975). *Liquid Thermal Conductivity – A Data Survey to 1973*. Edinburgh: HMSO.
- Stephan K. & Heckenberger T. (1988). *Thermal Conductivity and Viscosity Data of Fluid Mixtures*. DECHEMA Chemistry Data Series, Vol. X, Part 1. Frankfurt: DECHEMA.
- Washburn E. W. (1926). *International Critical Tables*, National Research Council. New York, McGraw–Hill.
- Timmermans J. (1959). *Physico–chemical Constants of Binary Systems*, Vol. 2. New York: Wiley–Interscience.
- Tyrell H. J. V. & Harris K. R. (1984), *Diffusion in Liquids*. London: Butterworths.
- Johnson P. A. & Babb A. L. (1956). Liquid Diffusion of Non–electrolytes. *Chem. Rev.*, **56**, 387–453.
- Hicks C.P. & Young C. L. (1975). Gas–Liquid Critical Properties of Binary Mixtures. *Chem. Rev.*, **75**, 119–171.

All issues of the *Journal of Chemical and Engineering Data* and the *Journal of Chemical Thermodynamics* were searched for relevant articles. Current literature was scanned for any data within the scope of this project as a part of the routine operations of TRC.

After a candidate journal article has been identified for retrieval and evaluation, subsequent processing of this reference was recorded and tracked with a computer filing system to avoid duplication in both retrieving abstracts and in making copies of journal articles. The acquired reprints of journal articles were classified into the following categories for filing:

(i) Data to be evaluated now
(ii) Data to be entered
(iii) Data to be evaluated at later date
(iv) Data entered, and paper contains secondary references to be obtained
(v) Data entered, and secondary references obtained
(vi) Review papers and papers containing secondary references but no data
(vii) Cited secondary references containing neither data nor relevant references

Papers in (i) were assessed as soon as possible after receipt and transferred to (ii) or (iii). After data entry, if secondary references were cited for relevant properties of any mixture, the paper was transferred to (iv). When all secondary references were obtained, the paper in (iv) was transferred to the completed section (v). Papers in (vi) became a part of the body of intensive sources for frequent rescanning. Category (vii) was maintained to avoid repetitious consideration of such references; a dismaying number of references not containing pertinent information were often cited with implications to the contrary.

The processes chosen for the literature retrieval and off–line evaluation evolved from trying some methods that proved to be inefficient. The initial searches using Chemical Abstract Services On–Line were expensive and inefficient for the following reasons:

(i) Repeat finding of same abstract.
(ii) New searches required for different mixtures sets.
(iii) Many hits were irrelevant, and formats for visual review were cumbersome, for example, large printouts or extensive scrolling.

Using hard copy from decennial, pentennial and annual indices was much more efficient because for a particular compound all relevant properties of its mixtures within the scope of this project were scanned at one time.

17.4.4.2 Data assessments and entry preparations

The original journal articles were reviewed to determine the quality of the data. For papers containing data under current purview of the DIPPR 882 project, all relevant data in each paper, including mixtures that were not under current purview, were assessed. Estimates of the measurement precision and the accuracy of the values for each variable, for example $(T, p, x,$ viscosity), associated with the data were expressed as equations. In exceptional cases, imprecisions for individual data points were recorded with the body of the reported data. Assessments of the purity of the substances contained in the mixtures were recorded using the format developed for the TRC SOURCE Database. The purity assessments were used to place subjective confidence in the data. Only in the rare instances where the nature and concentration of impurities were given could quantitative use be made of purity information. In many instances only a small portion of the foregoing information was explicitly recorded in the original papers and judgments were made based upon previous experience from both making experimental measurements and from evaluating literature data. Less expert personnel highlight the data essential for the evaluations; however, these efforts were time–intensive and required final assessment by more expensive labor. Evaluation of data published in the future could be greatly eased by ensuring that the authors present such information fully in an easily identifiable form.

Assessment sheets containing the auxiliary information and instructions for entering the data directly from the reprint without transcription to data forms were prepared. The formats were chosen to minimize the preparations required for data entry, the instructions required for training data entry personnel and the keystrokes required to enter the data.

Upon completion of entry of information from a selected reference, it was immediately checked to rectify entry and format errors. The full literature citation from the bibliographic data file was returned for comparison with that on the reprint. It was determined if the compounds were in an extensive file containing CASRN and synonyms; if not, an error message was issued to permit spelling checks. Names that presented persistent problems were automatically placed in a special file for subsequent examination by a nomenclature expert who made the necessary revisions in the master file of compounds. The files were also checked to ensure data of the proper dimensionality had been entered and that there were no problems in the formats of the records; where there were problems, explicit error messages were issued describing the entry deficiency.

Unit conversions and simple data transformations were accomplished with table look–ups for unit conversion factors and by algebraic manipulation of the data. As an element of the conversion process, the values were tested to see if they were within reasonable physical limits, for example, were mole fractions greater than 0 and less than 1. Where molecular weights were required, they were determined from molecular formulas in the list of compound names and registry numbers. The requisite data from the DIPPR Project 801 and other compilations such as that by Viswanath & Natarajan (1988) were used for testing zero mole fraction results and for possible data renormalization.

Data in a wide variety of tabular formats and numerical ranges were displayed with special programs developed at TRC. The programs provide formatted tables and graphical display of data with standard ASCII characters.

In the analysis of the data, checks were made to determine if the values of the properties at the limits of mole factions of one were consistent with the values given in the DIPPR Project 801 compilation, TRC Thermodynamic Tables and other reliable data sources.

The routine for tables contains codes to select the variables in the order that requires the least amount of vertical space on a page or screen, but which give the results in legible forms. The graphical display routine scales the data by factors of 10 so that no more than three characters were required to label the axes, and the labels were incremented by 'rational' one– or two–digit steps. A constant was subtracted from a coordinate where necessary to limit the number of characters for the axis labels to 3. The linear translation constants and the multiple of 10 scaling constants were shown with the units associated with each axis. The labels to distinguish between data sets were chosen automatically and were displayed with ASCII descriptors passed with the data. A modified version of the program was used to prepare the camera–ready copy manuscript of the two books.

17.4.4.3 Products of DIPPR 882 Project

The DIPPR 882 database DIPMIX on the Transport and Thermodynamic Related Properties of Mixtures – Version I is now available (DIPMIX Database 1993). This database employs the LogicBase software, which is a knowledge–base application generator capable of handling a wide variety of database applications, developed by Shell Development Company. It displays experimental data on viscosity, diffusion coefficients, thermal conductivity, critical constants, solubility and density of binary mixtures in both tabular and graphical form. The first version contains 430 mixture–property tables on the compound classes summarized in Table 17.13.

The accuracy and precision of each of the measured variables, as well as the estimated purity of the materials, are displayed. The database can be searched on all or certain properties for selected classes of compounds. Version II will contain all the tables in Version I and an additional 420 mixture–property tables. In selecting the additional tables, emphasis was placed on selecting new compound classes and binary mixtures that were not included in Version I.

The additional binary mixture classes contained in Version II are listed in Table 17.14. The hard–copy versions of the tables are being issued in multiple parts. The initial volume is entitled 'Transport Properties and Related Thermodynamic Data of Binary Mixtures, Part I' (Gammon *et al.* 1993). Part II of the series, containing an additional 420 mixture–property tables, was published recently (Gammon *et al.* 1994). It is planned to issue two more printed volumes and two additional versions of the database.

Table 17.13. *Compound classes contained in DIPPMIX – Version I.*

Class	Class
alcohols + water	organic acids + water
alcohols + hydrocarbons	organic acids + hydrocarbons
aldehydes/ketones + water	amines/amides + water
heavy hydrocarbons + light hydro– carbons (gases: H_2, N_2, CO_2)	ethers + hydrocarbons
polyalcohols (polyols) + water	aromatic 2,3 fused rings + hydrocarbons

Table 17.14. *Additional compound classes contained in DIPPMIX – Version II.*

Class	Class
hydrocarbons + hydrocarbons	alcohols + acids
hydrocarbons + acids	alcohols + amines, amides
hydrocarbons + inorganic gases	alcohols + polyfunctional
hydrocarbons + alcohols	polyols + inorganic gases
hydrocarbons + peroxides	acids + inorganic gases
hydrocarbons + amines, amides	acids + amines, amides
hydrocarbons + ethers	peroxides + inorganic gases
hydrocarbons + polyfunctional	amines, amides + inorganic gases
aldehydes, ketones + inorganic gases	amines, amides + amines, amides
aldehydes, ketones + alcohols	amines, amides + polyfunctional
aldehydes, ketones + acids	polyfunctional + inorganic gases
alcohols + inorganic gases	

References

DIPMIX Database on Transport Properties and Related Thermodynamic Data of Binary Mixtures, Version 1.0 (1993). Thermodynamics Research Center, Texas A&M University System, College Station, TX 77843–3111, USA.

Gammon, B.E., Marsh, K.N. & Dewan, A.K.R. (1993). *Transport Properties and Related Thermodynamic Data of Binary Mixtures*, Part I, Design Institute for Physical Property Data (DIPPR), American Institute of Chemical Engineers, 345 East 47th Street, New York, NY 10017.

Gammon, B.E., Marsh, K.N., & Dewan, A.K.R. (1994). *Transport Properties and Related Thermodynamic Data of Binary Mixtures*, Part II, Design Institute for Physical Property Data (DIPPR), American Institute of Chemical Engineers, 345 East 47th Street, New York, NY 10017.

TRC Databases in Chemistry and Engineering, TRC THERMODYNAMIC TABLES, Version 1.0 (1994). Thermodynamic Research Center, Texas A&M University, College Station, TX 77843–3111, USA.

Viswanath, D.S. & Natarajan, G. (1988). *Databook on the Viscosity of Liquids*. New York: Hemisphere.

17.5 Overview of data banks of CIS

A. A. VASSERMAN
Odessa Institute of Marine Engineers, Odessa, Ukraine

17.5.1 Introduction

In connection with the requirements of science and technology, the number of investigations of thermodynamic properties of substances has increased in different countries,
including the USSR, during the last 30 years. The program of the International Union
of Pure and Applied Chemistry on the creation of International Thermodynamic Tables
of technologically important gases and liquids has helped to promote this increase.

In the former USSR, investigations of thermophysical properties were carried out
in the Research Institutes of the Academy of Sciences, in the Research Institutes of
different branches of industry and in the higher schools. The aim of these experimental and theoretical projects was to obtain reliable data on the properties of substances
and materials in a wide range of temperatures and pressures. Since the 1960s, computers have been used extensively for the automation of experiments, for the treatment
of measurements and for the calculation of properties. More recently, the necessity
of computer usage for data collection, treatment, storage, retrieval and dissemination
has grown because the number of investigations of thermophysical properties and the
resulting quantity of data has increased greatly.

The creation of a new type of information system – the data bank – acquired special
importance. In contrast to the documentary type of information retrieval system, data
banks can be used not only to abstract data but also to calculate new results on the basis
of these data with the help of logical and mathematical operations.

In this section, the most familiar Thermophysical Data Banks, created in the big research institutes of the USSR, are described briefly. These banks supply higher schools,
research and planning institutes and enterprises of power engineering, cryogenics,
aerospace, chemical, gas and oil–refining industry, metallurgy and other branches of industry with the data on properties of materials (Gurvich 1983; Trakhtengerts & Lehotski
1987; Labinov & Dregulyas 1976). The information service is set up in accordance with
the principles of permanent subscription and of single inquiries as well. The banks also
provide service for subscribers from some East European countries, the former members of the Council of Economic Mutual Aid. The scientific and organizational control
of the work of data collection and evaluation was undertaken by the Commission on
Thermophysical Tables of Gases and Liquids of the Council through a comprehensive
project 'Thermophysics and Heat–and–Power Engineering' of the USSR Academy of
Science.

17.5.2 Institute of High–Temperatures Data Banks

Among the Russian academic institutes dealing with thermophysical problems the Institute of High Temperatures in Moscow is the leading one. The Thermophysical Center was created there. The bibliographic bank THERMAL and the bank of numerical data BATEDA are functioning in this center. THERMAL started about 20 years ago. It provides the users with information on thermophysical properties of substances by making typed reviews of published work and copies of the originals (Gorgoraki *et al.* 1977). These reviews are the result of systematic study of original publications by the employees of the center. Thanks to the scientific contacts with foreign research institutes, the Thermophysical Center receives foreign literature: papers of conferences, scientific reports of research organizations, dissertations *etc.* – and makes corresponding reviews. The computerized store of the bank has tens of thousands of the originals: Soviet, starting from 1970, and foreign, starting from 1975. THERMAL contains information about the class of substances, chemical formula, aggregate state, investigated properties, limits of temperature and pressure as well as bibliographic data (Trakhtengerts & Lehotski 1987).

The bank THERMAL is a set of information blocks (bibliographic and factual data) and packages of information and scientific programs that give an opportunity to retrieve, analyze and treat the data. It works in two regimes (Baibuz 1984):

- regime SELECT permits a comparison of new information with the inquiries of permanent subscribers and selection of the corresponding reviews;
- regime RETRO provides the retrospective search of abstracts throughout the whole store for new inquiries from permanent subscribers and newly connected subscribers.

Bank THERMAL gives a wide range of methods for searching the records on behalf of users. The search can be carried out for a single substance, several substances and all substances of a definite class. It is possible to search for any work relating to one or more specified properties for definite substances, for works of named authors or for papers published during a certain period of time and also to select works corresponding to given intervals of temperature and pressure. THERMAL contains elements (in their ionized state also), inorganic compounds, simple organic compounds (paraffinic and aliphatic hydrocarbons up to butane and but–1–ene, and halogenated hydrocarbons), oxides and sulphides of carbon, carbides, carbonyls and cyanides of different elements, simple organic acids and their salts and some other technically important organic compounds (Gorgoraki *et al.* 1977). Substances whose properties depend on the method of production or treatment are not included. Also excluded are mixtures, solutions, alloys and minerals (with the exception of air and mixtures of its components, certain aqueous solutions and solutions of substances in some standard solvents). The information kept in THERMAL includes thermodynamic properties as well as others: magnetic, electrical and optical. There are data in the bank on thermochemical quantities, thermodynamic functions of ideal and real gases and condensed phases, P–V–T relationships

and equations of state, vapor pressure, surface tension, parameters of triple and critical points, temperatures and heats of phase transitions, and speed of sound. Among
transport properties there are data on viscosity, thermal conductivity, self–diffusion and
thermal diffusion coefficients. Some molecular characteristics necessary for the calculation and generalization of thermodynamic and transport properties are stored in the
bank also.

The substances in solid, liquid and gaseous states (including dissociated and par-
tially ionized states) are considered at temperatures up to 10000 K and pressures up to
10^6 MPa. The two–phase region and phase equilibrium lines are also included.

The bank of numerical thermodynamic data BATEDA stores, analyzes, treats and
delivers to the users the data on thermophysical properties of pure inorganic and simple
organic substances as separate values or tables. This bank includes data for about 100
substances which are of special interest for power engineering and some new branches
of technology. It is planned to increase the number of substances in the bank to 500
(Baibuz 1984). BATEDA has three regimes of work: dialogue, storage of data and
data selection on the order of a user program. It also includes a library of scientific
programs for the analysis, treatment and calculation of thermophysical data (Baibuz &
Trakhtengerts 1984). Except for the general databases, BATEDA can hold databases
for individual users in memory.

The main logical unit of the stored information in BATEDA is the description of
a document. It includes a brief bibliographic description, sufficient to find a source,
text comments, the names of substances and properties and numerical values of the
properties. The data can be represented also as equations, with a note on their range of
applicability (Trakhtengerts et al. 1984).

The main part of the data stored in BATEDA was selected by the scientists of the
Thermophysical Center. The bank includes also the most reliable data from published
reference books as well as standard and recommended reference data certified and approved by the USSR State Committee on Standards. BATEDA is available for scientists
and thermophysical experts. They can take part in accumulation and treatment of data,
add and correct the data in the bases, put comments on the stored data, use the data
for their own programs and for scientific studies on obtaining the reliable values of
properties. All that gives an opportunity for continuous improvement of the databank.

While preparing the reference books (Medvedev et al. 1965–1981; Gurvich et al.
1978–1982), the scientists of the Institute of High Temperatures selected numerous
data on thermodynamic properties of substances in the condensed and ideal–gas state.
The bank IVTANTHERMO (Gurvich 1983) was created in 1981 on the basis of these
data. This bank produces intercoordinated data on thermodynamic properties, with measures of their acceptability based on the source, analysis and treatment of the primary
data. The given data are the result of measurements or theoretical calculations. The fundamental physical constants, the basic thermodynamic quantities and other constants
recommended by international scientific organizations are used in the treatment of data.

IVTANTHERMO consists of several databases and a set of algorithms and programs. The databases contain: auxiliary data, for example, fundamental physical constants, atomic masses of elements *etc.*; constants necessary for the calculations of thermo-dynamic functions; primary experimental data on equilibrium constants of chemical reactions and on saturated vapor pressure of substances; thermochemical constants; tabulated values of thermodynamic properties for wide temperature intervals.

A set of programs works in an interactive mode and gives an opportunity to carry out the following operations:

- to handle experimental data on the isobaric heat capacity and on enthalpy changes in the condensed state, on saturated vapor pressure and on equilibrium constants of reaction;
- to calculate thermodynamic properties (isobaric heat capacity, Gibbs energy, entropy and enthalpy changes) in the condensed and gaseous states;
- to estimate the accuracy of the calculated values of properties and determine the coefficients of equations describing these properties;
- to reproduce the tables of properties;
- to calculate equilibrium constants of chemical reactions, and the composition and properties of multielement and multicomponent systems in the ideal–gas state.

Thermodynamic properties of ideal gases can be calculated over wide temperature intervals: from 100 up to 6000 K and, for a number of gases, up to 10,000 and even up to 20,000 K. For substances in the condensed state, the calculations are stored from 100 K to the temperature at which the saturated vapor pressure reaches 10 MPa. One can get from the bank a complete table of thermodynamic properties of a substance or the data for a specific temperature and also the equations approximating these properties over the whole temperature interval. The properties of 1200 substances were calculated by IVTANTHERMO in 1983. Oxides, hydrides, hydroxides, halides, sulfides and nitrides were included among these substances. It was planned to increase the number of sub-stances to 2200 with a large increase in the number of organic compounds (Gurvich 1983). Recently, methods for creation of reference data of transport properties of com-bustion products were implemented in the Institute of High Temperatures. These data are the supplement of the database of IVTANTHERMO. A store of reference data on collision integrals was created also (Baibuz 1990).

The data of IVTANTHERMO are used mainly by the branches of industry connected with high temperatures, and by research institutes which deal with calculation of real–gas properties and need data on their properties in the ideal–gas state.

17.5.3 Information System of the Russian Center for Standardization

The Automated Information System of authenticated data on thermophysical properties of gases and liquids (AIST) (Kozlov *et al.* 1978; Voronina *et al.* 1980) was created by the

USSR National Standard Reference Data Service (NSRDS). The system was developed in 1976–1980 in the All–Union Research Center of NSRDS (now Russian Research Center on standardization, information and certification of raw materials, materials and substances) in Moscow. It provides specialists with attested databases, formed on the basis of standard and recommended reference data. The data of the IUPAC Commission on Thermodynamics, the International Association for the Properties of Steam, the U.S. National Bureau of Standards and other authenticated foreign data are used in the system as well. The information blocks of the system are sets of program modules, being the mathematical models of substances, and the blocks of numerical data for each substance. The basis for the model of a substance is a unified equation of state for gas and liquid in the form of a double power expansion of the compressibility with respect to density and temperature. The principles of the molecular–kinetic theory and the dependence of the excess viscosity and thermal conductivity on density and temperature are used for the calculation of the transport properties.

Coefficients of the equation of state and of the equation for transport properties are stored for each substance. Parameters of the critical point and coefficients of equations for calculation of the ideal–gas functions, the saturated vapor pressure and the melting pressure are kept also. The thermal properties in the single–phase region and on the phase–equilibrium lines can be calculated on the basis of well–known relations with use of these coefficients. The system contains data for 30 reference substances: monatomic and diatomic gases, air, water and steam, carbon dioxide, ammonia, paraffin hydrocarbons (up to octane), ethylene (ethene), propylene (propene), benzene and toluene. The system can calculate the thermophysical properties of poorly investigated gases and liquids and of multicomponent mixtures also on the basis of data for reference substances.

AIST can calculate the values of density, compressibility, enthalpy, entropy, isochoric and isobaric heat capacity, speed of sound, adiabatic Joule–Thomson coefficient, thermal pressure coefficient, saturated vapor pressure, enthalpy of vaporization, heat capacities on the saturation and solidification lines, viscosity and thermal conductivity. Values of properties can be determined at temperatures from the triple point up to 1500 K and pressures up to 100 MPa. The system generates the following databases with appropriate algorithms and programs for their calculation:

- basic tables of standard and recommended reference data;
- extended tables of data containing more properties;
- tables of data transformed by independent variables;
- tables of reference data with simplified representation (local equations of state for limited regions of parameters);
- tabulated data on properties of poorly investigated substances;
- tabulated data on properties of multicomponent mixtures;

- program modules for calculation in computer–aided design systems on the basis of standard and recommended reference data.

As mentioned by Trakhtengerts (1986), AIST is one of the few systems producing tables for any combination of independent variables, not only as functions of pressure and temperature.

The characteristic features of AIST are reliability and internal consistency of the produced data. This is conditioned by high precision of the initial data and by principles put in the analytical basis of the system. The unified equations for gases and liquids are used for calculation of thermodynamic and transport properties. The equations for calculation of the properties of mixtures are formed by combining coefficients of equations for components. An analogous approach is realized on calculation of properties of poorly investigated substances with use of equations for reference substances and the extended principle of corresponding states. The existence of an intercoordinated analytic base ensures the compactness of the program part of the system and the unification of programs.

The system uses original methods, algorithms and programs proposed by its elaborators. These methods include, in particular: a method of calculation of thermodynamic properties of substances at arbitrary combinations of independent variables (Kozlov *et al.* 1982a), a method of compilation of generalized equations on the basis of a combination of equations with different precision (Kozlov *et al.* 1982b), a method of producing equations based on an approximation of the analytical functions describing the thermophysical properties (Kozlov *et al.* 1984) and a method of obtaining the properties of mixtures on the basis of the properties of components (Kozlov *et al.* 1983). AIST was included in the system of Standard Reference Data of the Council of Economic Mutual Aid as a thematic store on thermophysical properties of gases and liquids.

17.5.4 *Information system of the Thermodynamic Data Center at Moscow Power Institute*

The Center on Thermodynamic Properties of Gases and Liquids, which coordinated the thermophysical investigations in higher schools of the USSR, was established at the Moscow Power Institute in 1977. The computerized information system of this center provides users with data on properties of gases and liquids (Sychev & Spiridonov 1984; Sychev *et al.* 1985a,b).

The databases determining the thermodynamic behavior of substances and a package of applied and operating programs form the basis of this system. The information database of the substances, being a structural unit of the common database of the system, is a successive file containing the coefficients of equations and a collection of quantities which are characteristic for a given substance (critical parameters *etc.*). The system can generate output data in accordance with a user's inquiry on the basis of limited input information about a substance. The equation of state, presented through the independent variables – temperature and density, and equations describing thermodynamic properties in the ideal–gas state and the values of pressure along the saturation and solidification curves as functions of temperature are used for calculations of the thermodynamic properties. The data in the ideal–gas state are taken from the bank IVTANTHERMO.

The system contains information on the properties of monatomic and diatomic gases, air and paraffinic and aliphatic hydrocarbons. It provides the calculation of thermodynamic properties in the single–phase and two–phase regions, on the saturation and solidification curves, on the Boyle, inversion and ideal–gas curves, on the spinodal and quasi–spinodal, on the lines of the isochoric and isobaric heat capacity extremes at temperatures from the triple point up to 1000–1500 K and pressures up to 100–300 MPa (Sychev *et al.* 1985a,b). The system calculates the same thermodynamic properties as AIST and additionally the humidity of moist vapor, the changes in properties at the transition from the single–phase into the two–phase region and a broad spectrum of partial derivatives and derivatives along the isolines. The software of the system, having a modular structure, makes it possible to introduce additional thermodynamic and transport properties during its operation.

The information system of the Moscow Power Institute, as well as AIST, can calculate the values of thermodynamic functions at different combinations of the independent variables. It can determine also the analytic dependencies as explicit functions of these variables. The initial data for compiling such functions are formed inside the computer with the help of the basic set of equations. Therefore, the system gives information on the thermodynamic properties of substances in the form of tables or equations, approximating part of the thermodynamic surface in terms of given independent variables (Sychev & Spiridonov 1984; Sychev *et al.* 1985b).

Recently, in the Thermodynamic Data Center of Moscow Power Institute, a database on thermodynamic properties of gaseous and vapor–liquid binary mixtures was created on the basis of results of experimental investigations carried out in this institute. Freons of the methane series, carbon dioxide and sulphur hexafluoride are the first component of these mixtures, with nitrogen, hydrogen and helium as the second component. The corresponding software product was developed for the information system (Baibuz 1990).

17.5.5 *Information System of the Thermodynamic Data Center at Ukrainian State Company UKRNEFTECHIM*

The Automated Unified System for Thermophysical Subscribing (AVESTA) was created (Labinov & Dregulyas 1976) in the Thermodynamic Data Center at the All–Union Research and Planning Institute of the Petrochemical Industry (now the Thermodynamic Data Center at Ukrainian State Company UKRNEFTECHIM) in Kiev in 1972–1976. The system stores and generates data on the thermophysical properties of individual substances, mixtures and oil products.

The list of substances contained in the system takes into account, in the first place, the needs of the petrochemical, gas and pharmaceutical branches of industry and chemical engineering. There are different hydrocarbons, halogenated hydrocarbons, alcohols, aldehydes, ketones, ethers, phenols, nitro–compounds, heterocyclic compounds and oil products – in total, about 1600 individual substances and about 40 oil products. AVESTA provides users with:

- analytical reviews of subjects at the Thermodynamic Data Center;
- tables of experimental data on thermodynamic properties of substances;
- tables of reference data (with indications of their accuracy level);
- algorithms and programs for calculation of thermophysical properties (with an indication of their region of application and the precision of the results);
- a system for generation of data on the properties of gases and liquids for automatic design and for the operation of technological processes (Labinov & Dregulyas 1976);
- possible access to the system by the communication lines.

The software of the system includes a library of programs for calculation of thermophysical properties of substances, a data bank containing the values of basic physical and chemical properties of individual substances, an experimental data bank, a bibliographic database and a set of operating and service programs. There is a library of computational programs for the calculation of thermophysical properties of pure substances, binary and multicomponent mixtures in the liquid and gaseous states, including the phase equilibrium lines, and the calculation of properties of oils and oil products also. The same properties are included in the system AIST but here, in addition, there are values of the Prandtl number, diffusion coefficients, thermal conductivity, surface tension and parameters of vapor–liquid equilibrium for binary and multicomponent mixtures, calculated at temperatures from 90 to 800 K at pressures up to 200 MPa (Labinov *et al.* 1975).

The data bank on basic physical and chemical properties of individual substances stores the following data: the name of the substance, molecular mass, the structural formula of the molecule and the number of carbon atoms, melting and boiling temperatures at atmospheric pressure, critical parameters, acentric factor and polarity, parameters of the Lennard–Jones and Stockmayer potentials, the coefficients of an equation for calculation of isobaric heat capacity in the ideal–gas state and some other properties. The content of the bank is enlarged continuously by input of information on new substances.

The experimental data bank was formed by analysis of results of thermophysical–properties measurements. It contains more than 700,000 numerical values of thermophysical properties of individual substances, their mixtures, oils and oil products. The bibliographic database contains more than 180,000 descriptions of papers dedicated to investigations of properties. The operating programs implement the following functions:

- they organize the reception of inquiries and analyze their contents;

- they select a computational method which gives an optimum level of accuracy in the calculated values;
- they organize the computation within the limits of the program library;
- they determine in accordance with a subscriber's inquiry the coefficients of equations which approximate the results of calculations.

The service programs permit the delivery of results on the printer, screen or subscriber's terminal. These programs allow users to interact with the bank on basic physical and chemical properties of substances, with the experimental data bank and with the library of computational programs.

AVESTA plays a great role in the rapid provision to many organizations and enterprises associated with the petrochemical industry of data which are necessary for equipment design and for control of technological processes. The large section of calculation procedures used in this system was certified by the State Committee of Standards (Kozlov 1980). AVESTA has been demonstrated at international exhibitions and fairs.

17.5.6 Data bank of the Institute 'Giprokautchuk'

A data bank for thermophysical properties of individual substances and their mixtures was created in the planning and research institute 'Giprokautchuk' in Moscow (Bogomol'ny *et al.* 1984; Bogomol'ny & Rumyantseva 1987). This bank is oriented toward the needs of the synthetic rubber industry. It includes a bank of published data and a system for calculation of properties. The number of substances in this bank is greater than 1400. The majority of these are members of various classes of organic compounds, with the number of carbon atoms in a molecule ranging from 1 to 20. Binary and multi-component mixtures of these substances are also included. Substances whose properties are dependent on methods of production or treatment are not included (as in the bank THERMAL). Solutions of electrolytes are not investigated, nor are mixtures of indefinite compositions or chemically reacting mixtures.

The properties of substances and mixtures are considered in the gaseous and liquid states and in the region of vapor–liquid mixtures. It is assumed that the substances are physically and chemically stable.

The following groups of properties are collected in the bank:

- constants (standard enthalpy of formation, melting and normal boiling temperatures, critical parameters);
- temperature–dependent properties (saturated vapor pressure, orthobaric densities, enthalpy of vaporization, surface tension, isobaric heat capacity, viscosity and thermal conductivity of saturated vapor and liquid);
- temperature– and pressure–dependent properties (density, viscosity and thermal conductivity of gas and liquid in the single–phase region);

- phase equilibrium data of mixtures (vapor–liquid, liquid–liquid and vapor–liquid–liquid).

Data on the fire and explosion hazards and on the toxicological properties of substances are collected in the bank also. There are flash points, concentration limits for ignition and the maximum permissible concentration in air. The data in the bank cover the temperature range 50–2500 K at pressure up to 10 MPa (Bogomol'ny & Rumyantseva, 1987). The database is a collection of blocks of identical structure. One block contains information or numerical data, having identical sense and form of presentation. Names of substances, names of properties, dimensions of quantities and coefficients of their mutual conversion, categories of data quality, coefficients of equations for calculation, numerical values of quantities and literature sources are included in separate blocks. The constants, one– and two–dimensional tables and data on phase equilibria of mixtures, occupy different blocks in the database and are treated by independent groups of programs. The software of the bank includes programs for storage, blocks and data printing, removal of faulty records, control of structure and degree of filling of the database, delivery of answers to inquiries and, finally, a set of programs for data treatment. This set is used by specialists who analyze the data and select the recommended values (Bogomol'ny & Rumyantseva 1987).

The data bank can select all the data for any given substance or data just for a given property according to the users' inquiries. In the second case, all the data on the indicated property which are stored in the bank are selected, a statistical analysis is made and the data are delivered to the user with a recommendation of the most reliable published value. The bank also can give information on the phase equilibrium of binary mixtures consisting of components of any multicomponent mixture indicated by the user (total number of components is not more than 40) (Bogomol'ny *et al.* 1984).

The software of the bank also includes a set of programs for calculating properties of gases and liquids. This system is intended for estimating data which are absent in the literature. Organizations connected with synthetic rubber production and enterprises and planning institutes of other branches of chemical industry are the users of data of this bank.

There are also some lesser known data banks in Russia, Ukraine and Azerbaijan, serving a comparatively narrow circle of users. Brief details of these banks are given in reviews (Baibuz 1984; Trakhtengerts 1986). It must be noted, in conclusion, that data banks in the former USSR were developed primarily using BESM computers and later on ES computers (IBM–360). Now they have been transferred onto IBM PCs, which ensures their compatibility with contemporary banks created in other countries.

References
Baibuz, V.F. (1984). Issledovaniya teplofizicheskikh svoistv veshchestv i banki teplofizicheskikh dannykh v Sovetskom Sojuze (Investigations on thermophysical

properties of substances and thermophysical data banks in the Soviet Union). *Obzory po teplofizicheskim svoistvam veshchestv (Reviews on thermophysical properties of substances)*. Moscow: IVTAN, No. 3, 47, 3–25.

Baibuz, V.F. (1990). Issledovaniya teplofizicheskikh svoistv veshchestv v Sovetskom Sojuze (Investigations on thermophysical properties of substances in the Soviet Union). *Obzory po teplofizicheskim svoistvam veshchestv (Reviews on thermophysical properties of substances)*. Moscow: IVTAN, No. 1, 81, 3–136.

Baibuz, V.F. & Trakhtengerts, M.S. (1984). Main conceptions of the IVTAN thermophysical data bank. *Dialogovye i informatsionnye sistemy (Dialogue and information systems)*. Moscow: IVTAN, No. 6, 49–60.

Bogomol'ny, A.M. & Rumyantseva, A.V. (1987). Data banks on properties of organic compounds and their mixtures. *Pryamye i obratnye zadachi khimicheskoi termodinamiki (Direct and reverse tasks of chemical thermodynamic), collection*. Novosibirsk: Nauka (Siberian department), 101–107.

Bogomol'ny, A.M., Rumyantseva, A.V. & Mozzhukhin, A.S. (1984). Data bank on vapour–liquid equilibrium of binary mixtures. *Informatsionnyi bjulleten' po khimicheskoi promyshlennosti (Information bulletin on chemical industry)*. Moscow: SEV, No. 2, 35–36.

Gorgoraki, E.A., Kraevskii, S.L., Trakhtengerts, M.S., *et al.* (1977). Avtomatizirovannaya informatsionno–poiskovaya sistema teplofizicheskogo centra IVTAN (Computerized information–retrieval system of the IVTAN Thermophysical Center). *Obzory po teplofizicheskim svoistvam veshchestv (Reviews on thermophysical properties of substances)*. Moscow: IVTAN, No. 4, 3–128.

Gurvich, L.V. (1983). IVTANTHERMO – computerized data system for thermophysical properties of substances. *Vestnik AN SSSR*, No. 3, 54–65.

Gurvich, L.V., Veits, I.V., Medvedev, V.A., *et al.* (1978–1982). *Termodinamicheskie svoistva individual'nykh veshchestv (Thermodynamic properties of individual substances)*, 4 volumes. Editors Glushko, V.P., *et al.* Moscow: Nauka.

Kozlov, A.D. (1980). Activity of NSRDS on providing the national economy with data for properties of substances and materials. *Informatsionnyi bjulleten' GSSSD (Information bulletin of NSRDS)*, No. 8–9, 7–12.

Kozlov, A.D., Kuznetsov, V.M. & Mamonov, Ju.V. (1982b). Method of obtaining reference data on the transport coefficients of gases and liquids. *Teplofizicheskie svoistva veshchestv i materialov (Thermophysical properties of substances and materials), collection*. Moscow: Izd. Standartov, No. 17, 148–161.

Kozlov, A.D., Kuznetsov, V.M. & Mamonov, Ju.V. (1983). A method of obtaining calculated data on thermophysical properties of mixtures on the basis of data for the pure components. *Teplofizicheskie svoistva veshchestv i materialov (Thermophysical properties of substances and materials), collection*. Moscow: Izd. Standartov, No. 19, 145–154.

Kozlov, A.D., Kuznetsov, V.M. & Mamonov, Ju.V. (1984). Analytic description of the behaviour of thermophysical properties of substances by means of modified least squares method. *Teplofizicheskie svoistva veshchestv i materialov (Thermophysical properties of substances and materials), collection*. Moscow: Izd. Standartov, No. 21, 126–134.

Kozlov, A.D., Kuznetsov, V.M., Mamonov, Ju.V. & Rybakov, S.I. (1982a). Method of calculation of thermodynamic properties of substances on computer at arbitrary combinations of arguments. *Teploenergetika*, No. 3, 71–73.

Kozlov, A.D., Mamonov, Ju.V. & Voronina, V.P. (1978). Automatized information–searching system of reliable data on thermophysical properties of gases and liquids. *Informatsionnyi bjulleten' GSSSD (Information bulletin of NSRDS)*, No. 7, 10–11.

Labinov, S.D. & Dregulyas, E.K. (1976). The main tasks and forms of activity of the data center of NSRDS on thermophysical properties of hydrocarbons and oil–products. *Informatsionnyi bjulleten' GSSSD (Information bulletin of NSRDS)*, No. 4, 22–23.

Labinov, S.D., Soldatenko, Ju.A., Dregulyas, E.K. & Bolotin, N.K. (1975). Data center on thermophysical properties of hydrocarbons, their mixtures, oils and oil fractions. *Informatsionnyi bjulleten' GSSSD (Information bulletin of NSRDS)*, No. 1, 7–8.

Medvedev, V.A., Bergman, G.A., Gurvich, L.V., *et al.*. (1965–1981). *Termicheskie konstanty veshchestv (Thermal constants of substances)*, 10 volumes. Editors Glushko, V.P., *et al.*. Moscow: VINITI.

Sychev, V.V. & Spiridonov, G.A. (1984). Thermophysical information–solving systems. Data center on thermophysical properties of gases and liquids. *Vestnik AN SSSR*, No. 10, 28–38.

Sychev, V.V., Spiridonov, G.A. & Grankina, L.N. (1985a). Thermophysical information–solving system on the base of small computer SM–4. *Doklady AN SSSR*, 280, 622–627.

Sychev, V.V., Spiridonov, G.A. & Kasyanov Ju.I. (1985b). Thermophysical information–solving system on the base of high–performance computer. *Doklady AN SSSR*, 280, 869–874.

Trakhtengerts, M.S. (1986). Banki teplofizicheskikh dannykh (Thermophysical data banks). *Obzory po teplofizicheskim svoistvam veshchestv (Reviews on thermophysical properties of substances)*. Moscow: IVTAN, No. 1 (57), 5–44.

Trakhtengerts, M.S., Laut, L.I. & Jufereva, G.N. (1984). Structure of the bases in a thermophysical data bank. *Dialogovye i informatsionnye sistemy (Dialogue and information systems)*. Moscow: IVTAN, No. 6, 61–74.

Trakhtengerts, M.S. & Lehotski Laslo (1987). Sozdanie bazy dannykh po teplofizicheskim svoistvam veshchestv (Creation of a data base on thermophysical properties of substances). Moscow: Preprint IVTAN, No. 1–225, 3–28.

Voronina, V.P., Kozlov, A.D. & Mamonov, Ju.V. (1980). Computerized information system of reliable data on thermophysical properties of gases and liquids. *Informatsionnyi bjulleten' GSSSD (Information bulletin of NSRDS)*, No. 8–9, 30–32.

Index